WINE ANALYSIS AND PRODUCTION

The Chapman & Hall Enology Library

Wine Analysis and Production by Bruce W. Zoecklein, Kenneth C. Fugelsang, Barry H. Gump, and Fred S. Nury

Principles and Practices of Winemaking by Roger B. Boulton, Vernon L. Singleton, Linda F. Bisson, and Ralph E. Kunkee

Forthcoming Titles
Winery Utilitites by David R. Storm
Wine Microbiology by Kenneth C. Fugelsang

WINE ANALYSIS AND PRODUCTION

Bruce W. Zoecklein
Virginia Polytechnic Institute & State University, Blacksburg

Kenneth C. Fugelsang
California State University at Fresno

Barry H. Gump
California State University at Fresno

Fred S. Nury
California State University at Fresno

SPRINGER-SCIENCE+BUSINESS MEDIA, B.V.

Art direction: Andrea Meyer, emDASH inc.
Cover design: Saeed Sayrafiezadeh, emDASH inc.

Originally published by Chapman & Hall in 1995
Softcover reprint of the hardcover 1st edition 1995

1 2 3 4 5 6 7 8 9 10 XXX 01 00 99 97 96 95

Library of Congress Cataloging-in-Publication Data

Wine analysis and production / Bruce W. Zoecklein...[et al.].
p. cm.
Includes bibliographical references and index.
ISBN 978-1-4757-6980-7 ISBN 978-1-4757-6978-4 (eBook)
DOI 10.1007/978-1-4757-6978-4

1. Wine and wine making—Analysis. I. Zoecklein, Bruce W.
TP548.5.A5W54 1994
663' .2—dc20 94-17776
CIP

British Library Cataloguing in Publication Data available

CONTENTS

List of Laboratory Procedures

PREFACE

Winemaking as a form of food preservation is as old as civilization. Wine has been an integral component of people's daily diet since its discovery and has also played an important role in the development of society, religion, and culture. We are currently drinking the best wines ever produced. We are able to do this because of our increased understanding of grape growing, biochemistry and microbiology of fermentation, our use of advanced technology in production, and our ability to measure the various major and minor components that comprise this fascinating beverage.

Historically, winemakers succeeded with slow but gradual improvements brought about by combinations of folklore, observation, and luck. However, they also had monumental failures resulting in the necessity to dispose of wine or convert it into distilled spirits or vinegar. It was assumed that even the most marginally drinkable wines could be marketed. This is not the case for modern producers. The costs of grapes, the technology used in production, oak barrels, corks, bottling equipment, etc., have increased dramatically and continue to rise. Consumers are now accustomed to supplies of inexpensive and high-quality varietals and blends; they continue to demand better. Modern winemakers now rely on basic science and the systematic application of their art to produce products pleasing to the increasingly knowledgeable consumer base that enjoys wine as part of its civilized society.

The process of making wine involves a series of concerns for the grower, as well as the winemaker. The first concerns are viticultural, including delivery of sound, high-quality fruit at optimal maturity. Upon arrival at the winery, fruit quality is assessed, the grapes are processed, and fermentation is begun. Almost immediately, and in many instances simultaneously, chemical and microbiological stability of the young and/or aging wine becomes important. Finally, problems occur on occasion requiring utilization of remedial techniques to produce an acceptable product.

Production considerations serve as the framework in organizing this book. Within each chapter is information culled from the authors' collective years of experience, as well as from the literate wine community around the world. These chapters provide numerous practical, as well as fundamental, insights into the various aspects of the process. Winemakers will benefit from these insights while still maintaining (and gaining further insight into) their own.

Analytical techniques have become valuable tools of modern winemakers wishing to better understand their product. These analytical tools are another major feature of this text. The authors have gathered numerous procedures commonly used for grape, juice, and wine analysis. These procedures are presented as they are generally practiced in the industry around the world. We have formatted them into an easy-to-follow recipe-style to make them more useful to the winery technician. Our procedures provide instructions for preparing required reagents, stains, and media, and then outline the analyses in detail. To make these procedures more accessible, we have gathered them into a single chapter at the end of the "text" chapters. In addition to the "standard" laboratory procedures we have included a section of rapid "diagnostic" tests to assist in identifying problems encountered during winemaking. Several frequently used conversion and correction tables have also been collected into one section for the reader's convenience. Finally, we have provided some information on the safety aspects of the various reagents employed in conducting the laboratory analyses.

In developing material for this text, the authors have emphasized analyses as they would be carried out in a production laboratory. Realizing that different laboratories have different analytical capabilities, personnel, and equipment, we have in many instances provided several different approaches to the same analysis. Throughout this book we have given special attention to practical considerations and their importance in the total spectrum of winery operations. We have done the same with the laboratory procedures. It is the authors' wish that the book's format will satisfy the interests of laboratory personnel as well as winemakers. It is assumed that the reader has some basic preparation in the fields of chemistry and microbiology. A novice reader would be advised to acquire a basic textbook in quantitative analysis for descriptions of fundamental laboratory skills.

In writing this text we not only surveyed the literature of the winemaking world, but have solicited direct assistance from several guest authors. To help present the subject of laboratory chemical and microbiological measurements from both the historical and future perspective, we have selections by Gordon Burns, President of ETS Laboratory (St. Helena, CA), and Lisa Van de Water, Director of The Wine Lab (Napa, CA) and Pacific Rim Oenology Services (Blenheim, New Zealand). The issue of the health aspects of wine has never been a more important subject than it is today. To provide a perspective on this subject, Dr. Carlos Muller, Director of the Enology Program at California State University (Fresno, CA) has contributed a Chapter.

A rational approach to the uses and benefits of sensory analysis is always a complement to the subject of laboratory measurements. The sensory organs are, for the most part, our most sensitive analytical tool for monitoring certain wine components and microbiological processes. A chapter on sensory techniques is offered by Dr. Susan Duncan of the Department of Food Science and Technology, Virginia Polytechnic Institute (Blacksburg, VA). Finally, in adding a current perspective to the use of enzymes in juice and wine processing, we have solicited the efforts of Katherine Haight, Research Associate at the Viticulture & Enology Research Center (Fresno, CA). We are greatly appreciative of the efforts of these contributors, who have added an extra dimension to this text.

WINE ANALYSIS AND PRODUCTION

CHAPTER 1

Introduction

Overview of Wine Analysis

Gordon H. Burns
Director, ETS Laboratories, St. Helena, CA

REASONS FOR ANALYSIS

Throughout the history of wine production, analytical techniques have became increasingly important with the development of technology and increased governmental regulation. Analyses of grapes and wines are performed for a number of reasons (Table 1–1).

COMMON ANALYTICAL COMPONENTS

The components of wine and must can be broken into classes that are analyzed during the production process (Table 1–2). Soluble solids, or "sugar" at harvest is not truly a measure of sugar at all, but rather a refractive index or densitometric measurement of grape solids presumed to be mainly sugar. This presumption generally serves the practical needs of the industry. Glucose and fructose analysis is used to determine the dryness/fermentability of a wine. Sugar free extract is routinely determined for export purposes.

Table 1–1. Reasons for analysis of grapes and wines

Quality control: ripening, processing, and aging
Spoilage reduction and process improvement
Blending: precise analyses leading to more precise blends
Export certification: European Economic Community, Pacific Rim, Canada
Global regulatory requirements

Acidity in wine is usually expressed as total acidity, or, more correctly expressed as titratable acidity. Volatile acidity (almost exclusively acetic acid) is determined as a measure of spoilage. Wine pH is perhaps the most important analytical parameter for wine due to implications concerning sensory characteristics and several forms of stability. Individual wine acids of interest to the analyst include tartaric, malic, and traces of citric acid in the berry and juice before fermentation, and tartaric, malic, lactic, acetic, citric, succinic, and traces of other organic acids in finished wine.

Table 1–2. Wine and must components.

Soluble solids: "sugar," extract, glucose, and fructose
Acidity: total, volatile, pH, individual acids
Alcohols: ethanol, methanol, fusel oils, glycerol
Carbonyl compounds: acetaldehyde, HMF, diacetyl
Esters: ethyl acetate, methyl anthranilate (labruscana)
Nitrogen compounds: NH_3, amino acids, amines, proteins
Phenolic compounds: total, phenolic fractions including anthocyanins
Chemical additions: SO_2, sorbic and benzoic acids, illegals
Other: common and trace metals, oxygen, CO_2, fluoride

Ethanol is obviously the alcohol of foremost importance in the wine industry. It not only determines the body of a wine and affects its flavor, but also affects the taxes imposed on a wine. Other alcohols of interest are methanol, glycerol, and various fusel oils.

Carbonyl compounds of primary significance in wine production include acetaldehyde, which occurs at elevated levels after oxidation of a wine; hydroxy methylfurfural, often present in baked wines; and diacetyl, a sensory component resulting from malolactic fermentation.

Esters present in wine include, among many others, ethyl acetate, which can be increased by hot fermentations and uncontrolled microbial activity. Methyl anthranilate is a component of grapes that are not *Vitis vinifera,* such as *V. labruscana* and some crosses.

Nitrogen compounds of interest to the analyst include ammonia, various amino acids, amines, and proteins. Much of the desirable and unique characteristics of wine can be attributed to phenolic compounds. Those measured routinely are total phenols, usually based on a colorimetric method. Also determined less frequently are individual anthocyanins, which are the color components of red wines.

Whether or not *Brettanomyces/Dekkera* infections are treated as spoilage or "characteristic enhancers" at a particular winery depends on the winery's current characterization of such infections as malignant or benign. In all situations, however, it is desirable to be able to rapidly, reliably, and quantitatively determine the *Brettanomyces* status of individual cellar lots. Analysis of 4-ethyl phenol provides this capability.

Chemical preservatives often added to wines include sulfur dioxide and sorbic and benzoic acids. The latter are typically monitored using high-performance liquid chromatography (HPLC).

Scandals have occurred occasionally in the wine industry with regard to the addition of compounds such as methanol and diethylene glycol. These situations have demanded prompt and accurate analysis.

Common metals of interest to the winemaker or analyst include potassium, calcium, and traces of copper and iron. There has been some focus recently on heavy metals such as lead and cadmium, which exist in wine just as they do in other natural foods.

During the wine production process, oxygen levels are monitored to prevent consequent oxidative problems that may occur when levels in wine are high. Carbon dioxide levels are measured in sparkling wines and are often important as stylistic tools for other table wines and wine-based products.

Another compound routinely analyzed in wines is fluoride, and the international community has placed limits on acceptable fluoride levels.

CURRENT ANALYTICAL TECHNIQUES

Analytical capabilities vary widely among wineries and service laboratories (Table 1–3). The minimum BATF requirements are the capability to measure ethanol and fill level. For simple alcohol measurement, the ebulliometer is a suitable device. Also needed is a calibrated vessel to measure the

Table 1–3. Analytical techniques and current applications.

"Wet chemistry": manual vs. automated methods
HPLC: acids, sugars, phenolics, microbial metalolites
AA: Cu, Fe, Ca, K, other trace metals including Pb
GC: ethanol, methanol, higher alcohols, esters, DEG
GC/MS: ethyl carbamate, procymidone, sulfides, 2,4,6-TCA, pesticide residues, contamination, 4-ethyl phenol NIR: ethanol, "residual sugar"

fill level of the bottles. At the other end of the spectrum, some wineries and service laboratories use techniques as sophisticated as gas chromatography/mass spectrometry (GC/MS).

The Association of Official Analytical Chemists (AOAC) and the Organization Internationale du Vin (OIV) define analytical reference methods that are often used within the wine industry. The OIV is the worldwide organization of wine-producing countries, and the European Economic Community (EEC) typically adopts OIV regulations as its own. In addition to the methods of these official bodies, the wine industry adopted other accepted routine methods by consensus.

Traditionally, wet chemical methods have been used and improved over the years so as to produce acceptable results. Many modifications are currently being made to wet chemical methods in order to automate them.

HPLC is used in a few large winery laboratories for analysis of various compounds. One of the first commonly accepted applications of HPLC in the winery laboratory was analysis of organic acids. HPLC techniques also exist for the determination of sugars, principally glucose and fructose. HPLC is also applied, mostly on a research basis, for the analysis of various phenols.

Flame atomic absorption and emission are routinely used in wine analysis to detect the presence of metals such as potassium, sodium, calcium, copper, and iron. Various trace metals may be determined with flame or electrothermal atomization (graphite furnace) techniques. The principal trace metals of interest today are lead, cadmium, and arsenic.

In recent years, the analytical community has made improvements in atomic absorption (AA) techniques for analysis of trace metals in wine. These improvements coincide with the use of more reliable platform surfaces in electrothermal vaporization chambers and better autosampling techniques. There are also ongoing attempts to improve the detection limits currently available with inductively coupled plasma (ICP) techniques. The primary attribute of a sensitive ICP method is the ability to perform simultaneous determinations of several metals in wine. Currently, detection limits are not adequate for the determination of certain important metals in the wine matrix.

Gas chromatography (GC) is used routinely by a number of wineries for the determination of ethanol. Methanol and the higher alcohols and fusel oils are also determined routinely by GC in wines and wine byproducts, such as brandy. Various esters and other compounds, such as diethylene glycol, are analyzed by GC for regulatory purposes.

GC-MS is used for the determination of ethyl carbamate, procymidone, and many other compounds as well as contaminants. Ethyl carbamate (urethane) is formed whenever urea and ethanol come into contact. Therefore, this compound exists naturally at minute levels in wines and other products such as bread and beer.

Wine is not immune to the current domestic and international regulatory concern regarding pesticide residues. Pesticides, fungicides, and other agricultural chemicals are certainly used on wine grapes. However, it is very rare that significant residue levels remain in the finished product. Consequently there is very little rational reason other than regulatory concern to analyze for pesticide residues in wine. One pesticide residue of current interest in the wine industry is procymidone, a fungicide, widely used throughout the world on grapes. Because it has not, to date, been registered for use on grapes in the United States, wines containing any trace amounts of this compound have been banned from importation.

GC-MS techniques have been developed for the analysis of organic sulfides and other sensory compounds in wines. This is done to validate sensory examinations of wines with more specific, compound-based analytical information. The compound 2,4,6-trichloroanisol can contribute to a sensory characteristic known as "corkiness." The sensory threshold of this compound in wine is around 2 parts per trillion, and this presents a significant analytical challenge to the wine analyst.

A recently developed technique using GC/MS technology allows a precise analysis of 4-ethyl phenol, which is specifically and quantitatively associated with *Brettanomyces/Dekkera* metabolism.

Near infrared spectroscopy (NIR) techniques can determine ethanol in wine with equal or better accuracy than that attained using GC. NIR methods are also being developed for fermentable or reducing sugars. Additional NIR methods are being investigated for total phenolics, and methanol. This development is particularly important given increasing hazardous waste disposal requirements for laboratories, because NIR uses no reagents and consequently generates no waste.

FUTURE ANALYTICAL TECHNIQUES

Some groundbreaking future applications now on the horizon in the area of wine analysis concern authentication of wine and grapes using ^{13}C NMR and isotope-ratio mass spectrometry (IRMS). Researchers already claim the ability in some cases to differentiate not only sources of wines, but even vintages based on IRMS and ^{13}C NMR techniques. Near-infrared (NIR) techniques for total phenolics, and even possibly phenolics fractions, may be available soon. Although considerable work remains to be done with regard to NIR analysis for determination of dryness or completion of fermentation, this technique holds considerable promise.

Also forthcoming are improved and simplified analytical techniques required for the analysis of 2,4,6-trichloroanisol and other compounds that may be implicated in causing cork taint. Sulfides present another analytical challenge in need of a solution. Wines may be bottled with no apparent sensory problems and develop perplexing and annoying sulfide-type aromas some months later. Although some analytical methods are available for the determination of these compounds at the detection limits required, such methods are generally too complex for use on a day-to-day basis.

Another issue of concern in the wine industry is increasing demands for regulatory compliance, thus increasing the demand for analyses of trace metals and agricultural chemical residues. Attempts to characterize wines using "signature" or "fingerprint" techniques have captured the attention of those in the wine industry. The signature concept holds the potential for allowing the analyst to trace authenticity of a wine to grape variety (including clonal selection), appellation, and vintage.

Historically, characterization of wines has been (and still is) conducted using sensory analysis. The nose remains the most sensitive analytical tool. However, at present quality control and regulatory requirements are forcing wineries to perform more analyses for both major and minor components. The necessity of testing for residues will be a powerful driving force toward more sophisticated instrumental methods in the industry. Finally, it is the interest in producing even higher quality products, and the search for more definitive information that cause winemakers to ask for better analytical measurements. It is safe to speculate that the number and quality of wine analyses performed will increase and these analyses will be accomplished using increasingly advanced analytical techniques.

A Technical Revolution in Winemaking

Lisa Van de Water
Director, The Wine Lab, Napa, CA, and Pacific Rim Enology Services, Blenheim, New Zealand

When I entered the California wine industry in 1974, it was different in many ways. Certainly, there were fewer wineries; the winery boom was yet to come. As an industry we felt young, eager, and ready to grow, and grow we did.

As amazing as it seems now, the opening of a new winery was a newsworthy event, of intense interest to wine devotees, who devoured everything that was written about premium wineries and their wines, true or not. Fame came easily; wine gluts and neo-prohibitionists were things of the future.

In the Napa Valley, vineyards and wineries replaced prunes and walnuts, changing the land and the community forever. However, the most profound changes in 20 years are in wine production. Quietly, gradually, a technical revolution in winemaking has been occurring. Far from subsiding, this revolution continues, with worldwide effects. I do not mean technology for technology's sake, but an intense interest in research accompanied by basic shifts in the understanding of the processes involved in wine production.

RELATING pH AND SO_2

Some wineries measured pH in 1974 but most did not, and few had any idea what the purpose was. I began to have an inkling that pH was a key measurement after tasting many wines the pH of which I had tested as well. Some of those wines were the celebrated 1970 Cabernets, many of which fell apart years too soon. Not coincidentally, the wines with the highest pH tended to fade soonest, "died like a dog," in the words of one winemaker. Wineries now pay close attention to pH, and even home winemakers know that pH is probably the most important test that can be made on juice or wine.

We also suspected that pH and SO_2 had something to do with each other, but it was 1982 before the relationship was elucidated, bringing "molecular" SO_2 into common usage. Until then, wineries sought to achieve a certain level of free SO_2, not necessarily adjusting the level according to pH, or even wine type.

Measurement of SO_2 has also changed. Over the last 20 years, enologists have moved away from the classic Ripper method, which has significant interferences, choosing the greater accuracy of the aeration-oxidation (vacuum aspiration) test for free SO_2. As a result, spurred by consumer awareness and TV scare stories, SO_2 use has plummeted. Twenty years ago winemakers might add 75 to 100 ppm SO_2 to white juice at harvest, and a bit less to reds. In the late 1970s, no-SO_2 fermentations of white wines were introduced, and by the end of the 1980s, California wineries seldom added any SO_2 at all to reds before fermentation.

Since then, the pendulum has swung back a little because of an explosion of devastating spoilage from spontaneous *Lactobacillus* growth during yeast fermentation. It was quickly realized that this problem could be controlled by resuming small SO_2 additions.

GRAPES ARE IMPORTANT

At the ASEV convention in 1982, a breakthrough in viticulture commanded everyone's attention. It was explained that vine structure influences grape composition: not just sugar level, but potassium and pH and vegetative aromas and more.

This information opened a whole new world for those of us who had tended to ignore grapes until they arrived at the winery. The notion that vine shape influenced wine quality was probably no surprise to old-timers, but for many enologists it was a revelation.

We were not sure what to do with this knowledge until the 1988 Cool Climate Symposium in Auckland, New Zealand. Suddenly, scattered information fell into place and started to be meaningful. Canopy management was no longer just a phrase, it was linked with wine chemistry. Now we recognize that "fine wine begins in the vineyard" is not just a cute marketing slogan; it is essential for viticulture and enology to go hand in hand.

Vine diseases inspire many current discussions about their effect on ripening and grape composition. Our involuntary experimentation with phylloxera will no doubt continue. In another 10 or 20 years our grasp of the relation of vine physiology to wine composition will seem as antiquated as our old ideas do now.

ATTENTION ON YEASTS

No enological field has moved so quickly as that of microbiology. Twenty years ago, for most wineries yeast selection was like choosing between vanilla and chocolate: Pasteur Champagne and Montrachet were the only dry wine yeasts made on a large scale. And glad we were to get them, too; many winemakers remembered that prior to 1962 there were no dry wine yeasts at all.

Trying different yeasts was unusual then, but during the 1970s a few wineries began fermenting the same must with different yeast strains. It was a lot of work; some strains had to be cultured from slants, others imported with great difficulty from Europe. As the years went by, most wineries branched out and tried different strains; some have used spontaneous fermentations to assess indigenous vineyard yeasts. By now a much greater sophistication toward yeast has developed, so winemakers choose yeast strains for the fermentation characteristics they wish.

Scientific information about wine yeast metabolism has increased tremendously in the past few years. Because of some truly revolutionary research work, we are beginning to get inside the yeast cell to find out what is really going on in there.

After many years of stuck fermentations (starting in 1977, the first full drought year, and becoming endemic the next), Californians knew that adding nutrients to already stuck wine usually did not help, but we were not sure why, or what to do about it. Research now is beginning to explain, in detail, the phenomenon of nitrogen uptake by the yeast cell, including inhibition by alcohol in the latter stages of fermentation.

Winemakers also noticed that some yeast strains tend to ferment to dryness more easily than others; lo and behold, they use nitrogen differently, as well. Disagreements about how yeasts act is becoming less confusing now that we know about yeast behavior changes depending on must composition.

It is like a huge jigsaw puzzle, in which finally some pieces are fitting together, and at least some of the picture is starting to emerge. Every few months, more pieces are joined, and sometimes they join up with other ones. This is one of the most exciting aspects of enology today and in the future.

INOCULATION FOR MALOLACTIC FERMENTATION

The industry's approach to malolactic fermentation (MLF) has undergone an enormous transformation, and should continue to do so. Over the past two decades, MLF has started to become a science instead of a mystery, although much folklore and confusion still surrounds bacterial habits, preferences, and metabolism. I am astounded when I consider what strides have been made with ML bacteria, and daunted when I realize that we are just beginning to get a glimpse into their world.

Twenty years ago, most wineries let MLF happen or not and hoped for the best. Winemakers shrugged their shoulders at the unfathomable workings of the little "bugs." Many wineries, even some producing millions of gallons, did not have a paper chromatography setup, and the enzymatic method for malic acid was just being developed.

In the early work with ML inoculation, a microbiology laboratory was required. In 1976 an enology student and I spent all summer developing a suitable broth, offering the first user-friendly *Leuconostoc* cultures that year. Now, ML inoculation is practiced routinely in most wines intended for MLF, using a variety of strains and preparations.

Most current ML research worldwide is devoted to understanding differences between the various genera, and strains within these genera. What exactly do ML bacteria metabolize, and when? What do they produce, and how, and why? What inhibits them and what stimulates them? How does *Lactobacillus* differ from *Leuconostoc,* and how do *Leuconostoc* strains differ from each other?

Work on these topics is in progress by dozens of researchers. Scientists in California, New York, Australia, New Zealand, South Africa, France, Germany, Switzerland, Argentina, and other places around the world are feverishly working on the conundrum of MLF.

Recently the industry, and research groups around the world, have tackled the very complex issue of interaction of yeast and ML bacteria. It is easy to notice in the cellar that some yeasts seem to interact more favorably with some bacteria than others, at least in some situations. It is entirely another matter to make sense of it at all. Efforts to elucidate what goes on between yeast and bacterium will occupy many people for quite some time.

SPOILAGE MICROBES

We are fascinated by spoilage. We want to understand how it happens, and, more important, help winemakers avoid microbial disaster. A few kindred spirits are at work on spoilage research. An ongoing investigation of the strains and pathways involved in *Lactobacillus* spoilage will be of major importance to red wine producers. Further understanding of *Pediococcus* growth in cellared reds, especially Pinot Noir, is a topic for future work.

We are also learning more about *Brettanomyces* and *Dekkera.* Knowing that they metabolize barrel extractives explains why infections seem more virulent in new barrels. There has long been controversy about how the yeasts are transferred, how they grow, and whether or not spoilage will result, but with further study on their nutritional requirements and sensitivities, winemakers should be able to relate to these yeasts with confidence.

Complaints of exploding concentrate drums and sorbated wines inexplicably fermenting in the bottle led in 1985 to the identification of *Zygosaccharomyces bailii.* For several years many had mistakenly assumed that we were just seeing a rogue *Saccharomyces,* but it was a new wine spoilage yeast instead.

LOOKING TO THE FUTURE

Other areas of wine technological research that have initiated changes in winemaking procedures include chromatographic flavor and aroma analysis, phenolic chemistry, ethyl carbamate, sulfide and mercaptan relationships, and tartrate equilibria, and the list continues to grow.

An exciting development is the use of computer technology to facilitate transfer of information. The international LACTACID computer conference on Internet, in progress since May 1993, illustrates how the computer age links scientists together.

There are still many questions to answer, and those will give rise to more. Some experiments will be flawed, others brilliant. People will continue to share and withhold data, agree and refute, collaborate and quarrel, but on a global scale more than ever before. The result is that winemakers can make more informed choices about where they want to go with a wine and how to get there.

Someone once asked Paul Pontaillier, of Chateau Margaux, why he did not try out this or that new idea in his winery. He replied that to him, winemaking was like an elephant surrounded by dinner plates. "The elephant takes one step, he breaks many plates." His point was that he wanted to understand what would happen before he changed his winemaking procedures. This understanding is what the technological revolution has been about, not reducing winemaking to formulas or work orders or "painting by the numbers."

The charge is sometimes leveled that with technological advancements, the "romance" of winemaking is lost. It can happen, but it does not have to be that way. In our search for greater knowledge of the ancient activity of winemaking, maybe we can help the elephant step a little more consciously.

Wine and Health—It Is More Than Alcohol

Carlos J. Muller
Enology Program, California State University, Fresno, CA

Dedicated to the Memory of John Kinsella and Salvatore Lucia

There is no doubt that wine is a healthful beverage. Wine has been consumed through the ages as food and as a food adjunct. Since antiquity, the virtue of wine as a panacea has been widely exploited in folklore and in the medical arts and sciences (Lucia, 1954, 1963). However, most of the purported attributes of wine that contribute to health were, until only recently, anecdotal; there was no hard experimental evidence to substantiate any such health claims.

Recently, however, two powerful and often diametrically opposed forces have come into play. The first is the efforts of the antialcohol establishment to find fault, and thus do away through often onerous regulation and taxation, with any and all of those beverages considered 'sinful.' Their efforts have poured millions of dollars in private and tax monies for research directly aimed at finding anything and everything of a negative connotation in beverages that happen to contain alcohol. Much of this research has centered not necessarily on the beverages themselves, but on those individuals who, for whatever reason, abuse alcoholic beverages. Indeed many ills of society can be directly traced to individuals who are prone to abuse. However, a blanket indictment of alcoholic beverages is not rational. It is the individual's choice to abuse or not.

The second, even those researchers noted for their vituperous attacks on alcoholic beverages (their null hypothesis being alcohol is bad; therefore . . .), cannot rationally discard a growing body of evidence indicating that there are indeed health benefits associated with moderate consumption of alcohol. It is interesting and amusing that such evidence has been obtained in great part through efforts by the very same researchers who have benefited handsomely from antialcohol funding!

Further, recent rigorous epidemiological studies of the French, a population who might be at health risk because of factors in their diet and lifestyle, such as high consumption of cholesterol-laden foods and smoking, have shown that these people have lower rates of cardiovascular disease and death than comparable individuals whose normal diet does not include wine (Renaud and de Lorgeril 1992; Seigneur et al. 1990). These classic studies have resulted in the now household term "The French

Paradox." Surely, such a paradox is worthy of a logical explanation. Why is wine so good for you? Certainly, wine offers some health attributes above and beyond those provided by alcohol alone and that are not shared by other beverages containing alcohol. Which wine constituents are responsible, either directly or indirectly, for these real or putative health benefits?

The need to answer such questions has triggered a flurry of reports dealing with those constituents in wine that have, or might have, health significance (Fitzpatrick et al. 1993; Frankel et al. 1993; Gurr 1992; Kinsella et al. 1992; Muller and Fugelsang 1993; Sharp 1993; Troup et al. 1994; Waterhouse and Frankel 1993). Most reports have centered around antioxidants, and many have dealt with the effect of these antioxidants on low-density-lipoproteins (LDLs) and cardiovascular wellness.

ANTIOXIDANTS

Ever since the classic papers by Harman (1956), and Tappel (1968), in which they proposed the then provocative contention that antioxidants might provide longevity, their thoughts have been echoed repeatedly (Ames 1989; Ames and Shigenaga 1992; Bunker 1992; Cutler 1991; Gutteridge 1992; Mehlhorn and Cole 1985; Pratt 1993; Sohal and Orr 1992), and research in the area of these ubiquitous substances has grown logarithmically (Afanas'ev et al. 1990; Aruoma 1993; Aruoma et al. 1993; Bendich et al. 1986; Beutler 1989; Bors et al. 1990; Cuvelier et al. 1992; Frei et al. 1989; Freisleben and Packer 1993; Halliwell and Cross 1991; Jackson 1993; Jackson et al. 1993; Jenkins 1993; Laughton et al. 1989; LeWitt 1993; Maxwell et al. 1993; Miller et al. 1993; Poli et al. 1993; Pryor 1991; Rose and Bode 1993; Scott et al. 1993; Shahidi et al. 1992; Sinclair et al. 1990; Stocker et al. 1990; Torle et al. 1986; Tsukamoto 1993; Vanella et al. 1993; Yoshikawa 1993). Indeed antioxidants have been found to protect a plethora of systems, in vitro, ex-vivo, and in vivo, from the ravaging attacks of harmful substances, both endogenous and exogenous, to which we are exposed on a daily basis (Babbs 1990; Bowling et al. 1993; Church and Pryor 1985; Comporti 1989; Cross et al. 1987; Gutteridge 1993; Halliwell 1993a, 1993b, 1993c; Halliwell and Gutteridge 1985a, 1985b, 1990; Kanner et al. 1986; Kumar and Das 1993; McCord and Omar 1993; Pittilo 1990; Poirier and Thiffault 1993; Pryor and Stone 1993; Richardson 1993; Sanfey et al. 1986). Most of these harmful substances are free radicals.

Free Radicals

Free radicals are chemical entities that have single, unpaired, unshared electrons. Because of these single electrons, they are somewhat unstable, that is, short lived; they are also highly reactive by virtue of the fact that, in order to gain chemical stability, they seek to either lose or gain an electron (Halliwell 1993d). Thus, these species are very prone to interact with many molecules such as membrane lipids and DNA and, in the process, often damage such molecules irreparably (Ames 1989; Feeney and Berman 1976; Jain et al. 1991; Laughton et al. 1989; LeVine 1992; Mello-Filho et al. 1984; Pryor 1982; Rehan et al. 1984; Sies and de Groot 1992; Ward et al. 1986; White et al. 1994). Free radicals are formed by a variety of processes: chemical, physical, and enzymatic. In our daily lives, we are exposed to all such processes, often simultaneously. No tissue or cell is immune to the onslaught of free radicals thus produced.

Among the most damaging free radicals to which we are exposed are those that are oxygen based (Becker and Ambrosio 1987; Beckman and Crow 1993; Cadenas 1989; Cochrane 1991; Cross et al. 1987; Feeney and Berman 1976; Fridovich 1986; Granger et al. 1986; Halliwell 1993a, 1993b, 1993c, 1993d; Halliwell and Gutteridge 1985a, 1985b, 1986; Halliwell et al. 1985; Lee-Ruff 1977; McCord 1993; Rehan et al. 1984; Ryan and Aust 1992; Sies and de Groot 1992; Ward et al. 1986). We are exposed to this type of radicals because we have evolved to exist in an oxygen-based environment. We need oxygen, and its relatives, for our daily and common metabolic processes such as respiration and the generation of energy. Although oxygen from the air we breathe is in itself not excessively reactive, some of its relatives such as peroxide radical ion are.

Fortunately, we are also equipped, after eons of continuous challenge and evolution, with the mechanisms to take care of these metabolic insults (Jackson et al. 1993; Lind et al. 1982; Maxwell et al. 1993; Stocker et al. 1990; Yu 1994). However, we cannot defend ourselves effectively without help from the foods we eat (Ames 1983; Aruoma 1993; Bendich et al. 1986; Beutler 1989; Castonguay 1993; Kinsella et al. 1992; Laughton et al. 1989; Liu et al. 1992; Muller and Fugelsang 1993; Newmark 1987; Pratt 1993; Pryor 1991; Shahidi et al. 1992; Singleton and Esau 1969; Stich 1991; Stich and Rosin 1984; Teel and Castonguay 1992; Warner et al. 1986). The reason for this is that we, as opposed to plants, cannot synthesize certain compounds because we lack the systems to produce them.

Among those defense compounds that we are precluded from producing endogenously are ascorbic acid (Bendich et al. 1986), and those derived from the shikimic acid pathway and its many branches thereof (Bentley 1990). The latter compounds, which contain a benzene ring, or its analogs, are among the best suited antioxidants, often specifically designed by nature to effectively take care of damaging free radicals. They are also produced by plants to prevent attack by viruses, bacteria, fungi, and other pathogens (Malamy and Klessig 1992; Muller et al. 1994; Raskin 1992a, 1992b; Sieman and Creasy 1992; Yalpani and Raskin 1993).

Of these defense chemicals, phenolics, both simple and complex, are ubiquitous in plants; they are plethoric in grapes and wine (Bachman 1978; Cartoni et al. 1991; Gorinstein et al. 1993; Kanner et al., 1994; Kinsella et al. 1992; Mahler et al. 1988; Muller et al. unpublished data; Salagoity-Auguste and Bertrand 1984; Scholten and Kacprowski 1993; Shahidi et al. 1992; Singleton and Esau 1969; Steffany et al. 1988; Stich and Rosin 1984; Torle et al. 1986). However, defense is not the only function of some of these phenolics. Skin pigments, for instance, also have the specific function of attracting useful predators to the fruit for propagation of the species.

Antioxidant Function

Whether a compound functions as an antioxidant in biological systems depends not only on its intrinsic chemical structure but also, and perhaps more important, on the biochemical environment in which it operates. It is also crucial that a putative antioxidant be capable of reaching the target cells or tissues and, in addition, that once there, be present in sufficient concentration to effectively take care of the biological insult (Halliwell 1990). Further, it is also imperative that once it has performed its beneficial function, the now spent antioxidant be efficiently removed to prevent further interactions with surrounding species.

Thus, in characterizing an antioxidant in grapes or wine, one must be careful and cognizant of the above constraints before making claims on a compound as an antioxidant in humans simply because such compound might work in vitro or ex vivo (Halliwell 1993e; Muller and Fugelsang 1994c).

Antioxidant–Pro-oxidant

Another question that must be asked is: What happens to an antioxidant once it has performed its defensive duties? In most instances, unfortunately, our knowledge of their biological fate is not fully understood. Neither is the fate of other potential target molecules that could be attacked by the now spent (oxidized) antioxidant. Many antioxidants might thus become pro-oxidants (Aruoma et al. 1993; Freisleben and Packer 1993; Halliwell 1990; Laughton et al. 1989; Smith et al. 1992; Snyder and Bredt 1992; Stich 1991). Whether a compound acts as an antioxidant or as a pro-oxidant depends on its oxidation state and that of other species with which it can interact. In reactions of this type, electrons move according to a rigorous hierarchy of electrochemical potentials (Koppenol 1985, 1990).

It is therefore important that a spent antioxidant be deactivated immediately after performing its function; otherwise, it might conceivably do harm as an oxidized species. Deactivation is often carried out by other antioxidants, by destruction or scavenging by other species such as enzymes, or by rapid removal from the tissue involved (Halliwell 1990). Fortunately, there are many mechanisms by which pro-oxidants might be inactivated; unfortunately, they are not operant in all cases (Yu 1994).

Wine Antioxidants

In wine, many constituents fit into the category of biological antioxidants capable of functioning as such in vivo, as defined above (Fitzpatrick et al, 1993; Frankel et al. 1993; Kinsella et al, 1992; Muller and Fugelsang 1993; Muller and Fugelsang unpublished data; Waterhouse and Frankel 1993). Others are, at present, only potential or putative antioxidants inasmuch as their absorption from the intestinal lumen, and other in vivo studies, have not yet been carried out. Absorption from the gut is often dictated by the absence or presence of other foods within the gut, the intestinal flora and its ability to metabolize such foods, gut enzymes and hormones, and other intrinsic factors such as age and general health of the individual (Crotty 1994; Read et al. 1994). In the case of phenolics, it is a well-known fact that proteinaceous materials interact strongly with phenolics to form macromolecules whose absorption from the gut might not be so facile (Hurrell and Finot 1984). Absorption studies for wine constituents might prove to be a fruitful area of research. It should be pursued vigorously, as should research aimed at isolating and identifying other potential antioxidants present in wine that might strengthen contentions pertaining to wine's beneficial effects on human health.

Antioxidant Function

Chemically, antioxidants work in a number of ways. The simplest way is by being able to effectively scavenge reactive species (Halliwell 1993d). Another manner is by providing either a hydrogen atom (which contains an unpaired electron), or, in some cases, a hydride (which contains a set of two paired electrons) onto a target substrate; these are classic one- or two-electron reductions. In the first category, substances such as salicylic and benzoic acids, which are capable of scavenging the highly reactive hydroxyl radical to form hydroxy compounds, are typical (Afanas'ev et al. 1990; Aruoma et al., 1993; Bors et al. 1990; Cuvelier et al. 1992; Frankel et al. 1993; Freisleben and Packer 1993; Halliwell 1993a, 1993b, 1993d; Kinsella et al. 1992; Koppenol and Bartlett 1993; Laughton et al. 1989; Levy 1979; Maskos et al. 1990; Rose and Bode 1993; Sagone and Husney 1987;

Scott et al. 1993; Yoshikawa 1993). Substances such as thiols and hydroquinones (both *ortho-* and *para-*) and other compounds possessing hydroquinone-like structures (such as ascorbic acid), in addition to the coenzymes $FMNH_2$, $FADH_2$ and NAD(P)H, fit the second category (Asmus 1990; Beutler 1989; Bors et al. 1990; Chou and Khan 1983; Cuvelier et al, 1992; Frankel et al, 1993; Freisleben and Packer 1993; Halliwell 1993d; Kinsella et al, 1992; Laughton et al. 1989; Meister 1992; Rose and Bode 1993; Scott et al. 1993). There is indeed a long list of these. Some, are capable of functioning as both free radical scavengers, and as electron donors (reductants), depending on the environment in which they operate. Many of these compounds are conversant with each other in vitro and in vivo in a manner analogous to the interaction between FMN and NADH.

Grapes contain little ascorbic acid; most is metabolized into tartaric acid shortly after manufacture by the plant (Saito and Kasai 1969); however, grapes contain an impressive array of phenolic compounds. Among the grape and wine antioxidants there are many that fit the category of radical scavengers. Salicylic and benzoic acids, and their metabolites, fit this category. But there is also a multitude of others that possess the *ortho-* or the *para-*hydroxy structural features of the hydroquinones. Substances such as gallic acid; 2,3- and 2,5-dihydroxybenzoic acids (DHBs); epicatechin; quercitin; and the grape pigment cyanidin fit the second category, and in some instances, both.

COEXISTENCE (SYNERGISM) OF ALCOHOL AND WINE ANTIOXIDANTS

Wine is unique among beverages in that it contains both **alcohol** and **antioxidants**. This coexistence has profound health implications in that in wine, unlike many other systems, its antioxidants are protected once they enter the body by the very same processes that our body uses to detoxify ingested ethanol. In the liver, and to a lesser extent in other organs, two NAD^+-dependent oxidizing enzymes, alcohol dehydrogenase (ADH) (E.C. 1.1.1.1.), which converts ethanol to acetaldehyde, and aldehyde dehydrogenase (AldDH) (E.C. 1.2.1.3.), which converts acetaldehyde to acetate, produce NADH in each step (Bullock 1990; Kennedy and Tipton 1990). This NADH (two reducing equivalents produced per molecule of ethanol detoxified) is then capable of recycling spent antioxidants by reducing them and, in the process, regenerate NAD^+, which in turn is required to detoxify more ethanol. Since most table wines contain about 12% ethanol (vol/vol), a plethora of reducing equivalents are produced during the ethanol detoxification process. Excess reducing equivalents might be deleterious in cases of abuse by keeping moieties in a reduced instead of in an oxidized state required to maintain tissue homeostasis.

Thus, the health implications of coexistence of antioxidants and ethanol are (1) antioxidants are recycled, minimizing formation of pro-oxidants, thus preventing further oxidative damage to tissues; (2) antioxidants are available in their reduced state; and (3) excessive levels of NADH, generated by the detoxification of ethanol, are lowered, thus minimizing reductive damage to other systems. These are unique properties of wines that are not necessarily shared by other alcohol-containing beverages. These facts might help explain, at least in part, why wine drinkers do not become as easily inebriated as those who consume equivalent amounts of ethanol from other sources.

The often quoted premise: "ethanol, is ethanol, is ethanol," might not necessarily hold true for wine

Dealcoholized wine, while still plethoric with antioxidants, might not offer these combined benefits. Further, because no (or very little, e.g., 0.5%) alcohol is present, antioxidant absorption from the gut might not be so facile.

Metal Chelation

Vicinal enediol compounds, such as 2,3-DHB, cyanidin, epicatechin, and quercitin, are also capable of chelating metals, particularly transition metals such as iron (Brune et al, 1989; Graziano et al, 1974; Halliwell and Gutteridge 1985b, 1990; Laughton et al, 1989). These compounds also mimic, because of their similarity of structural features, many, if not all, of the amply proven health-related attributes of ascorbic acid (vitamin C) (Bando and Obazawa 1990; Bendich et al. 1986; Bielski et al. 1975; Blondin et al. 1987; Bulpitt 1990; Chou and Khan 1983; Frei et al. 1989; Lohman 1987; Martell 1982; Meister 1992; Rainieri and Weisburger 1975; Varma 1991). When iron is chelated as Fe(III) (the most common form of nonheme food iron) by an *ortho*-hydroquinone (or by ascorbic acid), the iron is capable of being reduced to Fe(II) by the intramolecular action of the one electron reducing ability of the same molecule that holds the metal (Halliwell and Gutteridge 1985b). The result is that the reduced Fe(II) is then capable of generating the powerful oxidant, hydroxyl radical, via Fenton pathways (Goldstein et al. 1993; Halliwell 1993d). The generated hydroxyl radical is then capable of exerting its action onto other molecules unless it is immediately quenched by excess antioxidant, or by an appropriate enzyme-mediated defense mechanism. Furthermore, some of these vicinal enediol-possessing compounds are ideally suited and capable of bringing chelated iron into the portal circulation from the gut, a property shared by many of these compounds such as ascorbic acid (Graziano et al. 1974; Horning 1975).

These compounds (and foods that contain them) might be helpful as dietary coadjuvants in the treatment of cases of iron-deficiency anemia, and for women during their procreative years when iron demands by the body are great. They might also be helpful for young people during growth and development when both hemoglobin and myoglobin are rapidly being manufactured. It must also be remembered that iron is a crucial component of a plethora of enzymes and systems in the body that carry out vital functions such as defense and respiration. Both white and red wine enhance iron absorption from foods (Bezwoda et al. 1985). The authors also indicate that iron absorption from the gut is enhanced by simple, but not by complex phenolics.

The ability to absorb iron from the gut helps prevent iron-deficiency anemia; in elderly men and women, however, these dietary iron chelators might increase the body iron burden. This is of concern only in those individuals whose body iron holding capacity by the iron-storage proteins ferritin and hemosiderin is at risk of being exceeded. This applies specifically to older men who, because of lifelong dietary habits, have managed to accumulate considerable iron stores. It must be remembered that men, as opposed to women (from menarche to menopause), cannot easily dispose of more than about 1 to 2 mg of iron per day; thus, dietary iron intake, which in man often exceeds the disposal amount by several orders of magnitude, accumulates continuously (Carpenter and Mahoney 1992). This inordinate accumulation of iron is often the cause of many ailments of old age (Halliwell and Gutteridge 1985b, 1986, 1990, 1992; Halliwell 1993d; Herbert 1994; Rice-Evans 1989; Ryan and Aust 1992; Sempos et al. 1994). It has even been suggested that excessive iron accumulation might be important in the oxidative stress etiology of dementia (Roche and Romero-Alvira 1993). Inordinate iron excess might be directly responsible for the somewhat shorter average life span of men as compared to women. It must be remembered however, that most dietary iron arises from sources other than wine, such as meats, grains, and vegetables.

The iron-transport protein transferrin, and the iron-storage proteins ferritin and hemosiderin, normally hold iron in the body in an inactive state (Halliwell and Gutteridge 1990). Aside from the small amount of iron normally released from hemoglobin when the molecule commits suicide in the body at about 120 day intervals (with the concomitant production of biliverdin, another powerful antioxidant), there is very little, if any, free iron in blood plasma and other fluids (Halliwell and Gutteridge 1986; Ryan and Aust 1992). This is not the case, however, in the case of trauma, when blood and its components extravasate and flood the surrounding tissues.

Under these circumstances, iron-initiated damage by hydroxy- (HO·), peroxy- (O_2·), nitric oxide (NO), and other active radicals might be considerable (Beckman and Crow 1993; Choi 1993; Cochrane 1991; Graf et al. 1984; Halliwell and Gutteridge 1992; Puppo and Halliwell 1988).

Antioxidant Effect on Glutathione

Antioxidants might also work by helping retain the integrity of certain crucial sulfhydryl (thiol) groups such as those in glutathione (Asmus 1990; Beutler 1989; Butler and Hoey 1992). Glutathione is required in a myriad of homeostatic mechanisms in the body as varied as protection of mitochondrial DNA in the human brain (Ames 1989), intestinal absorption of amino acids (Meister 1973), and prevention of cataract formation (Bando and Obazawa 1988). Further, glutathione is required in many detoxification mechanisms including those of medicines (e.g., acetaminophen) and other xenobiotics. Glutathione is easily oxidized by endogenous and exogenous substances such as activated oxygen and radiation (including light) to the disulfide, a compound that offers lessened, if any, protection against oxidative insults (Asmus 1990; Beutler 1989; Meister 1992; Yu 1994).

In the eye, glutathione protects by scavenging free radicals formed photochemically. Formation of free radicals there eventually causes cross-linking of sulfhydryl groups in proteins with resultant formation of opacities and eventually blindness, unless protection in the form of an antioxidant such as glutathione is present in both the lens and in the vitreous humor of the eye (Bando and Obazawa 1988; Blondin et al. 1987; Spector et al. 1993).

Youngsters (and adults) often watch endless hours of television. Their proclivity to remain in close proximity to a source of strong photic stimulation might endanger their visual elements prematurely, unless their diet includes significant amounts of antioxidants. Dietary antioxidant requirements may be higher for people who spend a significant portion of their working time in front of a computer screen.

By their action on glutathione, if not directly, wine antioxidants might help prevent diabetes and associated sequelae such as visual loss. Antioxidants might also protect against diabetes caused by iron overload (hemochromatosis), a condition to which several ethnic groups are genetically predisposed (Halliwell and Gutteridge 1985a, 1985b). Some antioxidants that possess vicinal enediol structures (such as ascorbic acid and 2,3-DHB) might be useful in removing excess iron in conditions of iron overload (Muller, unpublished data). In addition, antioxidants might help prevent certain forms of liver damage, including cirrhosis and cancer (Kennedy and Tipton 1990; Poli et al. 1993). Protecting the integrity of reduced glutathione is crucial in preventing DNA damage from oxidative insults. Most DNA is amply protected by several inherent defense mechanisms. However, not all DNA is protected in all tissues: the DNA of some organelles is actually unduly exposed to injury (Jain et *al.* 1991)

Antidegenerative Diseases (Parkinson's, Alzheimer's, and Rheumatoid Disease)

Recent reports have implicated free radicals in the etiology of degenerative diseases of old age such as Parkinson's, Alzheimer's, gout, rheumatism, cataracts and other visual impairments, and rheumatoid arthritis (Ames et al. 1993; Bando and Obazawa 1990; Gutteridge 1992, 1993; Halliwell and Gutteridge 1985a; Halliwell et al. 1985; Kopin 1993; LeWitt 1993; Poirier and Thiffault 1993; Richardson 1993; Sohal and Orr 1992; Taylor et al. 1993; Varma 1991; Yoshikawa 1993). In many of these diseases, it is the lifelong accumulation of oxidative insults that eventually presents itself with frank pathologies. This emphasizes the importance of early and continuous, but moderate, exposure to dietary antioxidants.

Anti-Inflammatory and Anti-Low-Density Lipoprotein

Some of the compounds possessing antioxidant properties also have the ability to interfere with the inflammatory response in humans (Esterbauer et al. 1991; Kanner et al. 1986). Wine antioxidants might also help combat the painful inflammation of arthritic tissues. Further, some wine antioxidants such as salicylic acid and the DHBs have proven and powerful antipyretic and analgesic properties (Brune 1974; Davison 1971; Grootveld and Halliwell 1988). The recently isolated and identified phytoalexin resveratrol is an example of wine constituents capable of preventing LDL oxidation in vivo (Frankel et al. 1993; Kinsella et al. 1992; Waterhouse et al., 1993). A similar function, known for more than a century, has been amply documented for salicylic acid and its metabolites gentisic and 2,3-DHB, which are present in grapes and wine in larger concentrations than resveratrol (Muller and Fugelsang 1993; Muller and Fugelsang, unpublished data; Waterhouse and Frankel 1993). However, it must be remembered that amount, by itself, is not necessarily a measure of biological activity. Some constituents have powerful pharmacological action in minute amounts.

Further, some of these compounds also have the separate ability to prevent the formation and deposition of low-density lipoproteins (LDLs), and their oxidation products, in vascular endothelium (Esterbauer et al. 1991; Fitzpatrick et al. 1993; Gurr 1992; Kinsella et al. 1992; Waterhouse et al. 1993). LDLs are the lipoprotein fraction directly responsible for narrowing of blood vessels through their deposition and therefore for eventual ischemia of the heart and other organs with often lethal consequences. Formation of LDLs is also prevented by ethanol (and its metabolite, acetaldehyde) in vivo. (Anon., 1993; Friedman and Klatsky 1993; Gaziano et al. 1993; Ohlin et al. 1991; Savolainen et al. 1987; Suzukawa et al. 1994).

Nitric Oxide Carriers

In examining the literature of salicylic acid and other antioxidants, there seems to be a certain measure of interrelationship and pharmacological similarity between these compounds and the recently elucidated functions of the physiological messenger, nitric oxide (NO). However, these interrelationships, although tacitly apparent, have not been explicitly formulated. Is it possible that their actions are somehow intimately related?

A Hypothesis

I hereby propose that wine antioxidants such as salicylic acid and its metabolites might have an even more important function, in addition to their amply proven antioxidant and other traditional tasks: *that of being able to bind, transport, release, and permutate NO, and its oxidation-reduction analogs, nitroxyl anion (NO^-), and nitrosonium cation (NO^+).*

Salicylic (and its metabolites: 2,3- and 2,5-DHB), and gallic acids are ideally suited to perform these crucial NO-related functions because they have proven ability to reach and pharmacologically act in virtually any tissue in the body, and because their chemical structure (phenols or benzoates, or both) provides them with the ability to accept, stabilize, and release any of the three oxidation state forms of NO. This is accomplished either at an allylically stabilized position in the ring, or on a semiquinone oxygen. Further, 2,3-DHB can, as mentioned above, chelate iron(III), which provides an excellent bonding site for nitric oxide with subsequent release as nitrosonium ion (McCleverty 1979; Stamler et al. 1992a). The vasodilating and LDL-controlling ability of some of these compounds might be intimately tied to their ability to act as carriers and releasers of NO and its redox congeners. Nitration of salicylic acid and other phenolics by peroxynitrite has been reported in vitro (Koppenol and Bartlett 1993), and in vivo (Beckman et al. 1992). NO was named "molecule of the year" in 1992, and rightly so (Culotta and Koshland 1992). No other molecule has attracted as much attention in recent years. Ever since Furchgott (1984, 1988) and Palmer, Ferrige, and Moncada (1987) suggested that NO might be identical to endothelium-derived relaxing factor (EDRF), a substance that allows blood vessels to relax and thus permits better blood flow, intensive investigation has resulted in the finding that this free radical (or its oxidation-reduction congeners nitroxyl- and nitrosonium ions) is not only responsible for blood vessel relaxation, but it has been implicated in a multitude of crucial functions and activities as diverse as neurotransmission (Choi 1993; Garhwaite et al. 1988; Kontos 1993; Lowenstein and Snyder 1992; Lowenstein et al. 1994; Snyder and Bredt 1991), the immune system (Collier and Valance 1989), control of coronary vascular tone (Kelm and Schrader 1990), prevention of leukocyte adherence (Gaboury et al. 1993), microvascular permeability (Kubes and Granger 1992), vasodilation (Casino et al. 1993), penile erection (Burnett et al. 1992), cerebrovascular and neuronal regulation (Kontos 1993), antioxidant protection (Mehlhorn and Swanson 1992), and many others (Butler and Williams 1993; Snyder and Bredt 1992).

NO might be the first of a probable series of putative small molecules involved in intra- and intercellular communication (Gall, 1993; Garhwait, et al. 1988; Ignarro 1991). NO is, in some instances, cytoprotective (Cooke and Tsao 1993), and might prevent cardiac reperfusion damage (Gelvan et al. 1991); however, it does not seem to be beneficial to its target cells or tissues at all times. Under some circumstances, NO can be damaging, just as in the case of highly reactive oxygen species (Beckman and Crow 1993; Choi 1993; Crotty 1994; Matheis et al. 1992). This is apparently due to its ability to operate at different redox potentials (Lipton et al. 1993; Stamler et al. 1992a), and in addition, to its ability to interact with superoxide to form peroxynitrite (or as it should better be known, oxoperoxonitrite) (Beckman et al. 1990; Hogg et al. 1993; Koppenol et al. 1992). This latter compound has been implicated in the severe cell and organ damage caused by organ reperfusion after ischemia (Adkinson et al. 1986; Matheis et al. 1992). Peroxynitrite is an excellent bactericide due to its powerful oxidant ability (Zhu et al. 1992). Its protonated form, peroxynitrous acid, is an endogenous source of free radicals that might have a function in the body's ability to ward off infections (Yang et al. 1992).

As knowledge in this field stands today, NO is most often manufactured in situ from L-arginine by the action of NO synthase, a ubiquitous (both constitutive and inductive) enzyme that requires, depending on the production site, FMN, tetrahydrobiopterin, calmodulin, and other cofactors (Ignarro 1990; Leaf et al. 1989; Moncada et al. 1989, 1991a, 1991b). In this complex reaction, molecular oxygen is involved. One atom of oxygen is attached to NO and another forms the carbonyl oxygen in citrulline (Palmer and Moncada 1989; Tayeh and Marletta 1989).

However, in some instances, NO (and its congeners) are not manufactured at the site of action but are produced elsewhere and then transported by a suitable carrier. This transfer function has been ascribed primarily to thiol-containing compounds, and to thiol-containing proteins such as albumin (Kontos 1993). The latter has also been implicated in interconverting "bad" NO into "good" NO, and vice versa (Stamler et al. 1992b). In these instances, NO forms a nitrosothiol adduct that is ideally suited to perform its ascribed duties. Albumin also exerts protective action against spent antioxidants. However, albumin does not easily cross all cell membranes; salicylic acid and its metabolites 2,3- and 2,5-dihydroxybenzoic acids apparently do.

Because of their molecular structure, wine constituents such as DHBs, cyanidin, epicatechin, gallic acid, quercitin, resveratrol, and salicylic and other hydroxybenzoic acids seem ideally suited to quench or inactivate oxoperoxonitrite (Koppenol and Bartlett 1993), as well as the other above-mentioned reactive and cell damaging species.

SALICYLIC ACID

This ubiquitous compound has a long history of health-related attributes. Its pharmacological activity ranges from common antipyretic and analgesic properties, to the implication that it might provide a certain measure of cardiovascular wellness because of its fibrinolytic activity in whole blood (Moroz 1977), a property which it shares, perhaps via different mechanisms, with alcohol (Hendricks et al. 1994, Ridker et al. 1994). Fibrin formation appears to have as much importance in the etiology of atherosclerotic plaque formation and lesions as LDL oxidation (Smith and Thompson 1994). Salicylic acid (SA) is also an outstanding antioxidant capable of quenching the damaging hydroxyl radical (Kaur and Halliwell 1994, Obata et al. 1993). In this regard, it may have antiatherogenic properties by ameliorating LDL oxidation. These actions however, should not be confused with the antiplatelet-aggregation (antithrombogenic) function of aspirin (acetylsalicylic acid) which is due to the latters' ability to irreversibly inhibit thromboxane (TxA_2) formation and subsequent release from platelets by acetylating a key serine residue on the enzyme cyclooxygenase (Roth et al. 1975). The apparent confusion arises from the thought that atherogenic and thrombogenic processes are the same; they are often intimately related in vaso-oclusion, however they are distinct (Loscalzo 1992).

Salicylic acid has little action on thromboxane formation and release from platelets (Roth and Mejerus 1975), and is only a weak inhibitor of prostaglandin synthesis (Kaur and Halliwell 1994). Further, there should not be detectable levels of aspirin in wine unless it has been purposely added, a practice which is not permitted by current regulations.

Lately, most studies involving aspirin have centered on the benefits derived from its ability to block thromboxane formation. Yet, aspirin is rapidly deacetylated in the body to yield salicylic acid (Rowland et al. 1972). Further, aspirin has a relatively short residence time, whereas salicylic acid (and its metabolites) remain in blood plasma longer than aspirin. Thus, it stands to reason that many functions commonly ascribed to aspirin might in fact be due to salicylic acid. Salicylic acid, for instance, is equipotent with aspirin in inhibiting the chemotactic generation of 12-HETE from 12-HPETE in the lipoxygenase pathway of arachidonic acid metabolism by leukocytes, a key reaction in prevention of inflammation (De Gaetano et al. 1985).

Salicylic acid is effective against some viral (e.g., endocarditis) and bacterial infections in humans. SA (and recently also, wine) has been shown to be protective against the common cold (Cohen et al., 1993). Thus wine, because of its content of SA and similar compounds, might be potentially helpful in providing a certain measure of wellness in addition of the feeling of well-being naturally produced by the serenogenic (relaxant) compounds of wine and of alcohol (Delin and Lee 1992).

SA Levels in Wine

Typical total (free plus bound) SA levels in white and red wines are 11.0 mg/L and 18.5 mg/L. Levels for 2,3-DHB are: 21.0 mg/L and 26.5 mg/L. For 2,5-DHB, the levels are 22.0 mg/L and 28.5 mg/L, respectively (Muller and Fugelsang 1993; Muller and Fugelsang 1994b; Muller and Fugelsang, unpublished data). As expected, levels are higher for red than for white wines. It should be noted that the levels reported here are only averages of a limited sampling. SA content of wines is dependent on variety, vine health, and processing practices (e.g., skin contact time, oak-wood aging, and temperature) among other factors. We are currently investigating such variations.

SA Production

SA is produced by plants both as a constitutive (that is, always present), and as an inductive (produced on demand) defense chemical (phytoalexin), and thus part of the plants' systemic acquired resistance (SAR). It is produced by either hydroxylation of benzoic acid, or by hydroxylation of caffeic acid, followed by oxidative degradation (beta-oxidation). Salicylic acid is highly effective against bacterial, fungal, and viral pathogens in plants (Raskin 1992a; Yalpani and Raskin, 1993). Salicylic acid performs so many functions in plants that it has been postulated to be a plant hormone (Raskin 1992b). Because of its ability to react with NO and its oxidation products (Koppenol and Bartlett 1993), it might offer the plant a measure of defense against environmental insults such as oxides of nitrogen produced from internal combustion engines.

SA Protection Against Smog and Tobacco Smoke

Vehicular emissions, smog, and tobacco smoke contain oxides of nitrogen and other powerful pro-oxidants (Church and Pryor 1985; Halliwell 1993c; Pelletier 1975). Although the lungs produce NO in minute amounts, particularly while subject to respiratory infections, they are particularly sensitive to irritation and persistent damage by oxides of nitrogen. Salicylic acid and other similar wine constituents might offer, through their antioxidant action, a certain level of protection against lung cancer, emphysema, respiratory tract infections, and other pathological states to which smokers are at increased risk. This protection, of course, is afforded not only to smokers, but to nonsmokers, who might be exposed to 'second-hand' smoke. SA, by itself, or perhaps through NO-mediated endothelial relaxation, might also be able to counteract the vascular narrowing and plaque-depositing action of inhaled carbon monoxide. A similar action might be ascribed to epicatechin, quercitin, and resveratrol.

CONCLUSION

Does this mean that we should still be able to enjoy a cigar after dinner if we also imbibe wine with the meal? Apparently, yes! Although synergistic effects amongst all wine antioxidants (and possibly other constituents) definitely play a role in the healthful attributes of wine, it is perhaps the presence and levels of SA (and its metabolites), and other compounds such as quercitin and epicatechin, together with alcohol, that provide the answer to the "French Paradox" for both wine (Renaud and de Lorgeril 1992), and smoking (Renaud and de Lorgeril 1993). The search for wine constituents with potential health benefits is only just starting and must continue.

An interesting note regarding alcohol consumption is that, in those who abuse alcohol, particularly while undernourished, liver levels of ethanol detoxifying enzymes might be severely impaired, resulting in shortened life spans (Kennedy and Tipton 1990; Mouschmousch and Abi-Mansour 1991). However, those who consume alcohol in moderation from an early age seem to live longer. This appears to be the case among ethnic groups whose ancestors came to this country from Europe, where wine is consumed with meals or as an adjunct to meals. In these fortunate people, their livers and other organs induce higher levels of alcohol and aldehyde dehydrogenases, and in some cases, also induce more effective microsomal ethanol oxidizing systems (MEOS) than people who are not exposed to alcohol (Kershenobich et al. 1993). What this means is that these individuals can better detoxify alcohol; thus, they develop a certain tolerance for alcohol that enables them to enjoy wine with trivial alcohol-related negative effects.

As with any other food, moderation, not gluttony and excess, is beneficial. The term "moderation," so vastly used and misused, is most ambiguous. If one relies on the "U" or "J" curves generated from epidemiological data for general cardiovascular wellness and longevity as a function of alcohol intake (Veenstra 1991), perhaps two 5-ounce glasses of wine a day, taken with meals, might be sufficient (Friedman and Klatsky 1993; Gaziano et al. 1993; Hendricks et al. 1994; Kershenobich et al. 1993; Klatsky and Armstrong 1993; Rimm et al. 1991). However, we are certainly quite varied in our backgrounds, metabolisms, and lifestyles; thus, some of us might indeed benefit from larger and more frequent intake of wine (e.g, smokers), others from less. However, what is patently clear is that all of us would benefit from daily imbibing of some wine throughout our lives (Maclure 1994; Marmot et al. 1994).

In summary, those of us who have followed in the footsteps of our elders and have enjoyed wine since an early age, might be blessed in that, in enjoying wine with meals on a daily basis, we have also been receiving, unknowingly perhaps, the milieu of compounds that provided us with a longer, healthier, and more enjoyable life than our brethren who have chosen to abstain.

Acknowledgment

The author wishes to express his appreciation to the California Agricultural Technology Institute (CATI) for generous support.

(Submitted for publication: October 31, 1994)

CHAPTER 2

Application of Sensory Evaluation in Wine Making

Susan E. Duncan
Assistant Professor, Department of Food Science and Technology, Virginia Polytechnic Institute and State University, Blacksburg, Virginia

This chapter is designed to provide a general background about sensory evaluation to individuals in the wine industry. It is intended as a general review of sensory evaluation, not a comprehensive resource, but will provide adequate information to help the reader understand the importance of sensory evaluation in the wine industry. The reader is referred to the references mentioned in this chapter for publications that will provide complete information about sensory evaluation implementation and operation.

OVERVIEW OF SENSORY EVALUATION

Sensory evaluation is a "scientific discipline used to evoke, measure, analyze, and interpret reactions to the characteristics of foods and materials as they are perceived by the senses of sight, smell, taste, touch, and hearing" (Institute of Food Technologists 1981). Without the proper sensory evaluation techniques, it is difficult to interpret sensory responses and make logical and sound decisions.

The wine industry has historically relied on "experts" for determination of sensory characteristics of wine. Evaluation of wine by such "experts," highly trained to evaluate the slightest nuances in flavors and aromas, may be valuable to a winery, providing detailed descriptions of differences in wines. However, problems can occur if too much reliance is placed on a few individuals making sensory judgments that influence production or marketing decisions. "Experts'" opinions may not be reflective of true sensory characteristics or those important to the wine consumer. External factors such as mental or physical fatigue or distractions can interfere with the "expert's" analysis of the product. If a few "experts" are the common means of sensory evaluation within a winery, limitations of these evaluations are significant. A progressive winery will understand the importance of basing decisions on sensory data that can be statistically evaluated and interpreted.

In a competitive environment, it is important to base decisions about wine, new products, and wine improvements on the best information possible. Sensory evaluation controls the external variables as completely as possible so only the variable of interest is being measured. Sensory results can be interpreted statistically, providing a basis on which decisions can be made. Effects of changes in processing conditions on sensory attributes can be measured and the impact of the processing decision weighed. Sensory evaluation of wines can also assist in relating sensory impressions to others in a meaningful way by establishing a standardized sensory vocabulary. Correlation of sensory analysis with chemical measurements of selected wine compounds can assist in interpreting chemical data relative to wine characteristics. Properly conducted sensory evaluations can lead to improved decision-making with less risk involved, a means to targeting and achieving goals, and a way of categorizing attributes.

The selection of a sensory evaluation method is determined based on the type of information that is needed. A new grape variety, variations in pressure during pressing, a change in yeast strain or supplier can lead to changes that may impact on the sensory characteristics of the wine. It may be important to know if there are perceptible changes, what those changes are, and the impact of those changes on the consumer's perception of quality or acceptance.

STANDARDIZATION OF SENSORY EVALUATION

Before a method can be selected, it is important to understand the parameters of sensory evaluation that must be standardized and controlled. Standardization of environment and sensory techniques provides control of factors that can influence and bias the results of the sensory test; without these controls, incorrect information and interpretation may result. Therefore, it is very important to control as many variables as possible, such as test environment, sample preparation and presentation, and type and number of sensory evaluations (ASTM 1968; Meilgaard et al. 1991; Munoz et al. 1992; Stone and Sidel 1985; Yantis 1992).

Standardization of Environment

Sensory evaluation should be completed in an environment conducive to concentration. Many large wineries have sensory evaluation areas located within the winery. The best are designed to minimize interaction among panelists and allow the individual to focus on the evaluation of the sample. Individual booths, which have side panels that prevent interaction between panelists, are equipped with individual lighting, adequate space for sample evaluation, a signal system for communicating with the sensory technician, and a hatch door for receiving samples. The technician passes samples to the panelist through the hatch, which minimizes personal contact and influence. Neutral colors on the booth walls avoid creating a mood response or altering appearance of the sample. Incandescent light within the booth, at an intensity similar to a typical office, provides the best indoor light with regard to true wine colors; fluorescent light adds to the perception of brown hues. Red-colored lights or dark-tinted wine glasses may be used to remove any bias from color differences in the wine samples.

Conditions within the evaluation room are designed for the comfort of the panelist. Control of temperature and relative humidity, at approximately 21°C (70°F) and 45 to 55%, respectively, provides a comfortable setting. A well-ventilated room, with positive pressure, assists in controlling interferences from odors. Extraneous or excessive odors can interfere with evaluation of aromas from the wine samples and cause a bias in sensory measurement. Therefore, sample glasses containing wine are covered with watch glasses.

It is important that the evaluation occur in a location convenient to the panelists. Although some facilities have space dedicated for sensory evaluation, not every winery can allocate space for such a specific use. It is possible to prepare an evaluation area for temporary use from another facility, such as a conference room or lounge. Collapsible booths, manufactured from plywood or cardboard, may be set up on tables to provide the privacy for the individual panelist during evaluation. Preparation of samples for the evaluation is completed in a room convenient to the evaluation room, with controls to prevent observation of sample preparation by the panelists, as well as to minimize noise, odors, and other disturbances.

Controls for Sample Preparation and Presentation

The preparation and presentation of the samples must be uniformly controlled to avoid any biasing of response during evaluation (ASTM 1968; Meilgaard et al. 1991; Munoz et al. 1992; Stone and Sidel 1985; Yantis, 1992).

Sample Preparation and Temperature

Samples, selected to represent the product under evaluation, are also selected to represent the production lot. Samples are served in a standardized fashion, considering serving temperature, serving size, etc. Glass is an appropriate serving vessel as long as it is clean and free of any soap or chemical residues that can be detected by panelists. During preparation of the samples, it is important to cover the serving glasses with a glass cover. If the wine glass is not covered, volatile compounds will be lost from the sample, filling the room with aromas that could bias the evaluation. It is wise not to pour all samples too far in advance of serving because the volatile aromas could become variable on serving to panelists Serving the wine at room temperature is appropriate, making it easy to control the serving temperature.

Sample Size and Number

The panelist must have an adequate sample size to complete the evaluation required. Sensory methods requiring numerous evaluations of flavor, aromas, and body require a sample size of approximately 30 to 40 mL. Other methods involve simple comparisons that may require less sample, perhaps only 15 to 20 mL. Given the complexity of wine flavors and aromas, it is recommended that only a limited number of wine samples be evaluated in one sitting. The exact number will vary depending on the test method, complexity of the wine, and panelist experience, but it is recommended that no more than six samples be evaluated in one sitting. It is, of course, important that wine samples are expectorated to avoid fatiguing the palate and biasing judgments.

Sample Coding and Presentation

Samples must be coded to eliminate bias. A three-digit code, chosen at random, is assigned to each product and used to identify the product sample to the panelist. Use of the alphabet or single- or double-digit numbers as codes is discouraged because some letters and numbers can have special meaning to panelists. Three-digit random codes may be easily obtained from a random numbers table available in many sensory evaluation or statistics books (Meilgaard et al. 1991; Stone and Sidel 1985).

The order in which the samples are presented to panelists must be balanced so the influence of such factors as panelist fatigue, positioning of a high-quality wine next to a lower quality product, etc. will have little impact on the outcome of the test. A prejudicial preference for one wine over another based on the order in which the samples are tasted may occur. Frequently the first sample is preferred over subsequent samples, particularly by inexperienced evaluators. Another bias may be created if greatly contrasting wines are evaluated in sequence; the sensory impression of the second wine can be greatly distorted by the response to the characteristics of the first wine evaluated. Balancing the sample order is used to overcome this potential bias. For example, in a test with three wines, each product should appear in each position an equal number of times. The six possible combinations for positioning of three products (A, B, C) would be ABC, ACB, BAC, BCA, CAB, CBA. If the panelist is to receive more than one set of three samples, the order of presentation of these sample combinations should be randomized. This may be easily accomplished by drawing sample cards representing each three-sample sequence from a hat until the six combinations have been randomly assigned. This arrangement of samples prevents any undue influence from positional bias or contrast effects between wines.

SENSORY PANELISTS

Control of the human aspect of sensory evaluation is one of the more difficult factors of sensory evaluation. This may be accomplished best by carefully selecting the people that will be participating in the test. Important qualities in a sensory panelist include availability, dependability, interest, objectivity, stability, and acute senses of smell and taste (ASTM 1968; ASTM 1981; Hootman 1992; Meilgaard et al. 1991; Stone and Sidel 1985). The selection of panelists is also dependent on the type of information desired about the product. Panelists may be classified as consumers, experienced, or trained. Experienced and trained panelists complete testing in the laboratory setting.

Panelist Selection

Panelists are selected based on the type of test needed. Consumer panelists or experienced/trained panelists are used to answer different sensory questions.

Consumers

The consumer group may be selected to represent a geographic area, age group, socioeconomic status, or other population criterion. Frequently, they are randomly selected from consumers that frequent a given location such as a grocery store or a shopping mall. Information from these consumers is valuable for demonstrating preferences, degree of acceptability, product use, or consumer opinions about the wine samples. This could be the most important information received about the wine in relation to marketing and sales. Consumer testing is time consuming and costly but important to understanding consumer attitudes.

Experienced Panelists

Experienced panelists are used to provide information about the product itself, not personal opinion or preference. Specific characteristics or differences of the wine products are evaluated. To accomplish this, experienced panelists are provided preliminary information as a means of educating the panelist about the product. Panelists are instructed about the product characteristics of interest so they can evaluate wines for specific characteristics or determine if differences are evident between wine samples. These panelists are not considered trained panelists because validation of panelist performance is not completed. It is wise of the sensory specialist to develop a pool of experienced panelists so a panel of 12 or more individuals may be assembled quickly if necessary. If an experienced panelist cannot participate in a test on a given day, an alternate panelist from the "pool" may be substituted into the panel to obtain a sufficient number of responses.

Trained Panelists

Most wineries want to know how wines differ. This requires descriptive analysis and panelists that have been trained to assess qualitative and quantitative differences between products (Hootman 1992; Meilgaard et al. 1991; Stone and Sidel 1985). The trained panelist does not permit personal bias regarding the product to influence judgment of product characteristics. The understanding of wine character is much more complete with a trained panelist as compared with an experienced panelist because of the extensive training provided. To determine confidence in the performance of a trained panel, validation of panelist performance and understanding is accomplished through statistical analysis prior to beginning a descriptive analysis test. Therefore, other panelists cannot be substituted into a trained panel and absenteeism, lack of motivation, inadequate performance, etc., by panel members can be costly to the project outcome. The sensory specialist is advised to select individuals for participation on a trained panel carefully and train an adequate number so absenteeism has a minimum effect.

Number of Panelists

The number of panelists is dependent on the type of testing required (ASTM 1968; Hootman 1992; Meilgaard et al. 1991; Stone and Sidel 1985). To obtain a good understanding of consumer opinions, a minimum of 50 panelists are needed and more are desirable. Tests for differences between products may be completed with as few as 10 to 12 experienced panelists; however, differences may not be observed with such a small number (see Discrimination Tests). Twenty or more panelists are recommended. Descriptive panels, in which the panelists have received extensive training and performance is well documented, may consist of five or more panelists. It is wise to have more than five people on a trained panel, however, because if one panelist misses an evaluation, the influence on the data can be extreme. Ten to twelve trained panelists on a trained panel is more satisfactory to provide more statistical confidence in the data.

Panelist Screening

Panelists are selected for participation on an experienced or trained panel by initially screening for motivation to perform the test and ability to concentrate and communicate. Further screening is completed to determine abilities to identify differences using dilute solutions that may represent, for example, the basic tastes of acid, bitter, salty, and sweet. Additional screening may include determination of threshold testing for each taste sensation. Candidates that successfully complete the screening tests are eligible for participation in discrimination testing.

Potential panelists for descriptive testing should have the characteristics already listed but should also exhibit verbal acuity, the ability to think abstractly, and sensitivity to the characteristics of importance. The panelist must be highly motivated to undergo many hours of training time. Additional screening tests, in addition to those needed for discrimination testing, might include determination of characteristic differences in a wine system, a ranking of reference wine standards containing differing levels of a single characteristic, and a test to assess abstract thinking by providing a verbal description of some unidentified but common aroma compounds (Hootman 1992; Meilgaard et al. 1991; Stone and Sidel 1985).

Panelist Orientation

Before beginning any discriminative sensory analysis, panelists must receive some instruction (Hootman 1992; Meilgaard et al. 1991; Stone and Sidel 1985). This instruction may be as brief as an explanation of the scorecard and familiarization with standardized testing procedures. If the characteristic to be evaluated is not clearly understood or not easily perceived, it should be defined. A reference sample that clearly illustrates the desired characteristic may be used to help the panelist distinguish the attribute. The sensory specialist should document that each panelist can identify the characteristic of interest in a wine system before beginning the test.

Panelist Training

Descriptive testing requires extensive training and validation of the panel performance before confidence can be placed on results. The actual training time requirements are dependent on the ability of the panelists to learn to identify and measure the selected characteristics within the wine. In the initial stages of training, the panelists should meet as a group to discuss the characteristics that contribute to the descriptive profile of the wine. The sensory specialist acts as facilitator for the group discussion as well as preparing samples and reference standards used during the training sessions. Using wines that represent the range of characteristics expected in the sample set of wines, the group develops a list of descriptors for those characteristics, facilitated by the sensory specialist. The descriptors are discussed to determine definitions, and reference standards are created that reflect those definitions. The aroma wheel (Fig. 2–1) represents descriptors that may be observed in a variety of wines. Training ensues, initially using standards in water to demonstrate the descriptors until all panelists can identify each characteristic. Training continues using reference standards in a wine. Usually the panel will begin by discussing a small number of wines for the most identifiable characteristics, adding additional characteristics for discussion as abilities and confidence build. Training sessions are continued until all panelists have developed confidence in identifying and rating intensity of each characteristic of interest in the wine. Limiting the discussion to six or fewer descriptors allows panelists to concentrate on those limited concepts without fatigue. In quantative descriptive analysis, panelists also learn to evaluate the intensity of the characteristics on a scaling system, usually reflected by an unstructured line scale.

Reference Standards and Descriptors

The numerous descriptors that are used to describe aroma and flavor of wines are listed in the aroma wheel (Fig. 2–1). Table 2–1 includes formulations for reference standards for many descriptors. This information provides the groundwork on which a descriptive profile may be derived.

Performance Evaluation

Panel performance is assessed for descriptive testing before initiating the evaluation on the test wines. To evaluate the performance of the panel, two

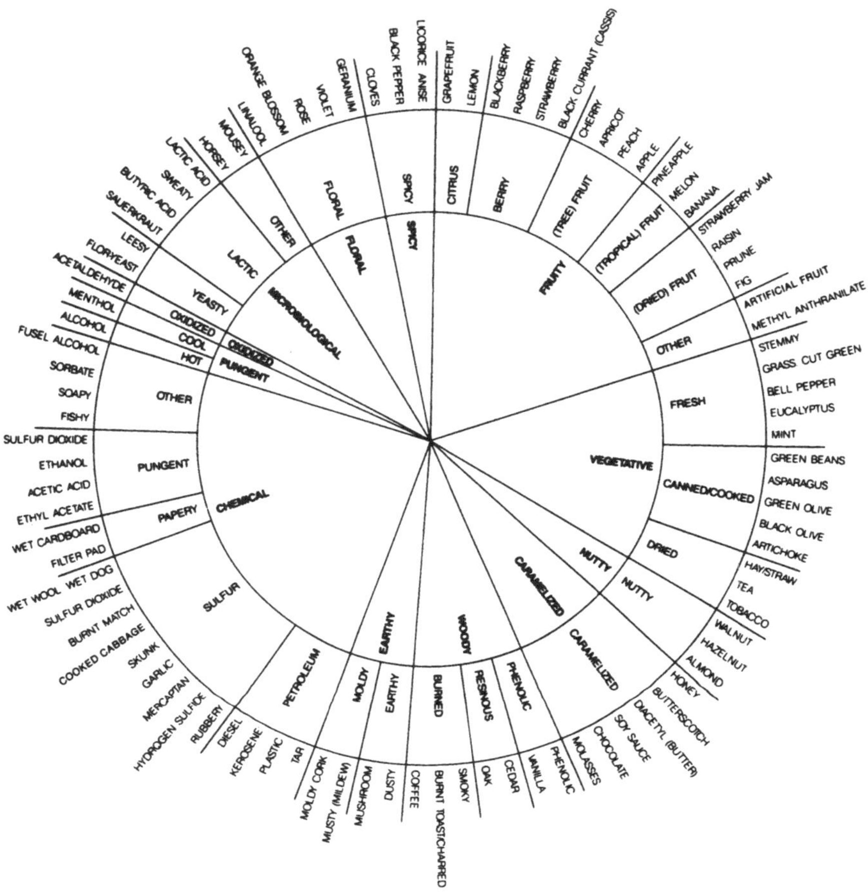

Fig. 2–1. Aroma wheel.

or three wines reflecting a variety of the characteristics of interest are selected for evaluation. The individual panelist evaluates each wine as in a testing situation (see Descriptive Analysis), identifying and rating each characteristic present in the wine. This evaluation is replicated three times to determine reproducibility among panelists and within each panelist, not on differences in the wines. Analysis of variance, a statistical method used to compare more than two mean values in a single study, is used to determine if differences exist among panelists for the characteristics measured (Meilgaard et al. 1991). Extreme differences among panelists' responses for a given characteristic suggest further training is needed to assist the panel in identifying and rating the characteristic of concern. If one individual panelist shows high variability in replication, the sensory specialist may work with that individual independently until greater skills are achieved or that panelist may be excused from the panel.

Panelist Motivation

Panelists must be motivated to contribute time and effort required for frequent testing situations. Discussion with each panelist about successes and improvement areas are helpful in directing and motivating the panelist to high achievement. The importance of the panelist's contribution to the test outcome should be stressed. Verbal and written expression of appreciation from the sensory specialist is important. Management should provide support of these motivational activities through some form of reward or commitment.

Timing of Panel

Training and testing time periods should be scheduled when the panelists will be most sensitive to product characteristics and have the attention to focus on the task at hand. Do not schedule the session immediately after a meal or coffee break as this may contribute to sensory error. Sensory sessions should be conducted at the same time of day.

METHODS OF SENSORY EVALUATION

Sensory evaluation methods may be divided into two broad classes: affective and analytical methods (IFT 1981). Affective methods use consumer panels or untrained panelists to answer questions such as Which product do you prefer? Which product do you like? How well do you like this product? and How often would you buy/use this product? Affective methods require much larger panel size than do analytical methods in order to have greater confidence about the interpretation of the results. The most common analytical methods of sensory evaluation used in the wine industry are discrimination and descriptive methods. Discrimination tests can be used to determine if products are different, if a given wine characteristic is different among samples, or if one product has more of a selected characteristic than another. Experienced panelists can complete discrimination tests (see Experienced Panelists and Panelist Orientation). Descriptive methods are used to provide more comprehensive profiles of a product by asking panelists to identify the different characteristics within the product and quantify characteristics. Trained panelists must be used for descriptive methods (see Trained Panelists and Panelist Training).

Discrimination Tests

There are several types of tests that may be used to determine if differences exist among products (ASTM 1968; Meilgaard et al. 1991; Stone and Sidel 1985). The most common methods include the triangle test, the paired comparison test, and the duo-trio test. As indicated earlier, discrimination tests may be completed by a small number of panelists (10 to 12) but statistical determination of differences is more enhanced with a greater number of responses. Analysis of these methods is made easy by the use of statistical tables from which results of the test may be quickly analyzed. These tests are relatively simple for the panelists if the panelist is knowledgeable about the product and characteristics of interest. In each method, the panelist is forced to make a decision or choice among the products. The amount of information drawn from these tests is limited to a detection of difference. It is not possible to know the degree of differences that exist among products, or if the change in the characteristic affects acceptability or preference of the product.

Triangle Test

The triangle test uses three samples to determine if an overall difference exists between two products. The three samples include two that are identical and one that is different. The samples must be coded with individual three-digit numbers (derived from a random numbers table) and presented at one time to the panelists (Fig. 2–2). The panelist is requested to identify the code on the scorecard representative of the odd sample (Fig. 2–3). This method requires the panelist to make a choice among the samples; the panelist has a 33% chance of simply guessing correctly. This test method has good applications in determining if a process change affects the overall product character. Fatigue is a factor as panelists usually must retaste several times. Adaptation may also occur as a result of retasting. It is recommended that no more than two sets (six samples) be evaluated at one testing session.

Interpretation is based on the minimum number of correct responses required for significance at a predetermined significance level, given the total number of responses received (Fig. 2–4). The minimum number of correct responses may be found in statistical tables provided in several publications (ASTM 1968; Meilgaard et al. 1991; Stone and Sidel 1985).

For example, two wines processed under identical conditions are fermented in the bottle for 6 months at two different temperatures, 7°C (45°F) and 13°C (55°F) in order to determine if the warmer temperature of fermentation results in a different product. Twenty-four panelists evaluate the two wines, using a triangle test (see Figs. 2–2 and 2–3 for scorecard

DATE ______________ TEST CODE ______________

Post this sheet in the area where trays are prepared. Code scoresheets ahead of time. Label serving containers ahead of time.

Products: ______________
Test method: ______________

Panelist No.	Order of Sample Presentation
1,7,13,19	A,B,B
2,8,14,20	B,A,B
3,9,15,21	B,B,A
4,10,16,22	B,A,A
5,11,17,23	A,B,A
6,12,18,24	A,A,B

Product Description	Research Identification	Sample Code
Wine fermentation at 7°C (45°F)	A	________
Wine fermentation at 13°C (55°F)	B	________

Fig. 2–2. Example of worksheet for preparation of samples and scorecards for triangle test.

Name ______________ Date ______________

Product ______________

Two of these three samples are identical, the third is different. Taste the samples in the order indicated and identify the different sample. Circle the code number of the sample with the odd/different taste. You may comment on any differences in the Remarks area. Rinse between series of tests.

	Codes			Remarks
Series 1	____	____	____	______________
Series 2	____	____	____	______________

For research use only. Do not write below this line.
Pair: ______________ Odd: ______________
Correct: Yes ________ No ________
Panelist No. ________

Fig. 2–3. Example of a scorecard for triangle test.

and worksheet). It is determined that a difference is needed at a 5% level of significance ($P < .05$). Fifteen of the 24 panelists correctly identify the odd sample. Is there an overall difference between the product fermented at 7°C compared with the product fermented at 13°C?

At the 5% level of significance and 24 panelists, a minimum of 13 correct responses were needed to determine there was a significant difference between the samples. The product fermented at 13°C is noticeably different from the product fermented at 7°C.

Minimum Number of Correct Responses Needed for Significance

	Significance level (%) (Meilgaard et al. 1991)		
Number of panelists	10	5	1
23	12	12	14
24	12	13	15
25	12	13	15

Fig. 2-4. Statistical interpretation of triangle test results.

Paired Comparison Test

The paired comparison method is another forced-choice procedure that uses only two samples. The test may be set up to determine an overall difference between samples, which is also called a simple difference test (Is there a difference between these two products? Yes or No), to determine if a difference exists for a single characteristic between the samples (Is there a difference in acidity between these two products? Yes or No), or if there is a directional difference for a single characteristic (Which product is more acid? Product 738 or Product 429) between the samples. Products are compared such that each sample is placed in the first taste position an equal number of times. This test causes less fatigue and is frequently used for strongly flavored or complex products. Whereas the triangle test provided only a 33% chance of guessing correctly, there is a 50% chance with the paired comparison test. Therefore, more panelists (at least 20) are recommended to complete this test. Interpretation is easily determined from a statistical table (ASTM 1968; Meilgaard et al. 1991; Stone and Sidel 1985).

Duo-Trio Test

The duo-trio test compares two products but has three samples within each test set. One sample is marked as a reference sample and is presented with two coded samples. The objective is to determine which coded sample is the same as the reference sample. The reference sample may always be set as the same product (constant reference) or may be randomly chosen so that each product is represented (balanced reference). This test method is less efficient than the triangle test, with a 50% chance of guessing correctly, requires a large amount of sample size, but is frequently used when a flavor is complex or intense. Interpretation is easily determined from a statistical table (ASTM 1968; Meilgaard et al. 1991; Stone and Sidel 1985).

Other discrimination tests that may be of interest include the two-out-of-five test, and "A–Not A" test, the difference-from-control test, ranking tests, and similarity testing. These are described fully in several of the references.

Descriptive Tests

Frequently it is important to know how a wine changes with a new vineyard site, how intense a characteristic is, etc. Discrimination testing, which is easy to use, easy to interpret, and easy for panelists to complete, is initially used to determine that a difference does exist. Such methods cannot provide information about the description of those differences. Descriptive evaluation methods are more difficult to complete and interpret but provide much more information. They provide a quantitative measure of wine characteristics that allows for comparison of intensity between products, and a means of interpretation of these results. Examples of descriptive test methods include quantitative descriptive analysis (QDA), flavor profile analysis, time-intensity descriptive analysis, and free-choice profiling (Hootman 1992; Meilgaard et al. 1991; Stone and Sidel 1985). QDA is frequently used because it requires less training time than several of the other methods.

Quantitative Descriptive Analysis

Descriptive analysis of wines using QDA requires products with relatively similar characteristics. If possible, 10 to 12 panelists are selected to participate based on ability to discriminate, communication skills, and task comprehension. Using fewer trained panelists is possible for descriptive analysis but individual responses have a greater effect on the mean scores. During the training, the group is provided with different wines that represent the range of characteristics that may be tested during the actual test sessions (see Panelist Training). Wine characteristics are identified, definitions or descriptions of the characteristics are determined, and references exemplifying the characteristics are established (Fig. 2–1, Table 2–1). The reference standards are used to further train the panelists to identify the desired characteristics and learn to rate intensity levels. A scorecard is developed by consensus of the panel that includes all characteristics of interest in the order in which they are to be evaluated (Fig. 2–5). Each characteristic is rated on a 6-inch (15.2-cm) line scale with descriptors of "weak" and "strong" as endpoint anchors. The anchor terms are located ½ inch (1.27 cm) from each end of the line; an alternate method places vertical anchors 2.5 cm from each end (Hootman 1992). Either method is satisfactory. It is important that panelists know they may use the entire line to mark their perception of intensity, even those regions toward the ends

Date ___________________

Sample Code __________

Panelist Number _________

Please smell the wine sample provided by removing the watch glass and taking one or two quick sniffs. Evaluate the sample for the first aroma characteristic listed. Mark the intensity of each characteristic by placing a vertical mark at the appropriate location on the line. Continue until all characteristics have been scored. You may sniff the sample as many times as necessary.

RASPBERRY

weak strong

PEAR

weak strong

ROSE

weak strong

GERANIUM

weak strong

OAK

weak strong

EARTHY

weak strong

CINNAMON

weak strong

Pass the scorecard and sample through the hatch and wait for the next sample. Please rest at least 1 minute between each sample. You have ____ more samples to evaluate today.

Fig. 2–5. Sample scorecard for a QDA evaluation of a wine sample.

of the line. Panelists continue training until the sensory specialist determines that everyone is demonstrating comprehension of the task and all panel members can identify and rate each characteristic. Preliminary testing should occur to determine the reliability and validity of the individual panelists before initiating the actual testing (see Performance Evaluation).

After the training is completed, the panelists function independently in the evaluation setting. No more than six or seven attributes should be evaluated at each setting to avoid fatigue. If intense aroma characteristics are evaluated, fewer than six wine samples should be evaluated in one testing. Products are evaluated for intensity of the characteristics on the scorecard. Panelists rate the intensity of each attribute by marking a vertical mark across the appropriate horizontal rating line. These marks are converted to numerical data by measuring the distance from the origin ("weak") of the line to the vertical mark. Usually no more than five or six products may be evaluated because wine is a very complex product.

Sometimes only one or two characteristics are of interest so an entire profile is not necessary. This reduces the effort required by the panelist and time needed for data handling and interpretation by the sensory specialist. A line scale as is used for QDA is appropriate for this, or a category scale with seven to nine verbal descriptors may be used. The category sacle is easier to use by panelists with less training than a QDA panelist.

Data analysis is completed using a mixed model analysis of variance for treatment-by-subject, with replication (Hootman 1992). To determine individual panelist's abilities to perceive differences among products, a one-way analysis of variance is completed for each panelist. This analysis can also be used to determine if an attribute is helpful in differentiating among wines. Subsequent analysis, using a two-way analysis of variance design, is needed to determine if any product differences exist and interactions by the panel (Hootman 1992). An independent statistical analysis is completed for each characteristic that is measured.

A graphic presentation, called a spider web plot, of all characteristics may be made to illustrate the differences and similarities of the descriptive profiles of the wine samples evaluated (see Fig. 2–6). This is accomplished by plotting the mean score for a given characteristic on an axis that represents the 15.2-cm line scale used on the scorecard (Hootman 1992). Each axis extends from a center point like spokes on a wheel and represents one characteristic. The center point is equivalent to the low-intensity origin of the line scale and the highest intensity is equivalent to the end of the axis.

Affective Test Methods

Commonly used affective methods include a paired preference test, a preference ranking test, and hedonic test method (ASTM 1968; Meilgaard et al. 1991; Stone and Sidel 1985). The test method must be simple and easy to understand so the consumers will know how to respond.

Paired Preference Test

The paired preference test is set up in the same manner as the paired comparison method for discrimination testing. Two samples are compared

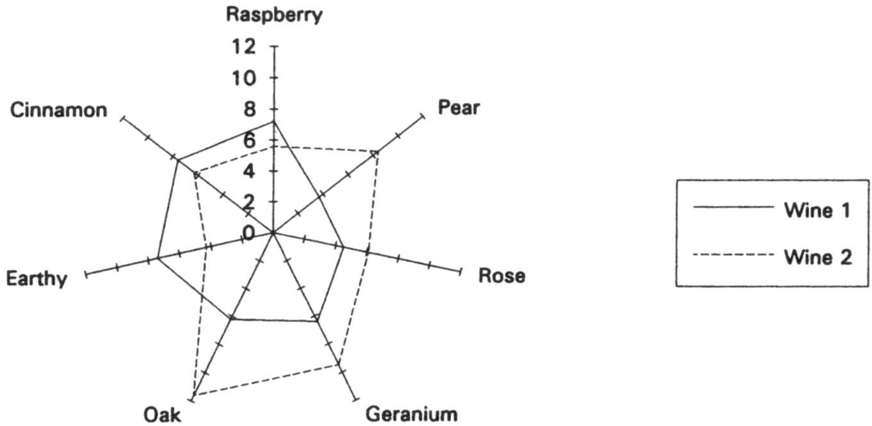

Fig. 2–6. Example of descriptive aroma profile of two wines using a spider web plot based on QDA analysis.

to determine which product is preferred. A large number of similar responses must be obtained to determine that one product is more preferred than the other. The minimum number of similar responses needed to determine if the preference is significant is based on the total number of responses obtained and may be determined from a statistical table.

Ranking Test

If more than two samples are evaluated, a preference ranking test may be completed. Usually three to five samples are the most that can be efficiently ranked by a consumer. This test asks the consumer to order the samples based on preference, with a ranking of "1" meaning most preferred.

Hedonic Test

The hedonic scale may be used to determine degree of acceptability of one or more products. This scale is a category-type scale with an odd number (five to nine) categories ranging from "dislike extremely" to "like extremely." A neutral midpoint (neither like nor dislike) is included. Consumers rate the product on the scale based on their response.

PRINCIPAL COMPONENT ANALYSIS

An advanced statistical method, principal component analysis, is frequently used in the wine industry to illustrate relationships between a reduced set of variables. Patterns in descriptive sensory data may be determined by analyzing the data by this multivariate statistical method. The number of variables is reduced using factor analysis such that the first principal component statistically identified explains most of the variability in the data (Meilgaard et al. 1991). The second component, which is not correlated with the first, explains the majority of the remaining variance. Additional principal components may be identified, up to the number of observed variables, until no significant variability can be explained from extraction of the principal component. Usually there are only two or three principal components of value because most of the variability will be explained in the first components that are extracted (Meilgaard et al. 1991).

A graphic portrayal of principal component analysis components may be used to illustrate relationships among principal attributes of different wines. Primary and secondary principal components are presented as axes at right angles with each other (Meilgaard 1991). The principal components for two wines are presented in Fig. 2–7. Emanating from the central origin are vectors representing each attribute. The length of the vector may be interpreted as an indication of influence on that principal component. Short vectors indicate attributes of relatively low importance. Close alignment of a vector with the principal component axis indicates a high correlation between the attribute represented by the axis and the variability

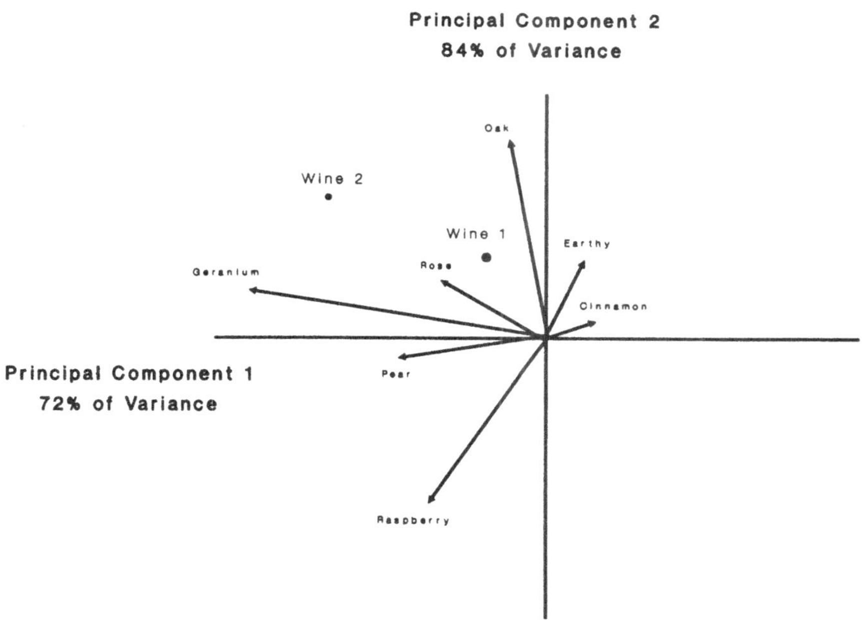

Fig. 2–7. Principal component analysis of two wines.

explained by the principal component. On this graphic representation, the principal component score of each wine is indicated as a point. The location of this point in relation to the principal components and the vectors representing attributes can provide information about the wine.

In Fig. 2–7, two components explain 84% of the variance. The fruity aromas of raspberry and pear and a geranium (floral) aroma are closely correlated with principal component 1. Oak and earthy aroma descriptors are closely aligned with principal component 2. The cinnamon characteristic appears to have relatively low importance in explaining the variance in the data analyzed. Oak and geranium aromas are high in wine 2 whereas wine 1 has some earthy and cinnamon aromas in moderate levels.

SUMMARY

Sensory evaluation in the wine industry is currently underutilized but can provide many benefits to the winery that employs the techniques correctly. With proper implementation of controls, careful selection of panelists, effective training (when necessary), utilization of appropriate sensory methods, and proper statistical interpretation, sensory evaluation can provide a basis on which decisions can be made. It is hoped that this review of sensory evaluation will encourage the reader to review several of the references for further information.

Table 2–1. Sensory formulas for descriptive analysis.

White wines are often described in terms of floral, spicy, and fruity odor notes and occasionally with terms such as vegetative, caramelized, and woody. Red wines generally have fewer floral, citric, tree, and tropical fruit odors and more berrylike, dried fruit, vegetative, and woody characters.

The following list of wine odors and ingredients is designed to aid in descriptive analysis of wines. The specific formulas needed may vary due to perception threshold differences and slight difference in the maturity and character of the natural ingredients. Formulas listed are for 25 mL of neutral base wine unless otherwise noted.

Aroma	Quantity	Composition (in 25 mL base wine)
Acetaldehyde	50 mg/L	In base wine
Acetic acid	1 drop glacial acetic acid + 2 mL wine vinegar	In base wine
Anise	1–2 drops	1–2 drops anise extract
Apple	4 mL	Canned apple juice
Apple cider	6 g	Fresh, sliced apple
Apricot	5 mL	Apricot nectar
Asparagus	To 5 mL	100% brine from canned asparagus
Astringent	200 mg/L	Alum (Aluminum ammonium sulfate)

Table 2–1. (*continued*).

Aroma	Quantity	Composition (in 25 mL base wine)
Astringent	0.9–1.1 g/L	Tannic acid
Banana	6 g	Crushed ripe banana steeped for 30 minutes in 20 mL of distilled water, spike base wine with 4 mL of solution
Bell pepper	2 pieces, 1.5 × 1.5 cm	Pieces of fresh bell pepper, soak for 25 minutes and remove
Berry	5 mL of crushed fresh or frozen fruit juice, or 1/2 tsp. jam	10 g fresh or frozen raspberries, blackberries, or strawberries; berry jam
Bitter	To 1.6 g/L	Caffeine in base wine
Blackberry	5 mL of juice	10 g crushed frozen or fresh blackberries
Black currant	10 mL	Liquor from canned black currants
Black (or Green) olive	4–6 mL	Brine from canned olives
Black pepper	10 pieces	Finely ground black pepper
Butterscotch	2.5 g	Butterscotch Life Saver, dissolve in 20 mL distilled water, spike 5 mL of solution into base wine
Buttery	0.25 mg/diacetyl	In base wine
Caramelized	5 mL honey, 1 mL soy sauce, or 1 crushed butterscotch candy	In base wine
Cedar	0.6 mL	0.5 g cedar wood shavings, soak in 1.5 mL of 200 mg/L cedar wood polish solution for 1 hour then remove
Celery	2 g	Celery, soak for 1 hour then remove
Cherry	6 mL	Liquid from canned cherries
Chocolate	1.5 g	Powdered cocoa
Cigar/tobacco	0.5 g	0.5 g of tobacco steeped in 20 mL of distilled water for 24 hours, then spike 3 mL of the solution into base wine
Citrus	4 mL of treated lemon juice	Macerate lemon skin in 25 mL of freshly squeezed lemon juice and 25 mL of base wine
Clove	0.4 g clove, 2 mL ethanol	Spike 1 mL into 25 mL base wine
Coconut	3 g	Shredded, dried coconut steeped for 2 hours in 20 mL of distilled water, spike 5 mL of solution into base wine
Coffee	30 mg	Fresh ground coffee
Cooked cabbage	2 leaves	Brine obtained by cooking 2 cabbage leaves in 200 mL of water for 1 hour
Cooked Fruit	2.5 g	Black currant preserves dissolved in 20 mL of boiling distilled water, spike 20 mL of solution into base wine

(*continued*)

Table 2–1. (*continued*).

Aroma	Quantity	Composition (in 25 mL base wine)
Corn	3.0 mL	100% brine from canned corn
Cut green grass	0.25 g	Freshly cut, crushed green grass, steeped for 2 hours in 20 mL of distilled water, spike 5 mL of solution into base wine
Dill	0.25 g	Dried dill, soak for 30 minutes in 20 mL of distilled water, spike 5 mL of solution into base wine
Earthy	5 mL canned potato liquor + 5 mL canned mushroom soup	In base wine
Ethanol	2.5 mL	Ethanol
Ethyl acetate	1 mL	Ethyl acetate in 20 mL of distilled water, spike 3 mL of solution into base wine
Eucalyptus	One leaf	Crushed, in base wine
Fig	5 mL	Liquid from canned figs
Floral	0.25 mL linalool + 0.5 mL geraniol in 100 mL of distilled water	Spike 1 mL of solution into base wine
Fruity	20 mL	5 mL bottled juice, 5 mL canned apricot juice, and 10 mL canned peach liquor
Grapefruit	3 mL	Fresh grapefruit juice
Green apple	4.5 g	Fresh, crushed Pippin apple steeped for 2 hours in 20 mL of distilled water, spike 20 mL of solution into base wine
Green bean	5 mL	Brine from canned green beans
Green olive	8 mL	Brine from canned green olives
Honey	0.25 g	Honey
Leather	4 cm^2	In base wine
Lemon	3 mL	Lemon juice
Licorice	1 mL	Anise extract in 20 mL distilled water, spike 1 mL of solution into base wine
Melon	2 g	Fresh melon in base wine
Methyl anthranilate	3 mL Welch's red or white grape juice	In base wine
Mint	0.5 mL	Mint extract
Molasses	1.5 mL	Molasses
Mushroom	One sliced mushroom	In base wine
Musty	damp canvas, 2×t2 in.	No wine
Nutmeg	0.6 g	Ground nutmeg soaked in 25 mL of base wine for 2 days, spike 2 mL of solution into base wine
Nutty	Ground almond, walnut, or hazelnut	No wine

Table 2–1. (*continued*).

Aroma	Quantity	Composition (in 25 mL base wine)
Oak	6 mL	Oak Mor, High Tone (Finer Filter Products, Newark, CA)
Onion	0.5 g	Soak macerated fresh onion in 25 mL of base wine, spike 1 mL of solution into base wine
Orange	1.5 mL	Orange juice
Oxidized	2 mL sherry + 50 mg/L acetaldehyde	In base wine
Peach	7 mL	Liquor from canned peaches
Pear	3 mL	Liquor from canned pears
Phenolic	1 mg	1 mg ethyl guaiacol
Pineapple	1 mL	Canned pineapple juice
Prune	2 mL	Prune juice
Pungent	4 mL	Ethanol
Raisin	5 raisins	5 cut up raisins, 8 mL of prune juice
Raspberry	2 g	2 g crushed fresh or frozen raspberries
Rose	1 mg	Phenethanol per 150 mL white wine + 1 rose petal
Rubber	5 g black rubber lab hose	Soak small pieces of black rubber lab hose for 2 hours in 20 mL of distilled water, spike 20 mL of solution into base wine
Smoke	0.5 mL	Liquid smoke
Soy/prune	12.5 mL canned prune juice + 1.5 mL soy sauce	In base wine
Soy sauce	2 drops	Soy sauce
Spice	212 μg/L	Eugenol solution
Spicy	12 mg	Eugenol, one ground black pepper and 0.5 mL aqueous cinnamon extract (120 mg cinn/L)
Strawberry	1	Crushed fresh or frozen strawberry
Sulfur dioxide	40 ppm	40 ppm of sulfur dioxide and 10% citric acid in distilled water, no base wine
Tea	1 tea bag	Soak for 1 minute in 20 mL of boiling, distilled water, spike 4 mL of solution into base wine
Vanilla	1.5 mL	Vanilla extract
Vegetative	2 mL brine from canned asparagus, 4 mL brine from jar of green leaves, or 3 mL brine of canned green beans	In base wine
Viscosity	3.04 g/L	Polycose (Ross Laboratories, Columbus, OH)

(*continued*)

Table 2–1. (*continued*).

Aroma	Quantity	Composition (in 25 mL base wine)
Wet dog/wet wool	Wet dog hair and piece of wet wool	No wine
Wood (raw)	2 g	Red oak wood shavings, steeped for 2 hours in 20 mL of distilled water, spike up to 5 mL of solution into base wine
Wood	1.25 g	Soak Nevers wood chips for 20 mins

SOURCES: Abbott et al. 1991; Andrews et al. 1990; Dozon and Noble 1989; Flores et al. 1991; Francis et al. 1992a, 1992b; Guinard and Cliff 1987; Heymann and Noble 1987; McDaniel et al. 1988; Noble 1984; Noble and Shannon 1987; Noble et al. 1987.

CHAPTER 3

Grape Maturity and Quality

WINE QUALITY

Quality is a subjective judgment that depends on the degree to which the wine is satisfying and balanced and reflects the character of the grape. It can be described in nine categories: color (hue, strength, purity, and stability), aroma intensity, vitality (purity), complexity, subtlety, palate strength, length, balance, and longevity. Hue refers to the dominant color wavelength, strength to the depth of color, and purity to the degree of "off" or tawny tones. Intensity refers to the magnitude of aromas and vitality to the quality and purity of those aromas. Complexity denotes the harmony of wine components. Delicate, refined flavors, strength of palate, length of finish and balance, or the entire integration of the wine, and longevity or conservation, are also important quality factors

Quality components are largely the results of fruit characteristics governed by the parameters shown in Fig. 3–1. Overlaid on basic grape quality is the mark of the winemaker, who can adjust grape growing and winemaking to emphasize or mute aromas, flavors, and textures to produce a well-balanced, integrated product. Wine styles differ because of the tre-

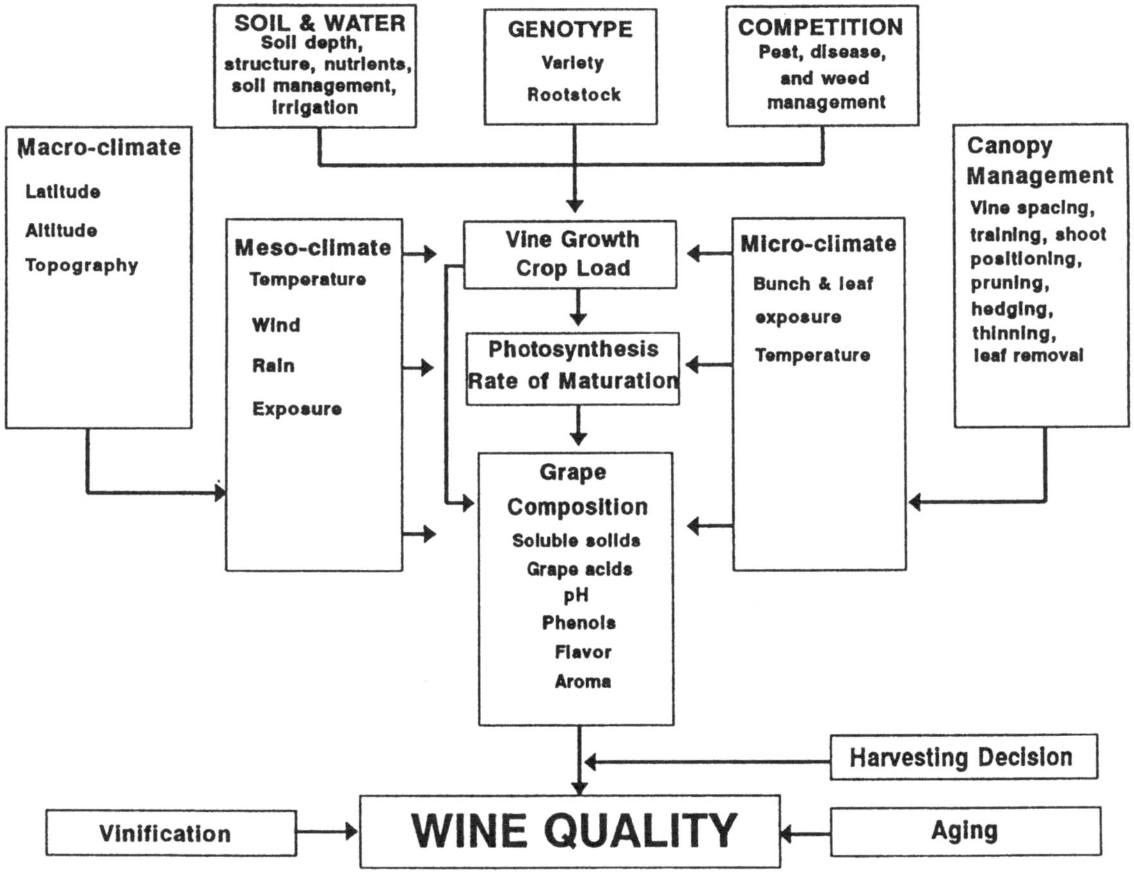

Fig. 3–1. Environmental and viticultural imports into grape composition and wine (adapted from Jackson and Lombard, 1993).

mendous number of variables in grape growing and in winemaking, as discussed below.

Clones

A clone is a population of plants all of which are descendants by vegetative propagation of a single parent vine. Many grapevine cultivars may not be clones, but mass selections. However, such a distinction may be academic. When a mutation occurs in a dividing cell, all cells derived from that individual cell carry that genetic change. All vines produced from the mutant shoot constitute a clone, yet are similar enough to their mother vine to have the same varietal name. Clonal variation can affect yield, setting rate, growth, clusters per vine, berry size, fruit rot susceptibility, and berry flavor components. The latter two are of particular interest.

Variations in vineyard and winemaking techniques likely obscure all but the most pronounced effects of clones on wine style and quality. However, the identification of superior grapevines within cultivars remains an important goal.

Climate

Climatologists recognize three levels of climate: macroclimate or regional climate, meso- or site climate, and micro- or grapevine canopy climate. Wine quality is influenced by the mesoclimate, particularly temperature, during the final state (stage III) of berry ripening. Jackson (1991) and Jackson and Lombard (1993) have divided grape-growing regions into two temperate zones: alpha zones with mean temperatures between 9–15°C (48°F–59°F) during stage III ripening; and beta zones with mean temperatures greater than 16°C (61°F).

The best variety for any region is one that matches the length of the growing season, so that maturation occurs even during the coolest portion of the season. The optimum choice of cultivars should allow for fruit maturity just before the mean monthly temperature drops to 10°C/50°F (Jackson 1991). In warmer climates, the length of the growing season easily allows for adequate fruit maturity but results in fruit development during the warmest part of the season.

Microclimate is determined by the presence of plant cover. Grapevine leaves are the major cause of microclimate variations; the presence of fruit, shoots, stems, and permanent vine parts are less significant (Smart 1985). In the sense that grapevine canopy influences microclimate, it is under the control of the viticulturist.

Canopy microclimate components include radiation, temperature, humidity, and evaporation. Berries maturing in densely shaded canopy interiors are generally associated with the following when compared with berries in open or exposed canopies: (1) low total soluble solids, (2) high titratable acidity, (3) high malate concentrations, (4) elevated pH, (5) high potassium, (6) low proline, (7) high arginine, (8) low total phenols, and (9) low anthocyanin concentration in red and high chlorophyll vs. flavonoid pigments in white cultivars (Kliewer 1980; Kliewer and Lider 1968; Smart 1976, 1985). These differences are due to exposure to sunlight and heat (Smart 1985) (see Tables 3.1 and 3.2).

Sunlight can affect vine and grape physiology through photosynthetic and thermal responses. The amount of diffuse solar radiation reaching the interior canopy leaves and fruit decreases geometrically as the number of leaf layers increases (Smart 1985), resulting in a reduction in photosynthetic rates (Winkler et al. 1974). Varying shoot numbers, reducing vine vigor, or adopting training and trellising systems that divide canopies into separate, thin curtains of foliage improve grapevine microclimate and enhance grape quality (Bledsoe et al., 1988; Smart 1985). For vineyards already planted to conventional trellis designs, alternative methods of decreasing canopy shade may be desired.

Canopy microclimate can be improved with selected fruit zone leaf removal which can result in reduced fruit rot incidence (Bledsoe et al. 1988; Zoecklein et al. 1992). Increased spray penetration, desiccation, and reduction of evaporation potential will likely contribute to rot reduction. For a review of grapevine canopy microclimate evaluation and management, see Smart and Robinson (1991).

Terroir

Although the precise contribution of each factor influencing the terroir (or soil-plant environment) is not known, experience suggests that a 'good' terroir encourages slow, yet complete maturation. Seguin (1986) suggested that a desirable terroir has adequate but not excessive nitrogen, the ability to moderate the influence of heavy rain, especially after veraison, as well as to moderate drought stress. The influence of soil on wine quality is widely debated. Rankine et al. (1971) concluded that soil composition is less important than climate. Jackson and Lombard (1993) reported that soil is known to have several direct influences on plant growth by affecting moisture retention, nutrient availability, heat and light reflecting capacity, and root growth as a result of penetrability. Soils that help promote excessive vegetative growth may negatively influence quality by altering canopy microclimate.

Table 3–1. Expected responses of a "normal" pruned and irrigated California vineyard trained on a 2-wire vertical trellis to various cultural practices

Change of Cultural Practices	Light to Cluster	Tartrate (g/berry)	Malate (g/berry)	K (g/berry)	Wine K	Wine Color	Crop Yield	TSS (°Brix)
U-Trellis	Increase	No Effect	Decrease	Decrease	Decrease	Increase	Slight Increase	Increase
Wye-Trellis	Increase	Increase	Decrease	Decrease	Decrease	Increase	Increase	No Effect
GDC Trellis	Large Increase	Increase	Decrease	Decrease	Large Decrease	Increase	Increase	No Effect or Decrease
Severe Pruning	Slight Decrease	No Effect	Increase	No Effect	Decrease	Increase	Decrease	Increase
Topping (Hedging)	Slight Increase	No Effect	Decrease	No Effect	Increase	Decrease	No Effect or Decrease	Decrease
Mod. Water Stress	Increase	Slight Increase	Decrease	Decrease	Decrease	Increase	Slight Decrease	Increase
Removal of Secondary Shoots	Increase	Increase	Decrease	Decrease	Decrease	Increase	Decrease	Increase
Ethephon (Growth Inhibitor)	No Effect	No Effect	Decrease	No Effect	Large Increase	Large Increase	No Effect To Slight Decrease	No Effect or Slight Increase

SOURCE: Kliewer and Benz 1985.

Table 3–2. Comparison of shaded and control[a] Cabernet Sauvignon fruit.

Parameter measured	Control	Shade	Significance level
Harvest date	17 Sept.	19 Oct.	
°Brix	23.3	21.1	+
T.A. (g/L)	5.5	6.4	+
pH	3.4	3.7	++
K^+ (mg/L)	2325	2510	+++
Malate (g/L)	1.65	2.84	++
Tartrate (g/L)	0.86	0.83	ns
Anthocyanins (g/g fruit)	0.98	0.56	+++

[a]Control treatment was a bilateral cordon 3-wire "T": trellis. Shaded treatment consisted of bunching the foliage around the fruit using bird netting.

SOURCE: Kliewer and Benz (Personal communication) 1984.

Vineyard Yield

Yield is an important economic and production consideration. The majority of vineyards producing quality wines tend to be those having low to moderate yields. Because high yield may delay maturity, direct effects are not easy to measure. The leaf-to-fruit ratio is considered to be an important factor influencing fruit sugar and the other components. For example, McCarthy et al. (1987) found that potential volatile terpenes (PVT) were higher in low-crop vines. Kingston and Van Epenhuijsen (1989) determined that 15 cm^2 leaf area per gram of fruit was optimum for soluble solids production. Yields can be estimated if both the average number of clusters per vine and the average cluster weight is known. Cluster counts may be determined on a vine-by-vine basis before bloom when they are easily visible whereas cluster weights are determined on a shoot-by-shoot basis near veraison when weights begin to stabilize (Wolpert and Vilas, 1991).

MATURITY SAMPLING

Because of the importance of fruit maturity on wine palatability, field sampling of grapes must be performed in an objective and statistically acceptable manner. Amerine and Roessler (1958a, 1958b) reported that berry sampling can provide an accurate and economical sampling technique. Variance in °Brix is a function of berry sample size. Theoretically, to be within ±1° Brix with a probability level of .05 (95 out of 100 samples), two lots of 100 berries each should be examined. To be within ± 0.5 °Brix 95% of the time, five lots of 100 berries should be collected. Kasimatis and Vilas (1985) reported that 10-cluster samples gave the same precision as two 50-berry samples for detecting differences of 1.0 °Brix.

The composition of berries can differ with their position on the rachis, the location of the cluster on the vine, the location of the vine in the vineyard, and the degree of sun exposure. Jordan and Croser (1983) recommend collection of berries from the top, middle, and bottom of the cluster. Terminal berries on the rachis may be less mature than others. Berry sampling in various locations on the cluster may be significant in the case of large clusters. Many sample by selecting berries only from the middle of the rachis, which may be acceptable in varieties with small clusters. Also, the side of the cluster must be randomized. One should also avoid collection from end rows or vines with obvious physiological or morphological differences. Ideally, sampling should reflect differences in soil type, topography, vine growth, and fruit sun exposure.

Fruit exposure may vary with the side of the row sampled because of variation in leaf cover. Smart et al. (1977) demonstrated that white and red grapes exposed to direct sunlight may be as much as 8°C (47°F) and 15°C (60°F), respectively, warmer than the ambient air temperature, this can affect fruit composition, as summarized in Table 3–2. Thus sampling technique must consider the effect of sun exposure. For consistency, some recommend that selected vines within each block be targeted for sampling.

There is a natural tendency to select samples based on eye appeal, often corresponding to the more mature berries. Such bias may yield results that are as much as 2 °Brix higher than measured by the winery after crushing the entire vineyard load (Kasimatis 1984).

Maturity Gauges

Grape maturity brings on a series of biochemical transformations; however, perfect synchronization among desirable components does not occur. Maturity is a multidimensional phenomenon that must be viewed in relative, not absolute terms and is dependent on the type and style of wine sought. Important harvest maturity considerations include:

1. General fruit condition.
2. Taste assessment of grape flavor and tannin maturity (reds) (see Chapter 7).
3. Assessment of varietal aroma and aroma intensity.
4. Soluble solids, titratable acidity, tartaric/malic ratio, and pH (see Chapter 4).
5. Berry softness.
6. Ability to ripen further.

Additionally, °Brix/acid ratio and °Brix x $(pH)^2$ have been suggested as maturity gauges (Cooke and Berg 1983). Soluble solids, titratable acidity, and pH are not specific indicators of physiological maturity or potential wine character and palatability. Harvest parameters vary considerably depending on the season, crop load, soil moisture, etc.

Table 3–3. °Brix and 100-berry weights for Barbera grapes harvested on two sampling dates.

	Sept. 5	Sept. 10
°Brix	22	22
Sample weight (g)	112	118

Vineyard uniformity is important to sampling and subsequently, to wine quality. In two California Cabernet Sauvignon vineyards, Long (1987) found that the one that consistently produced the best quality wine had less variability in replicated 400-berry samples. Extremes in fruit maturation reduce the quality even if the average ripeness seems satisfactory.

Sugar per Berry

Degrees Brix is defined as soluble solids per 100 g of juice and is a measure of all soluble solids including pigments, acids, glycerol, etc., and sugar. The fermentable sugar content of grape must accounts for 90 to 95% of the total soluble solids. Therefore, determination of °Brix provides an approximate measurement of sugar levels. However, this measurement is a ratio (wt/wt) of sugar to water and may change due to physiological conditions of the fruit. A potential problem encountered in °Brix or soluble solids occurs with changes in fruit weight. In time, °Brix may show no change, but in fact there may be major changes in the fruit weight (either increases or decreases). This information is of value to both grower and vintner in selecting harvest dates.

The concept of sugar per berry utilizes the same initial °Brix measurement, but takes into account the weight of a berry sample. For example, the data in Table 3–3 were taken from the same vineyard at 5-day intervals, with soluble solids (°Brix) by refractometry of both samples measuring 22°B. The grower concluded that there had been no change in the maturity of the fruit. However, as seen in Table 3–4, sugar per berry calculations lead to a different conclusion. Here, increases in soluble solids per berry and berry sample weight corresponded to maturation (see Table 3–5). Sugar per berry calculations yield considerably more information than that available by evaluation of °Brix measurements alone.

SAMPLE PROCESSING

Grape sugars, acids, and phenols are not uniformly distributed in the fruit, so sample preparation (i.e., degree of crushing or pressing) can have a

Table 3–4. Change in sugar/berry on two sample dates

	Sept. 5	Sept. 10
Soluble solids	$=\frac{22}{100}=\frac{X}{112}$	$\frac{22}{100}=\frac{X}{118}$
	X = 24.6 g/100 berries	X = 26.0 g/100 berries
Dividing by 100 berries/sample:		
	Soluble solids per berry = 0.246 g/berry (Sept. 5)	Soluble solids per berry = 0.260 g/berry (Sept. 10)

profound effect on analytical results. One example of this relationship is presented in Table 3–6, where the degree of pressing is seen to affect pH and titratable acidity. If sample preparation methods are not standardized, significant variations in fruit analysis will occur.

A variety of methods can be employed for sample processing. The simplest is to place berries in a cheesecloth bag and macerate. This is adequate only if the fruit has not raisined, and if there is not a high degree of uneven ripening. The analysis of juice derived from such a procedure corresponds closely to grapes processed with limited skin contact (i.e., most white wines).

Hand presses can be adequate if operated without seed breakage as may occur with the use of Waring-type blenders. Seed breakage causes an elevation of 0.2 to 0.3 pH units, corresponding to pH values expected when skins and seeds are fermented together for several days. The high sugar content of raisined berries may not be adequately represented by either procedure.

Table 3–5. Evaluation of results from sugar per berry data.

Changes in °Brix	Changes in berry weight		
	Decreases	No Change	Increases
Increases	Maturation and dehydration	Maturation	(a) Major increase: maturation and dilution (b) Minor increase: maturation
No change	Dehydration	No change	Dilution
Decreases	Dehydration and sugar export	Sugar export	Sugar export and dilution

SOURCE: Long 1984.

Table 3–6. Changes in grape juice chemistry with pressing.

Press Fractions	°Brix	TA (g/L)	pH	Total Phenols (mg/L GAE)	Absorbance (520nm)
1	19.5	13	3.0	200	0.25
2	19.5	11	3.2	250	0.62
3	19.5	9.5	3.4	320	1.10

SOURCE: Zoecklein 1986.

FRUIT QUALITY EVALUATION

At the winery, representative grape samples are collected and examined for material-other-than-grape (MOG; e.g., leaves, cane fragments), rot, fruit chemistry, juice aroma, and flavor.

Quantification of mold, yeast, and bacterial metabolites in collected juice samples can be used to help evaluate fruit quality. The nature and concentration of microbial metabolites differ as a function of biological and abiotic factors. Key indicators of fruit rot, such as the presence and concentration of ethanol, glycerol, gluconic acid, galacturonic acid, citric acid, lacasse, and acetic acid are discussed in the following sections.

Mold and Mold Complexes Associated With Grapes

Molds are saprophytic filamentous fungi. When conditions permit, their growth leads directly to fruit deterioration as well as exposing fruit to secondary activity of spoilage yeast/bacteria. Common molds involved in vineyard spoilage include *Penicillium, Aspergillus, Mucor, Rhizopus,* and *Botrytis.*

Mold growth on grapes is considered undesirable except for the association of *Botrytis cinerea* in the production of certain sweet wines. In some cases, mold growth and associated degradation of fruit stimulate the activity of native yeast and acetic acid bacteria, producing ethanol from yeast metabolism and its resultant bacterially-produced oxidation product, acetic acid. Populations of acetic acid bacteria (*Gluconobacter* and *Acetobacter* sp.) have been reported to reach levels of near 10^6 cells/g (Joyeux et al. 1984a, 1984b). In addition, gluconate and ketogluconate may be produced from bacterial oxidation of glucose (De Ley and Schell 1959). Ethyl acetate is often produced as a result of yeast growth on deteriorating fruit. Furthermore, ketoacids (e.g., pyruvate) and acetaldehyde, present as intermediates in the fermentation of sugars, and dihydroxyacetone from bacterial oxidation of glycerol, serve as important binding substrates for sulfur dioxide additions (see Chapter 10).

Penicillium and Aspergillus

After early fall rains, Penicillium may develop in berry cracks, making the fruit unfit for wine production. The penicillia are frequently referred to as "cold-weather molds," growing well at temperatures between 15°C and 24°C (59°–76°F).

Aspergillus niger is also common vineyard fungus found on damaged fruit. Initially the white mycellium of the mold resembles *Penicillium,* but as conidiospores develop, the colony becomes black. This mold is more abundant in warmer climates, and can metabolize sugars to produce citric acid, increasing the acid content of the juice.

Botrytis cinerea

Botrytis cinerea is unique in its parasitology. Frequently its development results in a decrease in grape quality referred to by the French as *pourriture grise,* (grey rot), or *graufaule* by German winemakers. Only under certain conditions does *Botrytis* produce an overmaturation termed *noble rot* or *edelfaule,* indispensable in the production of the great sweet Sauternes in France, Trockenbeerenausleses in Germany, and Tokay Aszu in Hungary. The most favorable conditions for *Botrytis* growth are when the mean temperature is between 15°C and 20°C (59°–68°F) with 90% relative humidity (Bulit and Lagon 1970). Zoecklein et al. (1990) listed the approximate moisture periods required and the effect of relative humidity on the percentage of *Botrytis*-infected fruit. *Botrytis cinerea* is found in all regions, but under conditions of high temperature and low relative humidity, little or no growth occurs. Penetration of the berry and growth of the fungus loosens the berry skin. After infection, depending on climatic conditions, two different effects may be seen:

1. In rainy weather, the infected grapes do not lose water, and the percentage of sugar remains nearly the same or may decrease. Although *noble rot* develops regularly and uniformly, *pourriture grise* or grey rot is normally heterogeneous. Secondary infection by other microbes may follow. Under cold and wet conditions *Penicillium, Mucor,* and *Aspergillus sp.*, as well as other fungi and yeast may overgrow *Botrytis* (Nair 1985), and is referred to as vulgar rot (*pourriture vulgaire*). Breakdown of the grape integument provides a substrate for the growth of native ("wild") yeasts and acetic acid bacteria and may produce a condition called *pourriture acide,* or sour rot.
2. In contrast to the above, *Botrytis* infection followed by warm, sunny, windy weather causes berries to lose moisture by evaporation. With dehydration, shriveling occurs, and the sugar concentration increases; this is called *pourriture noble,* or noble rot. Growth of the mold and associated acetic acid bacteria consumes a portion of the grape sugar. However, the utilization of sugar is countered by increases in sugar due to dehydration.

Factors Influencing *Botrytis* Growth

Grape berries have variable number of stomata, depending upon the cultivar and clone. These are functional only during the initial stages of berry development and become necrotic before veraison (Doneche 1993). These peristomatic areola cracks result in microlesions of from 10 to 100 μm wide, thus providing entry ports for *Botrytis* even in apparently healthy grapes (Purcheu-Planté and Mercier 1983).

Skin thickness and/or toughness may play an important role in limiting susceptibility and may also explain differences between varieties and clones. Tighter clustered varieties with berries that rub together are frequently more prone to *Botrytis* degradation due to disruption of the integrity of the grape cuticle wax (Rosenquist and Morrison 1989). The most common susceptible white varieties in the United States (California) are Chardonnay, Riesling, and Sauvignon blanc. Hill (1987) demonstrated that fruit exposure to UV light increased the production of stibenes, phenolic compounds that toughen the grape berry. English et al. (1990) showed that a reduction in the evaporative potential within the grapevine canopy caused by selective leaf removal significantly reduced the incidence of fruit rots. Using open canopies with low evaporative potential, increased spray penetration and desiccation reduces fruit rot. Additionally, the spores of *Botrytis* overwinter, and the removal or disking of prunings reduces infection.

Synthetic fungicides that have been used to help limit the incidence of *Botrytis* include Rovral, Dithane, Captan, and Benlate. Copper sulfate (Bordeaux mix) is also used and aids in *Botrytis* control by hardening the grape skin. Because agricultural chemicals are perceived by the public to pose health risks, it may be difficult to rely heavily on their use.

Botrytis cinerea and Fruit and Wine Chemistry

The largest quantitative changes occurring in the fruit as a result of *Botrytis* growth are those of sugars and organic acids. From 70 to 90% of the tartaric acid and from 50 to 70% of the malic acid is metabolized by the mold (Doneche 1990). Tartaric is more completely metabolized than malic acid because the latter is assimilated during the external phase of fungal development (Doneche 1993). However, the concentration effect resulting from berry dehydration tends to obscure these changes. Change in the tartaric to malic acid ratio leads to a reduction in titratable acidity and elevation in fruit pH.

Botrytis uses ammonia nitrogen, reducing the levels available for yeast metabolism. Additionally, thiamine (vitamin B_1) and pyridoxine (vitamin B_6) are depleted. Wines produced from *Botrytis*-infected grapes generally require supplementation with nitrogen and vitamins to help avoid fermentation sticking and possible H_2S formation (see Chapters 8 and 9).

Like other fungi, *Botrytis cinerea* produces laccase, (Dubernet et al. 1977), which catalyzes phenolic oxidation. The main nonflavonoid phenolic compounds of grapes are caffeic and *p*-coumaric acids, both free and esterified with tartaric acid. These are transformed to quinones by laccase, with resultant polymerization responsible for browning of the fruit. Ewart et al. (1989) report significant reduction in total anthocyanins in Pinot noir infected with *Botrytis* even when the laccase activity was low. High levels of grape sugar limit the dissolution of oxygen and partly inhibit laccase activity.

Laccase is resistant to sulfur dioxide, cannot easily be removed with bentonite, and is active in the presence of alcohol. Somers (1984) reported laccase activity in wines after 12 months in storage. The activity of laccase is temperature dependent with a maximum around 50°C (122°F) and pH activity between 2.5 and 7 (Ribereau-Gayon et al. 1976). Plank and Zent (1993) reported the use of laccase for the removal of phenolic compounds most susceptible to oxidative browning prior to bottling. In theory, this would facilitate subsequent removal of oxidized phenols by fining and/or filtration (see Chapter 16).

In addition to laccase, pectolytic enzymes and esterases, produced by the mold, break down grape tissue by cleaving methoxy pectins thus increasing the concentration of methanol. *Botrytis* causes an increase in the galacturonic acid content as a result of enzymatic hydrolysis of cell wall pectic compounds (Sponholz and Dittrich 1985). Galacturonic acid may be transformed to mucic (galactaric) acid by enzymatic oxidation and may reach must levels as high as 2 g/L (Boillot 1986; Wurdig 1976). This acid can combine with calcium to form insoluble calcium mucate.

One of the greatest impacts of *Botrytis* growth is the formation of polysaccharides that create clarification problems. Pectins are hydrolyzed by mold-produced polygalacturonase, with the formation of *beta*-1,3- and 1,6-glucans. In wine, ethyl alcohol causes the glucan chains to aggregate, thus quickly inhibiting filtration. Commercially, several glucanases are available to minimize these clarification problems (see Chapters 5 and 16).

Elevated levels of acetic and lactic acid are frequently seen in wines made from *Botrytis*-infected fruit. These spoilage acids arise from growth of yeast and bacteria associated with the mold.

Aspergillus, Botrytis, and *Penicillium* sp. oxidize glucose to produce gluconic acid. Since gluconic acid is not utilized by yeast or bacteria (Benda 1984) it may be used as an indicator of fruit deterioation. Ribereau-Gayon (1988) reported gluconic acid levels in "clean" fruit and in wines made from clean fruit to be near 0.5 g/L, whereas in wines produced from fruit infected with *B. cinerea* levels range from 1 to 5 g/L. In the case of sour rot or vulgar rot, where bacterial growth occurs along with the mold growth, levels may also reach 5 g/L.

Botrytis cinerea also produces significant amounts of polyols, of which glycerol is quantitatively the most important. Quantities produced may be as high as 20 g/L (Ribereau-Gayon et al. 1980). Glycerol may be metabolized by bacteria before harvest and sour rot berries often are emptied of their contents by insects. Infected fruit then develops high levels of acetic (40 g/L) and gluconic acid (25 g/L). Ribereau-Gayon (1988) suggested that the ratio of glycerol to gluconic acid indicates the "quality" of the rot. Higher ratios indicate the growth of true noble rot, whereas lower ratios suggest sour rot. *Gluconobacter oxydans* and *Acetobacter aceti* are known to oxidize glycerol to dihydroxyacetone. In this regard, *Gluconobacter* is more important than *Acetobacter* (see Chapters 6, 18). Thus, the presence of these species in significant numbers may impact the final concentrations of glycerol in the must.

Other polyols are formed by the fruit and their concentration increases by mold degradation. Mannitol, erythritol, and *meso*-inositol are secreted by *Botrytis cinerea* (Bertrand et al. 1976). Wines produced from grapes severely degraded by *Botrytis cinerea* have as much as 100 times more mannitol than from sound grapes (Boillot 1986). This compound can impart a rather sweet and sour character to wines.

Sensory considerations

In both sour rot and *Botrytis cinerea*, several aroma modifications may occur. Generally, fruitiness disappears sometimes with the formation of unpleasant odors described as "phenol" and "iodine." Noble rot often imparts a "honey" or "roasted" component to the aromatic character of wine. Nishimura and Masuda (1983) suggest that the characteristic aroma of wines made from *Botrytis*-infected fruit is due to 3-hydroxy-4, 5-dimethyl-2-furanone.

Botrytis secretes esterases that hydrolyze esters produced during fermentation. *Botrytis* also destroys monoterpenes, which are, in part, responsible for the varietal aromas of Muscat, Riesling, and Gewurtztraminer. Glycosidases produced by *Botrytis cinerea* hydrolyze the terpenyl glycosides (Gunata et al. 1989) Reductions of up to 90% of the primary monoterpenes such as nerol, geraniol, and linalool are reported to occur with the growth of the mold (Ribereau-Gauyon 1988).

Quantification of *Botrytis*

The incidence and severity of *Botrytis cinerea* degradation can be estimated by visual inspection, analysis of laccase, or immunochemical assay. Test kits estimating laccase content are available. Laccase activity above 3 units/mL indicate significant activity of *Botrytis* (Dubourdieu et al., 1984).

The presence of laccase (oxidative casse) is occasionally estimated by winemakers by exposing a small quantity of juice or wine to air at 20°–30°C (68°–86°F) for several hours. Laccase, if present, will cause rapid loss of red color or formation of brown hues, and brown deposits in white wines (Somers 1984) (see Chapter 20).

An immunochemical method for the detection and quantification of *Botrytis* has been developed. This method, now available in kit form, allows for qualitative and quantitative determinations. Reactivity has been shown against all isolates of *Botrytis cinerea* and no cross-reactivity with other molds or yeast has been noted. In the United States, the California Department of Food and Agriculture (CDFA) has recently approved this method.

Processing Considerations for *Botrytis*

Methods of handling grapes degraded by *Botrytis cinerea* and laccase include removal of deteriorated fruit, thermo-vinification, light whole cluster pressing, segregation of press fractions, cryoextraction, and post-fermentation heat treatment (Ribereau-Grayon et al. 1976).

Traditional pressing techniques are ineffective for "dried" grapes degraded by *Botrytis cinerea* without excessively high press pressures. When the rot is not homogeneous, pressing preferentially extracts juice from berries that are least affected by rot and lowest in sugar. Cryoextraction has been used to freeze grapes lowest in sugar, followed by immediate pressing, thus allowing only the ripest grapes to release juice. Mechanical processes that grind skins can cause glucans to diffuse into the must, posing clarification problems. (See Chapter 20 for determination of glucans).

Fermentation difficulties may arise from native yeast and bacteria, low levels of assimilable nitrogen, high must sugar, and the presence of botryticine. Botryticine, a heteropolysaccharide, rich in rhamnose and mannose, stimulates the production of glycerol by yeast (Doneche 1993). Botryticine has been demonstrated to stimulate production of acetic acid by *Saccharomyces* towards the end of fermentation.

Once sufficient alcohol has been produced, fermentation should be stopped, because the highest levels of volatile acidity are produced from the last few grams of sugar (Doneche 1993). Additionally, botrytized grapes produce wines with high concentrations of compounds likely to combine with sulfur dioxide, decreasing its percentage in the free microbiologically active form (see Chapter 10).

Laccase activity has been reported in aged wines (Somers 1984). Therefore, some choose to remove or inhibit the enzyme. Ultrafiltration using a membrane with molecular weight cut off of 50,000 daltons will remove laccase. Since laccase activity is inhibited in the absence of oxygen contact must be avoided or kept to an absolute minimum.

PESTICIDES

When used as directed, organic pesticides do not influence fermentation or wine quality (Lemperle and Kerner 1974). Fungicide residue (as a result of late season spraying) can cause difficulty in initiation and/or completion of fermentation. Residues of systemic and contact fungicides differ. Systemic fungicides are likely found in much lower concentrations yet may persist longer in the juice and wine (Gnaegi 1985). The vinification method can influence the amount of residue remaining. Skin contact may enhance, whereas preclarification (with and without bentonite and/or carbon) tends to reduce concentrations of contact fungicides (Lemperle and Kerner 1974) as do absorption by yeast.

Residues of colloidal sulfur from sprays used against powdery mildew can influence wine quality as a result of reduction by yeast to hydrogen sulfide during fermentation. A final copper spray (Bordeaux mix) is occasionally applied at the end of the season to help neutralize hydrogen sulfide formation (Lemperle 1988) (see Chapter 9).

SENSORY CONSIDERATIONS AS AN INDICATOR OF GRAPE MATURITY AND QUALITY

Except for monoterpenes, little is known about the components responsible for grape aroma in *Vitis vinifera.* Some factors that challenge investigation include the extremely low concentration of components, the difficulty of relating odors of individual compounds to overall grape or wine aroma, and the chemical changes that may occur during grape processing, winemaking, and aging (Rapp 1988; Williams et al. 1988).

The measurement of aroma components and the use of such analysis as a criteria of fruit quality has, and will continue to receive, a great deal of attention (see Chapter 20). Wines from Cabernet Sauvignon and Sauvignon blanc grapes often have a characteristic aroma described as vegetative, herbaceous, grassy, or green attributed to methoxypyrazines. The floral aromas of Muscat, Gewurtztraminer, and Riesling grapes and wines are attributed to monoterpenes, considered character impact compounds for those varieties.

Grape Aroma Components

The aroma components of grapes depend upon a small group of free secondary metabolities that have escaped oxidation to polyols or conjugation. Glycosylation and oxidative steps, such as hydroxylation, are natural processes that remove components as direct contributors to flavor. Oxidative products have no direct flavor value. There is, however, some reclamation of flavor available through acid-catalyzed hydrolysis of polyols and/or enzymatic hydrolysis of glycosides. Hydrolysis can produce a wide range of volatile components that can add complexity.

The most widely studied grape aroma components are the monoterpenes. The general biochemical mechanisms utilized for monoterpene production in varieties such as White Riesling are also present in non-terpene varieties. While non-terpene varieties also derive their flavor from free volatiles, the majority of potential flavor is locked up in the form of conjugated products (see Fig. 16–4).

Terpenoid compounds all possess characteristic branched chain carbon skeletons, and can occur with a variety of functional groups including alcohols, ethers, aldehydes and poly-functional derivatives. Monoterpenes in particular tend to have pronounced potent, pleasant flowery, and fruity aromas even at very low concentrations.

A number of surveys have been made of monoterpene concentration in different grape varieties (Rapp et al. 1982; Ribéreau-Gayon 1972; Schreier et al. 1976). A general classification of varieties allows division into (1) intensely flavored muscat varieties with total monoterpene concentrations as high as 6 mg/L; (2) nonmuscat but aromatic varieties with monoterpene concentrations of 1 to 4 mg/L; and (3) neutral varieties not dependent on monoterpenes for their flavor.

Free monoterpenes (i.e., not bound to other molecules) referred to as free volatile terpenes (FVT) provide most of the varietal character of wines such as Rieslings, Gewürztraminer, and Muscat. Initially monoterpenes are glycosidically bound to disaccharides. In the bound state, they do not contribute to odor, and therefore methods of hydrolyzing these aroma precursors to release free floral terpenes are of interest. Riesling grapes, for example, have about 85% of their total monoterpenes bound. Bound compounds are referred to as potentially free volatile terpenes, or PVT. The analysis of monoterpenes is provided in Chapter 20.

A number of methods have been suggested for taking advantage of the aroma potential of bound terpenes. These include skin contact, pressing, muté production, and use of commercial enzymes and selected yeasts. Because most monoterpenes are found in the skins, skin contact can enhance juice concentration. Degree of ripeness, cultivar, and temperature as well as yeast strain selection, can influence monoterpene extraction. Phenol extraction will also increase with skin contact. Higher press pressure results in a higher content of components found in the skins, including monoterpenes.

A practice that is not uncommon is to fortify skins. Alcohol aids in extraction of monoterpenes from the skins and is used to fortify juice for the production of *mutés*. Commercial enzymes can cleave the glycosidic bond between the monoterpene and sugars to convert PVT into the free, aromatic FVT (see Chapter 16). Yeasts differ in their ability to cleave the bond between monoterpenes and sugars, increasing the free, aromatic monoterpenes. Genetically engineered yeast strains with enhanced *beta*-glucosidase activity are commercially available.

Monoterpenes are important aroma components of some cultivars but not all. Chardonnay, Sauvignon Blanc, and Cabernet Sauvignon, for example, contain monoterpenes but they are not major aroma components. Some aroma constituents from these cultivars are also bound to sugars.

SOLUBLE SOLIDS IN WINEMAKING

Near maturity (≥18° Brix) soluble solids levels are within 1% of the actual sugar content. However, before maturation the actual sugar content may be 4 to 5% lower than the soluble solids (Crippen and Morrison 1986). Such a discrepancy can arise with refractive index measurements due to compounds with refractive indices similar to glucose and fructose. The balance of nonfermentable species includes pectins, tannins, pigments, acids, and their salts. Soluble solids help provide an indication of fruit maturity and potential alcohol yield as well as a tool to monitor fermentation rates, as a relative indication of sugar content in blends, and as a legal standard for certain wine types. Winemakers in certain regions rely on soluble solids measurements to dictate chaptalization and amelioration considerations.

Internationally, soluble solids are expressed using °Brix, °Balling, °Baume, or °Öechsle scales. Saccharometers using °Brix or °Balling (which are identical) are calibrated to read concentration of sucrose in g per 100 g of solution. Baumé hydrometers (used in Australia) are constructed to read percent salt in a pure salt solution; a reading of 1.8 °Brix (or Balling) equals approximately 1 °Baumé.

In Germany, the degrees Öechsle (°Ö) scale is used to estimate sugar concentration. Degrees Öechsle is based on the difference in weight of 1 L of must compared with 1 L of water. The first three figures in the decimal fraction of the specific gravity equals the Öechsle equivalent. For example, a 20 °Brix solution has a specific gravity of 1.082 and therefore a °Ö of 82. (For interconversion of specific gravity, °Brix, °Baume, and °Öechsle, consult Table I-2 in Appendix I.)

Conversion of sugar-to-alcohol

Theoretically, a given weight of fermentable sugar should yield 51.5% (by weight) ethanol, according to the Gay-Lussac relationship:

$$C_6H_{12}O_6 \rightarrow 2\ C_2H_5OH + 2\ CO_2$$

Thus, an initial 180 g glucose should theoretically produce 92 g ethanol (51%) and 88 g carbon dioxide upon complete fermentation. The actual alcohol yield is generally less than theoretical. As an estimate of potential alcohol, many winemakers use the conversion factor of 0.55. However this factor may be valid only in warm climates.

The cooler the grape-growing region the higher the conversion factor. In upper Monterey County (California), for example, winemakers may use figures as high as 0.62. Jones and Ough (1985) found that alcohol conversions ranged from 0.54 to 0.61. Differences were noted between regions and growing seasons, as well as slight variations between varieties. Another factor influencing the alcohol/°Brix ratio is fruit condition. With raisined berries, initial sugar extraction is incomplete, yielding erroneously low initial °Brix readings. With more complete extraction during pressing and fermentation, additional fermentable sugar is liberated, yielding higher-than-expected final alcohol.

Amelioration and Chaptalization

In certain winemaking regions of the United States and in most European wine-growing regions, amelioration, or chaptalization, of must or wine is permissible. Amelioration is defined as the addition of sugar and/or water to juice or wine, whereas chaptalization refers only to the addition of sugar. Sugar additions may be carried out either before, during, or after fermentation. If the winemaker desires an alcohol level of at least 11.5% (vol/vol), and is working with a conversion ratio of 0.55, for example, he must have an initial Brix of 21.0, as seen in the following relationship

$$\frac{11.5\%\ \text{final alcohol}}{0.55} = 21.0\ \text{g}/100\text{g}$$

The winemaker must know the sugar to alcohol conversion ratio for the particular variety and region in order to accurately predict alcohol, and make amelioration calculations and stylistic decisions.

The quantity of dry or liquid sugar needed to raise the Brix of a must may be determined from tables of formulas (see Table I-1 in Appendix I). When dextrose is used instead of sucrose, the 8% water of crystallization in the dextrose must be taken into account. With anhydrous dextrose, this does not apply. Must sugar content in weight may be determined as follows:

$$°\text{Brix} \times \text{Specific gravity} \times 10 = \text{weight g/L}$$

For wines of a California appellation, amelioration is normally permissible only in production of fruit wines, special naturals, and sparkling wines. In other cases, deficiencies in must sugar may be made up by the addition of grape juice concentrate. In cases where fruit maturity exceeds 22 °Brix at crush, U.S. winemakers are permitted to add sufficient water to bring the soluble solids (°Brix) back to that level.

Monitoring Fermentation

Soluble solids measurements are routinely used in winemaking to monitor the progress of fermentation. With fermentation, grape sugar decreases and alcohol increases. As a result, the fermenting juice has a lower density than the original. Pure water at 20°C (68°F) has a °Brix reading of zero. The density (specific gravity) of a wine is less than that of pure water due to the ethanol content. Therefore it is possible to have a dry wine with a negative °Brix value. Although hydrometers cannot be used for an accurate measurement of the sugar content in fermenting must and fermented wine, °Brix levels are used as standards for certain wine types. American wine labeling regulations reflect fruit sugar concentration according to the following: early harvest (grapes harvested at minimum of 20°Brix)—equivalent to the German Kabinett; late harvest—somewhat equivalent to German Auslese and requires a minimum of 24 °Brix at harvest; select late harvest—equivalent to Beerenauslese with a minimum grape sugar of 28 °Brix; special select late harvest—the highest maturity level requiring grapes picked at a minimum sugar concentration of 35 °Brix, the same as the German Trockenbeerenausleses.

LABORATORY MEASUREMENTS OF SOLUBLE SOLIDS

Densimetric Procedures

Hydrometry

The absolute density of any substance, expressed in units of grams per cubic centimeter (g/cm^3) or grams per milliliter (g/mL), is defined as:

$$\text{Density} = \frac{\text{Weight of substance}}{\text{Volume of substance}}$$

Direct measurement of volumes may present a problem, especially where gases are concerned. As a result, it becomes convenient to use the ratio of the density of a substance to that of a recognized reference such as water. This relationship, known as specific gravity, is expressed as:

$$\text{Specific gravity} = \frac{\text{Weight of} \times \text{mL of substance}}{\text{Weight of} \times \text{mL of water}}$$

The density of water at 4°C is, for all practical purposes, 1.000 g/cm^3. Because the weight of any substance will change as a function of temperature, any complete definition of specific gravity must include the temperature at which the determination was made, as well as the reference temperature for water. The temperature of the measured sample is noted above that of the reference. For example, the notation 15°/4°C indicates that the specific gravity of the solution in question was made at 15°C relative to water at 4°C.

The concentration of dissolved substances in solution is related to the specific gravity but one should not, however, assume a simple and direct correlation between observed specific gravity and concentration in all cases because molal volumes of substances in solution may vary in a complex and unpredictable manner. Tables are available that relate concentration of dissolved substances to apparent specific gravity; those most commonly encountered in analysis of wine are for Brix, Baume, °Öechsle, and alcohol (Table I-2 in Appendix I).

Tables usually reference only one or two standard temperatures; one must either measure the specific gravity at the defined temperature or, alternatively, employ a temperature correction factor. For the most accurate work, it is recommended that the solution be brought to defined temperature prior to measurement.

Hydrometric determinations are based on the principle that an object will displace an equivalent weight in any liquid in which it is placed. The volume displaced by an object is inversely proportional to its density. Hence a solution of high density will show less displacement than one of lower density. This relationship defines the basic principle of hydrometry.

A hydrometer consists of a calibrated scale within a glass tube that is usually constructed with a mercury or shot-filled terminal bulb to maintain it in an upright position. Hydrometers are available to read either specific gravity or the concentration of some component in solution. Examples of the latter include the familiar saccharometer and the salinometer.

Refractometry

Refractometric determinations are used *in lieu* of, or in addition to, hydrometric determinations of soluble solids. In principle, the passage of a ray of light from one medium to another of different optical density (different number of molecules interacting with the light) causes the incident ray to undergo a change in direction, or refraction (see Fig. 3–2). The angle formed by the entering light beam is called the "angle of incidence". The angle formed by the refracted ray is termed the "angle of refraction."

The "index of refraction" is defined as the ratio of the sine of the angle of incidence to the sine of the angle of refraction. For a solution of defined composition, the refractive index varies as a function of wavelength and temperature. Therefore, a reference wavelength has been selected for refractive index measurement: monochromatic sodium light at 589 nm and 20°C. Thus, for a solution of constant composition at a defined wavelength of light and temperature, the index of refraction is constant. As with hydrometric determinations, shorthand notation frequently is used. The designation N_D^{20} refers to a reading taken at 20°C using incident light of the D-line of sodium. Using this convention, the refractive index of water is defined as $N_D^{20} = 1.330$. (see Chapter 20).

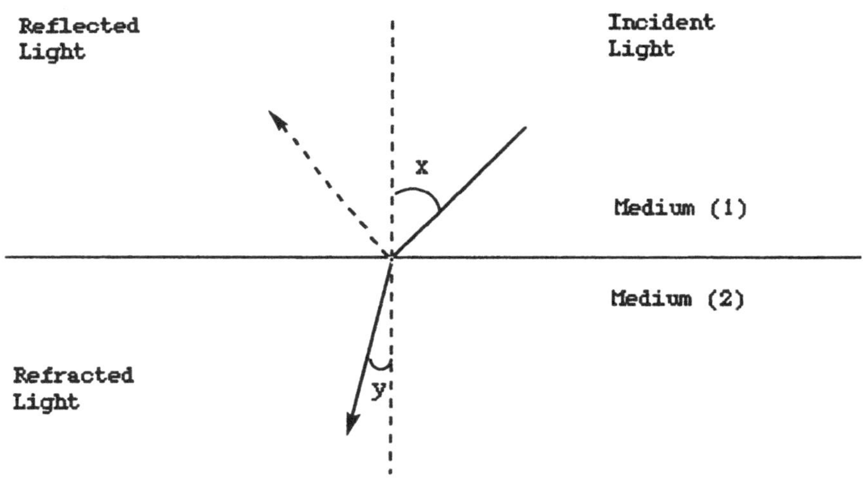

Fig. 3–2. Refraction of a beam of light resulting from passage from one medium into a second medium of greater density.

ANALYSIS

The following is a cross-reference of analytes and procedures found in Chapter 20: Aroma evaluation by sensory; monoterpene analysis by spectrometry; fungal metabolites (acetic acid, glycerol, ethanol, gluconic acid, galacturonic acid, citric acid) by HPLC and/or spectrometry; laccase by spectrometry; soluble solids and non soluble solids; sugar per berry by hydrometry and refractometry.

CHAPTER 4

Hydrogen Ion (pH) and Fixed Acids

Hydrogen ion concentration plays a major role in wine making. Its involvements range from physical-chemical and biological concerns to the sensory attributes and potentially, defects. Hydrogen ion activity usually is measured in terms of the logarithmic concentration term, pH. Initially, the winemaker is concerned with fruit and must pH. The pH values for white wines are often 3.4 or less, whereas higher values usually are observed for red wines, largely because of contact of juice and skins before and during fermentation.

ORGANIC ACID CONTENT OF WINE

The acid content of a wine is of importance, from the standpoint of flavor and, indirectly, effects on pH, color, stability, and shelf life of the product. Titratable acidity in grapes normally runs between 5.0 and 16.0 g/L; the acid levels obtained are influenced by variety, climatic conditions, cultural practices, and maturity of the fruit.

A number of organic acids are present in wines (tartaric, malic, citric, acetic, succinic, and lactic are common). In the case of di- and tricarboxylic acids, their several salts also may be present. These organic acids, along with others to be discussed in Chapter 11, are commonly grouped into the categories of "fixed acidity" and "volatile acidity," their sum being "total acidity." As such, tartaric, malic, citric, and succinic acids comprise the fixed acid contingent, whereas acetic, propionic, butyric, and sulfurous acids comprise the volatile acids in wine.

The organic acid content of a wine is traceable to three sources. The grape itself contributes tartaric, malic, and, to a much lesser extent, citric acid. By comparison to tartaric and malic acids, which are present at levels ranging from 0.2 to 1.0% (wt/vol) and 0.1 to 0.8% (wt/vol), respectively, citric acid is found in unfermented grapes at 0.02 to 0.30% (wt/vol) (Amerine and Ough 1980). Alcoholic fermentation results in formation of lactic, acetic, and succinic acids in addition to very small quantities of other Tricarboxylic Acid Cycle acids. Lastly, bacterial involvement may produce significant amounts of lactic and acetic, and, on occasion, propionic and butyric acids. Mold growth on the grape may result in gluconic acid levels of 0.15 to 1.0% (wt/vol) in a finished wine (McCloskey 1974).

INTERACTIONS OF HYDROGEN AND POTASSIUM IONS AND TITRATABLE ACIDITY

Historically, a number of studies have followed the development of the major organic acids and pH in grape berries during maturation (Amerine and Winkler 1942; Hale 1977; Johnson and Nagel 1976). Hydrogen ion concentration (pH) and titratable acidity (TA) in mature fruit cannot be explained simply in terms of the concentration of organic acid anions. Investigators have suggested that monovalent cations, such as potassium and sodium, enter the cell in direct exchange for protons (H^+) derived from organic acids (Boulton 1980a, 1980b; Hodges 1976). This exchange leads to grapes that have lower TAs and higher pH values than would be expected if one measured the acid anion composition (Boulton 1980a).

Boulton (1980c) proposes a membrane-bound ATPase responsible for potassium uptake. The evidence for concluding that ATPase-mediated transport occurs in grape berries comes, in part, from observations of the relationship between organic acid levels, titratable acidity, and pH in grape tissue. In grape juice, only about 68 to 74% of the expected protons are found by titration (Boulton 1980d). The missing protons can be accounted for by postulating an ion exchange process in the cell membrane that exchanges potassium and sodium for protons. The discrepancy between the protons expected from the organic acid anion content and those found by titration has been shown to be numerically equal to the molar quantity of potassium and sodium ions present (Boulton 1980a, 1980d).

The free proton pool in grape juice generally ranges from 0.001 to 0.01 mol/L, whereas the concentration of potassium plus sodium ranges from 0.05 to 0.10 mol/L, and the total acid concentration from 0.05 to 0.10 mol/L. As such, the pH value of juice is more sensitive to changes in the concentrations of potassium and sodium than similar molar changes in the principal organic acids, tartaric and malic. At typical juice pH values, an increase of 10% in the potassium and sodium concentrations in the fruit may lead to a shift of approximately 0.1 pH unit. In general, the concentration of potassium in grape berries is about 30 times that of sodium (Amerine and Winkler 1942). Sodium, therefore, usually plays a minor role in ATPase-mediated exchange in grapes (Boulton 1980b).

It has been proposed that potassium is taken up from the soil by hydrogen-potassium ATPase activity in the roots of grapevines. Therefore, the internal levels of ATP in the root tissues, as well as the root system size and the proportion of merestimatic tissues, operate in regulation of potassium uptake.

Once in the roots, cations move by bulk flow to the outer membranes of the leaf and berry cells. The presence of an ATPase in the plasmalemma of the cells enables cation transport across the membrane in exchange for internal protons derived from the organic acid pool. Cytoplasmic ATP levels are considered to act as substrate and the primary influence of this transport (Lehringer et al. 1979). Temperature also is considered important in ion transport (Boulton 1980c). The level of monovalent cations, such as potassium and protons in the phloem would be determined by the relative rates of uptake in roots, berries, and other tissues. Potassium uptake by the grape berry may be more rapid than by other tissues such as shoots, petioles, and leaves because of ATP availability (Boulton 1980c).

Freeman and Kliewer (1983) reported that shoots are net exporters of K^+ between anthesis (bloom) and fruit maturity. This export of potassium from the shoots was determined to be equivalent to 43% of the potassium content of the fruit.

The proposed mechanism for potassium accumulation is enzyme-mediated, so exchange is slowed when environmental conditions (i.e., low temperature) do not favor activity of the enzyme. In cool growing regions, the levels of malate in the grape's organic acid pool remain higher than in warmer areas where malate decreases because of respiration. In some instances, conditions may favor greater enzyme activity and uptake resulting in relatively high levels of berry potassium (>1,000 mg/L), high TA (>10 g/L), and high pH (>3.5) due to the exchange for hydrogen ions from malate. In very cool growing regions, the soil temperature may be sufficiently warm to stimulate enzyme activity and potassium uptake. In these instances, potassium levels and TA are expectedly high (>1,000 mg/L and >18 g/L, respectively). However, under such conditions, pH may still remain quite low (2.9–3.2) because of the relatively high concentration of malic acid (Boulton 1985).

In the case of grapes growing in warm regions, enzyme activity is greater because of the temperature, and the acid level is low because of malate respiration. Assuming potassium is available in the soils and aerobic conditions are present in the root zone, potassium is taken up and transported throughout the vine, and to the berries, where it exchanges for H^+ arising from malic acid. In this case, pH is high (>3.5), the potassium level is high (>1,000 mg/L), and TA is low (<6 g/L).

Some important parameters in potassium uptake and pH changes include soil, soil exchange capacity, rootstock, vine vigor, leaf shading, cultivar, and crop level, as well as seasonal variations. The rate of potassium uptake by roots appears to be somewhat independent of the concentration of available soil potassium, in that higher concentrations of potassium do not result in greater uptake rates (except when very high). However, if the concentration of potassium is less than that needed for saturation of the hydrogen-potassium ATPase, uptake will be related to soil potassium concentration (Boulton 1985). Waterlogged soils with low oxygen levels allow little respiration to occur in the roots. Thus not enough energy is generated to activate the enzyme, and little or no potassium uptake occurs.

Shallow soils have more potassium in the upper root system and thus may show greater potassium uptake. Additionally, high-pH soils permit greater potassium uptake. Soils high in magnesium and calcium may cause a reduction in vine potassium accumulation and may lead to potassium deficiency. However, it is difficult to control potassium uptake by soil management.

Activity that slows vine growth will also slow enzyme activity and thus may affect potassium exchange and pH elevation. Thus, devigorating rootstocks, or those that slow vegetative growth, and increased planting density to control vigor may also be used to control potassium uptake. Unfortunately, there may not be a good commercially available rootstock for vigor control.

There are differences among rootstock-scion combinations and hydrogen-potassium ATPase activity. Additionally, enzyme titer may vary among varieties. This is a major reason for the variation seen between pH imbalance and variety (Boulton 1980c).

A higher crop level results in a reduced accumulation of potassium principally because of a dilution effect. Higher crop levels in warm climates, however, delay maturity and, in combination with malic acid respiration, may produce fruit of low TA. Vines with excessive growth often have a significant amount of leaf shading. Beyond three leaf layers, light exposure is significantly reduced and shaded leaves are not photosynthetically active. When photosynthesis stops, no sugar is being produced, and ATP is channelled toward activation of the enzyme for potassium exchange. Thus, additional potassium is pumped into the berry. Malate still may be respired under these conditions, resulting in a decrease in the organic acid pool. As a result of the utilization of malic acid by the plant and uptake of potassium, the fruit has a low TA and high pH. Unfortunately, attempting to control potassium balance in the vineyard may not be practical. The relationships between leaf shading, vigor, management, and fruit and wine chemistry are demonstrated in Tables 3–1 and 3–2.

Temperature is the most important environmental factor influencing both the total acidity and the ratio of tartaric to malic acids (Winkler et al. 1974). Grapes ripening in cool day or night temperatures have higher total acidity and higher levels of malic acid than fruit ripened at warmer temperatures. Kliewer (1971) demonstrated that the amount of malic and tartaric acids found in grapes is negatively correlated with temperature during ripening and relatively independent of light intensity. The respiration quotient (RQ) is the ratio of oxygen absorption to carbon dioxide production. When sugars are respired, the quotient is approximately 1.0. The RQs for malic and tartaric acids are 1.33 and 1.60, respectively. Tartaric acid is respired only at temperatures above 86°F, whereas malic acid is respired at temperatures less than 86°F. In part, this explains the relatively low levels of malic acid in warm seasons and regions.

HYDROGEN ION CONCENTRATION AND BUFFERS

The molar concentration of hydrogen ion (H^+) in fruit juice and wine ranges from 10^{-2} mol/L, in the case of lemon and certain other citrus juices, to slightly less than 10^{-4} mol/L for grape juice and wine produced from grapes grown in warmer climates. These molar concentrations are equivalent to pH 2 and 4, respectively, using the classic relationship:

$$pH = -\log [H^+]$$

Hydrogen ions originate from the dissociation of parental acids. Weak organic acids, such as tartaric, malic, citric, lactic, and acetic, make up the major acid content of grapes and wine. These weak acids dissociate to only a small extent relative to strong acids:

$$HA \rightleftharpoons H^+ + A^-$$

The degree of ionization of a weak acid is denoted by the dissociation constant (K). These constants are generally on the order of 10^{-4} or smaller for typical wine acids, so the extent of weak acid dissociation in must and wine amounts to about 1%. That is, 99% of the acid content is in the parental form (Plane et al., 1980). These constants are frequently utilized in their logarithmic form (pK), and their values can be found in many handbooks.

Hydrogen Ion Concentration (pH) vs. Titratable and Total Acidity

There is a significant difference between pH and TA. The former is a measure of the free proton content of the solution, whereas the latter depends on the concentration of wine acids as well as the extent to which they dissociate. In the United States, TA is measured by potentiometric or indicator titration using standardized sodium hydroxide to the agreed-on endpoint of pH 8.2. It thus measures free $[H^+]$ plus any undissociated acids that can be neutralized by base.

There is also confusion between the terms titratable acidity and total acid content. The two are used interchangeably by the AOAC to refer to the same measurement (the AOAC procedure for measurement of titratable acidity appears under the title "Total Acidity"). Total acid content may be defined as the concentration of organic acids in the grape (Boulton 1982). It is a measure of the hydrogen ion concentration plus the potassium and sodium ion concentrations. TA on the other hand, is only a measure of hydrogen ions consumed by titration with standard base to an endpoint. TA is calculated (in the United States) in terms of grams tartaric acid per liter. The equation used to calculate this value uses the equivalent weight of tartaric acid (75 g/equivalent) in the formula. Thus this equation is constructed on the assumption that each tartrate species present has two protons to be neutralized by the NaOH titrant. Higher pH wine or juice samples with some monoprotonated titratable acids present would require less NaOH for the same absolute amount of tartaric or malic acid. Thus the values obtained by titration (titratable acidity) frequently will indicate less than the absolute concentration of wine or juice acids ("total acid content"), as determined by Boulton.

Buffering Capacity

Containment of pH within narrowly defined limits is critical to the physiological well-being of virtually every living system. The buffering capacity of a juice or wine is a measure of its resistance to pH changes. Essentially it is a measure of the organic acid pool (malic and tartaric) at winemaking pH.

A system with a high buffering capacity requires more hydroxide (OH^-) ions or hydrogen ions to change the pH than one of lower buffering capacity. Thus, buffering capacity can be defined, in practical terms, as the quantity of hydroxide or hydrogen ions needed to obtain a change of 1 pH unit (i.e., from 3.4 to 4.4).

Buffers occur as either weak acids and their conjugate bases, or as weak bases and their conjugate acids. The net result of buffering action is to create within the system resistance to changes in pH that otherwise would occur with addition of either acid or base. In the case of base addition, excess OH^- ions are consumed by H^+ ions of the buffer's acid component, to form water, whereas excess H^+ are consumed by the anion component.

SAMPLE PREPARATION AND REPORTING TA RESULTS

Many constituents, including acids, of the fruit are not uniformly distributed throughout the berry, so standardized berry sample preparation is an important consideration. The magnitude of these differences is readily seen in Table 3.6.

In the United States, results of TA analysis routinely are expressed in terms of the dominant acid contained in the sample, a value achieved by inclusion of the equivalent weight for the respective acid (see Chapter 20). The TA of grape wines in the United States is expressed in terms of tartaric acid. Use of tartaric acid for expression of TA is not a standard throughout the world wine community. In France, for example, TA is expressed in terms of sulfuric acid. By comparison, the German definition of TA is the volume of 0.1 *N* base needed to fully titrate a 100-mL sample.

In analyses of wine or musts, where one wishes to express results in terms of another acid, the appropriate equivalent weight must be taken into account (see Table 4–1). Therefore, if it is necessary to express results in terms other than as tartaric acid, one need only use the appropriate equivalent weight of the acid in question.

Table 4–1. Equivalent weight conversions for wine acids.

Acid	Milliequivalent weight (g/meq)	Molar mass (g/mol)
Tartaric	0.075	150.09
Malic	0.067	134.00
Citric	0.064	192.14
Sulfuric	0.049	98.08
Lactic	0.090	90.08
Acetic	0.060	60.05

ADJUSTMENTS IN TITRATABLE ACIDITY AND pH

Adjusting the acid content or pH of a must or wine may be necessary to produce a well-balanced product. Several of the most commonly employed methods of adjustment are discussed briefly below.

Organic Acid Additions

Several food-grade organic acids currently are approved as wine additives. Different regulatory agencies worldwide may place various restrictions on acids approved for addition to wine (see Table I-10 in Appendix I). The most commonly used acidulants in grape wine production are food-grade tartaric, malic, and citric acids. Fumaric acid is sometimes used for control of malolactic fermentations (see Chapter 18). As noted, the TA of grapes and grape wines is expressed as tartaric acid. Apple wines are expressed in terms of their dominant acid (malic), whereas all other fruit wines are expressed in terms of citric acid.

Many winemakers prefer to make necessary acid additions to the must rather than solely to the wine. Additions at this stage may help to maintain a low pH during fermentation, enhance color extraction, and generally produce a more desirable product. At this stage in wine production, tartaric acid is generally used. Several arguments can be offered to support this decision. First, tartaric is a relatively nonmetabolic acid when compared with malic and citric. Second, tartaric acid is a stronger acid than either malic or citric; one achieves a greater pH adjustment per pound of acid used. Third, when such additions can achieve a pH drop to less than 3.65, subsequent precipitation of potassium bitartrate during bitartrate stabilization produces one hydrogen ion per molecule of potassium acid tartrate formed (see Chapter 15) thus effecting bitartrate stabilization without increases in pH. A reduction in TA of as much as 2 g/L may occur. Some winemakers feel that acid additions are best made using combinations (ratios) of malic and tartaric acids that approximate those of grapes grown in their areas.

For additions to finished wine, citric acid also has been employed. It has the advantages of not forming insoluble precipitates with potassium or calcium in alcoholic solution, compared to tartaric acid. Furthermore, citric acid is active in iron chelation and thus helpful in preventing or reducing ferric phosphate casse formation. (The topic of metal casse formation is discussed in Chapter 12) Individuals wishing to express citric acid additions in terms of other acids are referred to Table 4–1. As a widely used rule of thumb, 1 pound of citric acid per 1,000 gallons (12.1 g/HL) raises the TA 0.13 g/L, expressed as tartaric acid.

Addition of citric acid to unfermented must should be avoided. Besides slowing the onset of alcoholic fermentation because of the inhibition of fermentative pathway enzymes, wine spoilage microbes—especially lactic acid bacteria—may rapidly metabolize citric acid, yielding acetic acid (Peynaud 1984). Furthermore, because citric is not a dominant acid in grapes, large additions may be expected to significantly alter the ratio of acids and thus the wine character. Typically, large quantities of this acid, when added to grape wine, result in what many would regard as citrus-like flavor. The OIV places a maximum limit for citric acid at 1.0 g/L.

Amelioration

In certain winemaking regions of the world, it is permissible to add dry or liquid sugar, water, or a combination of sugar and water, to reduce the acidity and/or increase the °Brix of grape must or wine. In California, this practice is permitted only in the production of sparkling wines, special naturals, and fruit wines. If the winemaker wishes to increase the Brix in low-sugar musts or to adjust the residual sugar level of table wines, grape juice concentrate is used. In some cases, amelioration of table wine musts may be permitted, but the resultant product may not carry a California appellation. Such wines may be designated as "American." BATF regulations define the limits of amelioration based on TA of the must or wine, provided that the final fixed acid content is not lowered to less than 5.0 g/L, expressed as the predominant acid. Further, if the acidity of the must does not exceed 5 g/L, no addition of water is allowed except that amount needed to reduce the soluble solids level to 22 °Brix. In any case, the maximum level of amelioration permitted in grape and most fruit wine is 35% (vol/vol).

Amelioration of must or wine with water has the obvious economic advantage of increasing the yield. However, the process also dilutes aroma, flavor, extract, color, etc. In certain intensely flavored varieties, this may be desirable. The expected reduction in extract values for ameliorated wines may be perceived as reduction in body, and so may detrimentally affect the character of certain wines. Reduction in TA due to amelioration varies with variety as well as level of amelioration. As a result of the unpredictable outcome of this technique, it is not a recommended procedure for deacidification.

Carbonate Deacidification

Neutralization of wine acidity by the use of carbonates is employed in both the United States and European wine communities (See Table I-10 in Appendix I). In both cases, the practice is restricted to use of potassium and calcium carbonate and potassium bicarbonate. Because of the relatively high pK values for malic acid and the solubility of reaction products, tartaric is the only wine acid to be neutralized under normal conditions of use. The net reaction involves carbonate neutralization of the two hydrogen ions available on tartaric acid and formation of carbon dioxide and water:

$$\underset{\text{carbonate}}{CO_3^{=}} + H^+ \rightarrow \underset{\text{bicarbonate}}{HCO_3^-} + H^+ \rightarrow \underset{\text{carbonic acid}}{H_2CO_3} \rightarrow H_2O + CO_2$$

Successful use of a particular carbonate depends on juice/wine chemistry. For example, high-acid/high-pH juices and wines are best deacidified with calcium carbonate (because of the relatively high tartrate ion concentration), whereas high-acid/low-pH juices and wines are best treated with potassium-containing salts to take advantage of potassium bitartrate precipitation (Mattick 1983). The addition levels needed to achieve the desired reduction in acidity vary for each carbonate compound. A reduction in TA of 1.0 g/L requires additions of 0.9 g/L of potassium bicarbonate and 0.62 g/L of potassium carbonate.

Potassium and calcium components of the salt may become involved in secondary precipitation reactions with bitartrate and tartrate anions, yielding potassium bitartrate and calcium tartrate, respectively. In the case of carbonates containing potassium, precipitation of the salt is improved by chilling the wine or must in advance of the addition. In the case of calcium carbonate additions, however, chilling has little, if any, effect on the rate of precipitation. There is evidence that lower temperatures may favor long-term reduction (Clark et al. 1988). Further, calcium tartrate precipitation is particularly troublesome, because salts may take weeks to months to form. Unless corrected during post-fermentation processing, calcium tartrate precipitation may well occur in bottled wine (see Chapter 15).

Double Salt Deacidification

Addition of calcium carbonate to grape juice or wine results in the formation of calcium salts of tartaric and malic acids. Although calcium malate usually is soluble at cellar temperatures, calcium tartrate precipitates from solution, creating an imbalance. This problem may be overcome by the addition to the must of such proprietary compounds as Acidex.

Theoretically, the ratio of tartaric and malic acids remaining in solution is constant. This represents the potential advantage of double salt precipitation over the use of calcium carbonate. The mechanism of reaction involves the use of calcium carbonate to raise the solution pH to a level at which precipitation of the double salt is favored.

To enhance precipitation, Munz (1960, 1961) included within the carbonate addition approximately 1% crystals of the double salt to serve as nuclei for subsequent crystal formation. In order for the reaction to operate, the hydrogen ion concentration must be above pH 4.5, or calcium tartrate formation will prevail. At this point, both acids are present as their dicarboxylate anions. Following addition of the Acidex to musts, there is an expected rapid increase of pH, followed by slight increases during the storage period. Steele and Kunkee (1978) reported an average decrease in TA in wines of 3.7 g/L and pH increase of 0.16. In red wines, by comparison, TA decreased approximately 4.5 g/L with increases in pH of 0.17. However, these workers were unable to demonstrate equimolar precipitation of malic and tartaric acids in their study, a situation they attributed to the lower initial concentration of malic acid in the musts. In this case, tartaric was the principal acid removed. Double salt deacidification is usually limited to grape musts, because of the lower percentage of free acid found in most grape wines.

Ion Exchange

Ion exchange technology has also found application in acid reductions in wine. In the case of anion exchange, weakly bound hydroxyl groups on the resin are exchanged for acid anions present in the wine. The hydroxyl groups combine with the dissociated proton from the acid to yield water. Because the acid anion remains bound to the resin, the exchanged wine is lower in acidity after the exchange. In the United States, the fixed acid level of grape wines may not be reduced to less than 4.0 g/L for reds and 3.0 g/L for whites. The process, although successful in achieving reductions in total acidity, generally results in decreased overall wine quality.

Biological Deacidification

Malolactic Fermentation (see Chapter 18)

Although the pathways and end products may vary, the overall result of this bacterial conversion is the decarboxylation of L-malic acid, forming L-lactic acid and CO_2. Conversion results in significant reductions in the TA and increases in pH of 0.20 or more (Bousbourus and Kunkee 1971). These changes depend on the species or strain of lactic bacteria used and the concentration of L-malic acid in the wine.

Use of Selected Yeasts

Yeasts of the genus *Schizosaccharomyces* use L-malic acid as a carbon source in production of ethanol. L-malic is first decarboxylated to pyruvic acid, which undergoes decarboxylation to acetaldehyde and subsequent reduction to ethanol (Temperli et al. 1965). The step from L-malic to pyruvic acid requires NAD-dependent malic acid enzyme whereas pyruvate decarboxylase and alcohol dehydrogenase mediate the second and third steps, respectively.

Schizosaccharomyces pombe has been examined as a means of must deacidification. It has been noted that, unless controlled, over-deacidification may result from its sole utilization. Because members of the group are susceptible to competition from other wine yeasts, Snow and Gallander (1979) used a secondary innoculation with *S. cerevisiae* near desired TA levels. Wines produced from musts deacidified by *Schizosaccharomyces* often exhibit undesirable sensory properties and pronounced imbalances. Some strains of *Saccharomyces cerevisiae* are also capable of partial utilization of malate via the same pathway (Rodriquez and Thornton 1990). The spoilage yeast *Zygosaccharomyces* may also utilize malate.

The utility of individual strains may become more important as the field of genetic engineering develops. Such potential includes inclusion of the fully expressed gene(s) for the malolactic fermentation within selected strains. Initial work in this area has been successful in implanting the appropriate genes, but the level of expression is far less than that needed to be of practical value to the winemaker.

Blending

Blending can be used to obtain proper acid balance by combining high-acid wines with those of lower acidity. Blending operations that result in pH shifts may affect both biological and chemical stability of the wine.

LEGAL CONSIDERATIONS

Citric, tartaric, and malic acids may be used to correct deficiencies in grape must acidity, provided the final TA, expressed as tartaric acid, does not exceed 8.0 g/L. The OIV permits the addition of tartaric but not malic acid. The maximun level of citric acid permitted is 1.0 g/L. On occasion, fumaric acid is used to inhibit malolactic fermentations (see Chapter 18). BATF limits for addition of this acid are (300 g/hL) (25 pounds per 1,000 gallons). The fumaric acid concentration in finished wine may not exceed 3.0 g/L. Fumaric acid is not approved for addition to wines by the OIV. Other pertinent regulations regarding such additions may be obtained by reference to appropriate government publications (see Table I-10 in Appendix I).

SENSORY CONSIDERATIONS

Although the acidic character of wine is due to hydrogen ion concentration, both pH and acidity play important roles in the total sensory perception of this stimulus. Amerine et al. (1965) noted that at equivalent levels of acidity, the order of perceived sourness of the common wine acids is malic, tartaric, citric, and lactic. Ethanol is effective in increasing the acid thresholds, and this increase is even more dramatic with the inclusion of sucrose. Phenols may also be active in increasing minimum detectable acid levels.

The perception of acidity is a function of palate balance according to the following relationship:

$$\text{sweet} \rightleftharpoons \text{acidity} + \text{astringency and bitterness}$$

This relationship suggests that acidity is magnified by the phenolic elements (astringency and bitterness) and muted by sweet elements including carbohydrates, alcohol, etc. (see Chapter 7).

ANALYSIS

The following is a cross-reference of analytes and procedures found in Chapter 20. pH (Hydrogen Ion Concentration); TA (phenolphthalein indicator or potentiometrically); individual organic acids by HPLC and enzymatic analysis, tartaric acid by metavanadate; microscopic examination of calcium tartrate and potassium bitartrate deposits; diagnostic identification.

CHAPTER 5

CARBOHYDRATES

Carbohydrates are polyhydroxy aldehydes, ketones, and their derivatives composed of carbon, hydrogen, and oxygen in the ratio $C_n(H_2O)_n$. On a molecular basis, carbohydrates exist as monosaccharides, such as glucose and fructose; disaccharides, such as sucrose; and long-chained carbohydrates, the polysaccharides. Polysaccharides may be hydrolyzed yielding lower molecular weight forms. Examples of polysaccharides include pectins, glucans, and dextrans, as well as alginates used in fining. Other compounds that qualify as carbohydrates include deoxy- and amino sugars, sugar alcohols, and acids.

REDUCING SUGARS (HEXOSES)

The six-carbon sugars, glucose and fructose, are utilized by yeast in alcoholic fermentation. These two sugars are also referred to as reducing sugars, which may be operationally described as those sugars containing functional groups capable of being oxidized and, in turn, bringing about reduction of other components (i.e., Cu II). Thus, certain pentoses are also

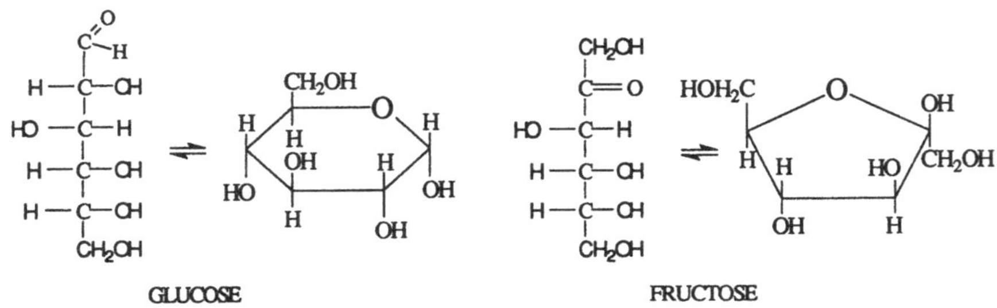

Fig. 5–1. Structures of two primary sugars in grapes.

classified as reducing sugars, even though they are not fermentable by wine yeasts.

Glucose and fructose may be differentiated on the basis of the location of their respective functional carbonyl group. As seen in Fig. 5.1, the carbonyl group of glucose is located on the first carbon and thus glucose is referred to as an *aldo*-sugar. In the case of fructose, the carbonyl function is located on the second carbon. Thus, fructose is an example of a *keto*-sugar. Intramolecular bond angles create molecular structures for these sugars such that they do not normally exist as straight-chained molecules but rather in cyclic configurations called hemiacetals (glucose) or hemiketals (fructose).

Because cyclization does not involve the gain or loss of atoms by the sugar molecule, the straight-chained and cyclic forms are isomers. The cyclic form represents quantatively the most important configuration. Glucose, for example, exists both in solution and crystalline form almost entirely as the cyclic hemiacetal (Fairley and Kilgour 1966). The fact that sugars display most of the reactions considered typical of aldehydes is the result of an equilibrium established between the open-chained and cyclic configurations present in solution.

Cyclization introduces another structural consideration into the chemistry of sugars. In solution, sugars can occur in rings composed of four carbons and one oxygen or five carbons and one oxygen. The former is termed a furanose ring and the latter a pyranose ring (see Fig. 5.1).

Sugars are optically active and can be assayed on the basis of optical rotation when measured polarimetrically. Glucose is dextrorotary and hence is frequently referred to as "dextrose." Fructose, by comparison, is levorotatory and is called "levulose." The dissaccharide sucrose is frequently referred to as "invert sugar." In its native configuration, sucrose is nonreducing and therefore cannot be measured by copper reduction techniques. However, upon hydrolysis, or inversion, the "reducing" monosaccharide components glucose and fructose can be measured.

Glucose-Fructose Ratios

In grapes, glucose and fructose occur in approximately equal concentrations, each contributing approximately 10 g/100 g to juice (Amerine et al. 1972). Sucrose is the third most abundant sugar, accounting for 0.2 to 1.0 g/100 g (Hawker et al. 1976). Although glucose and fructose are normally present in a ratio of 1:1 in the mature fruit, this may vary significantly.

Climatic conditions during the growing season may affect the glucose-fructose ratio. Kliewer (1967a) found that the ratio decreased in warmer seasons and increased during colder periods. Amerine and Thoukis (1958) reported ratios ranging from 0.71 to 1.45 in California's 1955 vintage, whereas Kliewer (1967a) cited ratios of 0.74 to 1.05 in *Vitis vinifera* wine varieties. During maturation, the ratio of glucose to fructose usually decreases.

During fermentation, yeasts utilize glucose and fructose differentially (Kunkee and Amerine 1970). At reducing sugar (RS) levels of 17 to 20%, glucose was reported to be fermented faster, whereas in the range of 20 to 25% RS both sugars fermented equally. At RS levels greater than 25%, the rate of fructose utilization was greater.

The ratio of glucose to fructose declines during fermentation from near 0.95 at the start to 0.25 near the end of fermentation (Peynaud 1984). Thus at this stage, fructose is usually present in greater amounts than glucose. In that fructose is nearly twice as sweet as glucose, the cited ratios explain why wines sweetened with grape concentrate or *muté* appear less sweet than wines with the same **analytical concentration** of reducing sugar produced by arresting the fermentation.

Reducing sugar analyses play multiple roles in wine processing. The quantity of fermentable sugar remaining in the wine upon completion of fermentation may be important in microbial stability as well as in potential blend preparations. Additionally, monitoring the fermentable sugar content in pomace, distilling material, etc., is of concern in overall plant efficiency.

Although it might be expected that "dry" table wines have 0% RS, typical analytical results are higher, due to the contributions of nonfermentable pentoses. Dry wines have been traditionally defined as having RS levels of less than 2.0 g/L (0.2%). Spoilage yeasts such as *Brettanomyces/Dekkera* as well as lactic acid bacteria can grow at sugar levels <2 g/L (Fugelsang et al. 1993). McCloskey (1978) defines a "dry" wine as ranging from 0.15 to 1.5 g/L (when determined by enzymatic assay specific for glucose and fructose). Because the primary reducing sugars in a dry wine are pentoses that are not fermentable by yeast, a dry wine (<0.02% RS) is generally considered stable with respect to yeast refermentation. However,

Smith (1993) reported that (theoretically) hexose concentrations of 100 mg/L (0.01%) can support *Brettanomyces* populations of 10^7 cells/mL (CFU).

PENTOSES

The five-carbon monosaccharides, commonly referred to as pentoses, may comprise approximately 28% of the reducing sugar content of a dry table wine (Esau 1967). Among the pentoses present in wine, arabinose is reported to occur in highest concentrations at 0.40 to 1.3 g/L. (Amerine and Ough, 1974). Analytically, this group of sugars is not easily separated and is, therefore, included in traditional analyses of reducing sugar. Certain pentoses may serve as energy sources in the malolactic fermentation (Doelle 1975). They may also be an important source of furfural in baked sherry (Amerine et al. 1972).

SUCROSE

The disaccharide sucrose serves as an important energy storage compound in most plants and vegetables. Although sucrose is itself unfermentable, the products of its hydrolysis, glucose and fructose, are readily utilized. In the case of grapes, upon translocation to the berry, hydrolysis by invertase enzymes yields glucose and fructose. Thus, sucrose levels in grape berries, at maturity, range from 0.2 to 1%. Because yeasts produce their own invertase enzyme, chaptelization of sugar-deficient musts with sucrose does not cause problems relative to fermentability.

POLYSACCHARIDES (AND ASSOCIATED INSTABILITIES)

Polysaccharides serve two important functions. They are either structural in nature (i.e., cellulose, pectin) or energy reserves (starch). They occur in wine as either a carryover from juice/must extraction and/or from microbial activity. Because of their size and colloidal nature, these macromolecules can present problems in clarification and filtration.

Cellulose and hemicellulose represent the primary structural polysaccharides of the plant cell wall. Upon acid hydrolysis, a portion may be released into the wine.

Pectins are naturally occurring heteropolysaccharides consisting chiefly of galacturonic acid linked via *alpha*-1,4 bonds. The carboxyl group of acid monomers may exist in either the acid form or as the methyl ester. In musts treated with pectic enzymes, the latter may be enzymatically cleaved to yield methanol (see Chapters 3, 6, and 16). Aside from polygalacturonic acid, native pectin incorporates several sugars as side chains, which include galactose, arabinose, rhamnose, and xylose. Gums present in wine are polymeric mixtures of arabinose, galactose, xylose, and fructose (Peynaud 1984).

Other important polysaccharides in musts and wine exist principally as polymers and shorter chained oligomers of glucose. These include glucans that arise from growth of *Botrytis cineria* on the grape berry. Two glucans are associated with growth of *Botrytis*. The first, of molecular weight 900,000, is important in clarification. The second, a heteropolysaccharide of approximately 40,000, is believed to inhibit yeast alcoholic fermentation. These polymers exist as branched chains of glucose linked via *beta*-1,3 bonds. Branching results from *beta*-1,6 linkages.

The extent to which glucans from *Botrytis*-infected fruit are extracted into juice depends on grape handling techniques. Careful pressing of whole clusters minimizes extraction of glucans into the juice (see Chapter 3). Glucans becomes increasingly insoluble in the presence of alcohol. Very low concentrations (2–3 mg/L) are sufficient to impede filtration (Wucherpfenning and Dittrich 1984).

Cellaring techniques such as fining are frequently ineffective in removing the polysaccharide. Correction requires use of glucanases, which bring about hydrolysis to mixtures of glucose and gentiobiose (Dubourdieu et al. 1981). Use levels vary with wine type and glucan concentration (see Chapters 3 and 16).

Diagnostic tests for the presence of glucans in wine utilizing alcohol precipitation were reported (Dubourdieu 1982). The reader is referred to Chapter 20. Polysaccharides arising from yeasts result from cell wall degradation during autolysis. These are *beta*-1,3 glucans as well as glycoprotein (chiefly mannoprotein) (Villettaz et al. 1980). Other polysaccharides, of bacterial origin, may occasionally be noted in wine. Lactic acid bacteria, particularly *Pediococcus cerevisiae* and *Leuconostoc mesenteroides,* may produce extracellular dextrans (polymers of glucose) when growing in fruit and other wines where sucrose is used in chaptelization. Additionally, some acetic acid bacteria also may produce extracellular polysaccharides when growing on glucose-containing media. These may include dextrans as well as polymers of mixed sugars.

In some instances, what is described as "protein haze" in wine is likely a complex of not only protein but also polysaccharide and polyphenols. Thus the presence of polysaccharide-protein-phenolic complexes may help to explain why heat and other protein precipitation tests do not predict or correlate well with analyses of total protein (see Chapters 8, 20).

Fermentation difficulty with must from botrytized grapes is due, in part, to the presence of botryticine, a heteropolysaccharide rich in rhamnose and mannose (Doubourdieu 1978) (see Chapter 3).

Sugars are involved in conjugated grape and wine aroma components. For example, monoterpenes can be bound to dissaccharide glycosides or as *alpha*-L-arabinofuranosyl-*B*-D-glycopyranosides or *alpha*-L-rhamnofuranosyl-*B*-D-glycopyranosides. Hydrolysis of conjugated aroma precursors yields free volatile aroma components (see Fig. 16-4).

MUTÉ PRODUCTION

Mutés (juice that is held or prevented from fermentation) are valuable blending tools. Mutés are used as sweetening agents, body enhancers, and to add life and freshness by increasing aroma intensity. Frequently aromatic juice from cultivars such as muscats are used for muté production by fortification, pressurized storage, refrigeration, or combinations. A common practice is to fortify pomace from aromatic varieties such as Riesling and Gewürztraminer and use that alcohol to fortify juice. Sulfur dioxide is added in limited concentrations (<200 mg/L) to help minimize oxidation.

SOLUBLE SOLIDS VS. REDUCING SUGAR VALUES

Confusion may exist between the concepts of soluble solids and reducing sugar. Reducing sugar is defined in terms of % wt/vol (g/L) whereas °Brix is defined as % wt/wt (g/100 g). A reducing sugar analysis, then, measures the amount of grape sugar in wine or must directly without interferences from alcohol, carbon dioxide, etc. On the other hand, °Brix relates the density of the entire must sample (assumed to be g sucrose/100 g sample) to that of pure water at 20°C. Interferences associated with this method are discussed in Chapter 3.

ANALYSIS OF REDUCING SUGARS

A reducing sugar may be defined as one that contains a free aldehyde or *alpha*-hydroxy ketone capable of being oxidized. Thus, sugars in the free *aldo-* or *keto-* form or that exist in equilibrium with these forms fit into this category.

Analytically, reducing sugars may be determined by chemical, enzymatic, and high performance liquid chromatographic (HPLC) techniques. The chemical methods generally involve reaction of reducing sugars with copper (II) in alkaline solution. As seen in the following equations, reducing sugars, such as glucose and fructose, reduce copper (II) to copper (I) oxide under alkaline conditions:

$$\text{Reducing sugar} + Cu^{+2} \rightarrow \text{Oxidized products} + Cu^{+}$$
$$3Cu^{+} + 3OH^{-} \rightarrow \underset{\text{(yellow)}}{CuOH} + \underset{\text{(red)}}{Cu_2O} + H_2O$$

In alkaline solution, sugars undergo decyclization to yield corresponding *aldo-* and *keto-* forms (Joslyn 1950a). This is followed by rearrangement and subsequent degradation. The color change associated with the reaction is believed to be due to enolization, with resultant double-bond formation producing color. Dependent variables for this reaction include the type and concentration of sugars present, the concentration of alkali, and temperature and time of reaction. Low temperatures and relatively high concentration of alkali favor formation of the characteristic red Cu_2O precipitate. Sodium-potassium tartrate is included in the reagent mix to facilitate separation of Cu_2O precipitate while maintaining sugar oxidation products and unreduced copper in solution. The copper (II) tartrate complex formed is stable even at the high temperatures of reaction. Reduced copper, however, does not complex with tartrate and readily precipitates from solution (Joslyn 1950a).

In the Rebelein method (see Chapter 20), potassium iodide and sulfuric acid are added to the cooled reaction mixture once the reaction between sugar and copper (II) is complete. Iodide reacts with the remaining Cu^{+2} ion to produce an equivalent amount of iodine, which is subsequently titrated with standardized sodium thiosulfate:

$$2Cu^{+2} + 2I^{-} \rightarrow 2Cu^{+} + I_2$$
$$I_2 + 2S_2O_3^{-2} \rightarrow 2I^{-} + S_4O_6^{-2}$$

McCloskey (1978) has reported an enzymatic procedure for the analysis of reducing sugars (see Chapter 20). Kits are now commonly available for this determination. The analysis requires the use of a spectrophotometer and the ability to pipette small volumes accurately.

HPLC procedures have also been developed (see Chapter 20). These procedures are generally constructed around a specialty HPLC column (sold by one or more manufacturers). Because column technology is rapidly improving, one should contact an HPLC supplier for columns and procedures.

Rapid Determination of Reducing Sugar

Rapid reducing sugar measurements may be run routinely by employing a variety of reducing sugar kits originally developed for use by diabetics. By reference to a color code, the reducing sugar content of a measured volume of wine is most accurately determined within a range of 0 to 1%. As previously mentioned, the presence of pentoses will prevent the reducing sugar level of wine from reaching zero. However, these pentoses are not fermentable by yeast (see Chapter 20-Reducing Sugar-Clinitest).

INVERT SUGAR ANALYSIS

The addition of sugar to fermenting and finished wines is permitted in certain countries. In California, these additions are restricted to formula wines such as champagnes and special naturals. In such cases, the sugars used are generally in the form of concentrates, dry dextrose, sucrose, or syrups of the latter two. Sucrose may create analytical difficulty because it is not a reducing sugar. Thus, preliminary sample treatment is necessary. This is usually accomplished by acid hydrolysis of the nonreducing disaccharide to its component-reducing monosaccharides glucose and fructose.

Once hydrolysis is complete, one may proceed with the reducing sugar analysis by one of the procedures described, keeping in mind that there will be a correction factor for dilution of the sample.

ANALYSIS

The following is a cross-reference of analytes and procedures found in Chapter 20. Glucose and fructose by HPLC and visual spectrometry; reducing sugar by visual spectrometry, titrametric and rapid estimation; Invert sugar, pectins and glucans by visual estimation.

CHAPTER 6

Alcohol and Extract

The alcohol content of a wine influences its stability as well as sensory properties. Wines are taxed, in large part, based on their alcohol levels. Careful monitoring of alcohol is important in stylistic wine production as well as in carrying out accurate fortifications, and in formulation of blends for bottling.

YEAST METABOLISM

Fermentation

Microorganisms have varying requirements for oxygen. At the extremes are those that require oxygen (aerobes), and those that cannot survive in its presence, the anaerobes. Between these extremes are facultative microorganisms such as yeasts, which are metabolically equipped to handle conditions where oxygen is either plentiful or limiting. Under oxidative (aerobic) conditions where concentrations of utilizable sugars are less than 3%, yeasts utilize respiratory pathways. Under these conditions, fermentative metabolism is repressed in that both pyruvate decarboxylase and alcohol dehydrogenase require glucose for induction.

Oxidatively, glucose is converted to pyruvate and, subsequently, one molecule of acetic acid (as acetyl CoA) enters the Tricarboxylic Acid Cycle and is oxidized to CO_2 and H_2O. The TCA cycle represents a key pathway for energy generation. From one molecule of acetic acid (as acetyl CoA) entering the cycle, 3NADH + H^+ are produced (which will eventually contribute electrons to the electron transport system), one molecule of the high-energy nucleotide, Guanosine Triphosphate (GTP), and two electrons from Flavin Adenine Dinucleotide (FADH). Further, organic acids produced in the TCA cycle serve as key intermediates in biosynthetic pathways (amino acids).

Alternatively, when oxygen is limiting, and/or glucose is present (at >4–5%), mitochondrial TCA cycle activity is repressed and fermentative metabolism ensues. Anaerobiosis is not the only condition required for fermentative metabolism. Glucose repression of respiratory growth (leading to fermentation) may be seen even under oxidative conditions. In either event, pyruvate decarboxylase and alcohol dehydrogenase are induced, resulting in the decarboxylation of pyruvate to acetaldehyde and subsequently reduction to ethanol.

In the case of recently rehydrated wine active dry yeast (WADY), growing oxidatively, the interval between inoculation into juice/must and glucose induction of pyruvate decarboxylase and alcohol dehydrogenase is biochemically critical. Because the aforementioned enzymes must be induced prior to activity, a lag period exists in which the yeast must maintain redox balance in terms of oxidized/reduced NAD. Fortunately, a mechanism exists to accomplish this need—glycerol formation (see Fig. 6–1 and related discussion). Until glucose repression of mitochondrial TCA cycle activity is fully established, the cycle remains active and intermediate organic acids (particularly succinate) accumulate.

Because the usual metabolic entry into the TCA cycle (via decarboxylation of pyruvate and linkage of acetic acid to coenzyme A) requires additional NAD^+ under conditions where the coenzyme is already limiting, an alternate donor must be utilized. During this transition phase, pyruvate

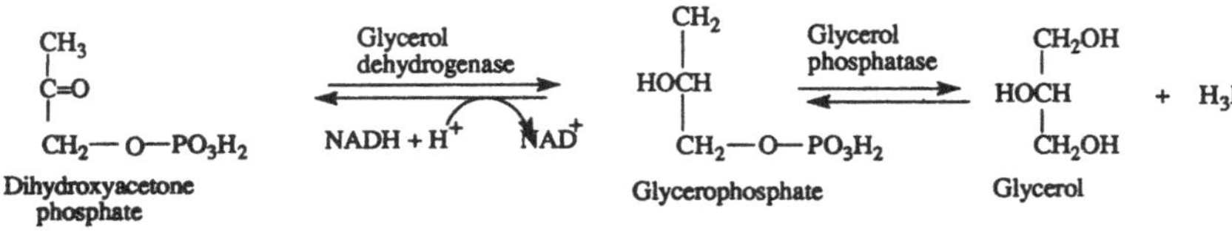

Fig. 6–1. Glycerol production from reduction of dihydroxyacetone phosphate.

carboxylase couples pyruvate + CO_2, yielding the TCA cycle intermediate, oxaloacetate. Once oxaloacetate is produced, partial operation of TCA Cycle (reverse phase) leads to formation and accumulation of succinate and regenerates NAD^+ from NADH + H^+ in the process (Oura 1977). As with glycerol, succinate is known to be produced from glucose during the early stages of fermentation (Thoukis et al. 1965).

According to Whiting (1976), cytoplasmic TCA cycle enzymes, with the exception of those needed for conversion of *alpha*-ketoglutarate to succinate, are present and operative during anaerobic growth. Subsequently, Heerde and Radler (1978) found *alpha*-ketoglutarate dehydrogenase to be present and active during fermentation. This enzyme catalyzes formation of succinate from *alpha*-ketoglutarate. Thus, although mitochondrial TCA activity may be repressed, there exists a vehicle or continued formation of intermediates.

Byproducts of Fermentation

The end- and byproducts of fermentation (ethanol and organic acids) are, themselves, at an intermediate stage of oxidation and as such, may be metabolically available to organisms (i.e., acetic and lactic acid bacteria as well as some yeasts). Additional alcohols of importance in winemaking include glycerol and other polyhydric alcohols such as fusel oils. Individually and collectively these may, on occasion, be of sensory and/or regulatory importance.

Glycerol

Glycerol is a normal byproduct of alcoholic fermentation resulting from reduction of dihydroxyacetone phosphate (Fig 6–1). Although the glycerol content of juice produced from sound grapes is low, less than 1 g/L, substantial amounts may be produced during fermentation. Amerine and Ough (1980) reported levels in U.S. wines ranging from 1.9 to 14.7 g/L (average of 7.2 g/L). Due to its relatively high specific gravity (1.26), glycerol may contribute to the overall sensory perception of body in certain wine types. However, it is questionable if this contribution is significant at alcohol concentrations of 10 to 12% normally found in table wine (Amerine and Roessler 1976).

The glycerol content of a wine produced from sound fruit may be affected by harvest and production parameters. Several of the more important considerations are the following.

1. Fermentation temperatures play an important role in glycerol formation. Workers have reported increased glycerol production with increased fer-

mentation temperatures in the range of 15°C to 25°C (60°–77°F) (Rankine 1955; Rankine and Bridson 1971).

2. Yeast strains are known to vary in their ability to produce glycerol. Radler and Schutz (1982) reported that glycerol levels produced by several strains of *Saccharomyces cerevisiae* ranged from 4.2 to 10.4 g/L. It has been known for some time that glycerol formation is most rapid before the onset of and during the initial stages of fermentation (Beckwith 1936). Ribereau-Gayon et al. (1956a, 1956b) also report that aeration enhances formation of glycerol.

   Certain of the native yeast flora may play a much more significant role in glycerol production than does *Saccharomyces cerevisiae.* Sponholz et al. (1986) report that *Kloeckera apiculata, Metschnikowia pulcherrima,* and *Candida stellata* may be particularly important in this regard. The osmotolerant spoilage yeast *Zygosaccharomyces rouxii* also produces increasing amounts of glycerol as the concentration of sugar increases (Spencer and Sallans 1956).

   Radler and Schutz (1982) suggest that glycerol formation results from competition between glycerol-3-phosphate dehydrogenase and alcohol dehydrogenase for available NADH. The activity of alcohol dehydrogenase appears to be fairly constant, whereas glycerol-3-phosphate dehydrogenase levels vary significantly between those strains producing larger amounts of glycerol compared with those producing lower levels.

3. Grape condition also affects the resultant glycerol content (see Chapter 3). Sweet wines, produced from botrytized grapes, are usually high in glycerol. Dittrich et al. (1974, 1975) found that wines produced from *Botrytis*-infected grapes yielded 15 g/L glycerol compared with 4.4 g/L from sound grapes. Amerine and Roessler (1976) reported that wine produced from moldy grapes contained more than 1.5% (wt/wt) glycerol when compared with 0.5% (wt/wt) from sound fruit.

4. Sulfur dioxide affects glycerol production by binding acetaldehyde (formed from decarboxylation of pyruvate) and thus preventing it from operating as a hydrogen acceptor for NADH (see Chapter 10). Under these conditions, dihydroxyacetone phosphate replaces acetaldehyde as a hydrogen acceptor, resulting in formation of glycerophosphate in amounts equivalent to the quantity of acetaldehyde bound. Hydrolysis of accumulated glycerophosphate then leads to the formation of glycerol (Neuberg's second form of fermentation). It has been reported that NAD regeneration occurs in this manner in the early stages of fermentation before the concentration of acetaldehyde reaches the levels needed for alcohol dehydrogenase activity (Holzer et al. 1963). However, utilization of this pathway does not provide the cell with either energy or biosynthetic intermediaries (Sols et al. 1971). The amount of sulfur dioxide needed to stimulate significant increases in glycerol is far in excess of that used in routine winemaking practices.

In some instances, glycerol may serve as a carbon source for microorganisms. Acetic acid bacteria, particularly *Gluconobacter oxydans,* are known to utilize glycerol oxidatively in formation of dihydroxyacetone. In the case of *G. oxydans,* this reaction is quantitative, and has been employed in commercial production of dihydroxyacetone. Although it has not been determined if this reaction can occur in wine, it is well established that formation occurs in infected grapes and must (Sponholz and Dittrich 1985). In their study, Sponholz and Dittrich reported levels of dihydroxyacetone reaching 260 mg/L in must infected with *G oxydans.* Upon fermentation, this level was reduced to 133 mg/L in wine. In addition to potentially altering the sensory properties of wine, the accumulation of dihydroxyacetone provides another important substrate for binding free sulfur dioxide. For more details regarding acetic acid bacteria, the reader is referred to Chapters 11 and 18.

Glycerol may also be utilized by lactic acid bacteria during the malolactic fermentation. Referred to as "acrolein taint," the pathway involves enzyme-mediated dehydration of glycerol to yield 3-hydroxypropionaldehyde and, upon continued reaction in acid solution, a second dehydration of the aldehyde yields acrolein. When glucose is present, 3-hydroxypropionaldehyde may also be reduced to the alcohol, 1,3-propandiol. *Lactobacillus brevis* and *Lactobacillus buchneri* as well as *Leuconostoc oenos* have been implicated in this problem (Hirano et al. 1962; Schutz and Radler 1984). Acrolein is sensorially detectable in wines with acrolein concentrations of 10 mg/L (Margalith 1981). For more detail see Chapter 12.

Higher Alcohols (Fusel Oils)

Alcohols of more than two carbons, commonly called fusel oils, are produced by yeasts during fermentation. The most frequently encountered fusel oils include isoamyl (3-methyl-1-butanol), "active amyl" (2-methyl-1-butanol), isobutyl (2-methyl-1-propanol), and *n*-propyl alcohols. This group of alcohols may present problems in distillation where they concentrate in the "tails" fractions of distilled spirits. In cases of stills without fractionization capabilities or of malfunction in column still operation, significant concentrations of these alcohols may appear in the product. Depending on the production objectives (i.e., brandy vs. neutral spirits), significant amounts may represent defects in the sensory interpretation of the distillate.

Quantitatively and qualitatively, fusel oils represent an important group of alcohols that may affect flavor. They may be present in wines at varying concentrations. Quantitatively, isoamyl alcohol generally accounts for more than 50% of all fusel oil fractions (Muller et al. 1993).

In table wines, the total fusel oil concentration is reported to range from 140 to 420 mg/L (Amerine and Ough 1980). Due to differences in processing, white wine fermentations typically produce lower final levels than red fermentations (Muller et al. 1993). In this range, fusel oils are believed to contribute to overall complexity whereas at higher levels they may become objectionable. In dessert wines, total fusel oil concentration may range from near 100 mg/L to over 1,000 mg/L. Values on the higher side reflect additions of brandy and/or lower-quality high proof.

Fusel oil formation was originally described by Ehrlich in 1907 as arising from transamination of the corresponding amino acid. As seen in Fig. 6–2, the Ehrlich mechanism involves initial transamination between an amino acid and an *alpha*-keto acid with subsequent decarboxylation and reduction. The final reduction step involves reoxidation of NADH + H^+. Fusel oils and their corresponding parental amino acids are seen in Table 6–1.

Because priority amino acids are taken up by the yeast early in fermentation and fusel oils are produced throughout the course of fermentation (Castor and Guymon 1952; Muller et al. 1993), the amino acid pool in juice is not quantitatively sufficient to yield the concentrations of fusel oils reported in wine (Reed and Nagodawithana 1991). Using radioactive tracers, Reazin et al. (1970) found that 35% of higher alcohols arose from carbohydrates. Fusel oils are known to arise from both amino acid and carbohydrate (glucose) sources (Thoukis 1958).

Physiologically, oxidative deamination provides the yeast a mechanism for obtaining more nitrogen when the pool has become depleted. The *alpha*-keto acids produced in the first step may be excreted by yeast or they may be decarboxylated and, subsequently, reduced to the fusel alcohol.

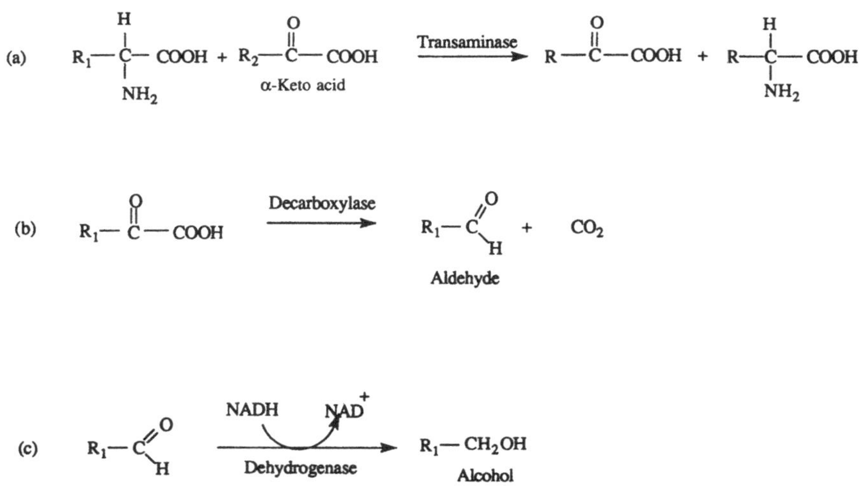

Fig. 6–2. Generalized pathway for formation of higher alcohols ("fusel oils") from amino acid and *alpha*-keto acid precursors.

Table 6–1. Origin of fusel alcohols from amino acid precursors.

Amino acid	Fusel alcohol
Leucine	3-methyl-1-butanol
Isoleucine	2-methyl-1-butanol
Valine	2-methyl-1-propanol
Threonine	Propanol
2-phenylalanine	2-phenylethanol
Tyrosine	Tyrosol

The reduction from aldehyde to alcohol involves reoxidation of the coenzyme NADH + H^+ to NAD^+ and, thus, helps to maintain redox balance within the cell. For additional information on juice and wine nitrogen see Chapter 8.

Factors Contributing to Fusel Alcohol Formation. Variables affecting the final concentration of fusel oils in wine include yeast strain (inoculated as well as native yeast flora), fermentation temperature, suspended solids level, oxygen level, nutritional status and pH.

Yeasts. Rankine (1967) examined several species of *Saccharomyces* as well as native yeast species with respect to higher alcohol production in wine. In the case of pure-culture *Saccharomyces* fermentations, significant differences in the concentration of various fusel oil fractions (most notably *n*-propanol) were observed. In mixed-culture fermentations, where cell number of the native yeast was low, fusel oil levels were not substantially different from those of pure culture *Saccharomyces* fermentations. Some native yeast species are capable of producing considerable levels of fusel oil. *Hansenula anomala,* for example, is capable of fusel oil production even in the absence of fermentative metabolism.

The fusel alcohol 2-phenylethanol (arising from 2-phenylalanine) has the unmistakable odor of roses and is also believed to play a sensory role in the perception of body. It is often found at much higher concentration in mixed culture native yeast fermentations (Sponholz and Dittrich 1974).

Temperature. Using several strains of yeast in pure culture, Rankine (1967) reports that increases in fermentation temperature from 15°C to 25°C (60°–77°F) stimulated formation of isobutyl by approximately 39%, whereas in the case of "active"-amyl, levels increased by 24%. Formation of *n*-propanol, however, decreased by 17%. Ough et al. (1966) reported similar results.

Oxygen Levels. Formation of fusel alcohols, particularly isobutyl alcohol, increases in aerated musts. Similarly, production of fusel oils increases with suspended solids levels and particle size. These increases most likely result from absorption and retention of oxygen within the solids' matrix (Guymon et al. 1961; Klingshirn et al. 1987). Under oxidative conditions, yeasts with limited fermentative capabilities (*Pichia* sp. *Hansenula anomala*, and *Candida* sp.) produce substantial quantities of fusel alcohols from fermentable sugars.

pH. Must/juice pH is also known to stimulate production of fusel oils. Rankine (1967) reports that in pure culture fermentations utilizing four yeast strains, increasing the pH from 3.0 to 4.2 resulted in increased production of "active"-amyl and *iso*-amyl (28%), isobutanol (85%), and *n*-propanol (11%).

Nutritional Status of Must. Studies have shown that, with the exception of *n*-propanol (a result of pyruvate-acetyl CoA condensation), increases in assimilable nitrogen resulted in lower levels of fusel oils in the wine (Ough and Bell 1980; Vos 1981). Large (1986) speculated that the transamination reaction may be repressed by assimilable nitrogen. In this regard, it has been reported that grape variety and climatic variations influence fusel oil content irrespective of yeast strain, must pH, and processing conditions (Rankine 1967).

Other Polyhydric Alcohols

Also referred to as "polyols" or, because of their sweet properties, "sugar alcohols," this group of compounds result from reduction of the aldo- or keto group of their respective parent sugar. Chemically, polyhydric refers to the presence of three or more hydroxyl groups. Glycerol and fusel oils fit this definition. Other examples that may play a sensory and, potentially, a stability role in winemaking include mannitol, erythritol, arabitol, sorbitol, xylitol, and *myo*-inositol.

With the exceptions of xylitol and *myo*-inositol, which occur naturally in the grape and whose concentrations are little effected by microbial activity, the group results from conversion of various parental sugars during the growth of mold, yeast, and bacteria present on infected fruit. In all cases, however, subsequent berry dehydration results in a concentration effect. Concentrations are typically much higher than can be accounted for by microbial formation. Because sugar alcohols possess sweet properties, they play potentially important sensory roles in balance and body, particularly in dessert wines.

Mannitol. Resulting from the direct reduction of either fructose or fructose-6-phosphate, mannitol is produced by molds, yeast, and heterolactic lactic acid bacteria. The levels reported in wine vary somewhat: whites

84 to 323 mg/L; reds, 90 to 394 mg/L; and in wines produced from moldy fruit, 452 to 735 mg/L (Dubernet et al. 1974). The presence and activity of *Botrytis cineria* and associated native vineyard yeast *Metschnikowia pulcherrima* plus subsequent dehydration of fruit results in very high levels reported in dessert wines such as Trockenbeerenausleses (4,365–12,884 mg/L) (Sponholz 1988). In this case, Sponholz reported that *M. pulcherrima,* by itself, was capable of producing 1,300 mg/L of the polyol.

Heterolactic LAB such as *Lactobacillus brevis* and *Leuconostoc oenos* may under certain conditions produce substantial amounts of mannitol in conjunction with lactic and acetic acids. The instability, referred to as "mannitol taint," is most commonly seen in sweet, lower alcohol, high pH grape and fruit wines. Heterolactic LAB lack the key enzyme fructose diphosphate aldolase, which catalyzes cleavage of the sugar-phosphate to yield triose phosphates glyceraldehyde-3-phosphate and dihydroxyacetone phosphate. This group utilizes the pentose phosphate pathway, which yields equimolar amounts of lactic and acetic acids, ethanol, and carbon dioxide. The second sugar usually present during fermentation, fructose, may then be reduced to mannitol directly or after phosphorylation.

Erythritol. The four carbon compound results from reduction of the sugar erythrulose. The latter is formed after isomerization and dephosphorylation of the pentose phosphate cycle intermediate erythrose-4-phosphate. Dubernet et al. (1974) report slightly higher levels of erythritol in wines produced from *Botrytis*-infected fruit (160–272 mg/L) compared with whites (33–100 mg/L) and reds (160–272 mg/L), suggesting that mold is not the most important microbe involved in formation. Sponholz et al. (1986) report substantial variation among *Saccharomyces* sp. with respect to formation. Wine strains of *S. cerevisiae* produced only 114 mg/L whereas one osmotolerant strain was capable of producing over 900 mg/L. By comparison, the native yeast species *Candida stellata* and *Kloeckera apiculata* produced 200 and 530 mg/L, respectively. As noted previously, fruit dehydration after mold growth may concentrate the metabolite.

Arabitol. Arabitol has been reported in white wines at levels of 13 to 59 mg/L, in reds at 32 to 111 mg/L, and in wines made from mold-damaged fruit at levels of up to 359 mg/L. Sweet, Germanic dessert wines produced from mold-infected fruit have levels as high as 2,353 mg/L (Dubernet et al. 1974). The origins of arabitol in wine vary depending on the organism involved. In the case of molds and many yeasts, it is produced from mannitol (Spencer and Spencer 1980). The osmophilic spoilage yeast *Zygosaccharomyces rouxii,* reported to produce 3,034 mg/L, uses ribulose (Ingraham and Wood 1961). Other native yeasts associated with *Botrytis* infection produce large amounts. *Candida krusei* and *Metschnikowia pulcherrima* have been reported to produce 2,321 and 1,552 mg/L, respectively. Again, dehydration contributes an unknown concentration factor to levels reported.

Sorbitol, Xylitol, and Myo-inositol. Xylitol and *myo*-inositol occur naturally in grapes. Their change in concentration reflects dehydration rather than formation by mold and yeasts. Sorbitol may be present in wines at varying concentrations reflecting the growth of molds and yeast on the fruit. Sponholz et al. (1986) reported formation by *M. pulcherrima* (257 mg/L), *C. stellata* (439 mg/L), and *K. apiculata* (120 mg/L). As noted, these yeasts are normally associated with the noble-rot ecosystem and dehydration effects should be considered.

ESTER FORMATION

Esters, in general, have fruity and floral impact characteristics that are important in the sensory properties of wine. A variety of esters are produced during the course of fermentation. A comprehensive listing of esters identified from wine is given by Rapp (1988). Wine esters may be categorized into two groups: (1) those arising from acetate and ethanol as well as fusel alcohols, and (2) those resulting from ethanol and straight-chained fatty acid precursors. Esters of the first group include ethyl-, isobutyl-, isoamyl-, 2-phenethyl-, and hexyl- acetate. Of those identified, ethyl acetate is generally present at the highest concentration (see Chapter 11). In some instances, ethyl lactate (arising from malolactic fermentation) may also be present in substantial amounts. Examples of the second group include the ethyl esters of hexanoic, octanoic, and decanoid acids. The fatty acid component of this latter group originates during formation or from degradation of longer-chained membrane fatty acids. Although their importance in wine (relative to the acetate esters) is less than those in the first group, this group plays an important role in sensory properties of distillates.

Other esters reported in wine typically are present at concentrations of less than a few parts per million. Because reported concentrations are generally near sensory threshold levels, esters play an important role (both positive and negative) in wine quality.

Nordstrom (1963, 64, 66) proposed the following enzyme-mediated reaction mechanism for volatile ester formation by yeasts:

$$\text{RCOOH} + \text{ATP} + \text{CoA}{\sim}\text{SH} \rightarrow \text{RCO}{\sim}\text{SCoA} + \text{AMP} + H_2O$$
$$\text{RCO}{\sim}\text{SCoA} + \text{R'OH} \rightarrow \text{R-COR'} + \text{CoA}{\sim}\text{SH}$$

The reaction presented above is catalyzed by enzyme alcohol acetyltransferase (AAT), which utilizes acetyl coenzyme A and the alcohol component as substrate (Yoshioka and Hashimoto 1981). Mauricio et al. (1993) reported that esterases, operating in a synthetic rather than hydrolytic mode, may also play a role. In the case of ethyl acetate, an esterase and AAT is required whereas with isoamyl acetate, only AAT is required.

Esters (specifically, ethyl acetate) may also be formed by simple chemical reaction catalyzed by hydrogen ions. However, such interaction favors the equilibrium toward the acid and alcohol. In addition, ethyl acetate might be formed by the NAD^+-catalyzed oxidation of the hemiacetal of acetaldehyde and ethanol. This system seems to occur within the yeast cell (Muller and Fugelsang, unpublished data). This process provides a means to regenerate NADH needed for continued reduction of acetaldehyde to ethanol.

Processing variables that affect formation and concentration of esters in wine include temperature of fermentation and fermentable sugar levels. Yeast strain has also been shown to be important (Soles et al. 1982). The presence of native yeast flora at the start of and during fermentation as well as during postfermentation processing is also important. Native species such as *Hansenula anomala, Kloeckera apiculate* (and its ascosporogeneous equivalent *Hanseniaspora uvarum*) produce substantial amounts of ethyl acetate as well as isoamyl acetate during the early phases of fermentation. During aging, *Brettanomyces* sp. (and *Dekkera* sp.) may also contribute to the formation of spoilage esters.

METHANOL

Methyl alcohol is not a normal product of alcoholic fermentation. Rather, it results from the hydrolysis of methylated pectin present in grapes by the enzyme pectin methylesterase (PME). Pectin is a polymer of galacturonic acid coupled via *alpha*-1,4-glycosidic linkage (see Chapter 16). In native pectin, approximately two-thirds of the carboxylic acid groups are esterified with methanol (Reed 1966).

Pectinases are used to enhance juice yield, color extraction, and clarification. They may significantly increase the methanol content in the resultant wine. However, in the case of grape wines, increases generally are still below the BATF limits of less than 1,000 mg/L (0.1%).

The methyl alcohol content of white wines ranges from 40 to 120 mg/L whereas the levels reported in reds are higher, ranging from 120 to 250 mg/L (Sponholz 1988). Lee et al. (1979) report methanol formation in Niagara and Concord varieties to be as high as 200 mg/L to 250 mg/L, respectively. Wines produced from grapes infected with the mold *Botrytis cineria* are also found to have higher methanol levels, up to 364 mg/L

(Sponholz 1988), compared to wines produced from sound fruit (see Chapter 3).

Gnekow and Ough (1976) found that in white wine fermentations methanol content reached an early maximum level whereas in the case of red varieties, which are fermented on the skins, methanol formation continues throughout fermentation, yielding much higher levels. Wines made from fruit other than grapes may be especially high in methanol content because of their higher relative pectin levels.

Distillates produced from plum and apricots have been reported to have methanol contents of 2,000 to 5,000 mg/L (Woidich and Pfannhauser 1974). It is not clear whether methanol plays a sensory role in wines. However, formation of methyl esters may be important. Methanol analysis is not a routine wine industry practice, but may be readily accomplished by gas chromatography (see Chapter 20—Fusel Oil Analysis by Gas Chromatography).

ETHANOL PRODUCTION

Yeast species and strains vary in their abilities to utilize carbohydrates in formation of alcohol and other byproducts, as well as to grow in varying concentrations of alcohol. Most strains of *Saccharomyces cerevisiae* are inhibited as alcohol levels approach 14 to 15% (vol/vol). However, several strains are more alcohol tolerant. In addition, many strains of wine yeasts can be acclimated to tolerate higher concentrations of alcohol through supplemented "syruped" fermentations.

The quantity of alcohol and carbon dioxide formed as well as the nature and concentration of byproducts vary with yeast strain, temperature of fermentation, and extent of aeration. Fermentation temperature plays an important role in alcohol tolerance. In general, yeast fermentations at lower temperatures result in higher alcohol yields, due in part to reduced losses of alcohol resulting from evaporation and entrainment. Work by Warkentine and Nury (1963) suggests that alcohol lost by carbon dioxide entrainment is relatively small (0.83% at 27°C/81°F). Generally, losses are correlated to fermentation temperature as well as surface-to-volume relationships in fermenters. Red wines often have lower alcohol levels than white wines with the same initial sugar content, primarily because of the practice of fermenting reds at higher temperatures in order to facilitate color extraction. Fermenting a must under strongly reducing (low oxygen) conditions, or in pressure fermentations will yield slightly higher alcohol levels (Amerine et al. 1972). Under such conditions, the yeast cell channels more energy to fermentative metabolism than toward cellular reproduction as normally occurs in the very early stages of fermentation.

DETERMINATION OF ALCOHOL CONTENT

The principal basis for wine taxation is alcohol content. United States regulations define wine as containing between 7 and 24% (vol/vol) alcohol. By comparison, the O.I.V. requires a minimum of 8.5% (vol/vol) alcohol to meet its standard as "wine." Within the U.S. federal definition, "table wines" must contain between 7 and 14% alcohol and "dessert wines" between 14 and 24%. In the dessert wine category, sherry must have a minimum alcohol content of 17%, whereas other dessert wine types (ports, muscatels, etc.) have 18% minima. Wines falling into the range of 14 to 18% are described as "light" sherries, ports, muscatels, etc.

The above regulations have recently been amended to reflect current industry interest in production of low-alcohol wines. Thus, wineries may produce wines of alcohol content less than indicated above, but such wines must meet all other standards for California table wines. Further, these wines must not contain less than 7% alcohol per federal definition.

The physical and sensory properties of wine are dependent, in part, on alcohol content. Thus blending or additions of wine spirits (WSA) that result in changes in the final alcohol content may subsequently result in change in the wine stability.

For the sake of discussion, procedures for the determination of alcohol in beverages may divide into methods utilizing the physical characteristics of solutions (colligative properties) and those based on the chemical properties of alcohol.

Physical Methods

Ebulliometric Analysis (see Chapter 20)

Ebulliometry is the most commonly encountered procedure for determination of the alcohol content of aqueous solutions. In principle, the analysis is based on the Raoult's law relationship of boiling point depression.

Raoult's law effects can be viewed from the perspective of a boiling point vs. composition diagram for the water-ethanol system (Fig. 6.3). The lower curve on this diagram represents the boiling point of various water-ethanol mixtures at 1 atmosphere pressure. As the percent ethanol increases, the boiling point decreases. Because one does not always have 1 atmosphere pressure in the laboratory, the ebulliometric method utilized a sliding scale that allows one to adjust (the left axis of Fig 6.3) for the actual boiling point of pure water.

The upper curve in Fig. 6.3 represents the composition of the vapor phase in equilibrium with the boiling liquid phase. One can readily observe that a 10% (vol/vol) ethanol-water solution boiling at ~199°F (point T on

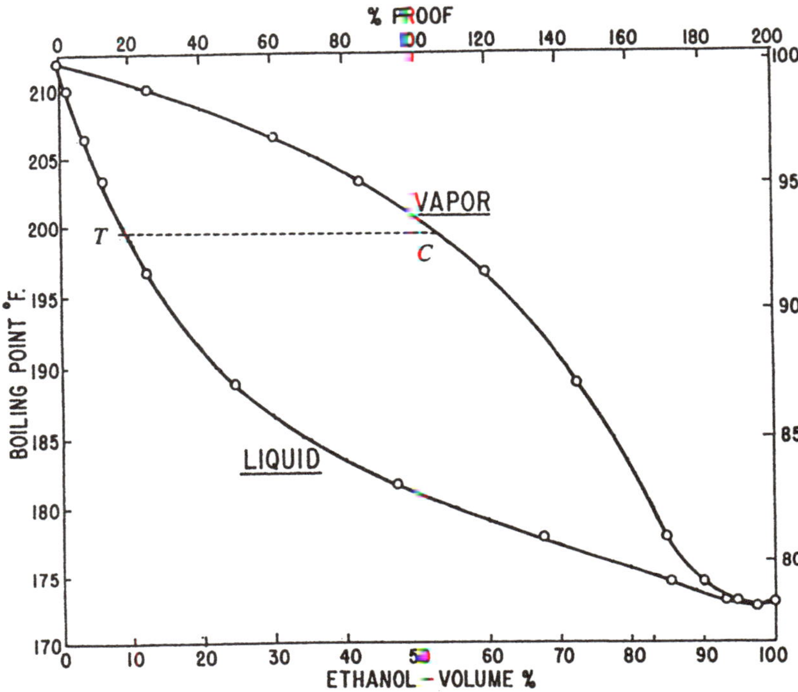

Fig. 6–3. Boiling point vs. composition for ethanol water mixtures.

Fig. 6.3) is in equilibrium with refluxing vapor of approximately 54% (vol/vol) ethanol (point C). This demonstrates why the ebulliometer must be operated under conditions of total reflux. Any ethanol-water vapor that leaks out of the condensor takes with it a proportionately large amount of ethanol. This causes the composition of the liquid phase to change (decreasing ethanol concentration) and the boiling point of the sample to gradually creep toward that of pure water.

Although simple in theory, several interferences may be encountered in routine laboratory application of ebulliometry; the most important of these is the effect of sugars. According to the colligative properties of solutions, sugar molecules would be expected to cause a boiling point elevation (hence lower apparent alcohol levels). However, observation contradicts this in that sweet wines usually boil at a temperature lower than expected, resulting in higher apparent alcohols. This is due to the sugar-water matrix squeezing out the ethanol (increasing its vapor pressure). To reduce the errors attributed to sugar, slightly sweet wines may be diluted with water to a sugar level of less than 2%, yielding a boiling point of 96°C to 100°C.

Amerine (1954), however, points out that it is doubtful if dilution avoids the errors because the result must be multiplied by the dilution factor, which, in turn, multiplies the relative error.

In contrast to the "sugar effect," aldehydes, esters, and acids tend to raise the boiling point of a solution. Thus the combined effects of solution components producing boiling point depression and elevation result in a compromise in which the boiling point, in practice, approximates that of the actual alcohol content.

Hydrometry (see Chapter 20)

The alcohol content of a distillate from an accurately measured volume of sample may be determined hydrometrically. Reducing sugars are not distillable and hence not a problem in distillate analysis, but two common interferences, sulfur dioxide and acetic acid, may cause problems. At levels approaching 200 mg/L, sulfur dioxide is steam-distillable as sulfurous acid. As such, it may affect hydrometric determinations of alcohol by producing results lower than expected (0.2–0.5% vol/vol); it is recommended that in cases where the sulfur dioxide content is known to be high, samples be neutralized prior to distillation.

Abnormal levels of acetic acid (greater than 1.0 g/ L) likewise present problems. Some workers suggest neutralization of such samples with 1 *N* NaOH prior to the distillation step. Young wines and sweetened wines may foam excessively during distillation. This problem can be avoided by the addition of small amounts of an antifoaming agent prior to distillation. However, antifoaming agents may coat glassware with residues that are difficult to remove.

Gas Chromatography (see Chapter 20)

Gas chromatography (GC) is a technique used to separate volatile components in the sample. For example, wine (or juice, distillate, etc.) is injected into a heated tube that is packed with a specialized absorbant, through which an inert gas flows. Ethanol and other volatile components are vaporized and carried through the tube (referred to as the GC column) toward a detector that senses their presence. Because of differences in their interaction with the absorbant, different compounds migrate or travel through the column at different rates, and are separated by the time they reach the detector. To quantitate ethanol, one may prepare standards of known concentration, inject them into the GC, and compare their detector responses to that of the unknown sample.

The same GC technique can be used for the analysis of fusel oils (see Chapter 20). For this particular analysis, a column with a different specialized absorbant is used, as well as a different column temperature. Other uses of gas chromatography in the winery laboratory are being developed. It is a powerful analytical tool that will see more and more utility in the future.

High Performance Liquid Chromatography (HPLC)

HPLC is also finding increased use in the winery laboratory. This technique can also be used to determine alcohol in harvested fruit as well as in finished wine.

Chemical Methods

Dichromate Oxidation

Although several chemical methods for alcohol determination appear in the literature, the most important currently used procedure is based on the quantity of acid dichromate required to oxidize the alcohol to acetic acid:

$$2Cr_2O_7^{2-} + 3C_2H_5OH + 16H^+ \rightarrow 4Cr^{3+} + 3CH_3COOH + 11H_2O$$

The excess dichromate remaining on completion of the reaction is titrated with ferrous ammonium sulfate (FAS):

$$Cr_2O_7^{2-} + 6Fe^{2+} + 14H^+ \rightarrow 2Cr^{3+} + 6Fe^{3+} + 7H_2O$$

The completeness of this redox reaction depends on the time of reaction as well as concentration of reacting components. The optimum temperature for oxidation is 60°C to 65°C. The reaction is also critically dependent on hydrogen ion concentration for the complete oxidation of alcohol to acetic acid rather than to mixtures of the acid and aldehyde intermediate.

Enzymatic Analysis

Ethanol can be oxidized in the presence of the enzyme alcohol dehydrogenase and NAD yielding the reduced form of the coenzyme, NADH + H^+. This is a stoichiometric reaction under proper experimental conditions, and the NADH produced can be determined spectrophotometrically at 334 nm. Reagents for this determination may be obtained from a number of chemical supply houses (e.g., Sigma), and complete kits for the analysis are also available (e.g., Boehringer-Mannheim). One drawback in using this method is that it is necessary to quantitatively transfer very small volumes of reagent and sample.

EXTRACT

Conventionally, extract, as measured by hydrometry, consists of the nonvolatile soluble solids left after dealcoholization of the wine sample (Amerine 1965). The OIV similarly defines extract as "the nonvolatile materials in wine," but further states that the physical condition for such a definition must be carefully stated. Thus extract includes sugars (or their condensation products), fixed acids, glycerin 2,3-butylene glycol, and phenols, as well as varying amounts of lactic and acetic acids. Because sugars present in the wine sample are nonvolatile, they contribute to "apparent extract." The sugar content should, therefore, be subtracted from the "apparent extract value" if one wishes to compare the sugar-free extract contents of different wines.

By international convention, extract is expressed in units of grams/liter (g/L). At present however, data are still frequently expressed as g/100 mL in the United States. Extract levels for dry white wines vary from 20 to 30 g/L (2–3%), whereas in dry red wine, values often exceed 30 g/L, the increase being attributed to higher levels of phenols. Thus extract content can be used to distinguish light-bodied from heavier-bodied wines. Dry table wines with extract values of less than 20 g/L (2%) often appear thin on the palate, while those with 30 g/L (3%) or more have a full-bodied character. Fermentation of low sugar musts and subsequent fortification can produce wines with low extract values. Furthermore, wines produced by amelioration may also have low extracts.

Extract values depend, in part, on variations in processing as well as the grape variety. Extract content may be affected by the type of press used as well as press pressures. Referring to Table 3–6 which compares press fractions separated in Methode Champenoise production of Pinot noir cuvees, it can be seen that, with increases in pressure there is an increase in extract components such as phenols. Other processing considerations that may increase extract values in wine include the use of pectolytic enzymes. Singleton et al. (1980) note that with increased pomace exposure and fermentation temperature, there are corresponding increases in pH, potassium, protein, and phenol extraction and, therefore, extract value. They also note decreased acidity under these conditions.

Processing activities, such as barrel fermentation and storage, increase the extract as a result of phenol extraction. Storage for 1 year in new 200 L European oak cooperage may contribute 250 mg/L gallic acid equivalents (GAE) of nonflavonoid phenols (Amerine et al. 1980). By comparison, American white oak contributes about half this level.

Extract Analysis

Densimetric Procedures

Extract can be determined from the difference in specific gravity of the original wine sample and the alcohol distillate removed from the wine. The specific gravity of the alcohol distillate may also be determined from an accurate value for percent ethanol (vol/vol) and an appropriate percent alcohol vs. specific gravity table (see Chapter 20). In another procedure given for extract analysis, the dealcoholized residue from the alcohol distillation is brought back to its original volume with distilled water and °Brix is then determined by hydrometer. It should be recalled that °Brix is defined in the units of **g/100 g**. To convert to **g/100 mL** (per United States definition of extract), one need only multiply the °Brix by the equivalent specific gravity (see Chapter 20). Final results are then expressed in units of g/L.

Several problems may be encountered in carrying out an extract analysis. Most obviously, the physical conditions of analysis (e.g., temperature, size of distillation flask) affect the product(s) formed, lost, or degraded. For example, water, lactic and acetic acids, ethylene glycol, and 2,3-butylene glycol may be differentially lost during distillation. Furthermore, dehydration products resulting from heating of sugars may form, changing the density of the solution formed upon dilution back to original volume.

Nomographs

In addition to densimetric procedures, nomographs may also be used as rapid means of approximating extract volumes in table and dessert wines. Historically, the Marsh nomograph (Appendix I, Fig. I-3) for dessert wines has been widely used. However, Vahl (1979) has presented an alternative scale for use with table wines (Appendix I, Fig. I-2). As shown in Figs. I-2 and I-3, nomographs consist of three scales, "degrees Brix," "extract," and "alcohol"; if two values are known, the third can be obtained directly. As seen in Fig. I-3, the extract content of a white dessert wine is defined as the reducing sugar content plus 2.0, whereas the extract content of a Port-style wine is reducing sugar plus 2.5.

ANALYSIS

The following is a cross-reference of the analytes and procedures found in Chapter 20: Ethanol by ebulliometry, dichromate oxidation, gas chromatography, hydrometry, spectrometry—visual and ultraviolet, titration, and HPLC; active amyl and isobutyl alcohol by gas chromatography; Fusel oils by gas chromatography; methanol by gas chromatography; *n*-propyl alcohol by gas chromatography; extract by specific gravity and °Brix hydrometry.

CHAPTER 7

Phenolic Compounds and Wine Color

Variations in wine types and styles are largely due to the concentration and composition of wine phenols. From the vineyard to production and aging, fine wines can be viewed in terms of management of phenolic compounds. Phenols are responsible for red wine color, astringency, and bitterness; they contribute to the olfactory profile; serve as important oxygen reservoirs and as substrates for browning reactions.

REPRESENTATIVE GRAPE AND WINE PHENOLS

Grapes and wine contain a large array of phenolic compounds derived from the basic structure of phenol (hydroxybenzene). Representative structures of the major classes are presented in Fig. 7–1. Two distinct groups occur in grapes and wine: the nonflavonoid and flavonoid phenols.

The total phenol content of wine is less than that present in fruit. Traditional fermentation following crushing and destemming leads to a maximal extraction of up to 60%. Microbial activity may lead to increases in the concentrations of certain phenols. Fermenting and/or storage in

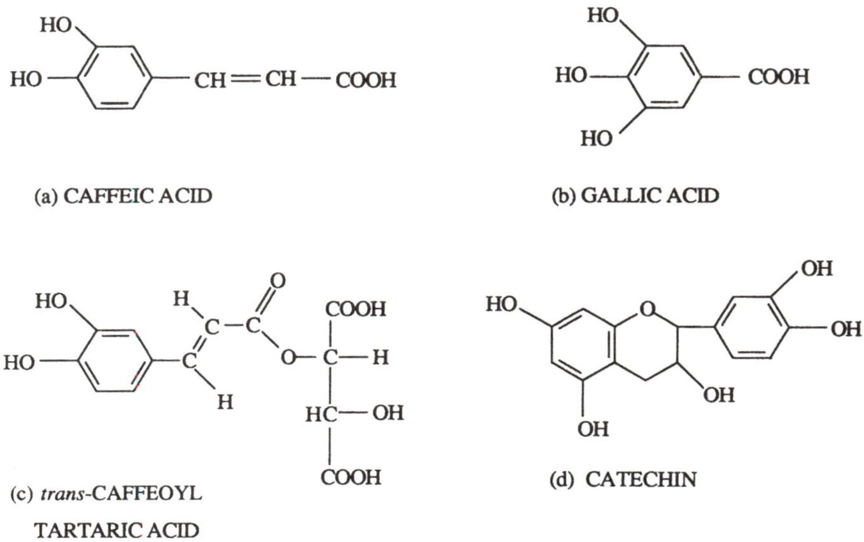

Fig. 7–1. Representative examples of nonflavonoid (a, b, c) and flavonoid phenolics (d) found in grapes and wine: (a) caffeic (dihydroxycinnamic) acid; (b) gallic (trihydroxybenzoic) acid: (c) *trans*-caffeoyl tartaric acid (the most common form in grapes); and (d) catechin.

oak provides additional sources of phenolics. Singleton (1980) reported the average values for phenolic fractions and total phenols (see Table 7.1). Due to the broad chemical diversity of phenolic compounds, total phenols in must and wines are usually presented in arbitrary units of a phenolic standard such as amount of gallic acid necessary to produce the same analytical response or gallic acid equivalents (GAE). As a result of changes in winemaking style, wines produced in the Unites States tend to be lower in tannin and with more supple tannins.

Nonflavonoid Phenols

In wines not stored in oak, the primary nonflavonoid phenols are derivatives of hydroxycinnamic and hydroxybenzoic acids, the most numerous of which are esterified to sugars, organic acids, or alcohols. The nonflavonoid component arises principally from juice extraction and secondarily from post-fermentation activity, including exposure to oak coopperage.

The levels of benzoic acid and its derivatives in red wines range from 50 to 100 mg/L and in whites from 1 to 5 mg/L (Singleton 1985). Salicylic acid is present at levels of more than 10 mg/L (Muller and Fugelsang, 1994), and other derivatives are present in only trace amounts.

Table 7–1. Gross phenol composition estimated in mg GAE/L for typical table wines from *Vitis vinifera* grapes.

Phenol class	Source[a]	White wine Young	White wine Aged	Red wine Young	Red wine Aged
Nonflavonoids, total		175	160–260	235	240–500
Cinnamates, derivatives	G,D	154	130	165	150
Low volatility benzene deriv.	D,M,G,E	10	15	50	60
Tyrosol	M	19	10	15	15
Volatile phenols	M,D,E	1	5	5	15
Hydrolyzable tannins, etc.	E	0	0–100	0	0–260
Macromolecular complexes					
Protein-tannin	G,D,E	10	5	5	10
Flavonoids, total		30	25	1060	705
Catechins	G	25	15	200	150
Flavonols	G,D	tr	tr	50	10
Anthocyanins	G	0	0	200	20
Soluble tannins, deriv.	G,D	5	10	550	450
Other flavonoids, deriv.	G,D,E,M	?	?	60?	75?
Total phenols		215	190–290	1300	955–1215

[a]D = degradation product; E = environment, cooperage; G = grapes; M = microbes, yeast.
SOURCE: Singleton 1980.

Most nonflavonoids are present at levels below their individual sensory threshold; however, collectively members of the group may contribute to bitterness and harshness.

Nonflavonoid Content of Juice

The phenol content of juice is largely nonflavonoid (Kramling and Singleton 1969). The levels of nonflavonoid phenols are relatively constant in red and white wines, because of their extractability from grape pulp. The major source of nonflavonoids from grape solids is the hydrolysis products of anthocyanins, hydroxycinnamic acyl groups (Singleton and Noble 1976). Hydroxycinnamate derivatives comprise the majority of this class of phenolics in both white and red wines. These compounds are present in juice and wine as the free acids and ethyl esters and in the form of tartrate or tartrate-glucose esters (Henning and Burkhardt 1960). Hydroxycinnamates serve as the primary substrate for polyphenol oxidase activity (Singleton et al. 1985).

Nonflavonoids Derived from Fermentation and Extraction from Oak

During alcoholic fermentation, slow (incomplete) hydrolysis of nonflavonoid esters occurs, resulting in free acid and ester forms. Caftaric and similarly bound or acylated phenols are hydrolyzed to varying degrees, yielding the corresponding free cinnamic acids. Somers (1987) reported that fermentation caused total hydroxycinnamates to decrease by nearly 20% due to adsorption by yeasts. Cinnamic acid may be involved in the formation of microbially produced compounds such as 4-ethylcatechol. This transformation involves decarboxylation of the acid to yield 4-vinylcatechol and subsequent reduction to 4-ethylcatechol. Similar microbially induced transformation of benzoic, shikimic, or quinic acids to yield catechol and protocatechuic acid has been reported (Whiting and Coggins, 1971). Ethyl phenols are important sensory compounds of red wines. Some produced by *Brettanonyces/Dekkera*, are responsible for phenolic or leathery off odors. Analysis of 4-ethyl phenol can be used as a marker for *Brettanonyces/Dekkera* (see Chapter 20). Tyrosol is produced by yeast from tyrosine during fermentation and is the only phenolic compound produced in significant amounts from nonphenolic precursors (Singleton and Noble 1976).

Nonflavonoid Component Arising from Oak

In wines not exposed to oak cooperage, the nonflavonoid phenol fraction is about the same as in the juice from which the wine was produced. Phenols extracted from oak are present almost entirely as hydrolyzable nonflavonoids. Vanillin, sinapaldehyde, coniferaldehyde, and syringaldehyde are reported to be the major species present in barrel-aged wines (Singleton 1985). (See section on Oak Barrel Components.)

Flavonoid Phenols

Much of the structure and color in wine is due to flavonoids that are found in skins, seeds, and pulp of the fruit. The base structure (aglycone) of flavonoids consists of two aromatic rings, A and B, joined via a pyran ring. (The base structure and standard numbering system are seen in Fig. 7.2.) Changes in the oxidation state result from variations in hydrogen, hydroxyl, and ketone groups associated with carbons 2, 3, and 4 leading to different members of the family.

Flavonoids may exist free or polymerized to other flavonoids, sugars, nonflavonoids, or a combination of these. Those esterified to nonflavonoids or sugars are referred to as acyl and glycoside derivatives, respectively. Polymerization of catechin and leucoanthocyanidin flavonoids produces the procyanidin class of polymers. Their classification is based on the nature of the flavonoid monomers, bonding, esterification to other compounds, or functional properties. The most common functional class of procyanidins is the condensed tannins.

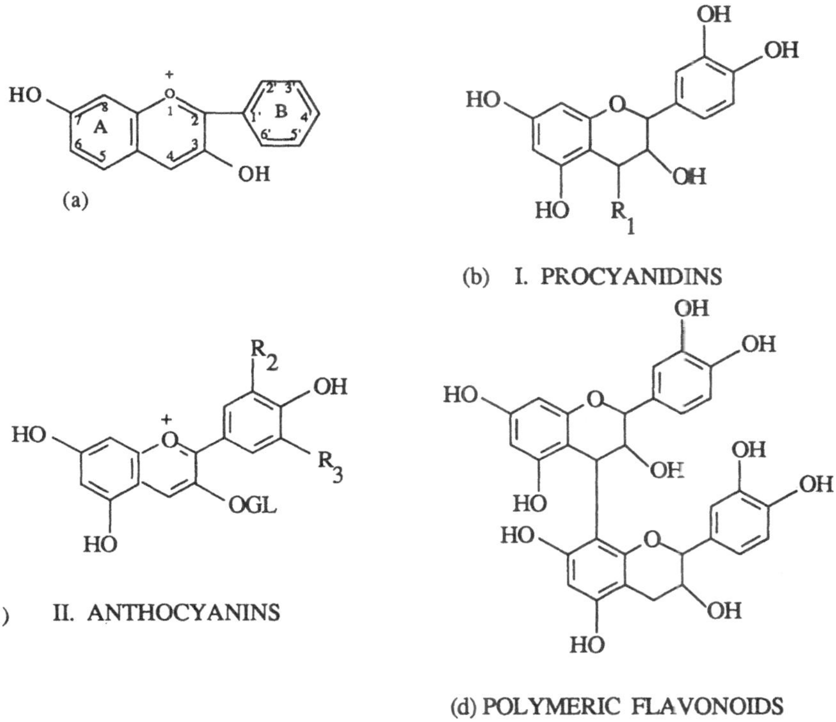

Fig. 7–2. (a) Base structure of typical flavonoid showing rings and numbering system. (b) Parent structures of wine flavonoids. I. Procyanidins: R_1 = H (catechins) ; R_1 = OH (leucoanthocyanidins); R_1 = =O (flavonols). II. Anthocyanins: R_2, R_3 = OCH_3 (malvidin); R_2 = OH, R_3, = H (cyanidin); R_2, R_3 =OH (delphinidin); R_2 = OCH_3, R_3 = OH (petunidin); R_2 = OCH_3, R_3 = H (peonidin). (d) Structure of typical dimeric flavan structures found in wine.

Monomeric flavonoids react to yield dimeric and larger forms, resulting in an array of heterogeneous structures. Polymeric flavonoids make up the major fraction of total phenolics found in all stages of winemaking. Polymerization, either oxidative or nonoxidative, yields tannins and condensed tannins, respectively (Ribereau-Gayon and Glories 1986). Further polymerization may eventually lead to precipitation.

Processing protocol will significantly affect the phenolic composition of wine. Flavonoids are derived primarily from the seeds, skin, and stems of grapes. Anthocyanins and flavonols (Fig. 7.2) are extracted mainly from the skins, and catechins and leucoanthocyanins from the seeds and stems. Increasing the skin contact time and temperature and the extent of berry breakage increases the flavonoid content. The distribution of phenols in a

Table 7–2. Phenolic levels in a "typical" *Vitis vinifera* red wine.

Phenol Type	Concentration (mg/L)
Nonflavonoids	200
Flavonoids:	
Anthocyanins	150
Condensed tannin	750
Other flavonoids	250
Flavonols	50

SOURCE: Singleton and Noble 1976.

red wine with 1,400 mg/L (GAE) is seen in Table 7.2. Flavonoid phenols usually account for 80 to 90% of the phenolic content of conventionally produced red wines and about 25% of the total in whites crushed but without skin contact.

Flavan-3-ols (Catechins and Epicatechin)

Flavan-3-ols flavonoids are primarily (+)-catechin and (−)-epicatechin. Their concentration in white wines ranges from 10 to 50 mg/L and may reach 200 mg/L in reds (Singleton and Esau, 1969).

The basic structure for the family is d-catechin (Figs. 7.1 and 7.2). Because of asymmetrism arising at carbons 2 and 3, four stereoisomers may occur: *d-* and *l-*catechin, and *d-* and *l-*epicatechin. Hydroxylation at the 5′ position yields *d-*gallocatechin and its isomer, *l-*epigallocatechin. Of these forms, *d-*catechin, *l-*epicatechin, and *l-*epicatechin gallate are the principal catechins found in grapes (Su and Singleton 1969). Compared to catechins, gallocatechins are present in much lower concentrations in mature fruit and wines (Singleton 1980).

Flavan-3-ols may exist in dimeric and larger groupings. Polymeric forms of flavan-3-ols are referred to as procyanidins or condensed tannins. In white wines made with limited skin contact, catechins account for most of the flavonoid phenols. Catechins are precursors to browning in white wines and to browning and bitterness in red wines. In white wines, catechins contribute significantly to the flavor profile (Singleton and Noble 1976).

Sulfur dioxide additions to must increase extraction of flavan-3-ols (Singleton et al. 1980), which may contribute to bitterness. This is a reason why many red and white wines are produced with little or no sulfur dioxide added before fermentation. Such practice allows for phenolic polymerization and precipitation, thus helping to avoid excess bitterness while enhancing suppleness.

The taste threshold of catechins in white wine is near 200 mg/L (GAE).

In contrast, catechins have a bitter threshold of 20 mg/L (Dadic and Belleau 1973) in 5% ethanolic model solutions.

Flavan-3,4-diols (Leucoanthocyanidins and Leucoanthocyanins)

Flavan-3,4-diols (leucoanthocyanidins) differ from catechins by an additional hydroxyl at carbon 4 (see Fig. 7.2). Three sites of asymmetry (C_2, C_3, and C_4) result in formation of eight possible isomers. Upon heating in acid solution, leucoanthocyanidins are converted to colored anthocyanidin forms. These compounds serve as precursors to larger polymeric forms. Leucoanthocyanidins differ from leucoanthocyanins in that the latter contain a sugar attached through a glucosidic linkage. Ribereau-Gayon (1974) suggested covalent bond formation between carbon 4 of the leucoanthocyanidin and either carbons 6 or 8 of a second reactive form to produce polymers. Depending on the nature of that second flavan component, several polymeric species are possible. If carbon 4 is available, the condensation reaction may continue, yielding a polymer of 3 to 4 members (Somers 1986). If reaction at C_4 is blocked, the condensation is stabilized at the dimer stage. Occasionally, this group is important in pinking reactions seen in white wines.

Flavonols

Flavonols are localized in the grape skin, occurring in glycosidic forms. The most commonly encountered sugar moiety in these flavonols is glucose. Ribereau-Gayon (1965) reported glucosides of kaempferol, quercetin, and myricetin present in the 3-glucoside form.

Hydrolysis of the glucose moiety occurs rapidly. Singleton (1982a, 1982b) reported trace flavonol concentrations in white wines, up to 50 mg/L in young red wines, and approximately 10 mg/L in older red wines. Quercetin comprises the majority of the flavonol fraction. In wines produced without skin contact, flavonols were reported at zero or only trace concentrations.

The bitterness threshold for kaempferol in 5% ethanol solution is reported at 20 mg/L. Quercetin and myricetin have bitterness thresholds of 10 mg/L (Dadic and Belleau 1973). The glucosides ordinarily are more bitter than the aglycones.

Complex Phenols (Tannins)

Polymers of both flavonoid and nonflavonoid phenols are referred to as tannins, a term derived from their ability to tan leather. Tannins are classified as hydrolyzable or condensed. Hydrolyzable tannins, based on nonflavonoid phenols, exist as esters and, as such, can be degraded or hydrolyzed. The condensed tannins, also known as procyanidins, cannot easily be decomposed by hydrolysis. Wine tannin is largely composed of polymers of leucoanthocyanidins and catechins (Ribereau-Gayon 1974), such as seen in Fig. 7.2.

Tannins exhibit the formation of blue-color complexes upon reaction with Fe^{3+}, react with protein and are described as astringent. Phenolic compounds eliciting such reactions, like dimeric flavonoids, characteristically have molecular weights ranging from 500 to 5,000. In young wines, tannins exist as dimers or trimers, but with further polymerization, 8 to 14 flavonoid units with molecular weights ranging from 2,000 to 4,000 may be formed.

The bulk of the polymeric flavan-3-ols (condensed or procyanidin tannins) are found in the seeds and to a lesser extent in stems and skins. Values determined by Kantz and Singleton (1991) for four varieties suggest 58.5% of the condensed tannins are in the seeds, 21% in stems, 16.5% in leaves, and 4% in the skins. Skin tannins are of greatest concern due to the ease of extraction. However, as skin contact time for red wines is increased, seeds play an increased role as a source of condensed tannins.

During aging, tannin levels in wines decrease as a result of oxidation and precipitation with protein. Tannins play a role in physical stability. Under conditions of high pH and high tannin levels, iron (Fe^{3+}) may combine with tannin after exposure to air, producing an insoluble colloidal instability known as ferric tannate or "blue casse." Tannins can also combine with high levels of protein, producing instability (see Chapter 8).

Wines deficient in tannin may be either blended or supplemented by addition of commercial tannin. United States government regulations state that tannin additions in white wines must be restricted so that the wine does not contain more than 0.80 g/L GAE after the addition. In red grape wines, the final tannin levels must not exceed 3.0 g/L (see Table I–10 in Appendix I). Commercial tannic acid is often used for such additions. Tannic acid differs chemically from the grape enological tannins, which were originally extracted from oak galls or grape seeds. For a further discussion on the addition of tannin to juice and wine, see Chapter 16.

Sensory Considerations

Arnold et al. (1980) found that all wine phenolic fractions studied, including monomers, were both bitter and astringent. Robichaud and Noble (1990) found that both bitterness and astringency increased with increased concentration of (+)-catechin, although the rate of bitterness increase was greater than that of astringency. Procyanidins have been reported to be both astringent and bitter (Arnold et al. 1980). On a per weight basis both bitterness and astringency increase with increased degree of polymerization (Arnold et al. 1978). Thorngate (1993) suggested that as polymerization increases, the number of possible hydrogen bonding sites increases, which would be expected to increase astringency. Bitterness is the result of access to membrane bound receptors and is likely limited by molecular size. Differences in the lipid solubility of trimers and tetrameres allow them to depolarize the taste receptor cells, thus increasing the perception of bitterness (Lee 1990).

The difficulties in assessing bitterness and astringency of various phenolic compounds involve the fact that astringency masks bitterness (Lea and Arnold 1978). As wines age, tannin polymerization and possibly precipitation occur. This, coupled with the reduction of tannin astringency resulting from protein fining agents, for example, may slowly unmask bitterness.

Degradation of wine flavonoids with age increases the production of more volatile phenols (Singleton and Noble 1976). These low molecular weight hydroxybenzoates and hydroxycinnamates are noted for their odors. Levels of volatile phenols in young white wines are reported to range from 1.5 to 3.2 mg/L (GAE). Higher levels have also been reported. Although the concentrations of individual volatile phenols in young wines usually are below threshold, their cumulative effects may contribute to the perception of smokiness and bitterness.

The major factors governing palate balance are the quantity and "quality" of tannins, alcohol, and acidity. These components interrelate, influencing the perception of balance according to the following.

$$\text{Sweet} \rightleftharpoons \text{acidity} + \text{bitterness and astringency}$$

The perception of sweetness derived from alcohol, polysaccharides and sugar (when present), must be in balance with the sum of the perceptions of acidity, astringency, and bitterness. This relationship suggests that the lower the acidity, the more tannin a wine can support. The palate balance formula is functionally analogous to the suppleness index described by Peynaud (1984):

$$\text{Supplenes index} = \text{alcohol (vol/vol)-(titratable acidity + tannin)}$$

Red wines are not considered supple unless the suppleness index is below 5.0. The palate balance relationship is important in both maturity assessments and processing. Methods of preparation of phenolic sensory standards are provided in Table 2–1.

Anthocyanins/Anthocyanidins

The color of red grapes is attributed to the presence of members of a large group of plant pigments, the anthocyanins. These compounds are present as glycosides (primarily glucosides). Glycosidation may occur as single sugar residues attached to multiple hydroxyls, as di- and tri-saccharides, or as any combination of these forms. In varieties of *Vitis vinifera,* pigments are present in the 3-glucoside form (bound at the carbon 3 position), whereas in other species, 3,5-diglucosides may occur (Singleton and Esau 1969). Except for Pinot noir, grape pigments are bound or acylated with acetic, caffeic, or *p*-coumaric acids (Fig. 7.3).

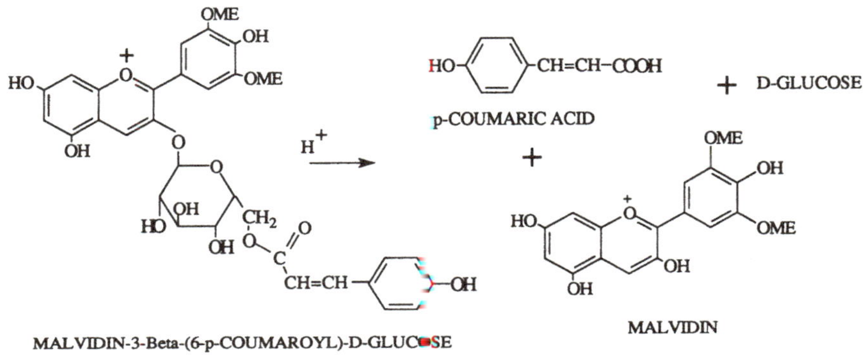

Fig. 7–3. Hydrolysis of typical grape anthocyanin yielding the aglycone (anthocyanidin), glucose, and coumaric acid.

Five anthocyanins are generally found in red grapes. These include malvidin-, delphinidin-, peonidin-, cyanidin-, and petunidin-3-D-glucoside (Fig. 7.2). Of these, malvidin (as the 3-glucoside) is the most common pigment in varieties of *Vitis vinifera.*

Upon acid hydrolysis (Fig. 7.3), anthocyanins yield one or more moles of sugar and the parent anthocyanidin (also referred to as the aglycone). The color of these pigments and their stability is a function of pH, the presence of various metallic cations, and other factors, which are summarized below.

pH

Anthocyanins are amphoteric, so their color is primarily pH-dependent. The reversible, pH-dependent equilibria that exist for malvidin-3-glucoside is seen in Fig. 7.4. At low pH values (pH <4.0), the main equilibrium forms are the red-colored flavylium ion and its colorless pseudobase. The pK for equilibrium between these forms (pK = 2.6) favors the colorless form.

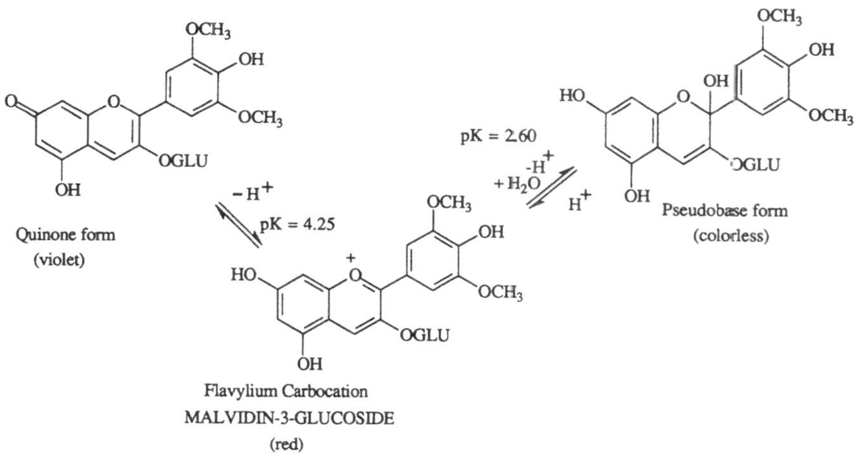

Fig. 7–4. pH dependency of malvidin-3-glucoside pigment.

Thus, at wine pH values greater than 3.0, less than 50% of the potential red color is visible.

Sulfite Bleaching

The addition of SO_2 can result in temporary color reduction based on a reversible reaction presented in Fig. 7.5 (Jurd 1964). Because the site of sulfite binding (carbon 4) is also the point at which reaction with other phenolics can occur, polymerized pigments or tannin-pigment polymers (with linkages between C_4 or C_6 and C_8) are resistant to decolorization by sulfur dioxide.

Temperature

In general, increased wine storage temperature results in an increased red (or brown) color, presumably due to accelerated rates of polymerization and other reactions occurring at the higher temperature.

Polymerization

A majority of the anthocyanidin pigment is incorporated into dimeric or larger units. Copolymerization may take place between pigments and tannins with bond formation occurring between C_4 of the anthocyanidin and either C_6 or C_8 of the tannin to yield a red-colored dimer. The presence of sulfur dioxide inhibits polymerization as a result of occupying the C_4 position of the flavonoid. This is an important red wine production consideration (see section on Red Wine Processing Considerations). Because C_4

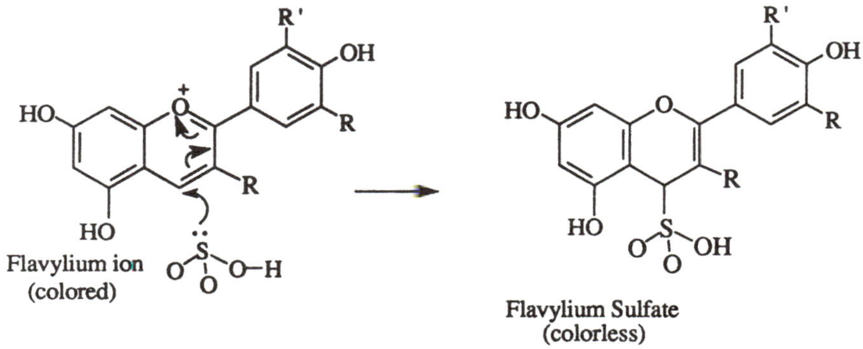

Fig. 7–5. Sulfur dioxide bleaching of pigments.

of the anthocyanidin is involved in bond formation, polymeric forms are not reactive toward sulfur dioxide nor are they responsive to changes in pH. By comparison to monomeric anthocyanidins, polymeric pigments show increased stability.

In the presence of acetaldehyde, rapid reaction may occur between anthocyanidin and tannin precursors. The extent to which these molecules are already polymerized determines the stability of the product. Further reaction with an already highly polymerized tannin leads to instability and precipitation (and decreased color) By comparison, in cases where polymerization has not reached this state, a stable complex results that is more colored than the anthocyanidin pigment by itself. This phenomenon may explain enhanced color and stability of color noted in barrel aged wines (see section on Oak Barrels). Direct condensation without involvement of acetaldehyde is slow.

Observations of enhanced color after addition of aldehydic spirits or sugar to red wines may be explained in light of these reactions (Singleton and Guymon 1963; Singleton et al. 1964). The increased color results from formation of pH-resistant polymers involving anthocyanidins and acetaldehyde. When the anthocyanidin component is present in the colored oxonium ion state (responsive to changes in pH), increased spectral color results from polymerization. Once formed, such polymers are resistant to changes in spectral color with decreases in pH. Conversely, oxidation of anthocyanidin monomers or precipitation of polymeric species may result in decreased color (Berg and Akiyoshi 1975).

Sugar additions can result in increased color due to formation of pH-resistant polymers. In this case, sugars are theorized to operate as polymerizing agents, either by hydrogen bonding or by binding water (Berg and Akiyoshi 1975).

GRAPE GROWING AND PROCESSING CONSIDERATIONS

Grape Phenols

Singleton and Esau (1969) reported the phenol content of red grapes average 5,500 mg GAE/kg, whereas white varieties average 4,000 mg GAE/kg. Table 7–3 compares the relative quantities of phenols in the grape. Expressed in terms of GAE, 46 to 69% of the total berry phenols are located in the seeds, 1% in the pressed pulp, and 5% in the juice. The balance, 50% for reds, and 25% for whites, is located in the grape skin. Total phenol levels significantly differ among varietal wines, as seen in Table 7–4.

Table 7–3. Total phenol levels in *Vitis vinifera* grapes expressed as Gallic Acid Equivalents (GAE mg/kg)

Component	Red grapes	White grapes
Skin	1,859	904
Pulp	41	35
Juice	206	176
Seeds	3,525	2,778
Total	5,631	3,893

SOURCE: Singleton and Esau 1969.

The phenol content of the fruit is primarily determined by variety, however, climate may exert an important influence. Flavonoids appear to vary more than nonflavonoids in regard to site, vintage, and climatic variations. For example, the anthocyanin content will be greater for a particular variety grown in a cool compared to a warm region. The relative ratio of pigments remains the same, but variations in the total anthocyanin and total phenol content may differ.

Grape Maturity and Wine Phenols

Canopy management is widely accepted as a means of influencing grape and wine composition including phenols by altering microclimate to influence vine physiology.

Leaf area-to-fruit weight is a critical factor in determining anthocyanin content. Research indicates that 10 to 15 cm^2 of leaf area is needed to adequately ripen 1 gram of fruit if ripeness is defined in terms of soluble solids without consideration of grape flavor and color (Kingston and Van Epenhuijsen, 1985). It must be noted that the same level of sugar may be achieved with wide variations in grape color and other quality parameters. The higher the exposed leaf area-to-fruit weight ratio the greater the grape

Table 7–4. Average phenol content in wines expressed as GAE mg/L.

Variety	Phenol Level
Refosco	2,300
Cabernet Sauvignon	1,520
Petite Sirah	2,120
Zinfandel	1,380
Pinot Noir	1,300

SOURCE: Singleton 1985.

color. Iland (1990) suggested a desirable ratio of 20 to 30 cm^2/g of fruit for maximum color formation.

Maturity affects qualitative changes in grape phenols. From veraison to harvest, there is a net increase in grape berry tannins and anthocyanins; however, with the exception of anthocyanins, this increase may be countered by increases in berry size. Overripe and shriveled fruit have lower phenols (including anthocyanins) levels possibly as a result of conversion to unextractable oxidation products (Singleton 1985). The total phenol concentration may show a twofold seasonal change. Therefore some measure mean berry weight for comparative purposes (Somers 1986).

As grapes mature so does phenol polymerization, resulting in a decrease in astringency. Polymerization along with the drop in acidity resulting from maturity increases suppleness of the berry tannins (Ribereau-Gayon and Glories 1986). Polymerized tannins with a molecular weight of 3,000 or greater are believed to be smoother on the palate than lower molecular weight tannins, considered to be "hard" and astringent. Fruit maturity is particularly important for red wine palate balance, a key quality component (see sensory section). The quantitative and qualitative differences in grape phenols are of great interest to red wine producers and a principal factor influencing fruit harvest. Good fruit and ripe tannins provide winemakers complete flexibility to make wine for early consumption or aging.

White Wine Processing Considerations

A white wine with 200 mg/L total phenols (GAE) can be expected to contain about 100 mg/L caffeoyl tartrate and related cinnamates, 30 mg/L tyrosol and small derivative phenols, and 50 mg/L of flavonoids such as catechins and tannins (Singleton 1982).

The catechin content accounts for most of the total flavonoids found in white wines made with little skin contact. Such wines contain few leucoanthocyanidins. As pomace contact time and temperature is increased, both catechins and leucoanthocyanins increase, with the polymeric tannins increasing faster than the catechins. Skin contact of crushed grapes is not uncommon in producing certain white wines. Seeds, stems, and skins provide equal quantities of phenol extractives at cool temperatures (DuPlessis et al. 1988a, 1988b). Ough and Berg (1971) reported that increased temperature during skin contact resulted in more deeply colored whites, and higher pH, potassium, proline, and total phenols. For temperature increases of 5°C to 15°C (41°–60°F), Pallotta and Cantarelli (1979) reported these increases: catechins of 160%; leucoanthocyanins 58%; tannins 31%; and total phenols, 42%. Elevated temperatures during skin contact (greater than 10°C/50°F) produce wines of deeper color, increased oxidative sensitivity, and coarser character. Also, these wines mature more rapidly during barrel aging. Retention of volatile components in must and subsequent wines is increased when pomace temperatures do not exceed 19.5°C (Ramey et al. 1986).

Red Wine Processing Considerations

There are a number of processing decisions that can affect phenol extraction, red wine palatability and style. These include the degree of cold soaking, thermal vinification, extent of berry breakage, whole cluster, berry and stem return, fermentation temperature, and length of time the juice remains in contact with grape skins. Further, the configuration of the fermenter (height to width ratio) affects the extraction process.

Crushing/Destemming

Vigorous crushing favors extraction of astringent and bitter tannins (Ribereau-Gayon and Glories 1986). Producers should avoid excessive crushing or handling, which might increase the nonsoluble solids level. Seeds contain a relatively high level of phenols; therefore seed scaring or breakage should also be avoided. The percentage of berries broken and destemmed varies, although many attempt to maximize flexibility by having systems that can convey uncrushed fruit to the fermenter. The interest in whole cluster or destemmed berry return resides in the perceived benefits of prolonged fermentation, lower phenol extraction, and increased fruit character.

Stem Return

The addition of stems to the fermenter is relatively uncommon, but is occassionally utilized with low tannin varieties such as Pinot Noir. Factors that may be influenced by the presence of stems in the fermenter are summarized. Color is decreased, and the phenolic content increased. The increase in phenols helps to stabilize color, which may be particularly useful for low color varieties. A high percentage of stems adds a "spicy" and possibly bitter character. Stem tannins are different from those of the skins, which tend to be rougher and more astringent (Ribereau-Gayon and Glories 1986). Stem return frequently increases must pH and can affect cap management and pressing efficiency. Winemakers using stem return must assure that stems have adequate "maturity" or lignification. In years with excessive fruit rot, grapes are dejuiced soon after the beginning of fermentation. About 20% stem return can help to make up the tannin deficiency resulting from early pressing.

Tank Size and Shape

Fermenter tank size frequently depends on vineyard plot size, speed of picking and filling, cooling capacity, cap management system, and pressing capacity. Fermentation and phenol extraction vary with tank size, height-to-diameter ratio, and cap management technique. Small tanks (2000 L or 529 gallons) can waste heat necessary for extraction, whereas large tanks cannot easily be managed by hand. High-profile tanks are considered to be less suitable for varieties with limited color. Most U.S. producers use stainless steel red fermenters with relatively low height-to-diameter ratios (1.0–1.3), which produces a cap relatively easy to manage using gentle management schemes.

In the United States, many winemakers use "closed" tanks because of the added flexibility for post-fermentation maceration, white wine production, and volume and floor space considerations. Some vintners believe that closed systems increase wine complexity by reducing the loss of volatile components. Alcohol loss due to entrainment with CO_2 can be as much as 0.5% (vol/vol). Another consideration is cap management which may be easier with opened vessels. The perceived aeration gained in open top fermenters may be preferable for the development of complexity, and enhanced phenol polymerization. Tank selection is also important in cold soaking (prefermentation skin contact).

Temperature

Temperature is an important factor affecting the extraction of phenolic compounds. Winemakers desiring to produce a fresh, fruity, aromatic wine generally use low fermentation temperatures (20°–25°C/68°–77°F) which favor retention of fermentation aromatics. Higher temperatures (30°C/86°F) increase the anthocyanin, tannin and total phenol extraction. Higher temperatures favor extraction but may affect the ability of the fermentation to complete. Ribereau-Gayon and Glories (1986) suggested a technique for high-temperature final maceration at the end of fermentation, where the wine and pomace are heated to 40°C (104°F) for 24 to 48 hours. With certain varieties enhanced polymerization occurs and suppleness is improved.

Short Vating

This processing technique calls for dejuicing wines early in the course of fermentation (16–18°Balling) and, occasionally, followed by barrel fermentation. To aid in early color development, pectolytic enzymes are frequently added at crush. Elevated fermentation temperature prior to dejuicing also facilitate color extraction.

The binding of pigments with tannins to form stable color complexes likely involves both oxidative and nonoxidative mechanisms. It is believed that acetaldehyde produced by coupled oxidation of ethanol (see Fig. 7–6) forms the bridge between tannins and pigments. Sulfur dioxide can inhibit or retard the formation of tannin-pigment complexes by binding free acetaldehyde (also see Chapter 10).

Cold Soaking

Prefermentation maceration (cold soaking) is a production tool used to increase complexity, color and color-stability in red wines. In the absence of alcohol, bonds are formed between anthocyanins and other phenols which are believed to help stabilize color in the resulting wine. Cold soaking is an aerobic process (frequently involving pump mixing) which en-

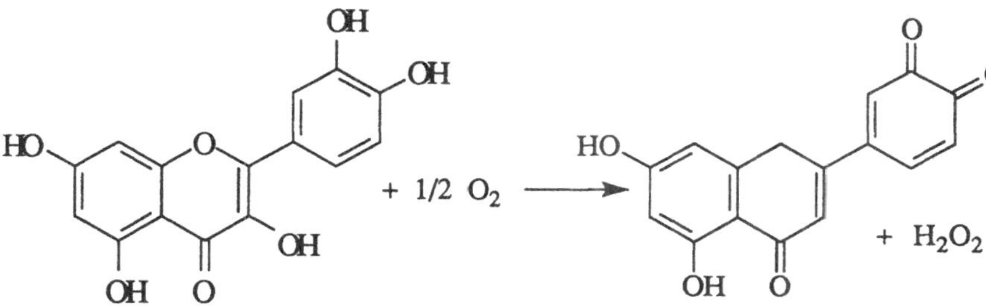

Step I

$$H_2O_2 + C_2H_5OH \longrightarrow CH_3CHO + H_2O$$

Step II

Fig. 7–6. Coupled oxidation of dihydroxyphenol yielding the corresponding quinone and hydrogen peroxide. Step II: oxidation of ethanol yielding acetaldehyde.

hances phenol polymerization and possibly increases suppleness. Relatively low temperatures (hence the term, cold soak) are required to prevent spontaneous fermentation. In a study of two Cabernet Sauvignon clones, Zoecklein (1994) demonstrated cold soaking for 48 hours lowers hues (A_{420}/A_{520}nm) while increasing color intensity ($A_{420} + A_{520}$nm) and total anthocyanins.

Cap Management

Pomace contact influences color, body, flavor, astringency, and the evolution and life of red wines. Cap management systems include pump over, punch down by hand or mechanically, sprinklers, rototanks or other specialty tanks, submerged cap, and thermovinification.

The ratio of skin to liquid is an important quality parameter. A high percentage of solid phase (as occurs with small berry varieties) increases the color and phenol content of the resultant wine. Dejuicing approximately 10% of the juice prior to fermentation is a well established method of increasing the overall character of some red wines by influencing the ratio of solid to liquid.

Cap management from fermenter to press has evolved to using the most gentle methods possible. Harsh treatment breaks the skins and increases the nonsoluble solids level (and phenols), which can detrimentally affect wine quality. Fire hose type spraying of the cap has given way to sprinkler and splash plates with many producers simply punching down. Many believe that gently punching the cap twice daily achieves a complete extraction of flavors without adding excess bitterness. Mechanical cap punches are not uncommon in the U.S. industry.

Pumping over or punching down is frequently done before the start of fermentation (cold soaking) to aerate the must. However, too much pumping before fermentation can be detrimental due to overoxidation and overstimulation of the subsequent fermentation. In the United States, winemakers are experimenting with using much smaller inocula and native yeast, which may allow for longer skin contact time, even at elevated temperatures (see Chapter 18).

A small percentage of U.S. producers use specialty tanks. Rototanks can provide more character, flavor, and 20 to 30% greater phenol extraction due to a thinner cap.

Post-Fermentation Maceration Period

Maceration concerns not only the extraction of phenols but solubilized polysaccharides, proteins, and peptides, which originate in the grape and yeast cell walls. Maximum color extraction is reached about halfway through the fermentation for most varieties (Berg and Akayoski 1956b). Anthocyanins are rather easily extracted, whereas other phenols such as tannins are extracted more slowly. This is a principal reason why dejuicing prior to dryness, particularly with mature fruit, produces wines with good initial color, relatively low astringency, low total phenols, and that are frequently floral, and light in body, complexity, and depth. Fruitiness is generally inversely proportional to the phenol level.

Extended cuvaison affects the evolution of tannins, and creates more body, complexity, palate depth, and enhanced color stability. In contrast to anthocyanins, tannin phenols are extracted throughout the skin contact period. This extraction has a significant effect on astringency, color, and particularly, color stability (Scudamore-Smith et al. 1990). Tannins derived from extended skin contact appear to stabilize anthocyanins by forming larger polymeric complexes with pigments. The presence of color in a wine 9 to 15 months old is directly proportional to the skin contact time, that is, the amount of tannin phenols extracted (Berg and Akayoshi 1956b; Sterns 1987).

As maceration time increases, so does the extraction and polymerization of phenols to form higher molecular weight tannins. This is reflected in the sensory analysis: reduced, yet stable color, and more, yet "softer" tannins. The increased extraction and polymerization of phenols during extended maceration produces wines that are "round" and "firm" in the palate often with considerable aging potential.

Cellarmasters in Bordeaux sometimes continue pump overs after the end of the alcoholic fermentation, but often close the tank and wait for the cap to "fall" into the wine. Extended maceration occurs in completely filled vessels or those rigorously blanketed with displacement gases. The length of post-fermentation skin contact is determined by a variety of factors, including fruit maturity (perceived presence of "ripe" tannins), source of fruit, history of wines produced from that source, stylistic goals, and taste.

Post-fermentation contact is generally monitored by taste. Samples from the racking and bottom valve are compared. Red wines are dejuiced after the tannins start to soften, and are said to be "well behaved," less "green," and less "raw." Maceration affects the evolution of tannins and for California Cabernets it can last for 6 weeks. Scudamore-Smith et al. (1990) found that although color densities of extended pomace treatments were higher than those of conventional fermentations after fermentation, by 13 months they were similar. Grapes from some vineyards benefit from extended maceration, but not others. This may represent qualitative and quantitative differences in grape phenols.

Controlled Aeration and the Use of Sulfur Dioxide

In addition to management of phenols in the fermenter, winemakers use controlled aeration of red wines as a production tool to help evolve and soften tannins. Aeration is thought to increase the reaction rate of tannin-anthocyanate components (-anthocyanins and -anthocyanidins), resulting in condensation and polymerization, which allow wines to mature as quickly as possible. Judgments are based on style, phenol content, and pH, with air exposure accomplished by aeration during tank and barrel to barrel rackings. Aeration is most effective in the months after fermentation when tannins are not completely polymerized. At this stage the condensation of tannins with anthocyanins in the presence of air will help stabilize pigments. Aeration occurring too late in the development of a wine can cause precipitation (Dournel 1985).

Sulfur dioxide is frequently not added until after aeration and after the malolactic fermentation is complete. Free sulfur dioxide levels are generally kept below 20 mg/L to help facilitate development of anthocyanin-tannin complexes (see Chapter 10).

FACTORS CONTRIBUTING TO WINE COLOR AND COLOR STABILITY

White musts often contain traces of chlorophyll, carotene, and xanthophyll (Amerine et al. 1972). Occasionally, certain varieties of white grapes, when grown in cool climates, retain some of their chlorophyll derivatives (Singleton and Esau 1969). Thus white wines may have a trace of green coloration in addition to the normal range of almost colorless to amber. White wine color considerations are primarily preventative, as the emphasis is on preventing or delaying browning.

Red color extraction into the juice and wine depends on the variety, region, season, maturity, physical condition of the fruit (berry softness, presence of mold, raisining, and processing). Grape varieties such as Rubired, Royalty, and Salvador have their principal anthocyanins present in the diglucosidic form, which enhances extractability. In addition, these three varieties are teinturiers with their color pigments present in the easily accessible pulp.

Grapes degraded by molds such as *Botrytis, Penicillium,* and *Aspergillus* are rich in oxidative enzymes, which unless inactivated or removed, may catalyze oxidation of phenolic substrates. The result is to produce brown or tawniness in reds and a yellow to amber shift in white wines.

Sulfur dioxide can significantly affect the development of color in wines.

The addition of SO_2 inhibits polyphenoloxidase activity. With mold-damaged fruit, sulfites may play a limited role in the inactivation of microbially produced oxidative enzymes (e.g., laccase; see Chapter 3). It is common practice, however, to add no or little sulfur dioxide prior to fermentation of sound grapes to encourage phenolic polymerization and precipitation. The reduction in phenols can be substantial enough to reduce bitterness and background astringency in white wines. Anthocyanins are generally extracted within the first 4 to 5 days of fermentation depending on the variety, fruit maturity, and temperature (Feuillat 1987, Powers et al. 1980; Sterns 1987). As fermentation proceeds there is a decline in anthocyanins and color intensity and an increase in total phenolics extracted, as depicted in Fig. 7–7 (Ribereau-Gayon and Glories, 1986). The reason for the decline in color following the color peak during fermentation is unclear but several theories have been suggested.

Somers and Evans (1979) demonstrated that, as alcohol levels increase, color at both 420 and 520 nm decreases. They suggested that destruction of colored co-polymers of anthocyanins present in the juice occurred as a result of increases in ethanol. The strong reducing environment of fermentation may be responsible for the decline in converting free anthocyanins to colorless flavenes (Ribereau-Gayon 1974). The fixation of free anthocyanins onto grape solids and yeast may also result in loss of color. Bakker et al. (1986) demonstrated that polymerization of anthocyanins with tannins is already occurring during fermentation. This polymerization may produce small noncolored polymers that may or may not be later oxidized back to colored forms.

Factors affecting the rate of color change include phenol composition and concentration, levels of oxygen and sulfur dioxide, temperature of fermentation and storage, concentration of metals, and pH. Following fermentation, decreases in red color may be up to 33% (Van Buren et al. 1974). Processing operations such as ion exchange, filtration, centrifugation, cold stabilization, and fining may significantly reduce red wine color. During malolactic fermentation, additional decreases in red wine color are often noted, resulting from the upward shift in pH.

Tannins appear to stabilize anthocyanins by combining to form larger polymeric pigments. The presence of anthocyanins increases the solubility of phenol-tannin complexes in red wines (Singleton and Trousdale 1992). These authors hypothesized that the incorporation of the flavylium salt and its attached sugars into the polymer increases the solubility of a given condensed tannin molecule. Low tannin levels may be responsible for poor color stability in Pinot Noir (Leglise 1980).

The ratio of anthocyanins to tannins influences both red color and color stability. For example, blush wines that go quickly from pleasing pink to

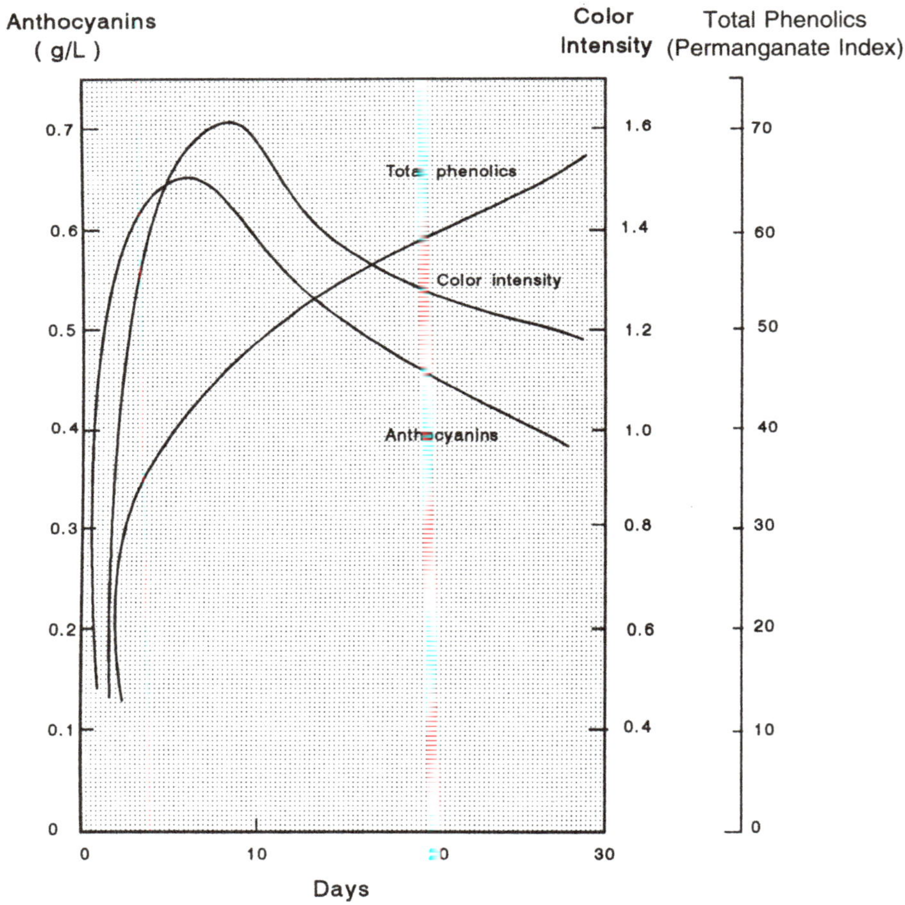

Fig. 7–7. Changes in color intensity during fermentation as a function of anthocyanin and total phenolics concentration. Adapted from Ribereau-Gayon and Glories (1987).

orange-brown may have too low of a ratio of anthocyanins to tannin. Such a ratio allows a large percentage of the tannins to bind with each other, forming brown hues (increased optical density in the yellow to brown range at 420 nm), rather than reacting with anthocyanins.

The shift from the pink of a blush wine to yellow or tawny does not usually occur quickly but results in the loss of desirable color. The ideal ratio of anthocyanin to tannin for color stability is around 1:10 (Peynaud 1984). If the free anthocyanins are in greater supply than tannins (greater than 1:10), the hue or dominant wavelength can change from red to yellow.

During red wine aging, phenolic compounds undergo a number of transformations that depend on the temperature, sulfur dioxide concentration, degree of oxidation, time, and the anthocyanate to tannin ratio. Phenolic reactions occurring in red wines have been summarized by Ribereau-Gayon and Glories (1986) and Recht (1993) and are depicted in Fig. 7–8.

With oxidation, procyanidins form tannins (T), which have a yellow hue, a molecular weight of 1,000 to 2,000, and maximum astringency. Procyanidins may also undergo oxidative or nonoxidative polymerization reactions forming condensed tannins (TC). A greater degree of polymerization results in the formation of very condensed tannins (TtC) whose astringency is diminished.

Condensation reactions can occur between procyanidins and other molecules. Polysaccharides and peptides are known to form tannin-polysaccharides (TP), which reduces astringency, thus helping to enhance wine suppleness. Various other condensation reactions occur between anthocy-

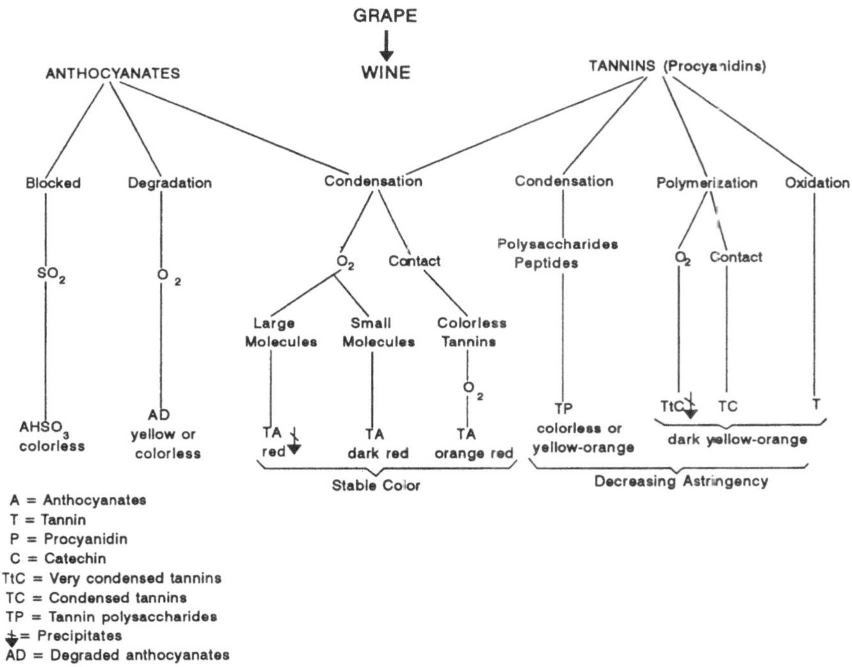

Fig. 7–8. Phenolic reactions occuring in red wines. Adapted from Rabereau-Gayon and Glories (1987) by Recht (1993).

anates and tannins forming tannin-anthocyanate complexes (TA). These condensation products lead either to intensification toward purple or tile-red hues.

Alternative Techniques for Color Extraction

Pectolytic Enzymes

Winemakers often wish to produce red wines with desirable color but reduced astringency. This generally requires short vating (shortening the time period for skin contact). Pectolytic enzyme additions with short vating facilitates increased color and yet low tannin extraction. Yield increases of from 3 to approximately 9% have been reported with the use of enzymes (see Chapter 16.)

Thermal Vinification

Various processes employing heat, collectively termed thermal vinification, have been used to maximize color extraction in red wine production. Composite mixtures of skins and juice may be heated together, or juice alone may be heated and then mixed with crushed grapes. After pressing, the juice is cooled and usually vinified in the manner of standard white wines. The heating time and temperature vary with region and producer. Joslyn and Goldstein (1964) reported that heating at 74°C (165°F) for a period of only 1 minute was sufficient for color extraction in Carignane and possibly most red varieties.

Most thermally vinified wines initially are deeper in color than their traditionally fermented counterparts, although clarification may be slower because of increased phenolic extraction. Thermal vinification may have an oxidizing or accelerated aging effect on red wine resulting from condensation and precipitation of phenolic constituents. Thus the harshness or astringency associated with young red wines is overcome.

OXIDATION

Enzymatic Oxidation of Musts

The extent and rate of color change (deterioration) depends on many parameters, including pH, the amount and type of phenolic substrate present, the temperature, and the amount of dissolved oxygen. The major viticultural parameters affecting browning in wines include variety, growing region, maturity, and condition of fruit at harvest. During processing, contact with air and the presence of metal ions, most notably copper and iron, may be principal factors accelerating browning rates in wine.

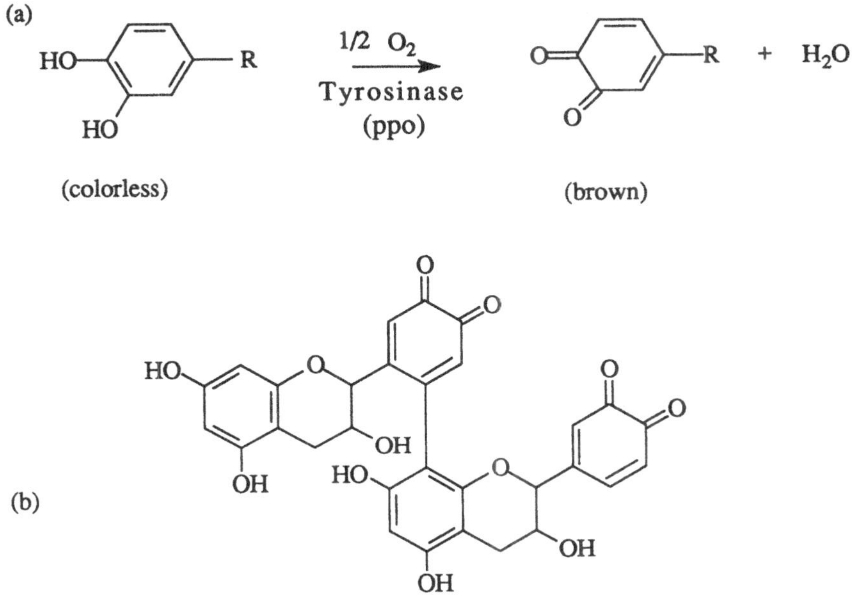

Fig. 7–9. Enzymatic oxidation of nonflavonoid species.

Enzymatic oxidation occurs primarily in freshly crushed fruit. The group of enzymes responsible for catalyzing oxidative reactions are the polyphenol oxidases, also referred to as phenolases or tyrosinases. Polyphenoloxidases catalyze oxidation of dihydroxyphenols to their corresponding quinones (Fig. 7–9a). The substrates for this reaction are the nonflavonoids (hydroxycinnamates and their derivatives).

The activity of this group of enzymes is greatest in the skin; their concentration in grape juice depends on grape variety and condition at harvest as well as the period of skin contact. Because of its general activity in binding proteins, bentonite addition prior to fermentation may reduce levels of the oxidative enzymes as well as enhance clarification (see Chapter 16).

Oxidized quinones may form polymers of the type presented in Fig. 7–9b. Bond formation occurs between carbon 6 of the first and either carbon 6 or 8 of the second quinone. As condensation continues, color intensifies from an initial yellow to brown (Hathaway and Seekins 1957).

Oxidation of Wines

In white wines, as compared with juice, browning is due largely to oxidation of phenolic compounds such as catechins and leucoanthocyanidins and generally proceeds by chemical means. The following three chemical mechanisms have been proposed to account for wine oxidation.

Carmelization

Upon heating hexose sugars in acid solution, rapid decomposition (dehydration) takes place. The reaction does not require proteins or amino acids (see Maillard reaction), and may occur in the absence of oxygen.

Carmelization occurs during production of baked sherry. Some producers of sherry in the United States heat shermat (with approximately 2% sugar) to 120°F to 140°F (49–60°C) to accelerate the process. As a byproduct of the reaction, a dark carmel color is produced. The same reaction takes place during excessive pasteurization of sweet wines. Similarly, browning overtones derived from carmelization may be present in red and white wines vinified from raisined berries.

Maillard Reaction

This reaction (in foods) involves condensation of sugars with amino acids and proteins. Active carbonyl groups, as would be present in reducing sugars, aldehydes, and ketones, are necessary for the condensation reaction to occur. Thus, blockage of these active groups, due to prior reaction with SO_2, renders the compound unreactive and serves to block the browning reaction sequence.

There has been little supportive evidence for the occurrence of the Maillard reaction in white wine browning (Caputi and Peterson 1966; De Villiers 1961). Nevertheless, because the necessary reactants are present in wine (amino acids, proteins, and reducing sugars), the possibility of its occurrence should not be overlooked.

Direct Phenolic Oxidation

The major nonenzymatic reaction causing browning in wine involves reaction of susceptible phenolic derivatives with molecular oxygen (see Chapter 14). Metals may act as catalysts in this oxidation (see Chapter 12).

Secondary Browning Reactions

Several nonenzymatic secondary browning reactions may accompany initial quinone formation, yielding yet darker products. These include coupled oxidations, condensation-polymerization reactions, and complexation of amino groups. Oxidation of susceptible phenolics to corresponding quinones is usually accompanied by darkened color. Concomitant with this oxidation reaction, hydrogen peroxide is formed (Wildenradt and Singleton 1974), which subsequently oxidizes ethanol to acetaldehyde (Fig. 7–6).

The acetaldehyde formed combines with tannins and red wine pigments to increase red wine color stability.

Phenol Instability

The phenol constituents of new wines are quite unstable. They are susceptible to degradation, condensation, and oxidative changes, although instability, the formation of haze or deposits caused solely by phenols, is rarely a problem in finished wines (Somers 1987). Catechins and procyanidins are likely to be present at low levels in white free run juice and are the principal substrates for oxidation. These components are responsible for the most common form of phenolic instability in white wine, oxidative browning (See Chapter 20 for a testing procedure). This type of oxidative instability is distinctly different than that seen in new red wines, which are supersaturated with phenolics and tartrates. These latter wines can be rapidly stabilized by refrigeration (see Fig. 15–2).

Pinking in White Wines

Development of a red blush in white wines, a reaction called pinking, is occasionally observed. Where there is an overall reduction in oxygen exposure, pinking is reported to be the result of rapid conversion of flavenes to the corresponding red flavylidium salts. With reference to Fig. 7–10, one can see that flavenes are formed in an acidic medium by slow dehydration of corresponding leucoanthocyanins.

In the presence of oxygen, flavenes and leucoanthocyanins are converted to brown pigments. Under reducing conditions, however, accumulations of flavenes may occur. Subsequent rapid exposure of wine to air, such as may occur during transfer, filtration or bottling, converts flavenes to their red flavidium salts, which confer a pink blush to the wine (see Chapter 16).

Flavonol Haze

Somers (1987) reported the occurrence of a flavonol haze in bottled white wine. The yellow deposits consisted principally of quercetin believed to have been extracted from leaves during fermentation.

Ellagic Acid Deposits

Hydrolysis of ellagitannins extracted from oak can yield ellagic acid. Ellagic acid can appear as a crystalline deposit in wines following bottling. This form of phenolic instability may be considered a normal occurrence during storage in oak (Somers, 1987). Actual bottle instability does not usu-

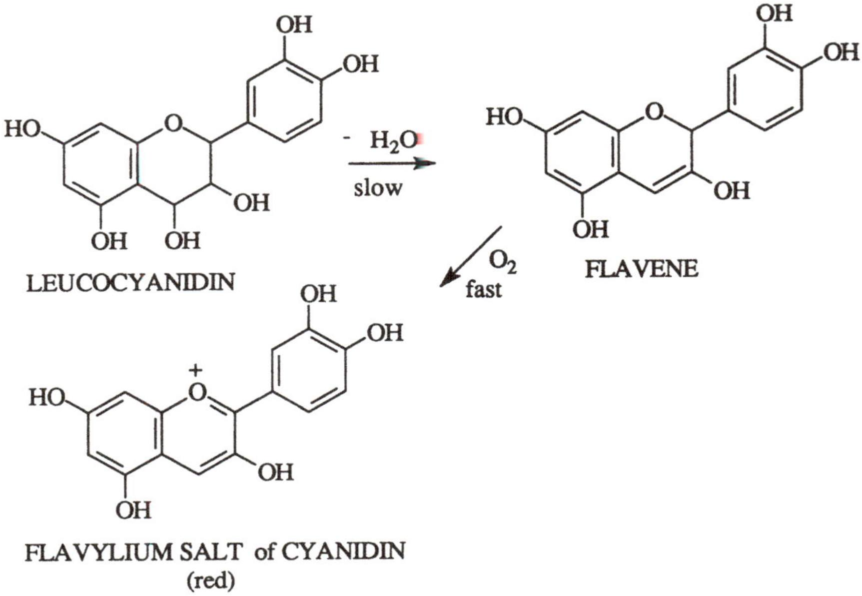

Fig. 7–10. Conversion of leucocyanadin to flavylium salts (colored).

ally occur due to the usual ample time for stability during bulk aging. Contact of wine with wood chips just before bottling should be avoided.

OAK BARREL COMPONENTS

Modern winemaking frequently involves fermentation and/or storage in oak, which involves the interaction of air, wine, and wood. During oak storage controlled oxidation results in decreased astringency and increased red wine color stability due to the combination between anthocyanin and tannin molecules (Pontallier 1987). Color stability is promoted by the presence of acetaldehyde produced by ethanol oxidation. This is facilitated by an oxidation catalyst such as gallic acid present in large quantities in oak wood. Additionally, wines extract volatile flavor components from the wood to support the aromatic panoply of the wine.

There are some physical and chemical variations between American white oak (*Quercus alba*) and the European species *Q. robur* and *Q. sessils*. Variations in wood composition also occur due to variations in climate, environment, and barrel processing. The sensory features of wines stored in oak are dependent on factors influencing wood composition, including

Table 7–5. Major constituents in oak.

Component	Amount (%)	Molecular type
Cellulose	45–50	Glucose polymer
Hemicellulose	20–25	Sugars
Lignin	25–35	Phenols
Tannin	5–10	Glucose and gallic acid
Minor constituents	0.1–0.5	Terpenes, lipids, etc.

Crum, 1993.

origin, barrel age and size, method of wood drying or seasoning, grain tightness, geographic location of seasoning, cooperage techniques, stave width and degree of toasting. Fermentation and/or storage duration also influence the sensory attributes of wines. Crum (1993) reported that cellulose undergoes little chemical alteration during wood processing (see Table 7–5). Hemicellulose is composed mainly of 5 carbon sugars such as xylose and arabinose with a limited amount of the 6 carbon sugars mannose and galactose. Products of hemicellulose brought about by heat treatment (e.g., toasting) include furfural, maltol, cyclotene, and ethoxylactone.

Oak lignin is a complex polymer mainly of coniferyl and syringyl alcohols that can generate a variety of volatile phenols based on the guaiacyl and syringyl nucleus. The most abundant of these are vanillin, syringaldehyde, coniferaldehyde, and sinapaldehyde (Sefton et al. 1990). The importance of these to wine flavor is unknown (Dubois and Dekimpe 1982). However, from initial seasoning to changes during stave bending and particularly during toasting, lignins undergo significant change.

Over 200 volatile components of oak have been identified (Boidron et al., 1988; Maga 1985; Masuda et al. 1971). These include furans, which contribute to toast-like properties; aldehyde phenols, like vanilla; volatile phenols, contributing spicy and barbecue. Among the volatile components from sources other than ligins are two isomeric lactones, the so-called oak or whisky-lactones, which have coconut-like aromas (Masuda and Nishimura 1971). It has been suggested that American oak contains higher levels of the oak lactones than French oak (Guymon and Crowell 1972).

Oak tannins consist of gallic acid esters of glucose and esters of dimers such as ellagic acid and trimers. Due to the rapid hydrolysis of galloyl esters in wine they are likely not of significance to wine flavor (Crum 1993). Complex oak tannins that degrade slowly significantly impact wine flavor. Levels of these tannins are decreased by both seasoning and heating.

Other components of oak wood include terpenes and lipids. Norisoprenoids are C_9, C_{11}, and C_{13} compounds that result from the breakdown of plant carotenoids (Wahlberg and Enzel 1987). In contrast to lignin-derived compounds that show little variation between oak type, the norisoprenoid compounds of American oak are very different from those of European oaks.

Processing Considerations

During stave heating, or toasting, wood is subject to relatively high temperatures (200°C), giving rise to chemical changes. Some volatile compounds such as methyloctalactone initially present in the wood and ellagitannins disappear quickly. On the whole, the 'woody' aromas are lessened as the degree of toasting increases. Some substances develop as a result of heat deterioration of wood. These include furaldehydes, phenol aldehydes, and volatile phenols. Francis et al. (1992a) found that heat was the most important barrel production parameter regarding influence on the sensory characteristics of oak wood extracts.

The type of oak affects the phenol content of barrel aged wines. The main phenols of both European and American oak are ellagitannins, the total content being higher in European oak (Quinn and Singleton 1985). Singleton et al. (1971) reported that new wines aged for 1 year in 200-L (53 gallon) French oak barrels contribute 250 mg/L phenols (GAE). By comparison, wines stored in American white oak contributed approximately one-half that level. Francis et al. (1992a) found that the differences among French woods were not as great as those between French or American wood with the later perceived to have less intense aroma properties.

American oak has gained popularity in the United States for white wines due, in part, to economics. Both American and French wood contribute tannin and aroma. American oak has a more aggressive and immediate taste and aroma and contains more vanillin, contributing to vanilla aroma. French oak contains more tannins yet has less obvious ''oaky'' flavors and aromas.

The ''woody'' character is generally less intense when the alcoholic fermentation takes place in oak. This phenomenon is linked to biochemical transformations of some aromatic components caused by yeast and MLF when present. Some components are altered and become less aromatic and some are absorbed into yeast cells (Chatonnet 1993). Yeast can also act directly on oak components. One example is the conversion of ferulic acid to vinyl-4-guaiacol which results in development of ''clove-like'' properties. An increasing number of wineries are experimenting with short-vating their red wines (at > 18° Balling) and continuing the fermentation in oak.

It is generally believed that the woody character is less pronounced and better integrated if newly fermented wine remains in contact with the lees for a certain period of time. Yeasts that are put back in solution by stirring are capable of fixing some of the volatile substances released from oak barrels. Additionally, enzymatic activity that starts during alcoholic fermentation reduces aromatic potential of vanillin and furanic aldehyde during lees contact (Chatonnet 1993). Almost one-third of the tannins released by the wood are adsorbed on to the yeast cell walls and half combine with the colloids in the wine, resulting in a natural fining (Chatonnet, 1993) (see Chapters 8 and 16). Lindbloom (1993) proposed a hedonic scale for the evaluation of oak aromas, and flavor. Descriptors applied to oak include coconut, vanilla, roasted, resinous/varnish, sweet, toasty, smokey, and spicy.

Phenolic Taint

Wine stored in barrels, particularly used barrels, may develop elevated levels of ethyl phenols (Chatonnet 1993). These compounds provide "phenolic odors" reminiscent of barnyard and sweaty saddles. Chatonnet (1993) suggested that *Brettanomyces intermedius* is the primary source of phenolic taint in aged wines (see Chapters 18 and 20 for determination of the presence of *Brettanomyces*).

Resveratrol

Seigneur et al. (1990) determined that consumption of red Bordeaux wines, but not whites, nor alcohol, induced platelet hypoaggregation and increased high-density lipoprotein (HDL)-cholesterol. Both of these features are believed to have cardioprotective value and are likely postulated to play a role in the so-called French Paradox. Resveratrol (3,5,4-trihydroxylstilbene) has been reported to induce platelet aggregation and lower lipids (Arichi et al. 1982; Kimura et al. 1985). Resveratrol has been found in leaves and grape skins (Creasy and Coffee 1988) and in grape wines (Siemann and Creasy 1992). Resveratrol in wine could be related to the therapeutic effects similar to those studies involving resveratrol in laboratory animals.

Siemann and Creasy (1992) reported ranges of resveratrol in red wine from 2.86 to less than 0.03 μmol/L and from 0.438 to less than 0.001 μmol/L for whites. They also found much higher levels of resveratrol in New York compared to California Chardonnays and Bordeaux reds compared to California Cabernet Sauvignon. However this may be explained by geography, variety, or winemaking differences. As a phytoalexin, more would be expected to be produced in an area of higher fungal disease pressure. Resveratrol is found in the grape skin, not the flesh, and therefore processing methods may influence wine concentration. In addition to skin contact, winemaking practices such as PVPP addition, for example, would bind phenolics and lower resveratrol concentrations. Following veraison and maturation, resveratrol concentration declines substantially (Jeandet et al. 1991). If the presence of this grape phenol in wine provides health benefits, increasing the concentration in the plant may be possible. The enzyme that produces resveratrol, stilbene synthase, has been identified in grapevines. It may be possible to alter enzyme activity to exercise control of resveratrol production by the vine (see Chapter 1).

EVALUATION OF COLOR BY SPECTROPHOTOMETRY

White wines are characterized by transmission over a broad range of wavelengths (Fig. 7–11). The absence of any strong absorbance bands in the visible spectral region accounts for their absence of color. With older or oxidized wines, lowered transmittance (more absorbance) in the 400 to 600 nm range produces some greenish color, which when mixed with the yellow yields a characteristic brown tone. Upon excessive oxidation the absorption maximum continues to shift toward the ultraviolet range with an increased transmittance in the area of the spectrum where brown overtones are seen.

Most young red wines have a transmittance minimum (absorbance maximum) at 520 nm and a transmittance maximum (absorbance minimum) at 420 nm (Fig. 7–12). These spectral characteristics are also common to blush style and port wines. As wine matures, there is a shift in absorption maximum to between 400 and 500 nm, usually near 450 nm. With excessive oxidation of red wines the spectral characteristics approach those of a sherry and the polymerized red pigments may precipitate.

The presence of metal ions or complexes can degrade color depending on the pH and the presence of complexing (chelating) agents such as citric acid (Jurd and Ansen 1966). Aluminum, copper, and iron form complexes with anthocyanins and other phenolics resulting in modification of red color by a shift toward the ultraviolet region. This shift is perceived by the eye as an increase in blue. It has been reported that anthocyanin-metal complexes may be responsible for the blue coloration of certain native American grape varieties (Jurd and Ansen 1966).

Humans perceive color as a characteristic of the wavelengths and intensity of light being reflected off the surface or being transmitted through an object. This occurs within the wavelength range of 380 to 770 nm of the electromagnetic spectrum, the part that is visible to humans. Color may be completely defined in terms of three fundamental attributes: (1) the dominant wavelength (hue), such as red, yellow, green, or blue; (2) brightness (luminescence), which defines the amount of gray in the color ranging

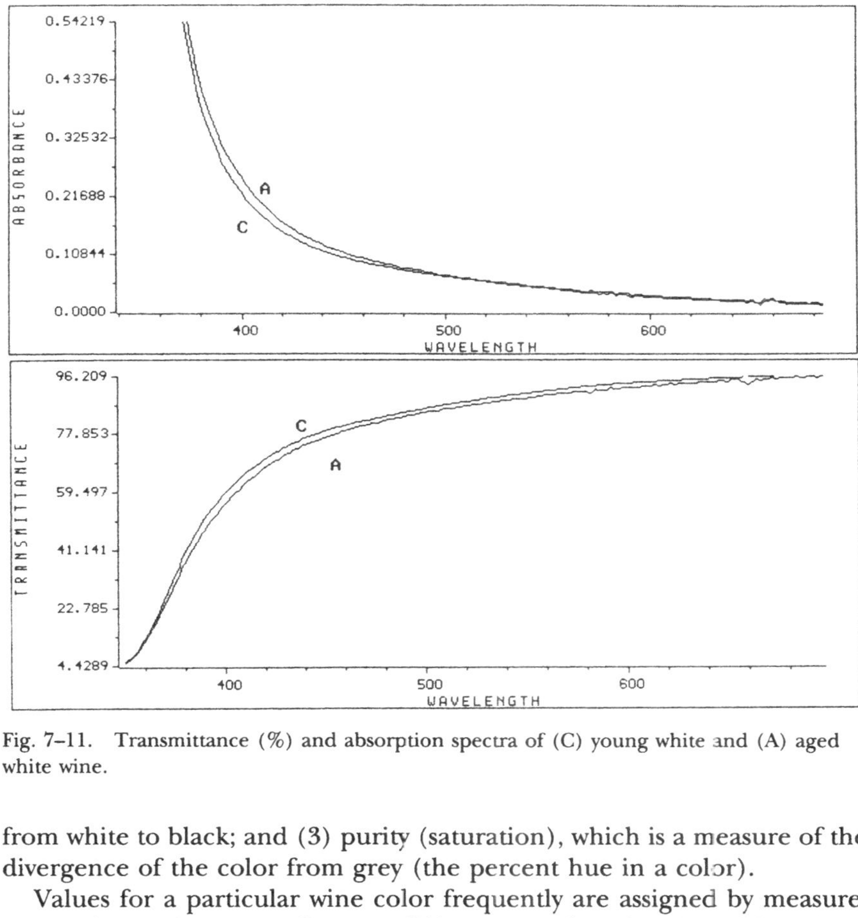

Fig. 7–11. Transmittance (%) and absorption spectra of (C) young white and (A) aged white wine.

from white to black; and (3) purity (saturation), which is a measure of the divergence of the color from grey (the percent hue in a color).

Values for a particular wine color frequently are assigned by measurement of absorbance readings at different wavelengths. For white wines, absorbance is usually determined at 420 nm. In the range of 400 to 440 nm, increases in brown coloration of white wines are detected readily. For red wines, data are collected by dilution, and absorbance measurements at 420 and 520 nm (Sudraud 1958). Wines are diluted with deionized/distilled water adjusted to the pH of the original wine. The extent of dilution is a function of the original intensity of wine color, and may range from 1+3 to 1+9. All particulate matter must be removed from the sample prior to analysis by filtration through a 0.45 to 1.2 μm membrane filter. A measure of color intensity or density can be achieved by summation of absorbance readings at 420 and 520 nm. By comparison, hue or tint is measured as the ratio of absorbance at 420/520. Spectrophotometric methods for

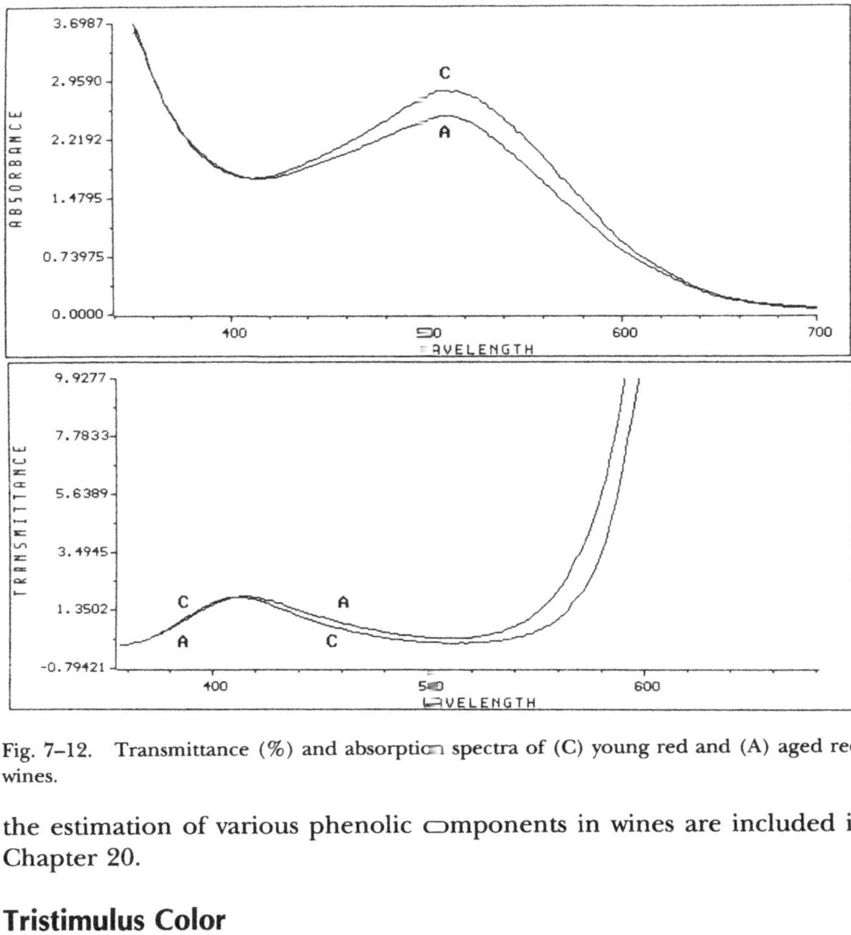

Fig. 7–12. Transmittance (%) and absorption spectra of (C) young red and (A) aged red wines.

the estimation of various phenolic components in wines are included in Chapter 20.

Tristimulus Color

The standard system for color measurement, against which all other systems are compared, is that of the CIE (Commission Internationale de L'Eclairage). This system specifies color in terms of three quantities, *X*, *Y*, and *Z*, called tristimulus values. These values represent the amounts of the three primary colors, red, green, and violet, required to match the color of the object under consideration. The system further defines chromaticity coordinates, *x*, *y*, and *z*, which are determined by dividing the appropriate tristimulus value by the sum of the three values [for example $x = X/(X+Y+Z)$]. Chromaticity coordinates give the proportion of the total color stimulus attributed to each primary color.

To specify the color of a wine using this system, the percent transmittance of the sample is measured over three series of wavelengths (516–656 nm for red, 507–640 nm for green, and 424–508 nm for violet). Ten %T (% transmittance) values in each series are summed and multiplied by a normalizing factor to obtain the tristimulus values of *X*, *Y*, and *Z*. The chromaticity values are then calculated as specified above and indicate the respective fractions of red, green, and violet in the sample. One major advantage of this system is that it specifies color in terms of what one sees, the *transmitted* electromagnetic radiation. The main disadvantage is that one is required to make 30 transmittance measurements in order to calculate the chromaticity values.

Spectral Estimations of Red Juice and Wine Phenols

The various phenolic constituents in grapes are known to absorb radiation in the UV and visible region of the electromagnetic spectrum. Absorbance of a red juice or wine at 280 nm is used as an index of total phenolics. The concept of (A_{280}-4) as a phenolic index for red *Vitis vinifera* wines (expressed as adsorption units, a.u.), is based on adsorption maxima showing mean values (with a 10-mm path length) for nonphenolics at 4.0 (Somers and Ziemelis 1985). This adsorption maximum is primarily due to the influence of flavonoids. Somers et al. (1983) surveyed over 400 young red Australian wines and found a range of 23 to 100 a.u. with a mean of 50. Young wines with (A_{280}-4) values less than 30 a.u. are believed to have a low capacity for aging (Somers and Verette 1988). Robust red wines with sufficient phenolic concentration to permit aging generally have (A_{280}-4) values greater than 40 a.u. with young red premium wines in the range of 45 to 65 a.u. (Somers and Verette 1988). During the first year the normal decline in absorbance at 280 nm is 10 to 20%.

Bakker et al. (1986) correlated (A_{280}-4) with the analysis of total phenolics measured by the Folin Ciocalteau method (see Chapter 20). They also showed that the total free anthocyanin contents of red table and port wines measured by HPLC were much lower than those measured by the spectral method of Somers and Evans (1977). The difference between methods decreased with wine age.

One can determine the relative amounts of simple and polymeric pigment forms by measuring the absorbance of a wine at 520 nm, before and after addition of excess SO_2. Anthocyanins decolorize upon addition of SO_2, whereas polymeric pigment forms do not. Treatment of young red wines with acetaldehyde releases anthocyanins that are bound to SO_2. The resultant increase in absorbance is a measure of the amount of total anthocyanins previously bound (and decolorized) with SO_2.

By shifting the pH of a young red wine to values less than 1.0, anthocyanins are converted entirely to their highly colored flavylium form yielding a large increase in absorbance at 520 nm. Because polymeric pigments are much less affected by low pH, this technique permits one to estimate the amount of monomeric anthocyanins present.

Spectral Estimation of White Juice and Wine Phenols

In white juice and wines, broad absorbance maxima occur in the ranges of 265 to 285 nm and 315 to 325 nm. These absorbance bands are produced by flavonoids (280 nm maxima) and hydroxycinnamate esters (320 nm maxima); other nonphenolic compounds absorb around 265 nm, and sorbic acid (if present) absorbs at 255 nm. The latter interference can be eliminated by extraction with isooctane to remove the sorbic acid (See isooctane procedure for sorbic acid analysis, Chapter 20). Estimates of the presence of nonphenolic interferences can be made by treating juices or wines with polyvinyl pyrrolidone (PVPP) to remove phenolic compounds and remeasuring absorbances at 280 and 320 nm.

Somers and Ziemelis (1972) demonstrated that the majority of white wines gave corrected absorbance values equal to ($A_{280}-4$) and ($A_{320}-1.4$), as measures of total phenolic components and total hydroxycinnamates, respectively. For wines made from free run juice, the difference spectra obtained after PVPP treatment is similar to that of caffeic acid, the major phenolic moiety of hydroxycinnamate esters in *Vitis vinifera*. The total hydroxycinnamates quantified as caffeic acid equivalents (CAE mg/L) on the basis of the molar absorptivity at 320 nm are as follows:

$$\text{Total hydroxycinnamates} = A_{320} - 1.4$$

$$\text{Caffeic acid equivalents CAE} = 10\ (A_{320} - 1.4)\ \text{mg/L}$$

Somers and Ziemelis (1985) surveyed 230 commercial white wines and found that the A_{320} ranged from 2.9 to 13.1 a.u., corresponding to a CAE value of 15 to 117 mg/L.

For juices and young wines in which the hydroxycinnamic acids are present as esters of tartaric acid, the factor 7/4 enables conversion of CAE to estimates of total hydroxycinnamoyl tartaric acids (Somers and Verette 1988). For fresh grape juice and young white wines there are no known interferences to these spectral estimates, which compare well with HPLC measurements.

Approximate assessment of the flavonoid extractives in white and rose wines can be made as follows:

$$(A_{280}-4) - 2/3\ (A_{320}-1.4)$$

This formula corrects for the absorbance of nonphenolic compounds and hydroxycinnamates at 280 nm.

ANALYSIS

The following is a cross-reference of analytes and procedures found in Chapter 20. Wine color density (intensity) and hue (tint) by spectrometry; total phenols by visible and UV spectrometry; phenol estimation by polyphenol (permanganate) index and pH 7 test; phenol fractions by spectrometry; tannins by spectrometry and gelatin index; pigments by HPLC; white wine oxidation; presence of red hybrid wine by fluorescence; 4-ethyl phenol by GC.

Quick screening methods for the determination of the presence of phenolic compounds in wines are also given in Chapter 20.

CHAPTER 8

NITROGEN COMPOUNDS

The nitrogenous components of must and wine play important roles in fermentation, clarification, and potential microbial instability. They may affect the development of wine aroma, bouquet, and foam characteristics in sparkling wines. Additionally, some health-related metabolic byproducts of fermentation are, in part, influenced by the type and concentration of nitrogen compounds in the must. The various nitrogenous constituents of wine include proteins, polypeptides, peptides, amino acid, amides, and ammonia.

NITROGEN COMPOUNDS OF GRAPES AND WINES

Nitrogen is taken up by the plant in the form of nitrate (NO_3^-), ammonia (NH_4^+), or urea. Nitrate is reduced to ammonia, incorporated into the amino acids glutamate, and glutamine and subsequently converted to other amino acids needed for protein synthesis. The total nitrogen content of grape must ranges from 60 to 2,400 mg(N)/L.

Proteinaceous compounds may impede clarification and stability in

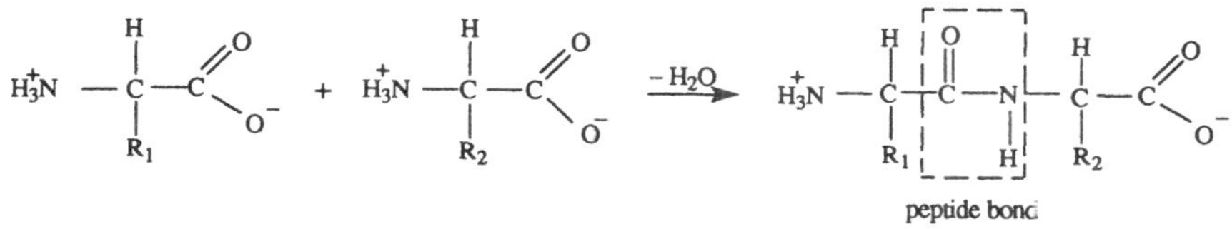

Fig. 8–1. Structure of typical amino acids showing peptide bond formation (leading to protein).

white table wine. The protein content of juice ranges from about 1 to 13% of the total nitrogen content (Correa et al. 1988; Moretti and Berg 1965). In wine, the levels are higher, approaching 38%.

Polypeptides exist as protein fragments composed of amino acids linked via peptide bonds (see Fig. 8–1). Polypeptides constitute a significant proportion of the total nitrogen content in wine. Ribereau-Gayon and Peynaud (1958) reported that polypeptides comprise between 60 and 90% of the total nitrogen, although Cordonnier (1966) estimated the value much lower, approaching 21%. Depending on processing techniques (such as thermal treatment of musts), the levels of polypeptides may be higher.

Ammonia (present as NH_4^+) serves as the primary form of available nitrogen for yeast metabolism. As grapes mature, ammonia decreases with increases in protein and peptide nitrogen (Peynaud and Maurie 1953). The concentration ranges from 24 to 209 mg/L in grapes and from a few mg/L to 50 in wine (Ough 1969).

Amino acids serve as building blocks for polypeptides and protein. Most of the 20 commonly occurring amino acids are found in must and wine. Although the amount of each varies with grape variety, cultivation, and processing techniques, proline generally is found in the highest concentration in must and wine. In must, arginine also is present in relatively high concentration.

Another group of nitrogen compounds present in wines are amines. Biogenic amines, such as histamine and tyramine, result from bacterially mediated decarboxylation of the corresponding amino acids. Species of *Lactobacillus* and *Leuconostoc* have been reported to produce histamine (Lafon-Lafourcade 1976). The amine content is dependent on winegrowing region, variety, and winemaking practices. Woller et al. (1981) reported that the histamine content of 459 wines ranged from 0 to 5 mg/L. In the majority (79%) levels ranged from 0 to 0.3 mg/L.

Histamine, tyramine, and *beta*-phenylethylamine have been tested on humans. Of these, only *beta*-phenylethylamine was linked to headache and nausea (Luthy and Schlatter 1983).

Nitrates (NO_3^-) and nitrites (NO_2^-) are present in wine at low levels, usually less than 0.3% of the total nitrogen. Amerine and Ough (1980) reported nitrate levels from less than 7 mg/L, in German wines. Italian white wines were lower, with an average of 1.65 mg/L. In some, nitrite levels were reported as low as 30 µg/L. Nitrates in table wine may not be of great importance, but concern is mounting about its role(s) in groundwater contamination resulting from discharge at the winery.

Nitrogen-containing flavor compounds are important in enology. For example, methyl anthranilate and o-aminoacetophenone are related to the "foxy" taste of *labrusca* grapes and related hybrids. Additionally, 2-methoxypyrazines are reported to be responsible for the vegetative, herbaceous aromas frequently noted in wines produced from Cabernet Sauvignon and Sauvignon blanc. Fermentation bouquet may be influenced by manipulation of nitrogen compounds (Rapp and Versini 1991).

EFFECT OF VINEYARD PRACTICES ON NITROGEN COMPOUNDS

Grape variety, rootstock, fertilization, harvest maturity, soil, climate, and fungal degradation may affect nitrogen content of the juice and finished wine. Amino acid content and concentration varies widely. Bely et al. (1991) found the free amino nitrogen concentration from 90 musts, representing 25 different varieties from four French viticultural regions, ranged from 28 to 336 mg N/L (avg. 120 mg N/L). Levels in berries increase during ripening. Glutamine is initially the main amino acid component of the juice, averaging 36% of the total (Sponholz 1991). The percentage of glutamic acid in juice decreases during maturation while proline increases. The content of arginine, an important storage amino acid, remains fairly constant during maturation representing around 20% of the total (Sponholz 1991). Amino acid concentration at harvest depends on climate and fruit maturity (Schrader et al. 1976). In cooler regions and seasons, higher amounts of amino acids are stored in the grape due to limited protein synthesis, although total nitrogen is not affected by climate. Fertilization and availability of water for transport can also influence amino acid content of the fruit. Sponholz (1991) reported that arginine and proline are the main amino acids if plant fertilization is low. With fertilization (more than 3 g N/plant), glutamine becomes the main amino acid transported in the xylem sap and subsequently into the fruit.

Free Amino Nitrogen (FAN)/Nitrogen Deficiency

Nitrogenous compounds in the grape that are metabolically available to yeast (amino acids and ammonia) are referred to collectively as free *alpha*-

amino nitrogen or FAN. Slow or incomplete fermentations are often linked to FAN deficiency. Nitrogen available for yeast growth varies among regions and vineyards. Mold growth on the fruit may dramatically change the qualitative and quantitative distribution of amino acids (Dittrich 1987). For example, decreases in the amino acid content caused by *Botrytis cinerea* range from 7 to 61% (Sponholz 1991) which may fall below that needed for complete fermentation. Little or no prefermentation addition of SO_2 may encourage the growth of native yeast populations, which compete for and deplete assimilable nitrogen levels needed by *Saccharomyces*. Bely et al. (1990) found that a minimum FAN concentration of 140 mg/L was required for satisfactory fermentation of table wine must. However, Henschke and Jiranek (1993) suggested that a maximum fermentation rate requires 800 to 900 mg N/L of which 400 to 500 mg N/L is assimilable. FAN deficiency in the juice is often corrected by the addition of ammonia in the form of diammonium phosphate (DAP).

Vos et al. (1980) established an optimal FAN/°Brix ratio of 43.9. Assuming a 20 °Brix must, this corresponds to a FAN level of 878 mg/L. Vos et al. (1979, 1980) reported the ammonium concentration needed for maximum fermentation rate as:

$$NH_3\ (\text{mg N/L}) = \frac{43.9 - \text{FAN}/^\circ\text{Brix}}{0.108}$$

Peynaud (1984) has suggested addition of ammonium nitrate when endogenous levels fall below 25 mg/L. In high-sugar musts, nitrogen supplementation may improve vigor, fermentation rates, and higher alcohol levels.

Must is frequently supplemented with assimilable nitrogen to prevent nitrogen deficiencies and fermentation problems. In the United States, the maximum addition of ammonium salts (such as DAP) for correction of nutritional defiences is 968 mg/L, whereas in EEC countries up to 300 mg/L is permitted. In Australia, additions are limited by the maximum wine phosphate and 400 mg of inorganic phosphate P_i/L are permitted (Henschke and Jiranek 1993). Although the addition of DAP corrects for nitrogen deficiency in the juice, it stimulates the increase in cell biomass and may not prevent the production of excess hydrogen sulfide (Vos et al. 1980). Several alternative nitrogen supplements are available including various "yeast foods." However, many of these are of limited value due to the variable concentrations of assimilable nitrogen (Henschke and Jiranek 1993). The involvement of urea in ethyl carbamate formation has led to its elimination as an approved wine additive in many countries. Yeast hulls or ghosts are thought to stimulate fermentation not simply by release of assimilable nitrogen and adsorption of toxic fatty acids, but by stimulating the mechanism involved in yeast cell membrane sterol formation as well (Munoz and Ingledew 1989, 1990; Wahlstrom and Fugelsang 1988).

The reduction of inorganic sulfur to sulfide via the sulfate reduction pathway occurs when the intracellular nitrogen supply is limited. In nitrogen-deficient musts, low levels of sulfur-containing amino acids (cysteine and methionine) forces the yeast to synthesize these from carbon/nitrogen precursors and sulfide (Juhasz et al. 1984; Ooghe and Kastelyn 1988). Sulfate and sulfite are reduced to sulfide (which accumulates) as part of this synthesis. (For an additional discussion on hydrogen sulfide production and control, see Chapter 9.)

Before amino acids and ammonia can be used by the yeast they must be transported across the cell membrane by amino acid permeases. This transport is dependent on the sterol concentration in the plasma membrane (David and Kirsop 1972) via oxygen-dependent synthesis (Aries and Kirsop 1977). Therefore, oxygen is as important as nitrogen, allowing transport of solutes such as amino acids and ammonia across the cell membrane (see Chapters 14 and 18).

During fermentation and growth, yeasts do not use amino acids equivalently. If the composition of amino acid in the must is sufficient for yeast, arginine is not used in high concentrates. Arginine accumulates in yeast vacuoles and is released and utilized with *alpha*-amino butyric acid (Kitamoto et al. 1988).

Glutamine, glutamate, asparagine, aspartate, arginine, serine, and alanine in addition to ammonium ion are excellent nitrogen sources for *Saccharomyces* (Cooper 1982). Other amino acids are used less or not at all (Large 1986). Jiranek et al. (1991) and Henschke and Jiranek (1993) highlighted the differences between strains of *S. cerevisiace* in the pattern of amino acid utilization.

Ough (1968) reported proline levels of 742 mg/L in grapes and 869 mg/L in wine. Unlike most amino acids in must, proline is not utilized by yeast under fermentative conditions for two reasons. First, the two enzymes involved in the uptake and utilization (proline permease and proline oxidase) are repressed by the presence of ammonium ion in must. Second, the oxidase enzyme requires oxygen for activity. Ingeldew et al. (1987) reported that wine yeasts may use proline as a nitrogen supplement and incorporating air into yeast starters may prove valuable in reducing the frequency of stuck fermentations. Under such conditions, sufficient oxygen may be incorporated to stimulate proline oxidase enzyme and yet not oxidize the wine (see Chapters 14 and 18).

Newly fermented wines have lower total nitrogen levels than do the musts from which they were produced. Gorinstein et al. (1984) reported that from 30 to 46% of the total nitrogen is assimilated by yeast during fermentation, the majority being rapidly utilized after the onset of full fermentation. Yeast strains with a strong demand for a growth-limiting amino acid may provide a practical method for stabilizing sweet wines (Schanderl 1959). Sparkling wine cuveés and fruit wines may be low in available nitrogen and benefit from nitrogen addition prior to secondary fermentation.

WINE PROTEINS

Soluble proteins in juice and subsequent wines increase with increasing grape maturity (Lee 1985; Murphey et al. 1989). Protein synthesis proceeds rapidly after *veraison* and parallels the rapid accumulation of sugar (Luis 1984). The protein level of the fruit is frequently higher in warmer regions. Low crop levels have also been associated with higher protein and higher total nitrogen (Ough and Anelli 1979).

The solubility of wine protein depends primarily on temperature, alcohol level, ionic strength, and pH. Changes in any parameter may affect the potential for protein precipitation.

At the isoelectric or isoionic point (pI) of a protein, positive and negative charges are equal. The pH of wine is very close to the isoelectric point for many wine protein fractions. Wine proteins are least soluble at their isoelectric points. If the wine pH is above the isoelectric point of the protein fraction, the net charge on the fraction will be negative, and the protein will bind electrostatically with positively charged fining agents. Conversely, if the wine pH is lower than the isoelectric point, the net charge on the fraction will be positive. In this case the protein will react with negatively charged fining agents such as bentonite (see Fig. 8–2). The greater the difference between the wine pH and the isoelectric point of the protein fraction, the greater the net charge on the protein and the greater its binding affinity toward charged fining agents.

Hsu and Heatherbell (1987) determined the effect of bentonite addi-

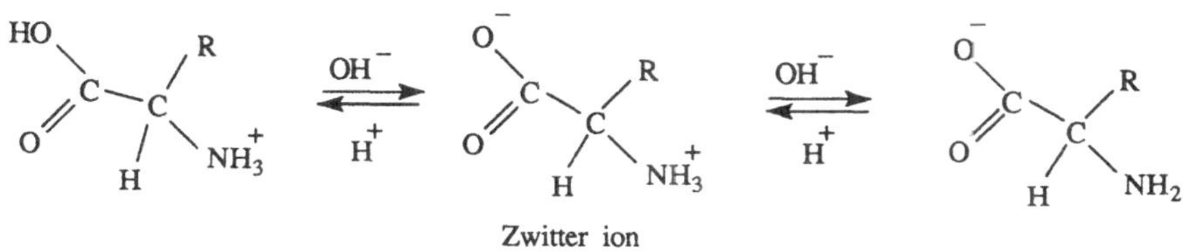

Fig. 8–2. Effect of additions of acid or base on proteins at their isoelectric points.

tion on protein removal and subsequent protein stability. Bentonite initially removed intermediate molecular weight (32,000–45,000 daltons) fractions (pI 5.8–8.0). To achieve stability as measured by a heat test (80°C/176°F for 6 hours followed by 4°C/39.5°F for 12 hours), they found it necessary to remove lower molecular weight (10,000–32,000 daltons) fractions with pI of 4.1 to 5.8. In this group, the glycoproteins represent the major fraction. Thus, heat-labile proteins appear to be comprised mainly of glycoproteins of less than 30,000 daltons. Lee (1985) also identified protein fractions that contribute to instability between 16,000 and 25,000 daltons with pI = 5.8 to 8.0.

PREFERMENTATION PROCESSING CONSIDERATIONS

The protein level extracted from the fruit is influenced by initial grape handling methodology. Juice produced from whole cluster pressing has lower protein levels compared with juice extracted from destemmed grapes (Dubourdieu and Canal-Lauberes 1989). Stems play an important role in limiting protein diffusion. Mechanical harvesting that largely eliminates stems may be considered an important factor in contributing to the protein load in white wines.

Skin contact generally increases juice protein concentration, depending on the variety. Dubourdieu and Canal-Llauberes (1989) found that 15 hours of skin contact at 18°C (64.5°F) doubled the protein concentration in Sauvignon Blanc, whereas in the case of Semillon the increase was 50%. Increased skin contact time of Muscadelle had little effect on protein concentration. Most of the protein extraction occurred during the first 10 hours of skin contact.

Settling and racking white grape juice before fermentation reduces the total nitrogen content by 10 to 15% (Koch 1963). Bentonite nonselectively removes proteins, peptides, and amino acids, and may adversely affect fermentation rates. As much as a 50% reduction in total nitrogen including reductions in amino acids may occur with bentonite fining (Ferenczy 1966). The amino acid content may be reduced by 15 to 30% with bentonite treatment of 1 g/L depending on the type of bentonite used (Rapp 1977).

FERMENTATION AND POST-FERMENTATION PROCESSING CONSIDERATIONS

Somers and Ziemelis (1973a) compared the effects of bentonite fining during and after fermentation on total nitrogen as well as nitrogen fractions in wines. The use of bentonite during fermentation reduced nonprotein nitrogen (approximately twofold) when compared with losses of protein nitrogen at each addition level to wine. Post-fermentation bentonite additions removed residual wine proteins and nearly equal amounts of protein and nonprotein nitrogen. Using bentonite during fermentation also reduces subsequent wine lees volume (see Chapter 16). Vos and Gray (1979), however, reported increased levels of hydrogen sulfide during fermentation in contact with bentonite. They speculated that this resulted from reduction in the FAN content of fermenting juice. The addition of nitrogen-containing fermentation adjuncts generally reduces this problem (see Chapter 9).

Following fortification, wines may precipitate large quantities of proteinaceous lees. However, alcohol levels of 10 to 12% are seldom sufficient to cause complete protein precipitation. The interaction between phenolic compounds and protein is important because phenol complexation in red and white wines removes/reduces the concentration of some proteins in solution. In white wines, relatively low phenol levels usually do not remove enough protein, causing instability. Leglise (1980) suggested the high protein levels in Australian Pinot Noir may lead to color instability by binding and co-precipitation with tannins and pigments. Pinot Noir occasionally requires bentonite fining to attain protein stability, presumably due to insufficient tannins. Wines fermented and/or aged in oak barrels frequently have lower concentrations of unstable protein and are much clearer than those held in stainless steel because proteins react with wood tannins and precipitate. Champagne producers take advantage of this interaction by occasionally adding tannic acid to their wines to help bind potentially unstable proteins resulting from both primary and secondary fermentation.

Effect of Aging on Nitrogen Components

After fermentation, and prior to first racking total nitrogen increases due to yeast autolysate (Gorinstein et al. 1984). During aging, nitrogen increases are primarily attributed to amine nitrogen, which reaches maximum levels after about 2 months of storage on the lees. The balance is in the form of amide nitrogen and protein. Proteins from yeast autolysate do not contribute to protein instability (Boulton 1980e).

Sur lie

The Burgundian practice of *sur lie* or extended lees contact (including stirring the lees) is a stylistic tool used in barrel fermented wines such as Chardonnay and Sauvignon blanc. Lees contact contributes complexity by integrating yeast characteristics with existing fruit and wood flavors adding greater depth of flavor. This practice is applicable only to wines made from "clean" sound fruit. During autolysis, cellular proteolytic enzymes bring about hydrolysis of cytoplasmic proteins, the products of which are released into the wine. Amino acids act as precursors to flavor compounds, resulting in subsequent development of aroma and flavor (Kelly-Treadwell 1988).

Keeping wine in contact with lees increases the soluble polysaccharide content as a result of enzymatic degradation of yeast cell walls by *beta*-1,3 glucanase (Chatonnet 1993). The polysaccharide-protein complex released into the wine is largely mannoprotein and strongly reacts with phenolic components (Chatonnet et al. 1992). Therefore, a wine aged *sur lie* in oak can have almost one-third of the tannins released by wood bound to yeast cell walls. Dubourdieu and Canal-Llauberes (1989) found that barreled wines aged and stirred *sur lie* had a phenolic content quite similar to wines aged in stainless steel.

Stuckey et al. (1991) reported a positive correlation between the sensory quality of Chardonnay and the level of total amino acids after 5 months *sur lie*. The total amino acid content was greatest in the nonstirred vs. stirred treatment. As lees contact time increases, fruity aromas change to a more muted "vinous" aroma. On the palate, *sur lie* wines often show increases in middle body when compared with conventionally produced lots. *Sur lie* increases aging potential, perhaps due to increases in available oxidizable substrate.

As nitrogenous components, primarily in the form of amino acids, are liberated into the product, it may become an excellent medium for microbial growth, especially for lactic acid bacteria. Thus, if the winemaker wishes to induce a malolactic fermentation, additional lees contact may be employed (see Chapter 18). The breakdown of sulfur-containing amino acids (methionine and cysteine) during aging *sur lie* may produce hydrogen sulfide (see Chapter 9).

Ethyl Carbamate

In the past several years much attention has been directed to ethyl carbamate (urethane), a compound suspected of being a mild carcinogen that may be naturally present in fermented foods as a consequence of the metabolic activity of microorganisms. Table 8–1 shows concentrations reported in wines by BATF in 1986 to 1987. The concern about ethyl carbamate may result in governmentally imposed limits and the testing and regulatory compliance that go with such limits.

The U.S. wine industry has established a voluntary target for ethyl car-

Table 8–1. Survey of Ethyl carbamate levels (μg/L) in commercial wines. (BATF 1986–1987).

Product	Range	Average
U.S. wines		
Table	0–102	10
Port	7–254	93
Sherry	18–209	82
Imported wines		
Table	0–80	12
Port/Madiera	17–108	55
Sherry	23–82	62

bamate of 15 ppb (μg/L) or less in table wines and less than 60 ppb in fortified wines. Additionally, the U.S. Food and Drug Administration has notified all countries exporting wines to the United States that they must meet targeted levels established by the American wine industry.

Known precursors of ethyl carbamate are urea, citrulline, carbamyl phosphate and n-carbamyl amino acids (Monteiro et al. 1989; Ough et al. 1988). Figure 8–3 shows the reaction sequence leading to ethyl carbamate.

The formation of ethyl carbamate is related to the concentrations of urea and ethanol, to time, and increases exponentially with increases in temperature. Because urea is the principal precursor of ethyl carbamate, controlling the urea concentration may be important in limiting ethyl carbamate levels. Factors influencing urea in wines include the arginine content of the grape, yeast strain, method of yeasting, fortification, timing of fortification, temperature, and duration of wine storage.

The amino acid arginine is the main precursor of urea. The majority of the urea formed comes from arginase-catalyzed degradation of arginine during fermentation. High urea levels can occur in wines produced from grapes of high (>400 mg/L) arginine content. Such grapes tend to come from vineyards heavily fertilized or displaying high vigor.

Urea is often formed during the early or middle stages of fermentation with subsequent yeast generations utilizing it during the latter stages. Fortified wines, which are made by arresting fermentation, may contain high concentrations of urea if the fermentation is stopped at the point of greatest urea production. With many yeasts the maximum extraction occurs at about 12 to 16 °Brix; followed by metabolism of the remaining arginine and reabsorption of the urea (Ough 1993b). Wineries that inoculate by pouring fresh juice over recently fermented tank bottoms may produce wines with elevated urea concentrations.

Yeast strains differ in their urea excretion and uptake during fermentation. Therefore, yeast selection may play a role in minimizing the poten-

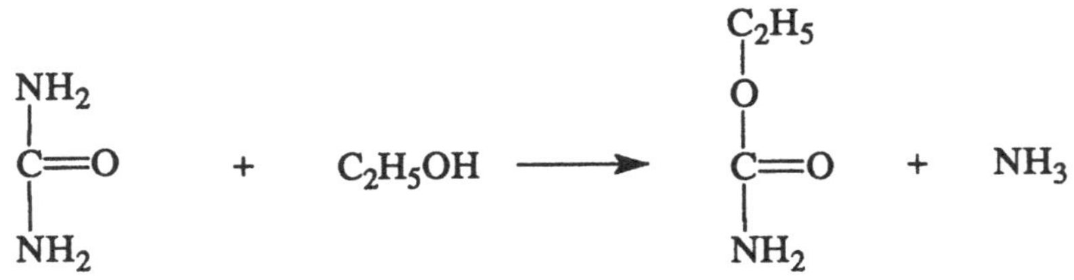

Urea Ethanol Ethyl Carbamate Ammonia

Fig. 8–3. Ethyl carbamate formation in wine.

tial for ethyl carbamate formation. The yeasts 71B (Lallenmand), SD 1120 (Red Star), and Prise de Mousse have been shown to release fairly low levels of urea during fermentation (An and Ough 1993). Yeasts that excrete little urea have slight but important differences in their arginine transport system and urea metabolizing enzymes.

Stevens and Ough (1993) and Ough (1993) reviewed several wine production factors influencing ethyl carbamate formation. Storage temperature is the single most important variable influencing the rate of formation, with wine type and pH having less effect. The concentrations of ethyl carbamate are proportional to the urea concentration during storage. Therefore, knowing the urea content and wine storage temperature allows for an estimation of the ethyl carbamate level that will be formed. The relationship between urea content and ethyl carbamate formation at 24°C (75.5°F) was established by Ough (1993). Storage of wine at temperatures greater than 24°C (75.5°F) with urea concentrations over 5 mg/L should be avoided.

Reductions in the level of urea formed can be achieved by minimal fertilization, utilizing yeast strains that release less urea, and by fortification when urea concentrations are low. Reducing the concentrations of urea immediately after fermentation, however, likely presents the best alternative to limiting ethyl carbamate production. (See discussion of ureases below.)

Ureases

Concerns about high ethyl carbamate levels in sake (a high urea rice wine usually heated prior to serving) resulted in the development of bacterially produced (*Lactobacillus fermentatum*) commercial urease enzymes. (Yoshizawa and Takahashi 1988). The products of urea are ammonia and carbon dioxide. A urease with activity within the pH range of wine has been approved for use in the United States. The enzyme is added after fermentation and before final filtration. Use levels are governed by alcohol, pH, urea concentration, enzyme contact period, and temperature. No perceptible effects on wine flavor or aroma have been noted at enzyme addition levels up to 500 mg/L; a level far greater than normally required (Caputi 1993).

Trioli and Ough (1989) reported minor inhibition of urease from excessive Fe^{+++}, Ca^{+++}, PO_4^{-}, SO_2, and phenolic compounds. Inhibition was more pronounced with L-lactic, acetic, pyruvic, and *keto*-glutaric and particularly L-malic acid.

The presence of fluoride at greater than 1 mg/L irreversibly inhibits the action of urease (Kodama et al. 1991). The mineral cryolite (sodium aluminum fluoride, AlF_6Na) is used in the United States (California) to help control the grape leaf skeletonizer (Archer and Gauer 1979). Famuyiwa and Ough (1991) showed that fluoride present in juice at 1 mg/L inactivates 10 mg/L added urease. Ureases can reduce ethyl carbamate formation, however, because urea is not the only precursor, complete elimination may not be seen.

EFFECT OF PROTEIN ON WINE STABILITY

Precipitation of soluble proteins in bottled wines creates an amorphous haze or deposit formed most frequently in white wines or wines of low polyphenol content. It is rarely encountered in wines with relatively high levels of flavonoid phenols, particularly tannins, which complex with and precipitate proteins. Proteins may also serve as nuclei around which soluble iron, copper, and other heavy metals may deposit (see Chapter 13).

The nature of protein instability in wines has been difficult to elucidate due to the many factors involved. Differences in proteins occur due to cultivar, maturity, climate, molecular size, electrical charge, as well as from interaction and precipitation with other components. Additionally, reliable methods of assaying soluble protein have not been developed.

The so-called protein haze in wine is likely a complex of protein, polysaccharide and polyphenols with minor amounts of inorganic ash (Ewart 1986). Unstable polysaccharide and polyphenol complexes may explain why heat and other protein precipitation tests as well as analysis of total protein are not completely effective predictors of potential instability. Estimates of soluble protein concentrations in wines range from 10 to 275 mg/L and may vary depending on method of determination (Boulton 1974). A wine's total protein content is not a good index of stability, and thus it cannot be used to predict protein instability.

Protein clouding is due not only to the precipitation of thermally labile proteins but also to formation of insoluble protein-tannin complexes. The grape is the major source of protein in wine. According to Somers and Ziemelis (1973b), about half of the total wine protein is bound to grape phenols, which are responsible for protein haze formation. Yeast cells may excrete small amounts of protein during fermentation (Bayly and Berg 1967), but much larger amounts end up in the wine upon completion of fermentation as a result of autolysis. Proteins originating from yeast autolysate are not thought to be involved in instability.

PROCESSING CONSIDERATIONS AND PROTEIN STABILITY

The treatment of fermenting and fermented wines with bentonite is a method of obtaining protein stability. Bentonite addition to wine has its disadvantages, including the formation of large volumes of lees, as well as possibly detrimental impact on flavor.

Murphey et al. (1989) found that both Gewurztraminer and White Riesling contained protein fractions with pIs below 3.5, regardless of fruit maturity. Such protein fractions would be potentially resistant to removal by bentonite fining. The problem worsens as pH increases with fruit maturity.

Alternative methods for protein stabilization have been investigated and include the following:

Immobilized Tannins

As previously stated, tannins interact with proteins resulting in precipitation. This is why wines stored in oak frequently clear readily. Weetall et al. (1984) demonstrated that an immobilized tannic acid derivative was effective in stabilizing a white wine.

Peptidases

Although peptidases have the potential to reduce protein content, they are not currently being utilized in winemaking. Studies have demonstrated that when applied in high concentration and at temperatures greater than 40°C (104°F), treated wines or juices showed reduced protein content (Heatherbell et al. 1985; Rokhlenko et al. 1980).

Ultrafiltration

Ultrafiltration (UF) is a tangential-flow membrane filtration process for separating molecules on the basis of size. Two flow streams are created: the permeate consisting of the portions passing through the membrane and the retentate of those fractions too large to pass. Depending on pore size, oxidized and polymerized phenols, proteins, yeast, and other compounds larger than nominal rating can be removed in a single step with UF.

Fermentation with Bentonite

Fermentation in contact with bentonite reduces the amount needed after fermentation (see Chapter 16).

METHODS FOR EVALUATION OF PROTEIN STABILITY

Winery operations may play a significant role in protein stability. Thus, bitartrate stabilization, malolactic fermentation, and acidification may render a previously stable wine protein unstable (as a result of pH shifts). Both heat and spirits additions may initiate denaturation and subsequent flocculation and precipitation of wine protein and/or protein complexes. Protein stability must be determined after all cellar operations are completed, and before bottling.

Determining Protein Stability

Cooke and Berg (1984) determined that there was no wine industry standard for protein stability in the United States (California). Tests performed for predicting protein stability generally involve heating, heating and cooling (at various temperatures and time durations), or the addition of a precipitation agent. The most common procedures call for subjecting wine samples either to heat or to a chemical oxidant, such as trichloroacetic acid, and subsequent examination for haze development.

Proteins are the most important foam-active components in champagne base wines. Hydrophobic proteins contribute more to foam constitution than hydrophilic proteins (Brissonnet and Maujean 1993). Thus *cuveé* protein levels must be adjusted such that there is minimal precipitation in the bottle while not detrimentally affecting carbonation.

In the case of sparkling wine, some choose to fortify *cuveés* in the laboratory by 1.1 to 1.5% before running a heat test. This may duplicate the additional alcohol achieved by bottle fermentations to produce a *Mousseuxs* (a product with a final CO_2 pressure ≥3.5 atmospheres).

Heat Stability Testing

Most predictive techniques involve some exposure of the wine to elevated temperatures for various periods of time. Precipitation of a colloid such as protein is affected by temperature and duration of heating. Virtually all wine protein may be precipitated by heat. Pocock and Rankine (1973) evaluated treatment temperatures and time over the temperature range 50°C to 90°C (122°–194°F). Precipitation of approximately 40% of wine protein occurred when a sample was held at 40°C (104°F) for 24 hours. By comparison, holding at 60°C (140°F) for the same time period precipitated 95 to 100% of the protein. The time necessary for haze formation decreases with increasing temperature. Pocock and Rankine (1973) recommended 80°C/176°F (6 hours) as a heat stability evaluation. Holding at lower temperatures, even for 24 hours, did not yield maximum haze formation in the wines tested. Increasing the temperature to 90°C (194°F) and decreasing the exposure time to less than 6 hours increased apparent haze formation even in samples where little protein was present. Ribereau-Gayon and Peynaud (1961) considered wines heated to 80°C (176°F) for 10 minutes to be stable if no haze developed upon cooling.

Chilling wine samples after heat treatment may increase visible haze formation. Berg and Akiyoshi (1961) recommended holding the sample at 49°C (120°F) for 4 days followed by cooling to −5°C (12°F) for 24 hours. Upon warming of the wine to room temperature, haze and/or precipitate formation is evaluated. Heat test procedures are given in Chapter 20.

Chemical Precipitation Tests

In addition to a wide array of laboratory methods involving heating, a number of chemical methods have been employed to predict stability. These include ethanol, ammonium sulfate, trichloroacetic acid, phosphotungstic acid (Bentotest), and tannic acid precipitation. Procedures are provided in Chapter 20.

DETERMINATION OF TOTAL PROTEIN AND NITROGEN-CONTAINING COMPOUNDS

Unfortunately, the analysis for total protein does not correlate well with observed protein instability in wine. Estimates of soluble protein concentrations in wines range from 10 to 275 mg/L and vary depending on the analysis method (Boulton 1980e). Many methods lack accuracy due to interference of nonprotein components. The Kjeldahl method for protein determination estimates the concentration based on total nitrogen. Because grapes and wines contain significant amounts of nonprotein nitrogen, Kjeldahl analysis of protein likely provides false high results (Koch and Sajak 1959). Kjeldahl analysis may be used to measure grape and wine proteins after selective precipitation of the protein with trichloroacetic acid, ammonium sulfate, alcohol, or phosphomolybdic acid (Berg and Akiyoshi 1961; Hsu and Heatherbell 1987; Koch and Sajak 1959).

A dye binding assay for the estimation of soluble proteins has been used for juice and wines. The method is based on the Bradford (1976) procedure using Coomassie brilliant blue G250. When Coomassie brilliant blue G-250 binds to protein the absorbance maximum shifts from 465 to 595 nm with the corresponding shift from orange to blue (see Chapter 20).

ANALYSIS

The following is a cross-reference of analytes and procedures found in Chapter 20. Protein stability by Bentotest, trichloracetic acid (TCA), heat test, ethanol test, ammonium sulfate, and tannic acid precipitation; protein by Coomassie Brilliant Blue; ammonia by spectrometry, ammonia (and ammonium ion) with selective ion electrode; *alpha*-amino nitrogen (FAN) by spectrometry; urea by spectrometry; proline by spectrometry; nitrite by ion selective electrode; fluoride by gas chromatography and ion specific electrode; ethyl carbamate; ability to ferment by fermentability test.

Quick screening methods for qualitative determination of protein in wine haze and precipitates are based on colorimetric methods such as Biuret, Folin-Ciocalteau, and Coomassie blue reactions outlined in Chapter 20.

CHAPTER 9

Sulfur-Containing Compounds

Sulfur, in its various forms, is important to the yeast in protein biosynthesis as well as vitamins and coenzymes. Formation of volatile sulfur compounds plays an important role in sensory properties of wines. These compounds may be variously described as rubbery, onion, garlic, cabbage, kerosene-like, and skunky. The somewhat undefined, but objectionable character arising from interaction of sulfur-containing compounds and wine components is sometimes described as "reductive tone." Wines exhibiting this property lack (or are deficient in) focused fruit and complexity attributes and are described as "reduced." Sulfur is available as sulfate (SO_4^{2-}), sulfite (SO_3^{2-}), amino acids (methionine, cysteine, and cystine), the tripeptide glutathione, as well as the vitamins biotin and thiamine (B_1), and acetyl-CoA and lipoic acid.

SULFATE (SO_4^{2-})

The concentration of sulfate in grape juice varies widely depending on variety, soil, and year. The sulfate content of California grapes ranges from

100–700 mg/L (Amerine and Ough 1980). In some instances, SO_4^{2-} may also be introduced to the must from addition of $CaSO_4$ in plastering.

Levels as low as 5 to 10 mg/L are reported as sufficient to support yeast cell growth (Eschenbruch 1974; Maw, 1960). Uptake of SO_4^{2-} requires activity of two permease enzymes whose synthesis is regulated by the cell's needs for methionine (Surdin-Kerjan et al. 1977). Once inside the cell, SO_4^{2-} must be reduced to a lower oxidation state to be biologically useful to the cell (see Fig. 9.1). Significant amounts of energy in the form of ATP are required to reduce SO_4^{2-} to the level of sulfide (S^{2-}).

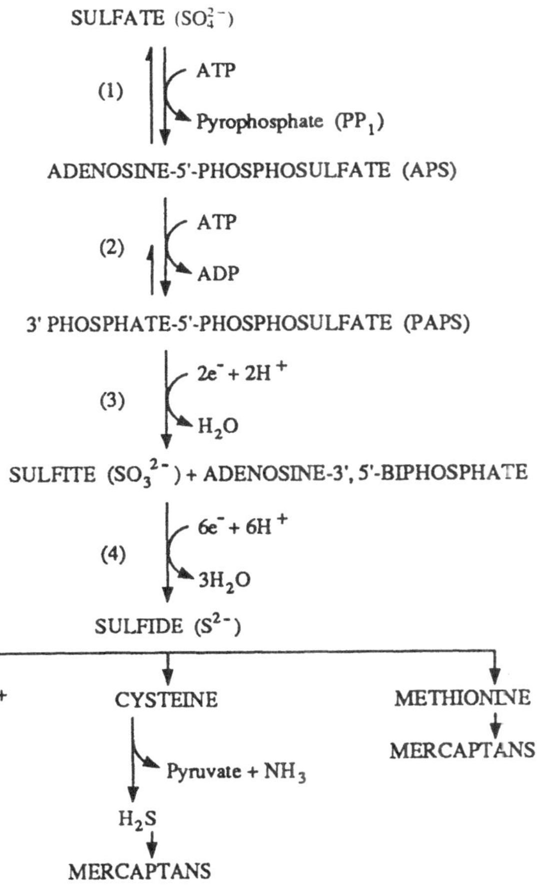

Fig. 9–1. Formation of hydrogen sulfide by yeasts.

SULFITE (SO_3^{2-})

As an intermediate in the reaction series forming methionine and cysteine, sulfite may be released into the wine where it contributes to total SO_2. Sulfate reduction is dependent on (regulated by) the requirement of yeast for the amino acids cysteine and methionine. Thus, if the content of these amino acids in must is sufficiently high, their formation by yeast is low and, correspondingly, SO_2 and H_2S production is suppressed.

Eschenbuch (1974) reported that 20 out of 250 strains of *Saccharomyces cerevisiae* produced more than 25 mg/L $S0_2$ during fermentation. Of these, five produced from 60 to 70 mg/L SO_2. Thus, *in situ* production of sulfite is an important issue in yeast strain selection.

HYDROGEN SULFIDE (H_2S)

At suprathreshold levels, H_2S is an undesirable odor-active compound reminiscent of rotten eggs. Sensory threshold levels are low, 50 to 80 μg/L (ppb) (Wenzel *et. al.* 1980). Dittrich and Staudenmayer (1968) reported that at lower levels (20 to 30 μg/L), H_2S may display 'yeasty' properties and play a role in complexity. If not corrected, H_2S may undergo reaction with other wine components to yield mercaptans, which also have undesirable properties and are difficult to remove.

Quantitatively, the most important source for H_2S is the assimilatory sulfate reduction pathway (Eschenbruch 1974) shown in Fig. 9.1. Henschke and Jiranek (1991) propose that in nitrogen deficient musts sulfite reduction is the major source of H_2S. Another source is chemical reduction of elemental sulfur. This requires the presence of compounds such as glutathione, which contain sulfhydryl (-SH) groups capable of donating a hydrogen (Maw 1960; 61). Sulfur candles, used to disinfect fermentation and storage containers, represent yet another source. Candles may not burn completely resulting in unburned sulfur entering the wine or juice. Use of dripless sulfur sticks and/or sulfur pellets or cups may effectively overcome this problem.

Because H_2S is an integral part of yeast metabolism, it is not possible to prevent its formation. However, vineyard management including selection and timing of spray applications, as well as wine processing techniques may effectively lower residual levels.

Hydrogen sulfide levels in wine are dependent upon (1) the kind and amount of elemental sulfur on the grapes, (2) yeast strains and their physiological condition during alcoholic fermentation; (3) juice/wine chemistry: pH, utilizable nitrogen levels, and including the vitamins B_6 (pyridoxine) and pantothenate, levels of sulfite, and sulfate, and ethanol concentration; (4) oxidation-reduction state of must and wine; (5) and physical parameters such as suspended solids and fermentation temperature. Details regarding each of these factors are presented in the following sections.

Elemental Sulfur

Elemental sulfur is used as a fungicide in vineyards worldwide. Residual elemental sulfur in must ranges from 0.3 to 8.9 mg/L and originates from vineyard spray formulations (Wenzel et al. 1980). Prefermentation sulfur levels of 1 to 5 mg/L are sufficient to produce objectionable quantities of H_2S (Thoukis and Stern 1962; Wenzel et al. 1980). The closer the time between the last application and harvest, the greater is the residual sulfur content in the juice. Eschenbruch (1983) recommended that sulfur not be applied for at least 35 days before harvest. The quantity of H_2S formed by elemental sulfur is inversely poroportional to particle size (Rankine 1963). Schutz and Kunkee (1977) compared H_2S formation from musts treated with dusting, wettable, sublimed, precipitated, and colloidal sulfurs. Colloidal sulfur brought about the most dramatic increase in H_2S. Wettable and dusting sulfur produced less H_2S followed by precipitated and sublimed sulfur. Many viticulturists are using micronized sulfur, which is produced as very small particles, ranging from 6 to 8 μm in size. Micronized sulfur is readily miscible in water and fumes at lower temperatures (72°F/22°C) than conventional formulations. Another advantage of micronized sulfur is that the application rate is less than one-third that of normal dusting sulfur for the same measure of control. Recommended addition levels for micronized sulfur range from 1 to 2.5 pounds/acre vs. 6 to 8 pounds for wettable formulations and 10 to 20 pounds for dust.

Redox State, Temperature, and pH

Hydrogen sulfide formation is also a function of oxidation-reduction (redox) state of the must during fermentation. Rankine (1963) found higher levels of H_2S produced from fermentations carried out in tall-form (height to diameter) tanks. The design of such fermenters was conducive to a rapid drop in redox potential. Fermentation temperature also affects formation of H_2S. Generally, less H_2S is produced at lower temperatures (Rankine 1963). The H_2S produced at lower temperatures is more soluble and less likely to be removed due to entrainment with CO_2.

High pH musts are prone to higher levels of H_2S (Eschenbruch and Bonish 1976; Rankine 1963; Rankine and Pocock 1969b). However, strain variation appears to be the most important variable in this regard.

H_2S Formation by Yeast

Yeasts differ significantly in their ability to produce H_2S with some strains producing more than 1 mg/L (Acree et al. 1972). Variations are influenced by both genetic and environmental factors, formation is linked to both sulfur and nitrogen metabolism by yeasts (Henscke and Juanek, 1991). One of the major problems associated with use of native yeast fermentations may be production of unacceptably high levels of H_2S. Furthermore, relatively large quantities of H_2S are produced by those yeasts that ferment more rapidly and, hence, bring about a more immediate drop in redox potential (Rankine 1963).

Among commercially available strains, Pasteur Champagne (UCD 595), Epernay 2, and Prise de Mousse are known to be low H_2S producers whereas Montrachet and Steinberg are known to produce higher levels. The reasons for these differences are not clearly understood. However, some yeasts have deficiencies in their sulfur metabolism that promotes increased production of H_2S. Such yeasts have an absolute requirement for the vitamins pantothenate and/or pyridoxine and substantial amounts of H_2S may result from deficiencies. Although grape juices are not normally deficient in these two vitamins, must treatment may result in depletion of one or both. Addition of methionine tends to suppress sulfate reduction and subsequent H_2S production (Eschenbruch 1974). However, other undesirable products can be formed.

Nutritional Status of Must

When assimilable must nitrogen is low, proteolytic activity of the yeast is stimulated; proteins and large peptides are degraded to assimilable forms in an effort to supplement this deficiency (Henschke and Jiranek 1991; Vos and Gray 1979). The result is liberation of H_2S from sulfur-containing amino acids. The free amino nitrogen (FAN) component of must plays an important role in H_2S formation. Specifically, assimilable free amino nitrogen content is inversely related to H_2S levels (Vos and Gray 1979).

The degree of juice clarification is also known to influence H_2S formation. Grape solids retain elemental sulfur and provide a source of protein nitrogen including sulfur-containing amino acids. More H_2S is formed in higher than lower solids fermentations. Centrifugation of juice may remove insoluble proteins. However, this may have only a limited effect in reducing hydrogen sulfide formation because insoluble proteins represent only a small fraction of the total protein content.

Bentonite addition and removal before fermentation is a frequently used step in white wine production. In addition to lowering solids levels, bentonite lowers H_2S formation by reducing levels of juice protein and amino acids (Ferenzy 1966). Complete suppression of H_2S by bentonite treatment of juice is unlikely. Although the activity of bentonite is not specific, it affects protein nitrogen and amino acids and may retard fermentation due to reduction in levels of FAN.

Wines fermented in contact with bentonite may exhibit increased levels of H_2S. Bentonite may cause conformational changes in protein including cleavage of protein disulfide bridges resulting in an increase in sulfhydryl groups. It is presumed that due to steric changes, the protein molecule is rendered more susceptible to subsequent protease attack. This problem is overcome by the addition of assimilable nitrogen before fermentation. Fermentation in contact with bentonite may be advantageous in obtaining protein stabilization (see Chapter 16).

ORGANIC SULFUR-CONTAINING COMPOUNDS

Dimethyl Disulfide

Dimethyl disulfide (DMDS) has been variously described as being onion-like or reminiscent of cooked cabbage. It is believed to arise from oxidation of methyl mercaptan (Thoukis and Stern 1962). Its concentration in wine is normally low, less than 2 μg/L (Leppanen et al. 1979). Goniak and Noble (1987) reported its threshold in wine at 29 μg/L. Spedding and Raut (1982) demonstrated that low concentrations of dimethyl disulfide had a beneficial effect on the quality of some wines.

Diethyl Disulfide

Diethyl disulfide (DEDS) arises from oxidation of methyl mercaptan (Thoukis and Stern 1962). DEDS's threshold in wine is reported to be 4.3 μg/L and at suprathreshold levels is described as burnt rubber or garlic (Goniak and Noble 1987).

Mercaptans

The mercaptans are another group of sulfur-containing compounds that are of sensory importance. Their name arises from the presence of a terminal -SH moiety. Although only a few such compounds exist in nature, they are very odorous. For example, *n*-butyl mercaptan is responsible for the objectionable odor of the skunk. Both ethyl and methyl mercaptan may rapidly oxidize to yield DEDS and DMDS (Thoukis and Stern 1962). Conversely, addition of SO_2 reduces the disulfides back to mercaptans, which can be removed by copper treatment (see Chapter 20).

Ethyl mercaptan is described as onionlike or rubberlike and has a sensory threshold of 1.1 μg/L (Goniak and Noble 1987). In wine, ethyl mercaptan is believed to be formed by reduction of H_2S with acetaldehyde via the reaction:

$$\underset{\text{acetaldehyde}}{CH_3CHO} \xrightarrow{H_2S} \underset{\substack{\text{thioacetaldehyde}\\ \text{(intermediate)}}}{CH_3CHS} + H_2O \rightarrow \underset{\text{ethyl mercaptan}}{CH_3CH_2SH}$$

Methyl mercaptan, considered to be of greater importance in wine, is thought to be produced from the amino acid methionine upon storage (DeMora et al. 1987; Thoukis and Stern 1962):

$$\underset{\text{methionine}}{CH_3SCH_2CH_2CHNH_2COOH} \rightarrow \underset{\text{methyl mercaptan}}{CH_3SH} + \underset{\textit{alpha}\text{-aminobutyric acid}}{CH_3CH_2CHNH_2COOH}$$

Its sensory threshold in wine is reported to be 0.02 to 2 μg/L. At detectable levels, methyl mercaptan is described as being reminiscent of rotten eggs or cabbage. Yeast autolysate may play an important role in the character and complexity of wine. However, where lees are not stirred, the process may result in the production of off flavors and aromas including H_2S and mercaptans (Rankine 1963; Schanderl 1955). This subject is also discussed in Chapter 8.

"Light Struck" Wines

In wines exposed to light, the amino acid methionine may undergo decomposition to yield several odorous compounds, including H_2S, methanethiol, dimethyl disulfide, dimethyl sulfide, and ethyl methyl sulfide. Commonly referred to as "light struck," these wines are characterized as having a cheese- or plastic-like aroma. The time of light exposure necessary to catalyze the reaction may be very short even in green glass bottles. This may be a particular problem in sparkling wines due to the magnifying effect that CO_2 has on aroma perception.

Other Sulfur-Containing Compounds

Thioesters have also been identified in wines. Their sensory properties are reminiscent of onions or cheese, often with burnt characteristics.

VINEYARD MANAGEMENT

Copper, manganese and zinc are components of many fungicides and insecticides. Vineyard management protocol may call for late season application of metal-containing pesticides to the grapes not only to control microbial growth but also in the belief that subsequent H_2S formation during fermentation will be diminished or eliminated (Lemperle 1988). Eschenbruch and Kleynhans (1974) reported a relationship between the use of copper-containing fungicides on grapes and increased incidences of H_2S formation in wines. In one case, there was almost a fourfold increase in copper present in must from fungicide-treated lots when compared with untreated controls. Hydrogen sulfide formation in resultant wines was also stimulated. In the case of the fungicide-treated lot, the final H_2S level was 52.7 μg/L as compared with 8.5 μg/L in the control. Rankine (1963) reported similar problems with hydrogen sulfide formation in wine and beer held in contact with these metals, particularly when SO_2 is present.

Additionally, it is not uncommon to find winemakers adding Cu(II) sulfite directly to fermenters hoping to achieve control of H_2S production. In that copper ions serve as powerful oxidants, these practices, in the case of unfermented juice, should be avoided. Several sulfur-containing fungicides and insecticides have been reported to produce sulfurous off-odors and flavors in bottled wines. These include tetramethylthiuramdisulfide (TMTD) (Schmitt, 1987), ethylenebisdithiocarbamate (EBDC) (Marshall, 1977), O,S-dimethyl-N-acetyl-amido-thiophosphate (Acephate) (Rauhut, 1990). Rauhut and Sponholz (1992) identify the following properties of pesticides which influence formation of sulfurous off-odors in wine: timing and frequency of application with respect to harvest date, application techniques, systemic activity and solubility in water, residual levels and stability during fermentation and in wine.

HYDROGEN SULFIDE AND MERCAPTANS IN WINE

The inverse relationship between FAN and H_2S production demonstrated by Vos and Grey (1979) may provide the best control. Suppression of H_2S has been obtained by exogenous addition of nitrogen in the form of ammonia (diammonium phosphate) and other fermentation supplements.

Formation of excessive H_2S in white wines can often be minimized by either settling, centrifuging, or filtering the must before fermentation, thus effecting removal of high-density solids along with associated elemental sulfur. In red wines with excessive H_2S some aerate at first racking, thus volatilizing the H_2S. Increased H_2S production will occur, however, if aeration is carried out too soon after the completion of alcoholic fermentation. In this case, elemental sulfur is believed to act as a hydrogen acceptor in formation of H_2S. Coincidental with H_2S formation are increases in yeast populations arising as a result of transient exposure to oxygen.

Additional techniques for reducing H_2S include sparging wines with nitrogen gas several weeks after the completion of alcoholic fermentation. Although this practice may be relatively effective in eliminating minor quantities of H_2S, desirable volatile wine components may also be swept away during sparging operations. In cases where methyl mercaptan appears to be the problem, carefully controlled aeration may bring about oxidation of methyl mercaptan to the less objectionable compound, dimethyl disulfide.

Some winemakers remove objectionable H_2S and mercaptans by direct contact with copper. Upon addition, copper reacts with hydrogen sulfide accordingly:

$$H_2S + CuSO_4 \rightarrow CuS + H_2SO_4$$

This may be accomplished by use of either "in-line" brass fittings or by addition of copper containing agents. In the case of exposure of wine to copper-containing alloys (brass), resultant copper uptake into the wine is difficult to predict and may lead to oxidation (see Chapter 13). For this reason, winemakers may elect to use copper-containing salts or fining agents. Addition of 4 g of cupric (II) sulfate ($CuSO_4 \cdot 5H_20$) per 37.85 hL (1,000 gal) raises the copper content by 0.2 mg/L. U.S. governmental regulations permit additions of up to 0.5 mg/L (as copper); residual levels in the wine cannot exceed 0.2 mg/L (as copper). The results of such procedures are often variable. Careful laboratory testing should precede any additions to wine. If residual copper levels are higher than 0.3 mg/L, instability may result (see Chapters 13 and 17).

It should be noted that although mercaptans react with copper, DMDS does not (see Chapter 20). Thus, if the wine in question has undergone any oxidation, it may be necessary to reduce DMDS back to the reactive species, methyl mercaptan. This can be accomplished by addition of ascorbic acid or SO_2. Addition levels of 50 mg/L or more of ascorbic acid are generally used. The reader is cautioned that SO_2 analysis by Ripper titration cannot be accurately performed in wines containing ascorbic acid, because the latter also reacts with the iodine titrant. Copper should not be added until the fermentation is complete and the yeast titer reduced by racking, filtration, etc. Yeast cells will bind copper ions to cell surfaces and may reduce reactivity with H_2S.

The addition of SO_2 to wines may reduce H_2S levels. The addition

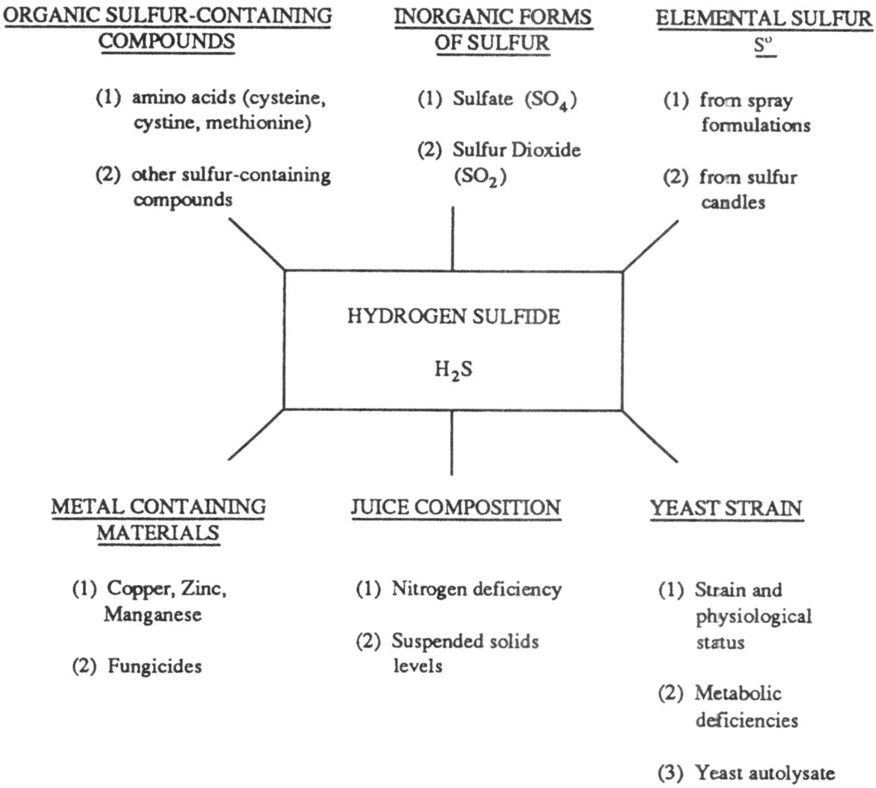

Fig. 9–2. Factors effecting H_2S formation (adapted from Eschenbruch 1983).

results in an SO_2-induced oxidation of H_2S to yield elemental sulfur, which, after precipitation, may be removed by centrifugation or filtration:

$$SO_2 + 2H_2S \rightarrow 3S^\circ + 2H_2O$$

Figure 9–2 summarizes the factors influencing hydrogen sulfide formation in wine. The reader is referred to the appropriate section(s) of this chapter for further discussion.

ANALYSIS

The following is a cross-reference of analyses and procedures found in Chapter 20. Sensory screen for determination of H_2S and mercaptans.

CHAPTER 10

SULFUR DIOXIDE AND ASCORBIC ACID

In its several commercially available forms, SO_2 is widely used in wine and related food industries as a chemical antioxidant and inhibitor of microbial activity. Although historically sulfites were generally recognized as safe, the U.S. Food and Drug Administration (FDA) has determined that the presence of unlabeled sulfites in foods and beverages poses a potential health problem to a certain class of asthmatic individuals. As a result, in 1987, the U.S. BATF implemented regulations requiring the declaration in labeling of sulfites present in alcoholic beverages at a level of greater than 10 mg/L (ppm) measured as total sulfur dioxide, by any method sanctioned by the international Association of Official Analytical Chemists (AOAC). The maximum permissible level of total sulfur dioxide for both BATF and the OIV is 350 mg/L.

There is an industry-wide trend toward reducing sulfur dioxide based on public health concerns, better quality fruit, desire for a malolactic fermentation, and the perceived delicacy of wines (lower levels of phenols in white wines and enhanced suppleness in reds). Reduction of sulfur dioxide levels, particularly before fermentation, is often consistent with such stylistic goals.

About a generation ago, the Pasteur Institute announced a 10,000 Franc prize to anyone who could identify a substance that could reproduce the desirable attributes of SO_2 and yet have fewer detriments. That prize remains unclaimed. This points out the effectiveness of sulfur dioxide and the difficulty in finding a viable substitute.

SULFUR DIOXIDE AS AN INHIBITOR OF BROWNING REACTIONS

Juice and wines contain many readily oxidizable compounds, including polyphenols. The antioxidative role of SO_2 in juice and wine lies in its competition with oxygen for susceptible chemical groupings. As a reducing agent, SO_2 can inhibit some oxidation caused by molecular oxygen. Aging may be considered a controlled oxidative process, where some oxidation by molecular oxygen is important in wine development. Ribereau-Gayon (1933) recommended that oxygen absorption be limited to the rate at which it could be catalytically reduced by oxidizable substrates such as flavonoid phenols and SO_2. Thus, in the case of white wines where oxidizable substrates are relatively low, SO_2 may play a more important role in consumption of oxygen.

The commonly observed browning phenomenon of freshly cut fruit (and crushed grapes) is the result of the activity of a group of plant enzymes, the tyrosinases (formerly polyphenoloxidases). These enzymes catalyze oxidation of nonflavonoid *o*-dihydroxy phenols (colorless) to their corresponding darkened quinones (see Fig. 7–9).

Although the exact mechanism of inhibition is not fully understood, it is known that SO_2 is active in destabilizing disulfide bridges that help maintain enzymes in their "native" or active form. The addition of 35 mg/L of SO_2 to must was found to inhibit completely oxygen uptake by tyrosinase (White and Ough 1973).

In addition to natural tyrosinases, grapes degraded with *Botrytis cinerea* contain the enzyme laccase. This enzyme rapidly oxidizes both *o*- and *p*-dihydroxyphenols (Peynaud 1984). Unlike tyrosinase, it does not hydroxylate monophenols and is also more soluble and significantly more resistant to SO_2 (see Chapters 3 and 16, and Chapter 20 for analysis of laccase).

The bisulfite ion species ($HSO3^-$) helps protect juices and wines from oxidative browning reactions, as well as scavenge hydrogen peroxide formed from oxidative reactions. For more information on browning reactions in wine see Chapter 7.

COMPOUNDS THAT BIND WITH SULFUR DIOXIDE

Several compounds present in juice and wine are active in binding with SO_2. In red wines, the major binding compounds are acetaldehyde and anthocyanins. Burroughs (1975) reported that at free SO_2 levels of 6.4 mg/L, 98% of the acetaldehyde present in solution was bound. This was followed by malvidin 3,5-diglucoside (63%), pyruvic acid (39%), and *alpha*-ketoglutaric acid (15%). Dihydroxyacetone, arising from oxidation of glycerol in grapes and musts infected with *Gluconobacter oxydans,* may additionally bind significant amounts of free SO_2 (Lafon-Lafourcade 1985).

Grapes degraded by *Botrytis cinerea* are rich in compounds that can bind sulfur dioxide. These include mainly the *keto* acids, pyruvic and *alpha*-ketoglutaric acid, which may be produced in higher concentrations (Doneche 1993). Glucuronic and galacturonic acids resulting from hydrolysis of plant cell walls also bind sulfur dioxide (Sponholz and Dittrich 1985).

Reactions with Acetaldehyde

Acetaldehyde is the principal aldehyde present in wine. Produced as an intermediate during alcoholic fermentation, the majority of acetaldehyde is reduced to ethanol during this phase. Upon storage of wine in the presence of air, nonenzymatic oxidation of ethanol may yield small amounts of acetaldehyde (Kielhofer and Wurdig 1960a). The majority of acetaldehyde formed, however, results from microbial oxidation of ethanol under aerobic conditions (Reed and Peppler 1973). As an intermediate in bacterial formation of acetic acid, acetaldehyde may accumulate even under conditions of low oxygen concentration (see Chapters 14 and 18).

Bisulfite reacts with acetaldehyde present in wine to yield the hydroxysulfonate addition product presented below. This reaction can be generalized to include other carbonyls.

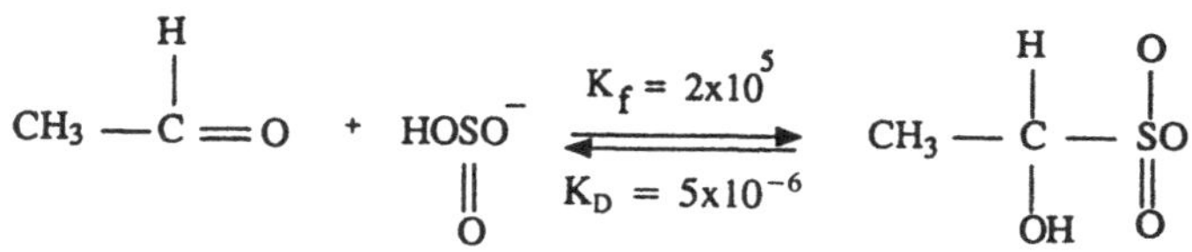

The small dissociation constant ($K_D = 5 \times 10^{-6}$) for this reaction indicates that the equilibrium favors formation of product (Burroughs and Whiting 1960). Compared to the dissociation constants for addition products of *alpha*-ketoglutaric acid (8.8×10^{-4}), pyruvic acid (4×10^{-4}), and glucose (6.4×10^{-1}), that formed with acetaldehyde is the most stable. Hydrogen ion concentration plays an important role in binding equilibria in that it

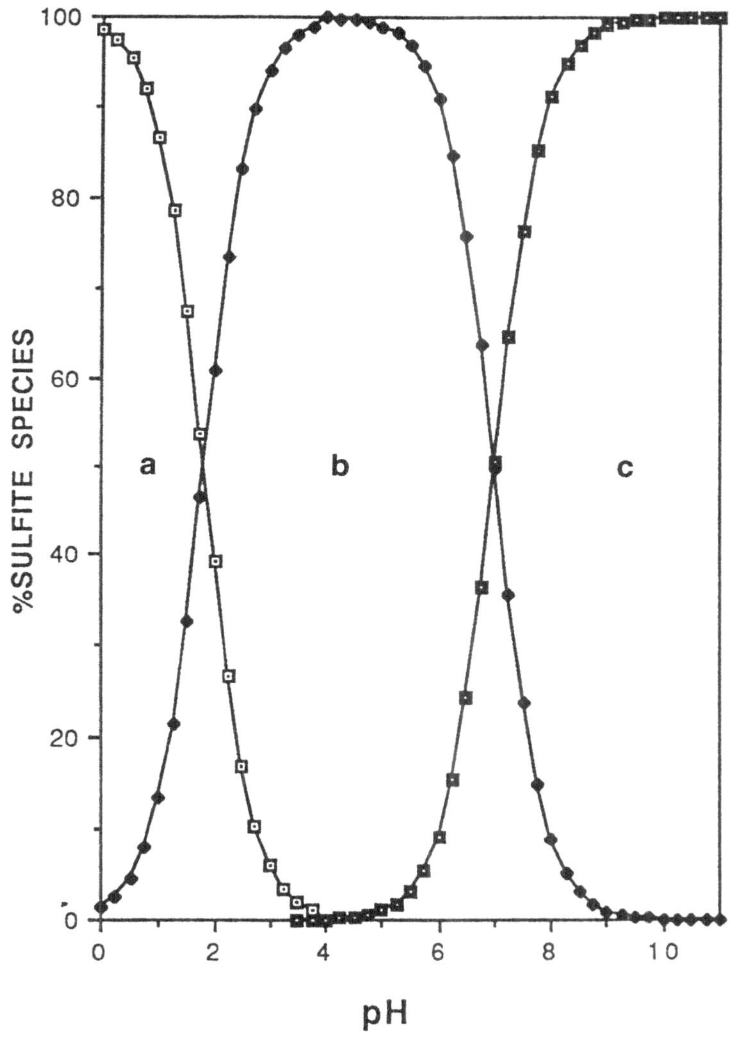

Fig. 10–1. Distribution of species (a) H_2SO_3, (b) HSO_3^-, and (c) $SO_3^=$, as a function of pH in dilute solution.

affects the concentration of bisulfite in solution (see Fig. 10–1). In cases where an aldehydic "nose" has developed in a wine, winemakers may exploit the favored reaction between bisulfite and acetaldehyde. With careful addition of SO_2, the more sensory-neutral addition product is formed (see Acetaldehyde Sensory Screen in Chapter 20).

Reaction with Pigments

Anthocyanin pigments have a strong affinity for bisulfite. However, the reaction of bisulfite and anthocyanin is not the same as the active carbonyl addition demonstrated by aldehydes, ketones, and sugars. Jurd (1964) has shown that the pigment-bisulfite reaction involves addition of the bisulfite anion to the colored anthocyanin cation forming (among others) the colorless species anthocyanin-4-bisulfite (see Fig. 7–5).

Timberlake and Bridle (1976) reported formation of the anthocyanin-4-bisulfite in young wines to be approximately 85% complete after the addition of only 15 mg/L SO_2. Such binding may significantly affect phenol polymerization, color stability and tannin suppleness in young red wines (see Chapter 7). By comparison, in older wines, Burroughs (1975), pointed out that more than one-half of the red wine color results from polymeric anthocyanins, which are resistant to SO_2 addition because of prior substitution. Furthermore, wines retain color longer in the presence of moderate amounts of SO_2 than in its absence. This topic is covered in greater detail in Chapter 7.

Reactions with Sugars

Several sugars present in must react with SO_2. These include glucose, xylose, and arabinose. From 50 to 70% of the total SO_2 addition to juice may be bound by sugars (Rankine and Pocock 1969b). The *keto*-sugar fructose does not form an addition compound with bisulfite ion (Braverman 1963). Fructose comprises approximately one-half of the reducing sugar content of must and this ratio increases as the fermentation reaches completion.

Sugars combine with bisulfite (HSO_3^-) at a much slower rate than do aldehydes and ketones, and the products formed are less stable. Joslyn and Braverman (1954) reported that in a mixture of acetaldehyde, glucose, and bisulfite, the aldehyde combined with bisulfite first, with 90 to 95% being bound after 2 minutes. Furthermore, addition of acetaldehyde to a glucose-bisulfite solution resulted in replacement of glucose by acetaldehyde.

DISTRIBUTION OF SULFITE SPECIES IN SOLUTION

The equilibria established upon dissolution of SO_2 in H_2O is seen in Fig. 10–1. As shown, the presence of any form(s) of SO_2 in solution is pH-dependent. At the pH of wine, bisulfite and molecular SO_2 are the dominant species. With increasing pH, the percentage of sulfite also increases.

BOUND AND FREE SULFUR DIOXIDE

Sulfur dioxide in wine occurs in two forms, bound (or fixed) and free, their sum equalling total SO_2. "Bound SO_2" refers to formation of addition compounds between bisulfite ion and other substances such as aldehydes, anthocyanins, proteins, and *aldo*-sugars. The quantity of SO_2 fixed and the rate of binding is a pH-dependent reversible reaction: the lower the pH, the slower the addition. Temperature also affects the equilibrium in a similar manner. A typical addition compound is shown on page 180.

Bound SO_2 has little inhibitory effect against most yeasts and acetic acid bacteria. However, at levels greater than 50 mg/L, bound SO_2 (as its acetaldehyde-bisulfite complex) is believed to be inhibitory towards lactic acid bacteria.

The undissociated molecular form of free sulfur dioxide is the most important antimicrobial agent. Within the pH range of juice and wine, the amount of free sulfur dioxide in the molecular form varies considerably (see Fig. 10–1). Attempting to control microbial growth with total or free SO_2 without reference to pH and molecular free SO_2 is of little value.

The level of molecular SO_2 needed for the control of microbial growth will vary with viable cell number, temperature, ethanol content, etc. Beech et al. (1979) determined that for white table wines, 0.8 mg/L molecular free sulfur dioxide achieved a 10,000-fold reduction (in 24 hours) in the number of viable *Brettanomyces* sp., certain lactic acid bacteria, and other wine spoilage organisms. The amount of free SO_2 needed to obtain 0.8 and 0.5 mg/L of the molecular form at various pH levels can also be estimated from Fig. 10–2. Depending upon wine pH, 30 mg/L free SO_2 may be excessive in one wine (lower pH) and a deficient quantity for control of microbial growth in another (higher pH).

Yeasts

SO_2 is probably taken up by yeasts as the molecular rather than anionic forms (Ingraham et al. 1956). Because cytoplasmic pH is generally near 6.5, intracellular SO_2 undergoing dissociation produces largely HSO_3^- and $SO_3^=$. Eventually inhibitory/lethal levels are accumulated by the cell.

Free molecular SO_2 is strongly inhibitory toward enzyme systems, especially those containing sulfhydryl (-SH) groups and disulfide linkages, as well as the coenzyme NAD. It catalyzes deamination of cytosine to uracil and may, subsequently, bind with and alter the activity of the latter. During fermentation, yeasts are more resistant to SO_2 due to rapid fixation by aldehydes. Therefore, attempting to stop fermentation by the addition of sulfur dioxide within OIV or BATF limits (350 mg/L) is not effective.

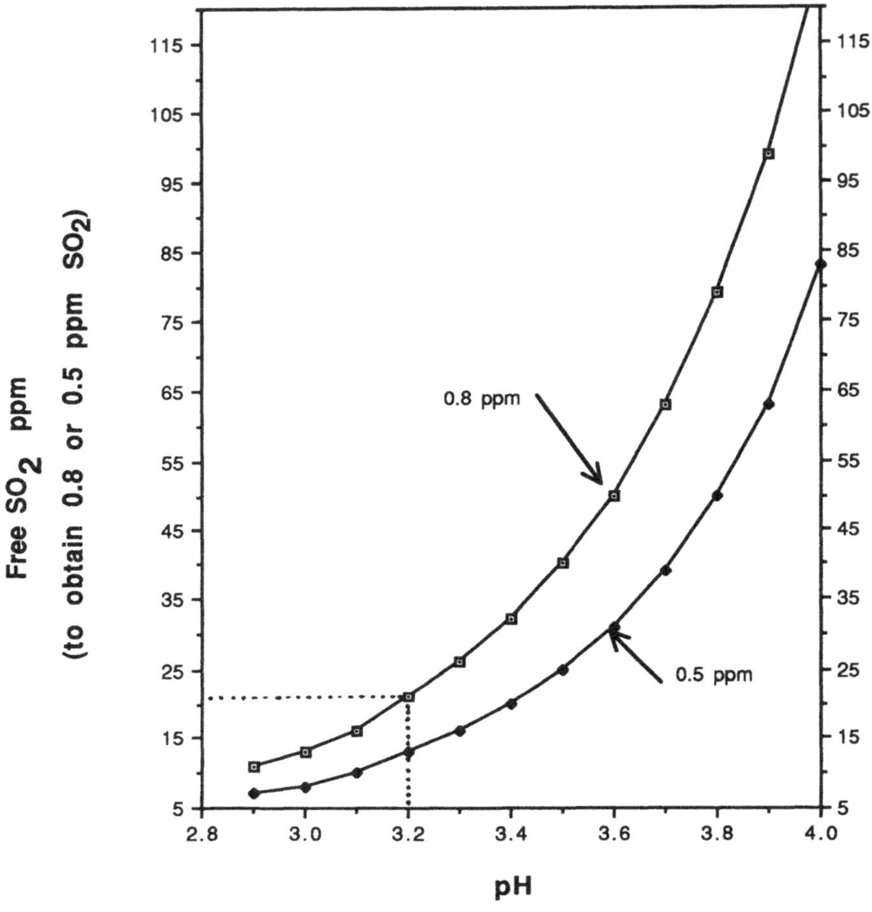

Fig. 10–2. Amounts of free SO_2 needed to obtain 0.5 or 0.8 mg/L (molecular) at various pH levels.

Some non-*Saccharomyces* yeasts approach and may exceed the resistance to SO_2 exhibited by wine strains of *Saccharomyces* (Heard and Fleet 1988a). At SO_2 levels of 50 to 100 mg/L, high population densities (>10^6 CFU/mL) of *Hansenula, Kloeckera/Hanseniaspora,* and *Candida* sp. have been observed after the onset of fermentation. Some spoilage yeasts (i.e., *Zygosaccharomyces bailii*) exhibit extraordinary resistance to SO_2. Thomas and Davenport (1985) report resistance to molecular SO_2 at levels of more than 3 mg/L. The trends towards little or no prefermentation sulfur dioxide addition enhances the likelihood of wild yeasts and LAB contributing to the fermentation.

Bacteria

In general, juice/wine bacteria are more sensitive to the effects of sulfur dioxide than are yeasts. Acetic acid bacteria vary in their sensitivity to SO_2. *Acetobacter aceti* and *A. pasteurianus* have been reported as frequent isolates in wines with more than 50 mg/L total sulfites. By comparison, *Gluconobacter oxydans* was only found in wines with less than 50 mg/L SO_2 (Drysdale and Fleet 1988).

Lactic acid bacteria also vary in their sensitivity to SO_2. In the case of *Leuconostoc oenos*, SO_2 as low as 10 mg/L is lethal (Wibowo et al. 1985). By comparison, bound levels of more than 30 mg/L delay the onset of growth (and MLF) and result in lower populations. Greater than 50 mg/L bound sulfite is strongly inhibitory toward LAB and MLF. *Lactobacillus* and *Pediococcus* sp. appear to be less sensitive to SO_2 than are strains of *Leuconostoc oenos* (Davis et al. 1988).

SULFUR DIOXIDE IN WINE PRODUCTION

There is an industry-wide trend toward reducing sulfur dioxide use whenever possible. Aside from health concerns, stylistic issues have sparked this interest. These include utilization of MLF, enhanced delicacy resulting from lower levels of phenols in whites, and more rapid polymerization in reds.

Grape Processing Concerns

Traditionally, sulfur dioxide additions were made at the crusher. Levels used depended on the cultivar and maturity as well as fruit integrity. In the United States, most white wine producers add sulfur dioxide either at the press pan, settling tank, or only after primary fermentation. The trend for red wine production is no or little sulfur dioxide until after primary and malolactic fermentations. If the goal is to maximize phenol polymerization in young red wines, the levels of free SO_2 generally should not exceed 15 mg/L (see Chapter 7). The effectiveness of the winery sanitation program will also be a consideration. Sulfur dioxide added at crush increases extraction of flavonoid phenols and therefore is generally not done or done sparingly. Processing in the absence of SO_2 increases enzymatic oxidation and subsequent polymerization and precipitation of nonflavonoid phenols.

Cellar Considerations

Monitoring the concentration of SO_2 following addition shows a rapid increase in the levels of the free form. This is followed by a decrease over 2 to 8 hours with corresponding increases in the percentage of bound SO_2. This is especially true in the case of sweet wines and those fortified with aldehydic spirits. Thus, SO_2 may be regarded as only a temporary preservative in wine because of its combination with carbonyl groups or oxidation to sulfate. There are some important limitations to the capabilities of sulfur dioxide as an antioxidant. Because it reacts very slowly with molecular oxygen, it is impractical to use SO_2 for scavenging oxygen from wines exposed to significant oxygen.

On occasion, SO_2 additions accompany controlled aeration for removal of H_2S (see Chapter 9). Elemental sulfur and water are produced (Tanner 1969):

$$2H_2S + O_2 \rightarrow 2H_2O + S^\circ$$

and

$$2H_2S + SO_2 \rightarrow 2H_2O + 3S^\circ$$

Several methods for sulfuring barrels are used, including sulfur wicks or pellets. Pellets frequently provide more sulfur dioxide than wicks (Chatonnet 1993a,b). When burned, oxygen is depleted and sulfur dioxide is released inside the barrel. Upon filling, residual SO_2 may be partially dissolved in the wine. Pressurized liquid sulfur dioxide is a common source for barrel sulfiting. Alternatively, use of 1% SO_2 in **acidified** water also is effective in controlling microbial growth. Chatonnet (1993a,b) reported the growth of *Brettanomyces* and thus high levels of ethyl phenols in barrels improperly sulfited.

Bottling Concerns

The concentration of molecular free SO_2 in wine is important in the control of wine microorganisms. Most winemakers increase the concentration of free SO_2 immediately before bottling. Although amounts may vary, this adjustment may be calculated to yield molecular free SO_2 levels of 0.5 to 0.8 mg/L (see Fig. 10–2). In the pH range 3.2 to 3.5, this corresponds to additions of 20 to 40 mg/L.

The loss of free SO_2 in wine is proportional to the dissolved oxygen content. Rankine (1974) reported that the headspace in bottled wine may contain up to 5 mL of air, which amounts to 1 mL (1.4 mg) of oxygen that is available to initiate oxidation. As a generality, 4 mg of SO_2 is needed to neutralize the effects of 1 mg of oxygen. Using this relationship, an additional 5 to 6 mg of free SO_2 is needed to reduce molecular oxygen in the headspace. This may represent a rather significant loss of free SO_2 that otherwise would be available as an antioxidant or antimicrobial agent. Purging bottles with inert gas before filling and/or vacuum filters and corkers are helpful.

SOURCES OF SULFUR DIOXIDE

Originally, sulfur dioxide was obtained by burning sulfur. Today, alternative sources of sulfur dioxide are available to the wine industry. Potassium metabisulfite and compressed sulfur dioxide gas, or solutions of the two are the most common. During storage, dry forms of SO_2 may lose their potency. Elevated temperature and high humidity can contribute to significant loss in strength. It is, therefore, desirable to monitor potency of these compounds, particularly if stored for extended periods.

Compressed gas has the advantages of being economical and avoiding the addition of potassium into the juice or wine. Whether one elects to use compressed gas or some salt of sulfur dioxide, it is necessary to establish the correct addition levels. Typical additions of SO_2 using (1) gas, (2) the salt, potassium metabisulfite, and (3) a concentrated solution of SO_2 are presented in the following examples where a winemaker wishes to increase the amount of total SO_2 in 1,000 gallons (37.85 hL) of wine by 28 mg/L (ppm).

1. *Addition of sulfur dioxide gas* (of 90% activity):
   (a) 28 mg/L = 0.028 g/L

   (b) 1,000 gal × 3.785 L/gal × 0.028 g/L = 106 g

   (c) $\frac{106\text{g}}{454\text{ g/lb}} = 0.234$ lb sulfur dioxide gas

   (d) the gas is 90% sulfur dioxide. Therefore, the amount of SO_2 gas needed is:

   $$\frac{0.234\text{ lb/1,000gal}}{0.90} = 0.260\text{ lb/1,000 gal (37.85 hL)}$$

2. *Addition of potassium metabisulfite (wt/vol):* Potassium metabisulfite (KMBS), is the water soluble potassium salt of sulfur dioxide. Theoretically, available SO_2 makes up 57.6% of the total weight of potassium metabisulfite.

Thus, the winemaker calculates the 28 mg/L SO_2 addition as follows:

$$\frac{0.028 \text{ g/L} \times 3.785 \text{ L/gal} \times 1{,}000 \text{ gal}}{0.576} = 184 \text{ g/1000 gal or (37.85hL)}$$

3. *Addition of sulfur dioxide (vol/vol):* Some winemakers choose to use solutions of sulfur dioxide in water for additions. Such solutions are usually created by bubbling liquid sulfur dioxide or dissolving KMBS into a measured volume of water, thereby creating a saturated solution of $SO_2 \cdot H_2O$. At 20°C (68°F) the solubility of SO_2 in H_2O is 11.28%
   (a) In a well-ventilated, isolated area, create a saturated solution of $SO_2{\cdot}H_2O$ solution. In cold water, solutions of 6–8% are readily achievable. Solutions of 15% SO_2 are commercially available.
   (b) Using the data in the following table (Wilson et al. 1943) and a piece of graph paper, plot concentration (% wt/vol) SO_2 vs. its corresponding specific gravity at 20°C.

Concentration of SO_2 (% wt/vol)	Specific gravity at 20°C
1.0	1.003
2.0	1.008
3.0	1.013
4.0	1.018
5.0	1.023
6.0	1.028
7.0	1.032
8.0	1.037

   (b) Using a specific gravity hydrometer, determine the specific gravity of the solution. Assuming that the reading was 1.028, this corresponds to a 6.0% (or 60,000 mg/L) solution of SO_2.
   (c) The volume of solution needed for a 28 mg/L addition to 1,000 gallons (37.85hL) is calculated as:

$$\frac{0.028 \text{ g/L} \times 3.785 \text{ L/gal} \times 1{,}000 \text{ gal}}{60 \text{ g/L}} = 1.766 \text{ L or } 1{,}766 \text{ mL}$$

Sensory Considerations

A number of sensory characteristics in wines have been directly attributed to the presence of sulfur dioxide. High levels of SO_2 impart a metallic (tinny) and harsh character to wines. Furthermore, excessive levels of free sulfur dioxide add a pungent aroma, a sharpness in the nose, and a "soapy" smell. Sulfur dioxide also may influence an important wine quality feature, longevity.

ANALYSIS OF FREE AND TOTAL SULFUR DIOXIDE

Because of the general concern for sulfites in foods, sensory and stylistic winemaking considerations, accurate analysis is essential. The internationally recognized official AOAC method for SO_2 in wine is the modified Monier-Williams method. This method involves a distillation of SO_2 out of the sample into peroxide, and the subsequent titration of H_2SO_4 formed. The method is cumbersome and only gives answers for total SO_2. A variant of the modified Monier-Williams is the aeration oxidation (AO) method. In this procedure, sulfur dioxide in wine or juice is distilled (with nitrogen as a sweeping gas or with air aspiration) from an acidified sample solution into a hydrogen peroxide trap, where the volatilized SO_2 is oxidized to H_2SO_4:

$$H_2O_2 + SO_2 \rightarrow SO_3^- + H_2O \rightarrow H_2SO_4$$

The volume of 0.01 *N* NaOH required to titrate the acid formed to an end point is measured and is used to calculate SO_2 levels.

The glassware involved is relatively inexpensive and the analysis is easy to perform. The AO procedure eliminates the interference from pigments and acetic acid.

The Ripper method for sulfur dioxide, which is more than 100 years old, uses standard iodine to titrate the free or total SO_2 in a sample. Although it is universally recognized that this method is somewhat inaccurate, the procedure is so simple that it is the most common method employed in the winery laboratory.

In this procedure standard iodine is used to titrate free sulfur dioxide. Free sulfur dioxide is determined directly. The completion of this reaction is signaled by the presence of excess iodine in the titration flask, which is complexed with added starch producing a blue black end point. Total sulfur dioxide can be determined by first treating the sample with sodium hydroxide to release bound sulfur dioxide. The Ripper procedure for free and total SO_2 suffers from several notable deficiencies: (1) volatilization and loss of SO_2 during titration; (2) reduction of the iodine titrant by compounds other than sulfite; (3) difficulty of end point detection in red wines; and (4) analysis cannot be accurately performed in juices or wines that contain ascorbic acid.

Commercial kits (ampules) for conducting the analysis of sulfur dioxide by the Ripper technique are available. These generally involve the same chemistry as is used in the Ripper titration. The accuracy, therefore, isn't any greater than the Ripper titration method and may be less. The Ripper titration for total SO_2 is accurate to $\pm$ 7 mg/L. The main problem with either the kit or titration form of the Ripper analysis occurs with red wines and white wines made from fruit that has been degraded by *Botrytis cinerea.*

With reds the problem is not simply the inability to see the end point, but is due also to phenolic compounds that react with iodine, producing a false high reading. In highly pigmented wines, this inaccuracy can be quite extensive. White wines generally possess fewer phenols and other interferences. Wines that have been made from grapes infected with *Botrytis* may contain metabolites that cause problems with the analysis. Improved accuracy, particularly in red wines, can be attained by other analysis methods such as the AO procedure.

There has been a renewed interest in developing simple, fast, and reliable methods of analysis for sulfur dioxide. Such new methods generally are instrumental rather than chemical, and thus lend themselves to automation. Techniques of current interest involve enzymatic analysis, gas and liquid chromatography (including ion chromatography), potentiometry and polarography, ultraviolet and visible spectrophotometry, atomic absorption, and fluorometric spectrometry (including flow injection analysis).

ASCORBIC ACID

Ascorbic acid, also known as vitamin C, can be isolated from a variety of fruits and ranges in concentration from 500 mg/kg in oranges to over 3,000 mg/kg in guavas. The ascorbic acid content in grapes is low compared with other fruits, ranging from 5 to 150 mg/kg of fruit (Amerine and Ough 1974).

Ascorbic acid is a monobasic acid with lactone ring formation occurring between carbons 1 and 4 (see Fig. 10–3). Because of asymmetrism at C4 and C5, four stereoisomers may occur: D- and L-ascorbic, D-isoascorbic (erythorbic acid), and D-erythro-3-keto-hexuronic acids. Of these D- and L-ascorbic and erythorbic acids are of interest to winemakers.

Modes of Action

Ascorbic acid appears to have no significant antimicrobial properties nor does it seem to play a major role in limiting enzymatic browning (Amerine

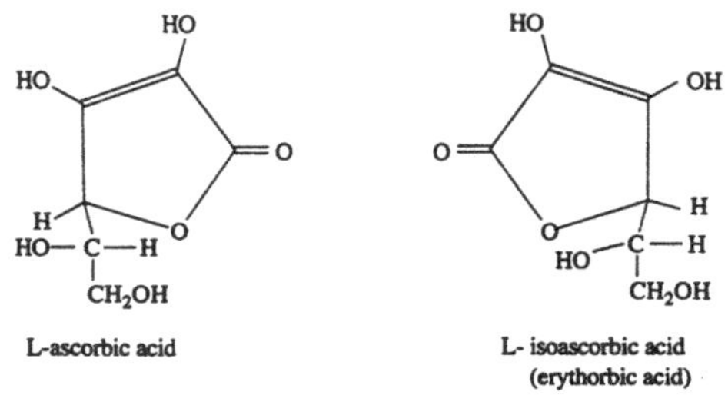

Fig. 10–3. Structure of ascorbic and erythorbic acid.

and Joslyn 1970). Oxidation is greatly accelerated in the presence of trace metals, especially copper, and the enzyme, ascorbic acid oxidase. Ascorbic acid readily undergoes auto-oxidation in the presence of flavone oxides (see Fig. 10–4). The latter are derived from oxidation of flavonoids in solution (Braverman 1963).

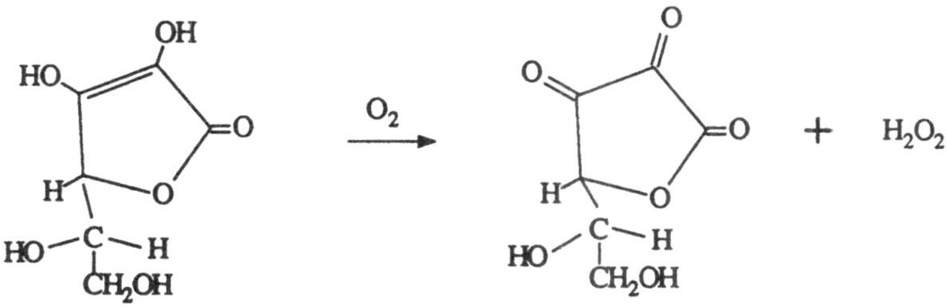

Fig. 10–4. Oxidation of ascorbic acid to dehydroascorbic acid.

Although its intended action is that of an antioxidant, Kielhofer (1960) demonstrated that ascorbic acid may not be effective and may, in fact, catalyze oxidation of some wine constituents. The reaction of ascorbic acid and oxygen generates hydrogen peroxide, which can, through coupled oxidation with ethanol, produce acetaldehyde (see Fig. 7–6). The latter binds with the free SO_2, making it unavailable as an antioxidant. As a result, more susceptible substrates (phenols) are oxidized. Thus, the effectiveness of ascorbic acid as a wine additive is disputed. In any event, because of its reactivity with molecular oxygen, use of ascorbic acid should be restricted to additions at bottling. Ascorbic acid and its optical isomer, erythorbic acid, first received governmental approval for use in wine and fruit juice production in 1956 and 1958, respectively. The OIV permits use of ascorbic acid at bottling at levels of 10g/hL (see Table I–10 in Appendix I). No maximum limit has been established by the U.S. government.

Because of its ability to catalyze oxidation of sulfurous acid, ascorbic acid should be used only in wines with low levels of SO_2. Fessler (1961) suggested that with the use of 60 to 182 mg/L (0.5–1.5 lb/1,000 gal) ascorbic or erythorbic acid, the total SO_2 level should be less than 100 mg/L. Ascorbic acid interferes with SO_2 determinations by Ripper titration.

ANALYSIS

The following is a cross reference of analytes and procedures found in Ch 20. SO_2 by Ripper titration (iodine and iodate); aeration-oxidation; Monier-Williams; and Enzymatic.

CHAPTER 11

VOLATILE ACIDITY

The total acidity of a wine is the result of the contribution of nonvolatile or fixed acids such as malic and tartaric plus those acids separated by steam volatilization. A measure of volatile acidity is used routinely as an indicator of wine spoilage (Table 11.1). Although generally interpreted as acetic acid content (in g/L), a traditional volatile acid analysis includes all those steam-distillable acids present in the wine. Thus, significant contributions to volatile acidity (by steam distribution) may be made by carbon dioxide (as carbonic acid); sulfur dioxide (as sulfurous acid); and, to a lesser extent, lactic, formic, butyric, and propionic acids. In addition, sorbic acid (added to wine as potassium sorbate), used as a fungal inhibitor, is also steam-distillable and should be taken into consideration when appropriate. The contributions of CO_2, SO_2, and sorbic acid interferences are discussed in Chapter 20 (Volatile Acidity).

MICROBIOLOGICAL FORMATION OF ACETIC ACID

The volatile acidity of a sound, newly fermented dry table wine may range from 0.2 to 0.4 g/L (Ribereau-Gayon 1961). Increases beyond this level,

Table 11–1. Legal limits for volatile acidity in wines. (Expressed as acetic acid)

Wine type	BATF (g/L)	California (g/L)	OIV (g/L) [a]
Red	1.40	1.20	0.98
White	1.20	1.10	0.98
Dessert	1.20	1.10	
Export (all types)	0.90	—	
Late harvest[b]			

[a]The volatile acidity of various specially fortified old wines (wines subject to special legislation and controlled by the governments) may exceed this limit.

[b]In the United States white wines produced from unameliorated juice of 28 °Brix (or more) volatile acidity can be 1.5 g/L. Red wines produced from unameliorated must of 28 °Brix (or more) volatile acidity can be up to 1.7 g/L.

however, may signal microbial involvement and potential spoilage. The principal source of acetic acid in stored wines is attributed to growth of acetic acid bacteria. Heterolactic lactic acid bacteria may also produce significant amounts of acetic acid in addition to lactic acid and CO_2 when growing on glucose (see Chapter 18).

Formation By Spoilage Yeasts

In some cases, high levels of volatile acidity may result from growth of wine yeasts. There is considerable variation in production of acetic acid and other byproducts among wine spoilage yeasts as well as strains of *Saccharomyces* sp. (Rankine 1955; Shimazu and Watanabe 1981).

Variations between strains of *Saccharomyces* with respect to acetic acid production has been related to acetyl-CoA synthetase. In this case, reduced levels of extracellular acetic acid is correlated with increased activity of the enzyme (Verduyn et al. 1990).

Among those yeasts involved in acetification of wine, *Brettanomyces* and its ascospore-forming counterpart *Dekkera* are known to produce relatively large amounts. Wang (1985) reported that acetic acid production by *Brettanomyces* in white wine after 26 days of incubation (28°C/82.5°F) increased from 0.31 g/L to 0.75 g/L. The impact of *Dekkera* on volatile acid formation is apparently less. Under identical conditions, Wang reported formation of 0.62 g/L acetic acid when compared with controls. In mixed culture fermentations, Fugelsang et al. (1993) also reported high levels of acetic acid when either spoilage yeast was present in coculture with *Saccharomyces cerevisiae* during fermentation (see Chapter 18).

Acetic acid is a normal byproduct of yeast growth and has its origin primarily in the early stages of fermentation. Several extrinsic factors may affect formation of acetic acid. These include pH, sugar, available nitrogen, and fermentation temperatures, as well as interactive effects of other microorganisms (see Chapter 18). Dessert wines produced from botrytized grapes often have higher levels of volatile acid than wines made from sound fruit. Ribereau-Gayon et al. (1979) report that extracts of the mold *Botrytis cinerea* may have a major impact on yeast and bacterial production of acetic acid during alcoholic fermentation. Addition of the extract was found to stimulate production of acetic acid and glycerol.

The effect of increased osmotic pressure resulting from high-sugar musts on volatile acid formation has also been reported (Nishino et al. 1985; Rose and Harrison 1970). Such fermentations typically have a longer lag phase with reduced cell viability and vigor. Generation time (budding) is also delayed. Cowper (1987) reported that at initial fermentable sugar levels above 20%, acetic acid increases with sugar level. He found acetic acid levels ranging from 0.6 to 1.0 g/L in musts of 32 to 42 °Brix (17.7–23.3 Baumé) when compared with controls at 22 °Brix (12.2 Baumé) with acetic acid of 0.4 g/L. Visually, yeast cells growing under conditions of high osmotic pressure appear stressed. This is generally manifest as a decrease or "shrinkage" in cell volume (Nishino et al. 1985). Concomitant with increased production of acetic acid is a proportionate increase in glycerol. This is not unexpected when organisms are grown in high solute (soluble solids) environments.

Must nitrogen levels may also play a role in acetic acid formation. When available nitrogen is low, higher initial sugar levels (as seen in over-ripe or mold-damaged fruit) may lead to increased production of acetic acid. Hydrogen ion concentration (pH) is also contributory to acetic acid production with more acetic acid produced at low (<3.2) pH (Copper 1987).

Fermentation temperature is also known to affect the levels of acetic acid produced by wine yeasts. Rankine (1955) found that volatile acid formation increased with increased fermentation temperature over the range of 15°C (59°F) to 25°C (77°F). Significant differences between yeast strains are also seen. Using two strains of *S. cerevisiae*, Shimazu and Watanabe (1981) noted that formation of acetic acid was maximum at 40°C (104°F) in one case, whereas maximum formation occurred at 10°C (50°F) in the second strain.

Unless controlled, the temperature of fermentation may rise to a point at which it becomes inhibitory to wine yeast. In practice, inhibition may be noted at temperatures approaching 35°C (95°F) or higher. Because acetic and lactic acid bacteria can tolerate temperatures higher than those needed to kill (inhibit) wine yeasts, stuck fermentations often are susceptible to secondary growth of these organisms. Pressure fermentations may also result in higher than expected volatile acid content (Amerine and Ough 1957), possibly due to selective inhibition of wine yeasts and growth of lactic acid bacteria.

Post-Fermentation Sources of Volatile Acidity

Cellar practices play an important role in volatile acid formation in stored wines. High levels of VA may result when headspace (ullage) is allowed to develop in wines. In this case, the combination of oxidative conditions and surface area may support rapid growth of both bacteria and yeast. Because acetic acid bacteria are aerobic organisms, depriving them of oxygen is a viable means of controlling further growth. However, wood cooperage does not provide the airtight (anaerobic) environment needed to completely inhibit growth.

Acetics may survive and grow at low oxygen levels present even in properly stored wines. Ribereau-Gayon (1985) reported viable populations (10^2 CFU/mL) of *Acetobacter aceti* present in properly maintained wines in wood cooperage. Survival of low numbers of the bacteria was attributed to slow exchange of oxygen (approximately 30 mg/L/year) into the wine. Transitory exposure to air, such as may occur during fining and/or racking operations, may be sufficient to stimulate growth. Although the exposure may be short term and the wine is subsequently stored properly, incorporation of oxygen was found to support continued growth of the bacteria. The problem becomes more apparent with increases in cellar temperature and wine pH.

During proper barrel storage, a partial vacuum develops within the barrel over time. Both water and ethanol diffuse into the wood and escape to the outside as vapor. In cellars where the relative humidity is less than 60%, water is lost from the wine to the outside environment and the alcohol content of the wine increases. Conversely, where a higher relative humidity exists, alcohol is lost to the outside environment. Diffusion of water and ethanol through pores in the staves creates a vacuum in the properly bunged barrel. Thus, even though some headspace may develop under these conditions, the oxygen concentration is very low.

Formation of a partial vacuum in the headspace requires tightly fitted bungs. Winemakers using traditional wooden bungs often store their barrels "bung over" or at the "two o'clock" position to ensure that the bung remains moist, and thus tightly sealed. Silicon bungs can provide a proper seal without being moist so barrels closed with such bungs are usually stored "bung up." Topping sealed barrels too frequently results in loss of vacuum and may accelerate both oxidation and biological degradation of the wine.

The volatile acidity of properly maintained barrel-aged red wines may increase slightly without the activity of microorganisms. An increase in volatile acidity of 0.1 to 0.2 g/L expressed as H_2SO_4, or 0.06–0.12 g/L as acetic acid is inevitable after 1 year in new wood, not as a result of biological degradation but due to hydrolysis of acetyl groups in the wood hemicellulose (Chatonnet 1993b). Additionally, acetic acid can result from coupled oxidation of wine phenolics (see Chapter 7) to yield peroxide, which, in turn, oxidizes ethanol to acetaldehyde and subsequently to acetic acid (see Fig. 7–6).

Although the practice is not recommended, winemakers forced to store wines in partially filled containers often blanket the wine with nitrogen and/or carbon dioxide. Nitrogen is the preferred blanketing gas because of its limited solubility in wine. Sparging of wines with carbon dioxide is also used. Upon standing, the gas escapes slowly from solution and, due to its density, remains at the wine's surface to offer a degree of protection against oxidative deterioration (see Chapter 14).

ACETATE ESTERS

The volatile character or "acetic nose" is not exclusively the result of acetic acid. Acetate esters, most specifically ethyl acetate, contribute significantly to this defect. For details regarding general ester formation, see Chapter 6.

Factors that can influence formation of acetate esters include yeast strain (as well as presence and population density of native yeasts), temperature of fermentation, and sulfur dioxide levels. Sponholz et al. (1990) report that growth of *Hanseniaspora uvarum* and its asporogenous counterpart, *Kloeckera apiculata,* during the early phase of fermentation results in significant production of ethyl acetate. These species frequently represent the dominant native yeast flora and their numbers may increase significantly even in fermentations inoculated with active *Saccharomyces* starters (Osborn et al. 1991). In their mixed culture experiments, Osborn and co-workers found levels of ethyl acetate increased as populations of *Hanseniaspora uvarum* increased. Radler et al. (1985) have identified "killer" strains of *Hanseniaspora uvarum* active toward sensitive strains of *Saccharomyces.* Other native yeast species that are known to produce substantial amounts of ethyl acetate (and other spoilage esters) include *Hansenula anomala* and *Metschnikowia pulcherrima* (Sponholz and Dittrich 1974).

Daudt and Ough (1973) found fermentation temperature to be the most important factor relative to the concentration of ethyl acetate produced. Maximum formation was reported over the temperature range of 10° to 21°C (50°–70°F).

Ethyl Acetate and Spoilage

Although high acetic acid content and the presence of ethyl acetate are generally associated with each other, they may not always be produced to the same extent. Ethyl acetate levels of 150 to 200 mg/L impart spoilage character to the wine (Amerine and Cruess 1960). Peynaud (1937) suggested a maximum ethyl acetate level of 220 mg/L be used rather than traditional analyses of acetic acid. This suggestion is based on the fact that high acetic acid content does not always confer a spoilage character to the wine. A volatile acid content of less than 0.70 g/L seldom imparts spoilage character, and, in combination with low concentrations of ethyl acetate, may contribute to overall wine complexity.

Acetic acid and ethyl acetate levels in unfermented must have also been examined as indicators of spoilage in grapes (Corison et al. 1979). The levels identified for "rejection," based on ethyl acetate in white and red musts, were 60 and 115 mg/L, respectively. Corresponding levels in wine were identified at 170 and 160 mg/L. By comparison, acetic acid levels needed for rejection in white and red musts were 1,190 and 900 mg/L. In white and red wines, rejection levels were lower: 1,130 and 790 mg/L.

SENSORY CONSIDERATIONS

Volatile acidity magnifies the taste of fixed acids and tannins but, itself, is masked by high levels of sugar and alcohol. This may help explain why VA can be sensorially detected in some wines at relatively low levels (<0.5 g/L) whereas in others it is not noticable at even higher concentrations. As discussed earlier, ethyl acetate is frequently associated with VA. The ester is frequently described as having "finger nail polish remover" properties and, at levels of 150–200 mg/L, can add spoilage "notes" to the wine. At lower concentrations, ethyl acetate may contribute to "fruity" properties of the wine.

REDUCTION OF VOLATILE ACIDITY

Both BATF and the OIV regulate the levels of volatile acidity (expressed as acetic acid) in domestic wines offered for sale. In California more restrictive regulations apply (see Table 11–1).

Reduction of high volatile acidity in wines is difficult. Attempts to lower volatile acid levels by neutralization generally yield undesirable results because of concomitant reduction in the fixed acid content. Similar problems (flavor and aroma stripping and modification) are encountered in the use of ion exchange. Reverse osmosis coupled with ion exchange has proven successful. In this case, only the permeate (acetic acid, ethanol, and water) is anion exchanged. Once the acetic acid is removed, permeate and retentate are recombined (Smith 1993 personal communication). Use of yeast for volatile acid reduction has also been studied; the application takes advantage of oxidatively growing yeasts using acetic acid as a carbon source. Utilization of acetic acid by active yeasts has led some winemakers to add high volatile acid wine to fermenting musts to lower volatile acid levels. However, such practices run the risk of contaminating the entire lot, and may have a detrimental impact on fermentation as well as on final wine quality. Judicious blending is probably the best practice to use in lowering the volatile acid content of borderline wines.

ANALYTICAL METHODS FOR VOLATILE ACIDITY

Steam distillation is frequently used in volatile acid analyses. The collected volatile acids present in the distillate are then titrated with standardized sodium hydroxide, and the results are reported as acetic acid (g/L). Although this is probably the most common procedure for VA determination, enzymatic, gas chromatographic, and high-performance liquid chromatographic procedures are becoming more popular.

ANALYSIS

The following is a cross-reference of the analytes and procedures found in Chapter 20. Acetic acid by gas chromatography, HPLC, distillation (Cash and Markham Stills), and enzymatic analysis; identification and isolation of acetic acid and lactic acid bacteria; ethyl acetate by gas chromatography.

CHAPTER 12

METALS, CATIONS, AND ANIONS

Grapes, musts, and to a lesser extent, wine contain trace amounts of heavy metals. The term "heavy metals" distinguishes those near the bottom of the periodic table (i.e., lead, mercury, and cadmium) and metal-like elements (such as arsenic) from the common "lighter weight" metals (such as sodium, potassium, calcium, and magnesium) present in significant amounts in grapes (see Chapter 15). Heavy metals are toxic to biological systems due to their ability to deactivate enzymes. As such, their allowable concentrations in foods is regulated. In terms of decreasing concentrations normally seen in wine, these include iron, copper, zinc, manganese, aluminum, lead, and arsenic (Table 12–1).

Unless exposed to significant airborne pollution, grapes accumulate only small amounts of heavy metals by translocation from the roots or by direct contact from vineyard sprays. These are normally absorbed onto the yeast cell membrane during fermentation, resulting in 0 to 50% of the original amount. Increased concentrations in wine result from contamination during post-fermentation processing. Sources include contact with nonstainless steel equipment and, potentially, as impurities in fining agents and filter media.

Table 12–1. Maximum acceptance levels for metals in wines.

	BATF	OIV	Australian[a]
		Concentration (mg/L)	
Arsenic	—	0.2	0.100
Bromine	—	1.0[b]	—
Cadmium	—	0.01	0.050
Copper	0.5	1.0	5.0
Fluorine/Fluoride	—	1.0	—
Lead	0.3	0.3	0.200
Sodium	—	60[c]	—
Zinc	—	5	5

[a]*Beverages, including wine.*
[b]*exception:* grapes from vineyards with brackish subsoil
[c]*exception:* wines produced from grapes of "exempt" vineyards

COPPER

Copper, in trace amounts, is an important inorganic catalyst in metabolic activities of microorganisms. The copper content of US musts and wines normally ranges from less than 0.1 to 0.30 mg/L. At higher levels, the metal plays an important role in catalyzing oxidation of wine phenols (see Chapter 7). Copper and copper complexes are more active than iron and its complexes. Copper at concentrations exceeding 1 mg/L may be sensorially detected (Amerine et al. 1972) and, along with cadmium and mercury, ranks among the most toxic of heavy metals. At levels above 9 mg/L, copper becomes a metabolic toxin that inhibits or delays alcoholic fermentation (Suomalainen and Oura 1971).

Sources of Copper in Wine

Copper in wine may be attributed to three sources: (1) vineyard sprays; (2) winery equipment; and (3) additions of copper salts for hydrogen sulfide correction in winery operations.

Use of copper-containing fungicides for mildew control may lead to significant increases in copper content of must (Eschenbruch and Kleynhans 1974). These workers demonstrated that timing of the spray application is crucial to residual copper levels. Applications of Bordeaux mixture (copper sulfate and lime) 1 week before harvest resulted in a twofold increase in copper (1.98 mg/L) in juice whereas spray applied 3 to 6 weeks before harvest did not materially differ from controls (0.98 mg/L). Use of copper oxychloride, however, resulted in significant increases even at 6 weeks before harvest (2.56 mg/L copper).

Elevated copper accumulations usually are due to the contact of must and/or wine with copper-containing alloys such as brass. Thoukis and Amerine (1956) reported a 40 to 89% reduction in the levels of copper present in must due to absorption onto the yeast cell. Yeast fining may be the best method of copper reduction in the United States where ferrocyanide (blue fining) is not permitted (see section on blue fining in this chapter and Chapter 16). Hsia et al. (1975) reported that wines produced from juices with 10 to 38% suspended solids had higher iron contents (10–20 mg/L) but were low in copper. They concluded that the suspended solids were rich in sulfates, which were reduced to sulfide with concomitant precipitating copper as copper sulfide.

Copper Instability

Instability, manifested initially as a white haze (in white wines) and later as a reddish-brown amorphous precipitate, may develop upon storage of bottled wine with excess copper. The precipitated "casse" develops only under the strongly reducing conditions found in bottled wine. Reoxidation by exposure to air or the addition of a strong oxidizing agent such as hydrogen peroxide causes the precipitate to disappear.

Ribereau-Gayon (1933) proposed that casse existed largely as cupric (copper II) sulfide in combination with wine colloids. Rentschler and Tanner (1951b), on the other hand, reported that the sediment from copper instability was high in cuprous (copper I) sulfide. Others have reported turbidity due to copper-protein and amino acid-copper complexes (Kean 1954) with the precipitate being high in protein nitrogen and rather low in sulfur-containing compounds (Kean and Marsh 1956a, 1956b). Thus casse formation resulting from Cu II reduction is not entirely the result of copper and sulfur interactions as once thought, but is instead a mixture of copper compounds and protein.

Protein levels in white wine may act as the limiting factor in cloud formation. Removal or reduction of protein may help prevent copper casse (see Chapter 8). Heat and light are known to accelerate casse formation. Peterson et al. (1958) found that under light conditions the copper complex formed was due to sulfite reduction and subsequent precipitation as copper (II) sulfide. Under dark conditions protein denaturation results from sulfite interaction. The copper-protein complex yields sulfate upon oxidation. Copper casse formation is also contingent upon low levels or absence of iron in the wine. Conditions that favor or hinder copper casse formation are summarized in Table 12.2.

Several compounds have been used to remove or reduce copper and iron in wine. These include potassium ferrocyanide ("blue fining") as well as the proprietary compounds Cufex and Metafine. The latter two are no

Table 12–2. Factors favoring and inhibiting copper casse formation in wine

Conditions necessary for copper casse formation	Preventive measures
Strongly reducing conditions (as seen in bottled wine)	Maintain copper levels at less than 0.3 mg/L
Iron absent or present in very low concentrations, protein present	Cold-stabilize and bentonite fine to reduce protein in white wines
Light and/or heat, which may hasten formation	Limit SO_2 additions

longer marketed in the United States. Because yeasts are effective in reducing metals during fermentation, their use alone or in combination with caseinate in post-fermentation metal reduction should also be considered (see Chapter 16). Screening procedures for identification of wine casses are given in Chapter 20.

IRON AND PHOSPHOROUS

At low concentrations, iron plays an important role in metabolism as an enzyme activator, stabilizer, and functional component of proteins. At higher-than-trace levels, iron has other roles: altering redox systems of the wine in favor of oxidation, affecting sensory characteristics, and participating in the formation of complexes with tannins and phosphates resulting in instabilities. The latter, also termed "casse," is seen initially as a milky white cloud and later as a precipitate. Two iron-containing casses may form in wines: "white" (ferric phosphate) and "blue" (ferric tannate) casse. The former represents the most commonly encountered iron-related casse.

Ferric tannate is not commonly observed. Resulting from tannin-iron complex formation, blue casse may be observed in white wines after tannic acid additions.

Ferric Phosphate Casse

White casse formation is dependent on (1) iron content, (2) pH, (3) redox potential, (4) phosphate content, and (5) nature and concentration of wine acids (Table 12–3). Normally, iron levels in grapes are low, even when grown on high iron soils. If no contamination occurs, a typical must has from 1 to 5 mg/L iron (Amerine and Ough 1974). Dupuy et al. (1955) could find no relationship between grape variety and iron content, although wines produced from different varieties grown on the same soil varied considerably. The most important source of iron in wine is contact with iron-containing alloys during processing. Modern wineries have largely eliminated this problem by use of stainless steel equipment.

Table 12–3. Factors favoring and inhibiting ferric phosphate casse formation in wine.

Conditions necessary for casse formation	Conditions impeding or inhibiting casse formation
Redox potential that favors the presence of Fe III over Fe II	Iron levels of less than 5 mg/L
pH 2.9–3.6	Clarification with bentonite and cold stabilization
Iron concentrations in excess of 7 mg/L	Citric acid additions 12–24 g/hL (1–2 lb/1,000 gal)

Iron instability, as ferric phosphate casse, is reported to occur only within the pH range 2.9 to 3.6. Thus, if iron is present at critical levels, any winery practice (i.e., blending) that alters the pH to fall in this range may affect the potential for casse formation.

Although iron in trace amounts is important in metabolic activities, levels exceeding 20 mg/L may inhibit fermentation. The yeast cell surface has a net negative charge that can rapidly and reversibly bind with exogenous divalent cations such as iron or copper. Thoukis and Amerine (1956) reported that 45 to 70% of the iron in must was removed during fermentation by adsorption onto the yeast cell membrane (see yeast fining in Chapter 16).

Most of the iron in wine is present in the ferrous, or Fe^{2+}, state. The ratio of Fe^{3+} to Fe^{2+} depends on the oxidation state of the wine, with the ferrous form predominating when oxygen levels are low. If oxidative conditions occur, Fe^{2+} is converted to Fe^{3+}. Subsequent reaction of Fe^{3+} with phosphates, normally present in wine at levels ranging from 135 to 200 mg/L (Chow and Gump 1987), may yield ferric phosphate casse, $FePO_4$. The addition of diammonium phosphate (DAP) to juice as a fermentation aid may significantly increase the phosphate content.

Iron may form complexes with several organic acids in wine, rendering it inactive in subsequent reaction with phosphate. Citric acid has a strong affinity for iron. When iron levels exceed 5 mg/L, some winemakers add citric acid at concentrations of 12 to 24 g/hL (1–2 pounds/1,000 gal). At concentrations significantly greater than 5 mg/L, special fining techniques may be required to remove (or lower the concentration) of the metal. (See section on blue fining and Chapter 16.)

ALUMINUM

The presence of aluminum in wine usually stems from contact with the metal in processing or storage. Aluminum contamination is rare. McKinnon et al. (1993) surveyed 267 (mostly Australian) wines for aluminum, and all but two had less than 1.99 mg/L. At wine pH, the rate of solubilization is rapid, and at levels exceeding 5 mg/L, instability, present as a haze and undesirable changes in color, flavor, and aroma may result. Aluminum in wine rapidly brings about reduction of SO_2 to H_2S. Aluminum haze is most pronounced at pH 3.8.

Bentonite may be an important source of aluminum in wine. McKinnon et al. (1992) also reported a twofold increase in aluminum after bentonite addition and a 54% increase attributable to filter pads and filter aids.

LEAD

Accumulations of lead in body tissues originates in the beginning of the food chain and from water supplies. Except for rare cases, lead is of chronic rather than acute importance. Prior to use of unleaded gas, auto emissions were an important source. On a daily consumption basis, the most important source of metal is drinking water. In the United States it is seldom found in water at more than 5 ppb (American Public Health Association et al. 1989). Reflecting the serious implications of chronic intake, the U.S. Public Health Service has lowered acceptable levels in drinking water to 20 ppb. In 1991 the U.S. Food and Drug Administration issued a temporary lead standard in wine of 300 ppb, which was the same acceptable limit imposed by the OIV. At that time Canada proposed 200 ppb. The OIV proposed a 250 ppb limit at their 1993 meeting. As of January 1, 1994, California imposed a limit of ≤150 ppb for all wines sold in the state.

In 1991, the U.S. FDA conducted a survey of lead in food. Levels in table wines ranged from 14 to 40 (median 30) ppb. In an additional survey conducted by BATF on 552 foreign and domestic wines, 14 of 435 imported wines had lead levels greater than 300 ppb whereas 3 of 117 U.S. wines exceeded this level (Higgens 1991). Reports by Henick-Kling and Stoewsand (1993) and Ough (1993a) summarize the sources of lead in wine.

Vineyard Soil

Soil, particularly soils in the immediate proximity of roads, represent a source for lead in wine. Henick-Kling and Stoewsand (1993) reported that lead in the soil of German vineyards ranged from 30 to 1000 ppb. Some French vineyards are much higher, 2.8 to 74 ppm (mean 14.1 ppm). Lead present in French/German must ranged from 10 to 570 ppb with higher concentrations (450 to 650 ppb) found in grapes adjacent to roads. Leaded gasoline has been used extensively in western Europe until recently.

Vineyard Sprays

Lead and arsenic in wine also arise from lead arsenate sprays used for control of lepidopteran larvae (Handson 1984). Lead and arsenic in white wines produced from treated grapes were present in concentrations 10 times (0.4 and 0.10 mg/L, respectively) that found in untreated grapes (0.03 and <0.01 mg/L). In the case of red wines (with longer extraction time), the difference increased to 14 times.

Wine Lees

Yeast and various fining agents are effective in lowering lead levels present in grapes by 50 to 90%. During fermentation, hydrogen sulfide reacts with lead producing lead sulfide, which is adsorbed by yeast. Correspondingly, lead in yeast and fining lees is relatively high (Ziegler 1990).

Brass Fittings

Brass also represents a source of lead and copper. Kaufmann (1992) reported that wine flowing through brass valves at 4.8 L/min resulted in lead increases of 25 ppb.

Fining Agents

Decolorizing/deodorizing carbons have been examined with respect to lead extraction (Enkelmann 1989). The lead content of carbons tested ranged from 37 to 79 mg/kg although analyses of wines treated with highest lead-containing charcoals showed increases of only 18.3 ppb.

Bentonites and filter aids represent negligible sources of lead in wine. Enkelmann (1990) reports increases of lead in wine attributable to filter aid of less than 1.5 ppb.

Bottles and Leaded Capsules

Because wine bottle manufacturing does not use lead or its salts, bottles do not contribute to lead in wine except as a contaminant in raw materials. An important source of lead in wine is lead foils. Analyses conducted by the U.S. BATF (Higgins 1991) compared samples taken directly (by pipette) from bottles having lead foils with samples of the same wine in which the foil was removed and wine poured without first wiping the bottle lip. Results showed lead concentrations significantly increased in wines poured from bottles with lead foils. However, wiping the lip of lead-foiled bottles before pouring is effective in eliminating this source of lead (Edwards and

Amerine 1977). As of Jan 1, 1992, California has banned tin-lead capsules in packaging wine.

Defects in cork or an imperfect seal leading to contact between wine and lead foil result in increased lead content. Gulson et al. (1990), using lead isotope methods that facilitated separation of lead arising from vineyard from that of foils, found that there was no movement from the foil into the wine. They attribute lead in wine to pouring without first wiping the bottle lip of lead capsuled wines.

METAL REMOVAL

Historically, German winemakers found that excess iron and copper could be removed from wine by the addition of ferrocyanide. Also known as *Blauschonung* or "blue fining," the practice has been used in Europe under strict governmental regulation for treatment of white and rosé, and in some cases, red wines. According to Peynaud (1984), 1 mg iron (III) requires 5.65 mg of potassium ferrocyanide. In practice, this may vary, ranging from 6 to 9 mg ferrocyanide/mg iron. Laboratory trials for each wine must be conducted to optimize fining additions (see Chapter 16). Because the major decomposition product of the reaction is cyanide, the practice is not permitted by BATF but is allowed by the OIV.

The reaction of ferrocyanide with iron (III) is slow, requiring up to 7 days (Castino 1965). Some workers recommend the use of ascorbic acid, at 50 mg/L, to reduce the iron present to the Fe(II) state, thus increasing the rate and completeness of the reaction. Properly used, ferrocyanide removes not only excess iron, but also most of the copper without adversely affecting the wine's flavor or bouquet.

The proprietary compounds Cufex and Metafine have been widely used for metal removal. However, currently neither product is being marketed. In the absence of chemical fining agents, yeast fining may be the best effective method to lower metal levels through absorption onto the yeast's cell membrane.

Reactions of Metals and Blue Fining Agents

Removal or reduction in the concentration of copper and iron by blue fining takes advantage of the fact that cyanide ion forms a very stable complex with transition metal ions. However, excess ferrocyanide remaining in solution may degrade, producing traces of free hydrocyanic or "Prussic" acid (HCN):

$$Fe(CN)_6^{4-} + H_2O \rightarrow Fe(CN)_5^{3-} + OH^- + HCN$$

The practice of blue fining is not permitted in the United States (see Chapter 16).

Because yeasts are effective in reducing metal levels during fermentation, their use in post-fermentation metal reduction should also be considered, especially in the absence of previously available agents. Langhans and Schlotter (1987) as well as our own trials have shown reductions of from 50 to 85% upon addition of recently rehydrated wine active dry yeast. Screening procedures for identification of wine casses are given in Chapter 20.

FLUORIDE

Fluoride in must and wine has received international attention. Because of exemption, the EEC allows importation of U.S. wines with 3 mg/L fluoride into member nations (Burns 1994, personal communication). This level also represents the maximum fluoride level allowed in U.S. drinking water. However, the EEC plans to reevaluate the status of fluoride in 1995 with the potential for reduction to 1 mg/L. The 1994 OIV maximum fluorine level allowed among member nations is 1.0 ppm (see Table 12–1).

In the warm climate of the San Joaquin Valley of California, fluoride-containing compounds such as the natural mineral cryolite, or proprietary formulations, have been used for years in control of vineyard pests such as larvae of the grape leaf skeletonizer and omnivorous leaf roller. These formulations are unsurpassed in terms of efficacy and benefit/cost ratio for the control. Both timing and rate of application are important in minimizing the levels of fluoride on fruit at harvest. Alternatives for control include the bacterial parasite *Bacillus thuringiensis* or "BT"; however, the relative cost is high.

Fluoride concentrations approaching 3 mg/L pose potential fermentation and post-fermentation problems. Wine yeasts are known to be variably sensitive to fluoride at 3 mg/L during fermentation (Fugelsang et al. 1994; Wahlstrom et al. 1992). Utilization of sensitive strains may result in stuck or protracted fermentations. Furthermore, fluoride is an irreversible inhibitor of acid urease, the enzyme used to hydrolyze residual urea upon completion of alcoholic fermentation (Famuyiwa and Ough 1991). Urea is a principal precursor to ethyl carbamate formation in wine (see Chapter 8). Fluoride concentrations of 1 mg/L inactivate 10 mg/L of enzyme (Famuyiwa and Ough 1991).

ANALYSIS OF METALS

Colorimetric analyses for both copper and iron are presented in Chapter 20. Copper reacts with diethyldithiocarbamate to form a colored complex measured spectrophotometrically at 450 nm. Iron as Fe (II) and Fe (III) quantitatively reacts with thiocyanate anion (SCN^-), forming a colored complex that can be measured spectrophotometriclly at 520 nm.

Atomic absorption is used in many laboratories as a rapid alternative to wet chemical procedures. Caputi and Ueda (1967) were among the first to report its value in heavy metals analysis in wine.

ANALYSIS

The following is a cross-reference of analytes and procedures found in Chapter 20. Copper and iron by atomic absorption, spectrometry and visual tests; lead by atomic absorption; phosphorus by atomic absorption; calcium by atomic absorption and selection ion electrode; Fluoride by ISE.

CHAPTER 13

Sorbic Acid, Benzoic Acid, and Dimethyldicarbonate

SORBIC ACID

Sorbic acid is a short-chained unsaturated fatty acid widely used in the wine and food industry as a chemical preservative. Because sorbic acid is not readily soluble, it is usually sold as the soluble salt, potassium sorbate. Effective primarily as a fungistat, sorbic acid is added to sweet wines. Although a generally effective inhibitor of fermentative yeasts, sorbic acid has little inhibitory activity toward lactic acid bacteria, acetic acid bacteria, or oxidative film forming yeast.

To be effective against microorganisms, sorbic acid must be incorporated into the cell. Antimicrobial activity resides in the undissociated molecule. The relative amounts present in this form vs. the ionized (negatively charged) form will determine ease of movement across the negatively charged cell membrane of yeasts. The pK_a of sorbic acid is pH 4.7 (the pK_a of an acid is the pH at which the ratio of undissociated to ionized forms is 1.0). Further decreases in pH bring about decreasing amounts of the ionized form and proportionately increase the concentration of the antimicrobial, undissociated form. Once incorporated into the cell the chemical may be operative against the dehydrogenase enzyme system of yeasts and molds (Desrosier and Desrosier 1977), interfering with oxidative assimilation of carbon.

The additive does not kill yeast but, if properly used, inhibits their growth. It has little inhibitory activity toward acetic and lactic acid bacteria. The activity of sorbic acid is directed toward controlling the growth of fermentative (*Saccharomyces* sp.) yeast in wine. Its effectiveness is dependent on several parameters including pH, sulfur dioxide concentration, alcohol content, and yeast titer.

The percentage of undissociated active form of sorbic acid decreases with increases in pH, and therefore the effectiveness of the compound is correspondingly reduced. The magnitude of this decrease is reported by Zoecklein et al. (1990). For example, at pH 3.0 the percentage of the undissociated form is 98.4% whereas at pH 3.7 it drops to 92.6%.

Sulfur dioxide appears to operate in a synergistic manner with sorbic acid. Ough and Ingraham (1960) reported inhibition of yeast growth in wine with additions of sorbic acid at 80 mg/L coupled with free sulfur dioxide levels of 30 mg/L. In a study of stability in white and red wines, Auerbach (1959) reported that inhibition in whites was achieved at a combination of sorbic acid at 75 mg/L and sulfur dioxide at 200 mg/L. In table wines, depending upon alcohol (% vol/vol) pH, yeast titer, and SO_2 level, as much as 100 to 200 mg/L sorbic acid is used.

Alcohol content is known to affect the activity of sorbic acid and is, therefore, another consideration in addition levels. Sweet wines of higher alcohol content require less sorbic acid for stabilization than those of lower alcohol levels. Table 13–1 compares wine alcohol content with the concentration of sorbic acid needed for inhibition of wine yeast. The yeast titer at time of addition is also important. Peynaud (1984) stated that cell concentrations must be less than 100 CFU/mL.

The relative inactivity of sorbic acid toward lactic acid bacteria may be due to formation of an addition compound between sorbic acid and SO_2 (Heintze 1976). Reaction between these components occurs in a 1 to 1 ratio. This may result in removal of enough free sulfur dioxide from the

Table 13–1. Inhibitory interaction of alcohol content and levels of sorbic acid in wine.

Alcohol content (% vol/vol)	Sorbic acid (mg/L)
10–11	150
12	100
14	50

Ough and Ingraham (1960).

system to support growth of lactic acid bacteria. Schmidt (1987) pointed to the importance of adequate levels of SO_2 in addition to sorbate for inhibition of lactic acid bacteria.

Internationally, legally permissible levels of sorbic acid in wine vary from 0 mg/L to several hundred mg/L. The BATF limit for sorbic acid in table wines is 300 mg/L, whereas for wine coolers, addition levels may not exceed 1,000 mg/L. In the case of coolers, benzoic acid, as potassium or sodium benzoate is a permissible additive in addition to sorbate. When used in combination with benzoates, the total concentration of both additives may not exceed 1,000 mg/L. However, unfavorable sensory responses to this concentration of either or both compounds would preclude use at this level. The OIV limit for sorbic acid is 200 mg/L.

Sensory Considerations

Attempts to establish threshold levels for sorbic acid in wines have resulted in a relatively wide range of values. Ough and Ingraham (1960) reported the threshold levels in wine to be as low as 50 mg/L, with the average 135 mg/L for trained panelists. By comparison, Tromp and Agenbach (1981) found the threshold to be 300 to 400 mg/L.

Because sorbic acid may be detected at use levels, winemakers should use only pure, fresh material. Thus, the use of old, yellow, oxidized potassium sorbate may result in detection at low levels. Oxidation of sorbate-treated wines also may lower the sensory threshold. Therefore, wines treated with the additive should be stored under low oxygen conditions with the SO_2 level kept high enough to prevent incipient growth of lactic acid bacteria. It is recommended that the preservative be added to sweet wine just before bottling. The use of potassium sorbate in wines destined for long-term aging is not recommended. De Rosa et al. (1983) reported development of an odor attributed to ethylsorbate in Charmat sparkling wines after 1 year of storage. The compound was reported to have a celery-pineapple-like odor.

An undesirable odor resulting from lactic acid bacterial (LAB) decomposition of sorbic acid may, on occasion, be noted. This odor, suggestive of geraniums, is commonly referred to as "geranium tone." Wurding et al. (1974) studied geranium tone and concluded that it resulted from the formation of 2,4-hexadien-1-ol and its lactate esters and acetates. It now appears that the significant portion of the off-character results from formation of the ether, 2-ethoxyhexa-3,5-diene, that is formed by rearrangement of hexadienol (Crowell and Guymon 1975) (Fig. 13–1). Of the lactic species present in wine, Edinger and Splittstoesser (1986) report only strains of *Leuconostoc oenos* carry out the conversion; neither *Pediococcus* nor *Lactobacillus* sp. used the acid in production of geranium tone.

SORBIC ACID —(H^+ / ETOH)→ ETHYL SORBATE
(2,4-hexadienoic acid)
↓
lactic acid
bacterial growth
↓
SORBYL ALCOHOL —(ETOH)→ ETHYL SORBYL ETHER
↓
Rearrangement
H^+
↓
3,5-HEXADIEN-2-OL —(ETOH)→ 2,ETHOXYHEXA-3,5-DIENE
("Geranium Tone")

Fig. 13–1. Microbiological formation of "geranium tone."

Probably the most successful way to deal with geranium tone is to blend. Unfortunately, the blend ratios of sound to defective wines are generally high, usually on the order of at least 11:1.

Due to the poor solubility of sorbic acid, the more soluble salt, potassium sorbate, is used in the wine industry. In using potassium sorbate, it is necessary to correct for the differences in molecular weights between the acid and its salt. This relationship is:

$$\text{Weight of potassium sorbate required} = \frac{\text{Molecular wt salt}}{\text{Molecular wt acid}} \times \text{Addition level of sorbic acid (mg/L)}$$

The calculated amount of potassium sorbate should be properly hydrated in wine or water before mixing into the wine.

BENZOIC ACID

The use of benzoic acid as the potassium or sodium salt has been restricted to the food industry in the United States. The product is not a BATF approved additive for table wines, but its use in wine coolers has been approved. Although the legal limit of addition is 1,000 mg/L, concerns relative to unpleasant sensory properties dictate much lower levels. Further, when used in combination with sorbic acid, the combined addition levels may not exceed 1,000 mg/L.

Like sorbic acid, antimicrobial activity of benzoic acid is linked to the unionized form. Thus, at pH values below its pK_a of 4.2, the percentage of active form increases significantly and the amounts needed for inhibition decrease correspondingly.

In addition to the effects of SO_2 and alcohol in combination with benzoates, carbonation level also acts to enhance antimicrobial activity. Schmidt (1987) reported linear decreases in the concentration of sodium benzoate required for inhibition with increases in carbonation levels in soft drinks.

In wine coolers, sodium or potassium benzoate is frequently used in combination with potassium sorbate and SO_2. The combination of sorbate and benzoate provides the needed level of antimicrobial activity at a concentration level that is generally not sensorially objectionable.

DIMETHYLDICARBONATE

Dimethyldicarbonate (DMDC) is the methyl analog of diethyldicarbonate (DEDC or diethylpyrocarbonate). Both DMDC and DEDC are active in inhibition of yeast at relatively low levels of addition (<250 mg/L). The mechanism of inhibition for both are similar; each brings about hydrolysis of yeast glyceraldehyde-3-phosphate dehydrogenase and alcohol dehydrogenase yielding inactive forms of the enzymes. Within the respective enzymes, the site of reaction appears to be the imidazole ring of histidine. DMDC remaining in solution breaks down to carbon dioxide and methanol (Fig. 13–2).

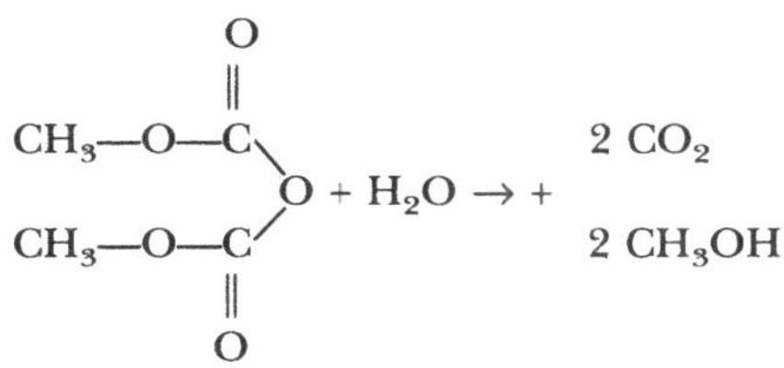

Fig. 13–2. Hydrolysis of DMDC to carbon dioxide and methanol (adapted from Porter and Ough, 1982).

DEDC was originally approved by BATF for use in wine in 1960 and revoked in 1972. Removal of DEDC was based on reports that under certain conditions, ethyl carbamate (urethane) may be produced upon decomposition of the sterilant in wine (see Chapter 8). Subsequently, it was proposed that the dimethyl analog replace DEDC. Because this compound does not contain the ethyl component needed in formation of urethane, it is believed to be an acceptable alternative.

With both DEDC and DMDC, temperature and alcohol act synergistically with the sterilant. Higher temperature and alcohol reduce the time needed for kill (Terrell, et al., 1993). Splittstoesser and Wilkinson (1973) reported the killing rate for *Saccharomyces cerevisiae* with DEDC to be 10 to 100 times faster at 40°C (104°F) than at 20°C (68°F). Using DMDC, Porter and Ough (1982) report that yeast cell count was reduced to zero (from 380 cells/mL) in 10 minutes upon addition of DMDC at 100 mg/L to wines (10% alcohol and 2% reducing sugar) held at 30°C (86°F). Review of their paper shows that viable cells were recovered only when the reaction was carried out at 20°C (68°F). Use of DMDC is permitted by BATF in wines as well as dealcoholized and low alcohol wines at levels not exceeding 200 mg/L.

ANALYTICAL DETERMINATION OF SORBIC AND BENZOIC ACIDS AND DIMETHYLDICARBONATE

Several analytical procedures are commonly used for determining the sorbic acid levels in wines. Because the acid is completely steam distillable (>99%), preliminary separation by distillation in a Cash or Markham still is an effective method to remove sorbic acid from the wine. The free acid (molecular form) is also extractable in iso-octane. Once isolated, sorbic acid may be derivatized to form a colored compound that can be measured spectrophotometrically with a simple visible colorimeter.

Molecular sorbic acid, due to its conjugation, can also absorb radiation in the UV region of the spectrum; this serves as the basis for several common analytical procedures. Samples containing sorbic acid can be diluted and analyzed using HPLC (see Chapter 20).

Caputi et al. (1974) developed a colorimetric procedure using thiobarbituric acid. The initial separation of sorbic acid is by steam distillation with subsequent oxidation to intermediate malonaldehyde. Reaction of the latter with thiobarbituric acid yields a highly colored condensation product that is measured at 530 nm.

Because of its state of conjugation, sorbic acid absorbs light in the UV area of the spectrum. Thus the distillate collected can also be analyzed directly at 260 nm without forming a colored derivative (Melnick and Luckmann 1954). It is necessary to acidify the distillate to prevent the sorbic acid from dissociating into its nonabsorbing ionic form.

Due to potential problems of component loss during the preliminary distillation step, Ziemelis and Somers (1978) developed a direct extraction procedure to extract a 0.25-mL aliquot of wine in iso-octane (2,2,4-trimethyl pentane) and absorbance was measured at 255 nm against an iso-octane blank. Iso-octane was chosen as the solvent of preference because of its limited extraction of wine phenolics, which also absorb in the UV.

Benzoic acid can be determined by measuring its absorbance in the UV region at 272 nm. The acid is first extracted into ether to separate it from other absorbing substances in the sample. Then the absorbance spectrum of the acid is scanned between 265 and 280 nm and a baseline constructed from the two absorbance minimums. Net absorbance is measured from this baseline to the peak of the absorbance maximum (at approximately 272 nm) and related to concentration of benzoic acid. As presented the method is applicable to samples containing from 200 to 1,000 mg/L benzoic acid.

Benzoic acid, as well as sorbic acid, can be determined using HPLC. A commercial organic acid column with acetonitrile-modified sulfuric acid mobile phase will resolve these two acids. Detection is best accomplished at 233 nm.

An alternative HPLC analysis of benzoic acid can be found in Official Methods (AOAC) section 12.018–12.021 (Fourteenth Edition, 1984). This procedure uses a C_{18} reverse phase column and a fixed wavelength (254 nm) UV detector.

Dimethyldicarbonate in wine reacts to form breakdown products of methanol, carbon dioxide, and ethyl methyl carbonate. The ethyl methyl carbonate residues in wine can be determined using conventional extraction and gas chromatographic techniques. An estimate of the original addition level of dimethyldicarbonate can be calculated from the ethyl methyl carbonate and percent ethanol analytical results (Stafford and Ough 1976).

ANALYSIS

The following is a cross-reference of analytes and procedures found in Chapter 20. Sorbic acid by visual spectrometry, by UV spectrometry, by direct extraction, extraction and UV spectrometry, by HPLC; benzoic acid by UV spectrometry and HPLC; DMDC.

CHAPTER 14

OXYGEN, CARBON DIOXIDE, AND NITROGEN

REDOX POTENTIALS IN WINE SYSTEMS

The oxidation-reduction (redox) potential of a chemical system such as wine is a measure of the tendency of the molecules, or ions, to gain or lose electrons. A compound with a large positive reduction potential (e.g., oxygen) will readily accept electrons producing the reduced form of that compound (e.g., water). Conversely, molecules (ions) with a lower (negative) redox potential (e.g., SO_2 or sulfite ion, $SO_3^=$) exhibit increasing tendencies to lose electrons, hence producing the oxidized form (e.g., sulfate ion, $SO_4^=$).

The various components in wine exist as mixtures of their oxidized and reduced forms (called a redox pair). Wine is a complex system made up of many such redox pairs. Examples may be found in Table 14.1 along with their standard reduction potentials (as described above, a measure of the tendency of the oxidized form of the compound to be reduced). Thus, reduction of one component causes oxidation of another until a final equilibrium point is reached, and net reduction equals net oxidation. Because pH affects the values of some redox potentials (see Table 14.1), the

Table 14–1. Typical redox pairs of compounds found in wine.[a]

Half-Reaction	Standard reduction potentials (volts) pH 3.5	pH 7.0
$1/2O_2 + 2H^+ + 2e^- \rightarrow H_2O$	1.022	0.816
$Fe^{+3} + 1e^- \rightarrow Fe^{+2}$	0.771	0.771
$O_2 + 2H^+ + 2e^- \rightarrow H_2O_2$	0.475	0.268
Dehydroascorbate + $2H^+ + 2e^- \rightarrow$ ascorbate	0.267	0.060
Fumarate + $2H^+ + 2e^- \rightarrow$ succinate	0.237	0.030
$Cu^{+2} + 1e^- \rightarrow Cu^+$	0.158	0.158
Oxaloacetate + $2H^+ + 2e^- \rightarrow$ malate	0.105	−0.102
Acetaldehyde + $2H^+ + 2e^- \rightarrow$ ethanol	0.044	−0.163
Pyruvate + $2H^+ + 2e^- \rightarrow$ lactate	0.027	−0.180
Acetyl−CoA + $2H^+ + 2e^- \rightarrow$ acetaldehyde + CoA	−0.203	−0.410
$SO_4^{2-} + 4H^+ + 2e^- \rightarrow H_2SO_3 + H_2O$	−0.244	−0.657
Acetate + $2H^+ + 2e^- \rightarrow$ acetaldehyde	−0.390	−0.600

[a]Standard conditions are unit activity for all components with the exception of H^+, which is 3.2×10^{-4} M or 10^{-7} M; the gases are at 1 atm pressure.

Recalculated from Florkin, M. and Wood, T., *Unity and Diversity in Biochemistry*. London: Pergamon Press.

position of this final equilibrium point in a wine is very dependent on the pH. The higher the pH the more negative the redox potential of many compounds, and therefore the better these compounds act as reducing agents.

"Oxidizing agents" are compounds (such as oxygen) that cause other compounds to be oxidized (that is, lose some of their electrons). A reducing agent, on the other hand, is a molecule or ion that causes other components to be reduced. The main reducing (or antioxidizing) agents found in wines are SO_2, ascorbic acid, and phenols. These compounds can react or bind with oxygen and lower the overall redox potential of the system (i.e., a lower oxygen level yields a lower redox potential for the system). Again, pH influences the reducing power of these agents.

The rate of reaction for different reducing agents is variable. Ascorbic acid is rapid, SO_2 is much slower, and phenols are even slower. As an example, after bottling, the oxygen content, and thus the redox potential of a wine, will be lowered over a 3 to 5 day time period due to relatively slow reactions of oxygen with SO_2 and phenols. Ascorbic acid is a more efficient antioxidant than SO_2 because it reacts with oxygen faster. Unfortunately, the reaction of ascorbic acid and oxygen may also generate hydrogen peroxide, which can react with ethanol, producing acetaldehyde (see Chapter 7). The latter binds with the free SO_2, making it unavailable as an antioxidant. As a result, more susceptible substrates (e.g., phenols) are oxidized.

It is the molecular form of a compound (e.g., acetic acid) that is volatile and responsible for aroma. Ionized forms (e.g., acetate), themselves, are nonvolatile. Thus shifts in the redox potential of a wine may produce the more volatile form of a number of compounds, causing a change or increase in the aroma. The opposite may also occur with aromas being masked or eliminated. Because pH affects the redox potential, it also determines the equilibrium state of a wine and the relative volatility of some aroma compounds.

Wine Buffering Capacity and Yeast Autolysate

Because phenolics and other compounds, such as SO_2, bind with oxygen, they contribute to the resistance of wine to oxidation. Collectively, these contribute to the buffering capacity of a wine. A poorly buffered wine at oxygen saturation has a high redox potential due to the lack of oxygen-binding compounds. A well-buffered wine having high capacity for oxygen uptake will have a much lower redox potential.

Winemaking practices may influence and enhance a wine's buffering capacity with respect to oxidative changes and thus, the rate at which it ages. The Burgundian practice of extended lees contact (*sur lie*) is a frequently used stylistic tool in production of certain white wines. Newly fermented wines are aged on yeast lees for varying periods of time. During yeast cell death, intracellular enzymes bring about hydrolysis of cytoplasmic components (proteins, peptides, amino acids) as well as the cell wall polysaccharides (largely mannoproteins). Mixing the lees periodically is reported to increase release of both groups of compounds (Ferrari and Feuillat 1988). In addition to improved aging potential resulting from the autolysate's contribution to the pool of oxidizable substrate, winemakers using extended lees contact report enhanced mid-body and complexity for both red and white wines. The sensory features of yeast autolysis (*sur lie*) were presented in Chapter 8.

OXYGEN

Oxygen contact with must and wine is of concern throughout the winemaking process. In some instances, such as in juice processing, and during barrel aging, controlled exposure to oxygen may play an important and beneficial role in wine quality. Controlled aeration may enhance phenol polymerization, influencing both color stability and suppleness in red wines. During bottling, oxygen levels should be as low as possible to prevent premature deterioration.

Since the early 1970s it has been increasingly reported that limited oxygen contact with the must (in absence of SO_2) before fermentation may not be as detrimental as once thought (Long and Lindbloom 1986; Muller-Spath et al. 1978). Controlled oxidation (and sulfite elimination) in juice processing has several advantages: (1) Oxidative polymerization and subsequent precipitation of astringent and bitter phenols in white juice reduce their levels in the wine. Enzymatic browning occurring in unsulfited juice is reversed during fermentation. (2) Prefermentation additions of SO_2 carry over into the wine as adducts of compounds such as pigments and acetaldehyde. Aside from having minimal or no activity with respect to oxidative and microbiological control, the reservoir of bound SO_2 creates a need for further additions in order to achieve the desired levels of molecular SO_2.

Achieving reductions in the levels of SO_2 used, while maintaining high quality, requires control of virtually every facet of production from the vineyard to the bottled wine. Concerns begin in the vineyard and include grape variety, climatological conditions, harvest maturity, and temperature at harvest, during transport, and processing.

Grape chemistry and integrity play important roles in oxidation. High pH musts and wines tend to oxidize at a faster rate than low pH lots. Grapes grown in warmer regions tend to darken faster than the same variety grown in cooler areas. Maturity may affect the tendency toward browning. Although a variety of enzymes are present in sound fruit, fermentation and processing (including juice fining) reduces their activity substantially. Thus, tyrosinase (polyphenoloxidase)activity is limited to prefermentation and does not play a major role in further oxidative degradation. Winemakers forced to deal with mold-damaged fruit expect to see more oxidation than with sound fruit. In these cases, use of SO_2 even at relatively high levels may not be useful in control of further deterioration. For example, *Botrytis*-produced lacasse is relatively insensitive to SO_2 and alcohol and its activity may continue in the presence of alcohol (see Chapter 3, and Chapter 20 for a procedure for the analysis of lacasse).

The color of wine is one of its most important characteristics. The potential for browning in wines may be closely tied to grape variety as well as mold growth that may have occurred. Oxidation in wines results from direct interaction of susceptible substrates with molecular oxygen. This happens normally during the course of controlled barrel aging. It may be accelerated when wine is exposed to air.

Oxygen and Yeast Metabolism

Oxygen plays important roles in the physiological status of yeast. Molecular oxygen is required in synthesis of lipids (principally oleanoloic acid) and steroids (ergosterol, dehydroergosterol, and zymosterol) needed for functional cell membranes. Steroids play a structural role in membrane organization, interacting with and stabilizing the phospholipid component of the membrane. It has been shown that yeasts propagated aerobically contain a higher proportion of unsaturated fatty acids and up to three times the steroid level of conventionally prepared cultures. This increase correlates well with improved yeast viability during the fermentative phase.

As fermentation begins, oxygen present in must is rapidly consumed, usually within several hours. After utilization of initial oxygen present, fermentations become anaerobic. Because yeasts are not able to synthesize membrane components in the absence of oxygen, existing steroids must be redistributed within the growing population. Under such conditions, yeast multiplication is usually restricted to 4 to 5 generations, due largely to diminished levels of steroids, lipids, and unsaturated fatty acids.

The need for oxygen supplementation may be overcome by addition of steroids and unsaturated fatty acids. Oleanolic acid, present in the grape cuticle (Radler 1965), has been shown to replace the yeast requirement for ergosterol (major steroid produced by yeast) supplementation under anaerobic conditions (Brechot et al. 1971). There is also evidence that exogenous addition of yeast "ghosts" or "hulls" may overcome the oxygen limitation, possibly by providing a fresh source of membrane components.

Methodology of starter propagation is important with respect to subsequent requirements for oxygen. Aerobic propagation has been demonstrated to significantly enhance subsequent fermentative activity. Yeast populations reach higher final cell numbers and fermentations proceed at a faster rate when starters are prepared with aeration (Wahlstrom and Fugelsang 1988).

Post-Fermentation Oxidation

Management of oxygen pickup in aging wine is crucial. Because air is 21% O_2, it is necessary to lower this concentration to the lowest point possible during movement or fining of delicate white wines. Transfer lines, pumps, and receiving tanks may be purged with inert gas before use. Commercially available racking devices use nitrogen gas to pressurize the headspace above the wine, thus facilitating transfer of wine from barrel to barrel without pumps. Controlled air exposure (splash racking, etc.) is frequently useful to help soften and evolve tannins and stabilize color in young red wines (see Chapter 7).

ACETALDEHYDE

As wines age, acetaldehyde levels increase due to chemical oxidation of ethanol, and in the case of improperly stored wines, growth of oxidative yeasts and bacteria at the wine's surface. The film (referred to in older literature as "mycoderma") exists as a mixed population of several species including *Pichia, Candida, Hansenula,* and oxidatively growing *Saccharomyces.* Ethanol represents the primary source of carbon in aerobic film growth. In addition to substantial production of acetaldehyde, film yeasts may produce acetic acid and ethyl acetate. Oxidative metabolism may be exploited in production of flor sherry where acetaldehyde levels may exceed 500 mg/L.

Control of film yeast is best accomplished by depriving them of oxygen needed for growth. Thus, minimizing oxygen contact during storage is essential in preventing population buildup. Cellar temperature is also known to impact development of oxidative yeasts. Dittrich (1987) reported that at 8°C to 12°C (47°–54°F) negligible film formation was seen in wines of 10 to 12% alcohol, whereas at higher temperatures, growth was observed at 14% alcohol.

In that some film-forming species are resistant to molecular SO_2 levels greater than 2 mg/L, attempting to suppress growth with SO_2 (once film formation is observed) is not feasible. In these instances, it is recommended that the wine be transferred (without disruption of the film) to sanitized cooperage (Baldwin 1993).

Acetaldehyde is also an intermediate in bacterial formation of acetic acid. Under low-oxygen conditions and/or alcohol levels greater than 10% (vol/vol), acetaldehyde tends to accumulate instead of being oxidized to acetic acid. Muraoka et al. (1983) report that aldehyde dehydrogenase is less stable than ethanol dehydrogenase.

Aside from chemical and microbiological formation, winemaking practices influence the level of acetaldehyde present in wine. Timing of SO_2 additions is important. Prefermentation additions as well as additions during the course of fermentation not only increases the concentration of acetaldehyde in the wine but lowers the concentration of free SO_2. Other parameters reported to result in higher levels of acetaldehyde include increases in pH and fermentation temperatures (Wucherpfenning and Semmler 1973).

The combination of anthocyanin and tannin molecules is promoted by the presence of acetaldehyde produced from ethanol oxidation during barrel storage. The anthocyanin-tannin complex is important to red wine color stability (see Chapter 7).

Oxygen uptake such as may occur during bottling may result in oxidation of ethanol to acetaldehyde. The muted varietal character of newly bottled wines, especially those with low SO_2 levels, reflects this transitory oxidation and accumulation of acetaldehyde. Although it is believed by some that SO_2 additions at bottling will limit short-term oxidation, the binding rate of $SO_3^=$ is a very slow process (see Chapter 9).

Sensory Considerations

Immediately after fermentation, table wines generally have an acetaldehyde concentration of less than 75 mg/L. The sensory threshold in wines ranges from 100 to 125 mg/L. Above this concentration acetaldehyde can impart an odor to the wine described as over-ripe bruised apples, sherry, and nut-like. An aroma screen provided in Chapter 20 takes advantage of the fact that acetaldehyde quickly binds with sulfur dioxide. Blending and refermentation are industry practices used to reduce acetaldehyde concentrations.

CARBON DIOXIDE AND NITROGEN

Carbon dioxide (CO_2) present in table wines may arise from several sources. Alcoholic and malolactic fermentations as well as refermentation of unstabilized sweet wines represent the important biological origins (see Chapter 18). Utilization of the gas in post-fermentation processing and bottling represents the principal abiotic source.

Newly fermented wines are saturated with CO_2. As a result of outgassing and post-fermentation processings (racking, fining, etc.), levels drop to near 1,000 mg/L during the first several months of aging. Depending on intrinsic properties of the wine and cellaring practices, additional decreases of 100 mg/L or more may be seen during the first year with negligible losses thereafter.

Because solubility is, in part, temperature dependent, fermentation and storage temperatures contribute to retention. The use of slow fermenting yeasts may also play a role in establishing soluble levels. In wine of 11% alcohol (at atmospheric pressure), increasing the temperature from 0°C to 20°C (32°–68°F) brings about a solubility decrease from 2.9 to 1.4 g/L (Jordan and Napper 1987).

Alcohol is also important in retention of CO_2. Because its solubility is inversely related to the concentration of alcohol, CO_2 retention at equivalent temperatures is greater in wines of lower compared with higher alcohol. Other contributing factors include reducing sugar levels and viscosity arising from polysaccharides and phenols. Originating from the grape as well as yeast autolysate, the polysaccharides exist as polymers of glucose, galactose, arabinose, and rhamnose. The concentration of polysaccharides in white wine ranges from 170 to 970 mg/L (Wucherphenning and Dittrich 1984).

Cellar processes play an important role in gas retention. For example, storing wine at low temperatures aids in retention, whereas warming accelerates degassing. Processes such as racking, fining, and filtration may reduce the amount present, whereas use of CO_2 in purging tanks and lines may contribute to increases. Nitrogen gas serves as an effective tool in lowering levels of CO_2. Compared with CO_2, which has a solubility in wine (at atmospheric pressure) of 1,500 mg/L, N_2 is relatively insoluble, with a solubility of only 14 mg/L. Because of this insolubility, N_2 in low concentrations can be effectively used to strip other volatile species (including CO_2), which are more soluble. The mechanism of action involves the equilibrium between the gaseous form and dissolved carbonic acid:

$$CO_2 \text{ (gas)} \leftrightarrow CO_2 \cdot H_2O \text{ (dissolved)}$$

During nitrogen sparging, N_2 molecules collide with and sweep some of the CO_2(gas) molecules to the surface. This, in turn, causes more $CO_2 \cdot H_2O$ to shift to the gaseous form where it too is swept from the system.

Exclusive use of CO_2 or N_2 may lead to either "spritzy" or "flat" character. Use of formulated CO_2/N_2 mixtures ("balanced mix") is reported to more closely simulate natural balance (Allen 1993).

Carbon dioxide and protein interactions are partially responsible for maintaining CO_2 solubility (Anderson 1959). The CO_2-protein interaction is electrostatic and not chemical. Due to its symmetry, CO_2 does not exhibit polarity. Carbonic acid, however, is asymmetric and its negative dipole is adsorbed to positively charged protein fractions. Thus, more CO_2 is adsorbed at lower than higher pHs.

Some wineries sparge wines with CO_2 before or during bottling. In the United States, CO_2 additions are permitted in still wines provided that not more than 3.92 g/L is present at time of sale. By comparison, OIV levels are set at 1.0 g/L.

Sensory Considerations

Carbon dioxide can provide a tactile sensation, magnify the sense of acidity, and enhance odor intensity. As such, white wines designed for early release may be produced using residual CO_2 to enhance the fruit character and enliven the palate.

Carbon dioxide is perceptible in water at 200 mg/L, and in wine at about 500 mg/L (Peynaud 1984). At levels of greater than 700 mg/L, CO_2 may be tactically perceivable and at greater than 1,000 mg/L, CO_2 bubble formation is frequently noted. It is the authors' experience that the difference threshold in wines of 900 to 1,300 mg/L is near 300 mg/L.

CO_2 enhances the sense of acidity, thus reinforcing tannin and bitter elements and reducing the sense of sweetness. As such carbon dioxide is an important stylistic tool influencing palate balance (see Chapter 7).

USE OF GASES

Carbon dioxide, nitrogen, and argon are used in wine production in three ways: (1) sparging, (2) blanketing, and (3) flushing. Sparging involves the introduction of very fine gas bubbles to help remove dissolved oxygen or CO_2 or occasionally to add CO_2. The solubility of a gas in a liquid is proportional to the partial pressure of that gas in the gaseous atmosphere in contact with that liquid. When fine bubbles are dispersed, a partial pressure develops between the sparging gas (usually N_2) and the dissolved gas (usually O_2). The difference in partial pressures causes the dissolved gas to leave the wine. The effectiveness of sparging is dependent on the wine, temperature, time, gas volume, and bubble size.

Blanketing attempts to maintain a gas layer above the wine surface in the hopes of minimizing wine-air contact. Both nitrogen and CO_2 are used for the purpose, although nitrogen is preferred because it has a very low solubility in wine, 14 mg/L. This is a principal reason why it is also an effective sparging gas. Carbon dioxide, on the other hand, has a solubility of 1,500 mg/L (Wallace 1980). In order to prevent the growth of aerobic microorganisms on the wine surface, the O_2 concentration must be reduced from the 20.9% O_2 found in air to 0.5% or less at the wine surface.

The reduction of O_2 to levels low enough to control biological growth for extended periods is difficult using blanketing gases. Some producers choose to lightly but regularly CO_2 sparge partial tanks. As the CO_2 comes out of solution it helps to displace the O_2 at the wine surface. Because CO_2 is "heavier" than either N_2 or air it may remain on the surface and help protect the wine. It is universally accepted that there is no better substitute for protection from O_2 than storage in completely full containers.

Wines stored in barrels that are properly sealed develop a partial vacuum over time (Peterson 1976). If the vacuum is maintained it helps to limit the growth of aerobic microorganisms even if the barrel is not completely full.

Gassing may be accomplished with N_2, CO_2, or "balanced mixtures" of the two, as well as argon. Gassing is best accomplished by introduction at the tank's bottom draw. This ensures more complete displacement of oxygen out the top. Properly carried out, O_2 levels less than 1% are achievable (Wallace 1980). Portable oxygen meters are useful to monitor the effectiveness . After movement of the wine, any headspace should be gassed prior to closing the access port.

The amount of gas needed to adequately displace O_2 in empty tanks and hoses or dissolved oxygen (DO) from wine is substantial. The following examples and figures are adapted from Wallace (1980).

Example: The volume of N_2 needed to lower the O_2 content of an empty 1,000 gallon tank from atmospheric to 1% can be determined with reference to Fig. 14–1b. Locate 1% on the "Percent Oxygen" axis. Follow across to the curve and down to the "Number of Volume Changes" axis. The value read on this axis is 3.25. Because there are 7.48 gal/ft^3, a 1,000 gallon tank has 133.7 ft^3. Knowing that 3.25 volume changes/ft^3 are needed to accomplish the goal, multiplying 133.7 ft^3 by 3.25 yields 434 ft^3. Since a standard nitrogen cylinder holds 224 ft^3, the job requires two cylinders. In like manner, a CO_2 cylinder holds 435 ft^3, and one is required.

Using Fig. 14–1a, one can determine the amount of N_2 needed to lower the O_2 level in wine. For example, assume it is necessary to reduce the DO in 1,000 gallons of wine from 2.2 to 1 mg/L (or by 55%). Find 55% on the "Percent Dissolved Oxygen To be Removed" axis. Trace a line across to the curve and then down to the Nitrogen-to-Wine Ratio (ft^3/gal) axis. In this case, the value is 0.05. Thus it is necessary to use 0.05 ft^3/gal × 1,000 gal or 50 ft^3.

Before bottling, O_2 should again be checked and lowered to less than 0.5 mg/L for whites and 1.0 mg/L for reds. During bottling, bottles are purged with N_2 or CO_2 before filling. The goal here is to reduce oxygen to 1 to 2%. During bottling O_2 should be checked regularly to ensure that there is no, or negligible, pickup between bottling tank and bottle.

Cuveé

If sparkling wine contains dissolved air or nitrogen under pressure, as well as carbon dioxide, gushing can occur. For this reason, nitrogen sparging and excessive aeration of the cuveé wine is undesirable. The solubility of air or nitrogen is very low under pressure. When bottles that contain air or nitrogen are opened, these gases immediately come out of solution as fine bubbles that then gather carbon dioxide and gush. These gases make the system unstable because their escape rates may be higher than that of the carbon dioxide. It is therefore imperative that cuveés not be nitrogen-sparged or undergo excessive aeration. There may be 15 psi or more of air in the wine at cuveé bottling (Miller 1966); if too much additional air is

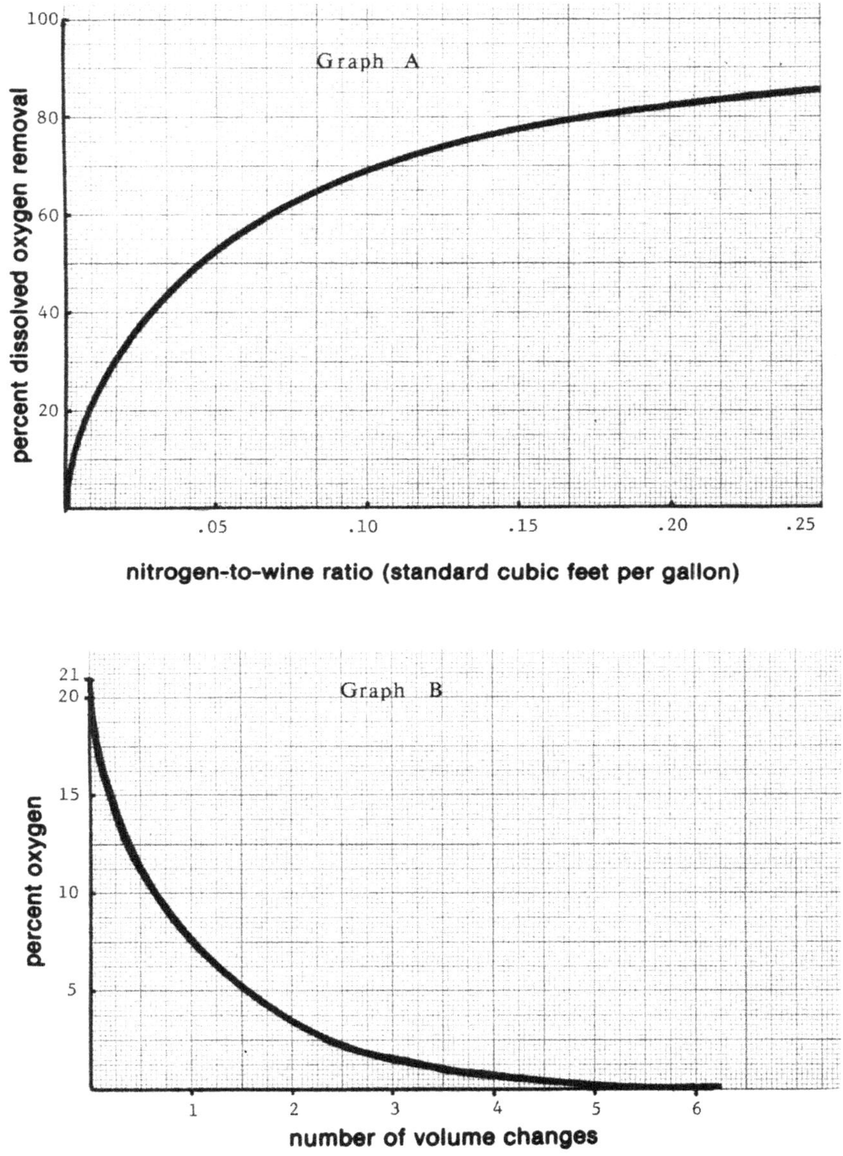

Fig. 14–1. Volume of nitrogen gas used in sparging/stripping operations.

dissolved in the wine, it may make the final bottle unstable or "wide" at the time of disgorgement and consumption.

MEASUREMENT OF CARBON DIOXIDE

Carbon dioxide in wine can be measured using a variety of analytical techniques and instruments, including the Carbodoser, titration, blood gas analyzers, and CO_2 specific electrodes.

The Carbodoser is a glass tube that measures the amount of CO_2 outgassed from a fixed quantity of wine. Comparing this volume to a calibration chart, one can read directly in mg CO_2/liter of wine. It is relatively easy to use and gives reproducible results. Because the Carbodoser method does not require the sample to be treated to facilitate "release" of CO_2, results may be variable depending on pH, temperature, etc.

The titration method requires two titrations (the original sample and a degassed sample)to an exact pH reading, and calculates the CO_2 content from the difference in the two titration volumes. The method is relatively time consuming, and requires the enzyme carbonic anhydrase. However, it is accurate.

The blood gas analyzer uses a small quantity of wine previously mixed with a CO_2-releasing agent (lactic acid). Gaseous CO_2 generated is transferred to a thermal conductivity detector (similar to those used in some gas chromatographs). The signal produced reads directly in millimoles CO_2/liter. This method is rapid (approximately 30 sec/sample), as well as accurate.

With the CO_2 electrode, one acidifies the sample and allows the gas to pass through a CO_2-permeable membrane into the electrode chamber. The electrode works with any standard expanded scale or "select ion" pH meter, and can provide accurate results. As with all electrode methods, there is some necessary "care" required to properly use the electrode, especially if it is not being used on a continuous basis.

ANALYSIS

The following is a cross-reference of analytes and procedures found in Chapter 20: oxygen by oxygen specific electrode; acetaldehyde by gas chromatography, visible spectrometry, and titration; acetaldehyde sensory screen by sulfite binding; carbon dioxide by enzymatic analysis, Carbodoser and specific ion electrode.

CHAPTER 15

TARTRATES AND INSTABILITIES

Tartaric acid (H_2T) and its salts, potassium bitartrate (KHT) and calcium tartrate (CaT), are normal constituents of juice and wines and important to stability. The formation of crystalline deposits is a phenomenon of wine aging, although it generally does not meet with consumer acceptance. Thus, winemakers strive to reduce the potential for bottle precipitation.

The tartaric acid content of grape must ranges from 2.0 to 10 g/L and varies according to region, variety, maturity, soil, and viticultural practices (see Chapters 3 and 4). In grapes and wines, tartaric acid is found in its ionized forms: bitartrate and tartrate. Depending on pH, the ratios of $H_2T/HT^-/T^=$ can vary greatly and thus significantly influence the potential for precipitation of insoluble salts. The relative distribution of each component as a function of pH is seen in Fig. 15.1.

POTASSIUM

The level of potassium in grape must ranges from 600 mg/L to over 2,500 mg/L in certain red varieties. During *veraison,* potassium from the soil is

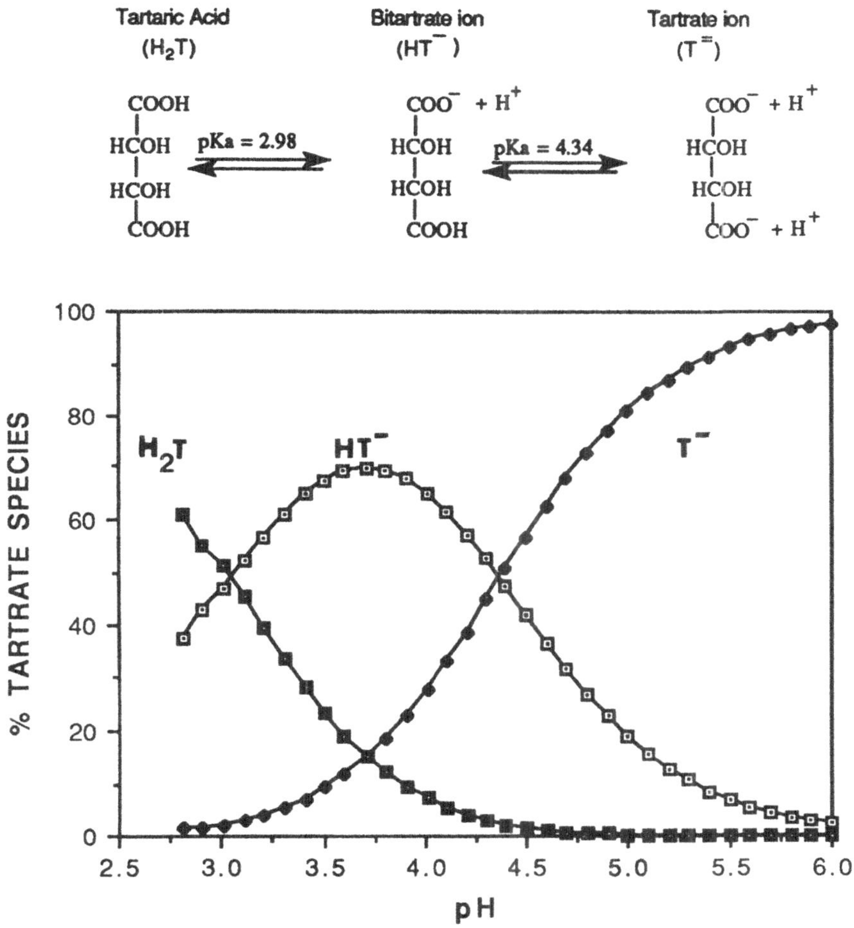

Fig. 15–1. Relative concentration of tartaric acid species in aqueous solution at different pH values.

translocated into the fruit where it forms soluble potassium bitartrate. Factors affecting potassium uptake are reported to include soil, cultivar, rootstock, etc., as well as cultural practices (see Chapter 4). Although potassium bitartrate is partially soluble in grape juice, alcohol and low temperature decreases its solubility, resulting in a supersaturated solution and subsequent precipitation. Methods for potassium analysis are given in Chapter 20.

CALCIUM

Calcium is present in wine at levels of 6 to 165 mg/L (Amerine and Ough 1980) and may complex with tartrate and oxalate anions to form crystalline precipitates. Several sources contribute to increased calcium in wine, including: soil, practices such as liming, fermentation or storage in concrete tanks, and use of calcium-containing fining material and filter pads. Furthermore, where $CaCO_3$ is used in deacidification or where "plastering" is used for adjusting the acidity of shermat material (sherry base wine), increased calcium levels occur. Because of changes in processing technology, CaT is generally not a problem in the United States. Calcium tartrate instabilities usually appear from 4 to 7 months after fermentation. Calcium tartrate can represent a problem due to its temperature independence and the difficulties in predicting instability.

All wines differ in their "holding" or retention capacity for tartrate salts in solution. If that capacity is exceeded, precipitation occurs, resulting in formation of "tartrate casse." In wine, solubility is largely dependent on alcohol content, pH, temperature (with KHT, not CaT), and interactive effects of the solution matrix created by the various cations (particularly potassium and calcium) and anions. Any changes in these parameters may influence stability (see Calcium Tartrate Stability).

BITARTRATE STABILITY

As seen in Fig. 15.1, the percentage of tartrate present as HT^- is maximum at pH 3.7; generally precipitation will be maximal at this point. Thus, any treatments causing changes in pH, such as blending or malolactic fermentation, may affect subsequent bitartrate precipitation. Therefore, winemakers are concerned with the potential for bitartrate precipitation and preventing "tartrate casse" formation in the bottle.

Crystallization depends on (1) the concentration of the salt and other components that may be involved in the crystallization equilibrium; (2) the presence of nuclei upon which crystalline growth may occur; and (3) the presence of complexing factors that may impede crystal growth. In general, a certain level of supersaturation is necessary for adequate nucleation. Once nucleation has occurred, further crystal growth results in precipitation.

During alcoholic fermentation, KHT becomes increasingly insoluble resulting in supersaturation. Potassium bitartrate stability is achieved by chilling (with or without seeding), ion exchange, or combinations of both. In conventional cold stabilization (chill proofing), wines are chilled to a selected low temperature in order to decrease KHT solubility. Perin (1977) calculated the optimum temperature needed for bitartrate stabilization:

$$\text{Temperature } (\text{-}^{\circ}\text{C}) = \frac{(\text{ethanol } \% \text{ vol/vol})}{2} \text{ -1}$$

KHT precipitation occurs in two stages. During the initial induction stage, the concentration of KHT nuclei increases due to chilling. This is followed by the crystallization stage, where crystal growth and development occur. The precipitation rate for KHT at low temperatures is more rapid in table than in dessert wines. Further, precipitation from white wines is faster than from reds (Marsh and Guymon 1959). During conventional chill-proofing, precipitation is most rapid during the first 12 days. After the initial period, KHT precipitation decreases due to decreased levels of KHT saturation. Temperature fluctuations during cold stabilization may have a significant effect on reducing precipitation rates because of the effect on the speed of nucleation.

Without crystal nuclei formation, crystal growth and subsequent precipitation cannot occur. Therefore, simply opening the cellar doors in the winter, although cost effective, may not be ideal for KHT precipitation. Because of the potential for increased absorption of oxygen in wines held at low temperatures, alternatives to conventional cold stabilization have been sought. (Problems relating to oxidation are discussed in Chapters 7, 10, and 14.)

Complexing Factors

Complexing factors can greatly affect KHT formation and precipitation. As a result, wine (especially red wine) may be supersaturated longer than a corresponding alcohol-water solution. As seen in Fig. 15.2, wine can support a supersaturated solution of KHT because portions of the tartrate, bitartrate, and potassium ions may be complexed and thus resistant to precipitation.

Metals, sulfates, proteins, gums, and polyphenols may form complexes with free tartaric acid and potassium thus inhibiting formation of KHT (Pilone and Berg 1965). The complexes are mainly between polyphenols and tartaric acid in red wines and between proteins and tartaric acid in whites. In a study of white Bordeaux wines, Peynaud et al. (1964) found that sulfate was the most important factor in stability next to potassium and tartrate. This influence would appear to be due to complex formation between sulfate and potassium (Chlebek and Lister 1966). Almost one-half of the sulfate in white wines and 100% of the sulfate in red wines is thought

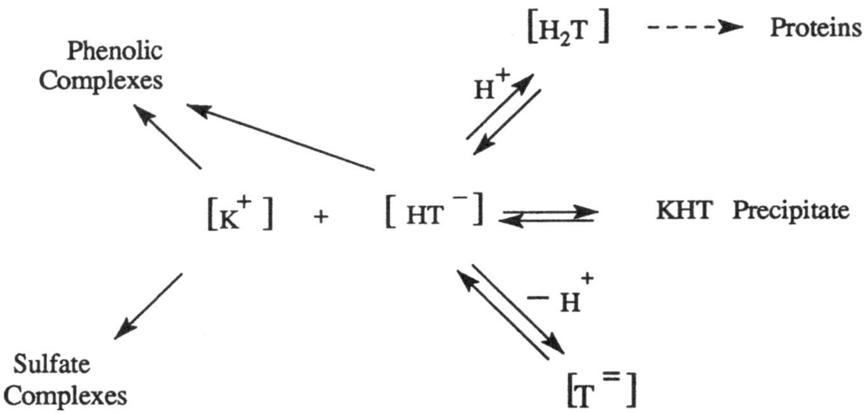

Fig. 15–2. Interaction equilibria between potassium bitartrate and "complexing factors" present in wine.

to form complexes with potassium ion to form K_2SO_4 or KSO_4^- (Bertrand et al. 1978). The extent to which tartrate complexes form and their relative ability to impede precipitation depends on the particular wine in question.

Red wine pigments can form complexes with tartaric acid (Balakian and Berg 1968). As pigment polymerization occurs, the holding capacity for tartaric acid diminishes, resulting in delayed precipitation of KHT.

Pectins and other polysaccharides such as glucans produced by *Botrytis cinerea* may inhibit bitartrate crystallization due to crystal adsorption, preventing further growth. However, in a study of German white wines, Neradt (1977) could find no inhibition of crystallization by either gelatin or acacia (gum arabic). Thus, each wine, because of its unique composition, will achieve unique solubility and equilibria under imposed temperature conditions.

Occasionally, winemakers choose to add complexing agents or inhibitors to help prevent KHT formation. However, none of the approved wine additives for such purposes are completely satisfactory.

Metatartaric acid is the hemipolylactide of tartaric acid. Although it has been approved as a wine additive in certain countries, addition is not permitted in the United States. Addition of 50 to 100 mg/L protects young wines from bitartrate precipitation even when stored at low temperatures for several months (Peynaud and Guimberteau 1961). Inhibition is due to the coating of growing bitartrate crystals by metatartaric acid (Peynaud 1984).

After addition, metatartaric acid is slowly hydrolyzed to tartaric acid with a corresponding loss of activity. The period of effectiveness is a function of the wine storage temperature. Peynaud (1984) reported that wines stored at 0°C (32°F) were stable for several years, whereas metatartaric acid disappeared after 2 months in wines stored at 25°C (77°F). The acid has its greatest application in wines that are expected to be consumed rather early.

Carboxyl methyl cellulose is another known inhibitor of KHT precipitation (Cantarelli 1963). Purified apple pectin and tannin can also inhibit crystal formation. Tannin addition at levels of 1 g/L strongly inhibits precipitation of bitartrate (Wucherpfennig and Ratzka 1967).

Wine Processing and Complexing factors

There is a close relationship between wine fining and KHT stabilization. For example, condensed polyphenols interfere with bitartrate precipitation, suggesting that removal of a portion of the polyphenols by addition of protein fining agents prior to cold stabilization may enhance subsequent KHT precipitation (Zoecklein 1988).

Cold stabilization (chill-proofing) may result in precipitation of both KHT and wine proteins. In white wines, proteins may affect tartrate holding capacity, thus inhibiting precipitation (Pilone and Berg 1965). As white wine phenols oxidize and polymerize, binding and co-precipitation with proteins can occur, affecting tartrate holding equilibria and stability.

Bentonite fining may decrease the tartrate holding capacity by reducing both proteins and phenolics (Berg and Akiyoshi 1971). Additionally, if a bitartrate unstable wine has a pH below 3.65, chill-proofing causes a downward shift in pH that may enhance protein precipitation. Some winemakers elect to bentonite fine during bitartrate stabilization which allows KHT crystals to help compact bentonite lees. (For additional information on protein stabilization and fining, see Chapters 8 and 16.)

Addition of 30 g/hL (2.5 pounds per 1,000 gallons) or more of bentonite has been found to reduce concentration product (CP) values of dry white wines from 15 to 18% and CP values in dry red wines from 25 to 32% (Berg et al. 1968). (see Concentration Product, below). Reducing sugars are also known to impact tartrate stability. Berg (1960) reported 20% higher CP values in sweet sherries (6 °Brix) when compared with sherries at 1 °Brix. Further, final tartrate deposition took 23% longer in the higher sugar group. A similar trend was demonstrated by Berg and Akiyoshi (1971) in examination of wines from California.

Bitartrate Stabilization and Changes in Titratable Acidity and pH

Wines with initial pH values below 3.65 show reductions in pH and titratable acidity (TA) during cold stabilization because of the generation of one free proton per molecule of KHT precipitated. The pH may drop by as much as 0.2 pH units with a corresponding decrease in TA of up to 2 g/L. By comparison, KHT precipitation in wines with pH values above 3.65 results in higher pH levels and corresponding decreases in TA. This is the result of removal of one proton per tartrate anion precipitated. The above values represent ranges seen in practice and may vary.

METHODOLOGY FOR ESTIMATING COLD STABILITY

There is no universal definition for cold stability, which is a relative term defined differently by different producers. Three common methods used internationally for evaluating cold stability are outlined: freeze tests, electrical conductivity, and concentration product (CP) values. (For specific test methodology, consult Chapter 20.)

Freeze Tests

These procedure rely on the formation of KHT crystals in a suspect wine held at reduced temperature for a specified time period. There is wide variety of times and temperatures for this analysis. Often a sample is frozen and then thawed to determine the development of bitartrate crystals and whether those crystals resolubilize. The absence of crystal formation or resolubilization indicates KHT stability.

As water in the sample freezes, there is an increase in the relative concentration of all species in the sample, including alcohol, thus enhancing nucleation and crystallization. It is difficult to accurately relate crystal formation in this concentrated wine sample with KHT instability.

Preliminary laboratory treatment of samples may call for filtration. However, filtration removes crystal nuclei which may affect test results and, therefore, should not occur.

The freeze test is essentially a measure of precipitation rate, that is, the formation of nuclei (nucleation), secondary crystal growth, and subsequent precipitation. Unless one provides seed crystals, precipitation over the relatively short period of the test is, in fact, a measure of the wine's ability to form nuclei and precipitate. Thus, the freeze test is of limited value in prediction of bitartrate stability in wine. Despite the deficiencies, however, it is probably the most widely used technique for attempting to predict cold instability. Frequently, a companion sample is held at refrigerator temperature and examined for protein-tannate or chill haze (see Chapter 20).

Conductivity Test

An accurate determination of KHT stability may be achieved by seeding a juice or wine sample with finely ground KHT powder. The oversaturated portion of tartaric acid and potassium present deposits on the added seed. Estimation of KHT stability is then based on the reduction in electrical conductivity of the juice or wine from the beginning to end of the test.

Changes of less than 5% in electrical conductance during the test period may be considered stable, although some define stability in more strict terms (e.g., <3% change or lower). Samples passing the test are stable only at (or above) the test temperature. In practice, most winemakers select 0°C (32°F) as stabilization temperature for whites and 5°C (41°F) for reds.

The conductivity test provides a value that is unique for the product tested. Complexing factors that may be present, and possibly affecting KHT crystal formation, are taken into account.

Concentration Product

By measuring the concentration of potassium, total tartrates (as bitartrate or tartrate), pH, and alcohol, one can calculate the solubility for KHT in terms of a CP value (Berg and Keefer 1958, 1959; DeSoto and Vamada 1963):

$$CP = [K^+ \text{ mole/L}]\ [\text{total tartrate mole/L}](\%HT^-)$$

The values for % bitartrate and/or tartrate at measured pH and alcohol levels are taken from Tables I-8 and I-9 in Appendix I.

As an example, typical data for a white wine are presented in Table 15–1. From Table 15–2 the minimum CP for a dry white wine is 16.5×10^{-5} at 0°C. Because the calculated CP, in this example, exceeds values considered "safe," the wine will precipitate bitartrate crystals. In addition to inherent problems of being unable to quantify the contribution of complexing

Table 15–1. Analytical data for a white wine.

Analysis	Result
K^+	1,100 mg/L
Ca^{++}	76 mg/L
Tartrate	1,600 mg/L
pH	3.74
Alcohol	12% (vol/vol)

Carrying out the calculations:

$$CP = \frac{1{,}100 \times 10^{-3}}{39.1} \times \frac{1{,}600 \times 10^{-3}}{150} \times 0.661$$

$$CP = 19.8 \times 10^{-5}$$

Table 15–2. Suggested maximum CP levels for potassium acid tartrate and calcium tartrate for California wines.

Wine type	Potassium Acid Tartrate ($\times 10^{-5}$)		Calcium Tartrate ($\times 10^{-8}$)	
	Highest level found	Suggested safe level	Highest level found	Suggested safe level
White table	18.5	16.5	230	200
Red table	34.7	30.0	590	400
Pale dry sherry	14.6	10.0	170	90
Sherry	15.0	17.1	202	125
Cream sherry	15.0	10.0	171	120
Muscatel	18.0	17.5	310	250
White port	11.0	10.5	175	155
Port	23.3	29.0	410	275
Dry vermouth	7.0	6.0	69.5	50

SOURCE: Yamada, H. and R. T. DeSoto (1963).

agents, component analyses may present problems in determining CP values.

Potassium bitartrate becomes increasingly insoluble during fermentation. Therefore, the ethanol content is a necessary chemical parameter in any determination of tartrate stability in wine using CP relationships.

As seen in Figure 15.1, the distribution of tartrate/bitartrate species is pH dependent. The percentage of tartrate as either bitartrate or tartrate was calculated by Berg and Keefer (1958, 1959) as a function of pH and ethanol concentrations in model solutions (see Tables I-8, I-9 in Appendix I). The resultant value must also be used in calculation of CPs.

Determination of CPs for wines yields a numerical value corresponding to the solubility for that particular wine. By comparison to solubility values provided in tables, the winemaker may calculate proportions of blend components that will yield a theoretically stable product. As with any blending operation, the winemaker should not assume that blending two or more stable wines will result in a final blend that is stable. Changes in alcohol, pH, tartrate, cation concentrations, and complexing factors may contribute to potential instability.

CORRECTION OF BITARTRATE INSTABILITY

KHT stability has been traditionally accomplished by chilling (with or without seed), ion exchange, or both. Other techniques reported including electrodialysis (Postel et al. 1977), reverse osmosis (Wucherpfennig 1978), and crystal flow (Riese 1980).

In conventional cold stabilization (chill-proofing), wines are chilled to a temperature designed to decrease KHT solubility to a level that, optimally, results in precipitation of the salt. Important variables affecting precipitation of KHT include (1) the concentration of reactants, specifically tartaric acid and K^+; (2) availability of nuclei for crystal growth; and (3) the solubility of potassium bitartrate formed. A formula for estimating stabilization temperature is found on page 231.

Conventional cold stabilization may last for several weeks or more. Dessert wines with higher alcohol and sugar levels require lower temperatures and longer holding times. Due to the energy costs involved, alternatives have been developed to accelerate the cold stabilization process.

Seeding

Seeding (contact seeding) techniques are used as a means of reducing time and improving efficiency. Addition of an excess of finely powdered KHT creates a supersaturated solution. The enormous surface area of powdered KHT reduces or eliminates the energy-consuming nuclei induction phase and allows for immediate crystal growth.

Several processing considerations are important in achieving stability using seeding; these are outlined in the following sections.

Quantity of KHT and Crystal Size

Because of the involvement of complexing agents in precipitation equilibria, the amount of KHT seed required depends on the wine. The amount of KHT seed used must produce a supersaturated solution. Table 15–3 compares changes in tartaric acid, K^+, and final CP values in wine treated with different levels of seed.

Tartaric acid, potassium, and CP values decrease with increased levels of KHT seed. Rhein and Neradt (1979) reported optimal levels of seed to be

Table 15.3. Influence of potassium bitartrate seed (40 μM) level on wine components at 0°C.

Seed addition level (g/L)	Tartaric acid (g/L)	K^+ (mg/L)	CP ($\times 10^{-5}$)
Control	1.58	920	15.1
1	1.11	808	9.3
2	1.03	794	8.5
4	0.93	765	7.6
8	0.78	754	6.2

SOURCE: Blouin et al. 1982.

4 g/L. In practice, stability may be achieved using lower concentrations. Reduction in the quantity of seed used prolongs the contact time needed for stabilization.

Seed particle size should range from 30 to 150 μm (Guimberteau et al. 1981). Rhein and Neradt (1979) report that maximum reactive surface area is achieved at a particle size of 40 μm. Because the reaction rate depends on available surface area, use of larger particles will slow the rate of crystal growth and extend processing time.

Agitation

Seeding should be conducted in tanks where adequate mixing can be accomplished. The density of KHT is greater than juice and wine. Crystal growth is dependent on available interactive surface area. Therefore constant agitation is essential. Table 15–4 compares a mixed with a static system. It is apparent that proper mixing is critical to producing a stable wine using this method.

Table 15.4. Effect of agitation on final tartrate, K^+ and CP values in white wine seeded with KHT at 4 g/L (0°C).

Wine	Tartaric acid (g/L)	K^+ (mg/L)	CP ($\times 10^{-5}$)
Control	1.58	920	15.1
Static system	1.38	870	12.5
Agitated system	1.17	805	9.8

SOURCE: Blouin et al. 1982.

Time, Temperature, and Filtration

Bitartrate seeding can occur at any temperature. However, treatment temperature should be identical to desired stability temperature. For example, many winemakers seed white wines at 0°C (32°) and reds, when stabilized, at 5°C (41°F). If stabilization is correctly performed, wines held at or above these temperatures will be stable with respect to KHT precipitation.

During the first hour of contact seeding, there is a rapid reduction in tartaric acid, potassium, and CP values. This reduction slows after the first hour and then levels off at the end of 3 hours in most wines (Blouin et al. 1982). It has been suggested, however, that using 40 μm KHT seed allows stabilization in 90 minutes (Rhein and Neradt 1979). For security, it is desirable to have a minimum contact of 4 hours (Guimberteau et al. 1981). As stated, reduction in the quantity of seed used prolongs the stabilization period.

Filtration of the wine after contact seeding is essential to prevent resol-

Table 15.5. Sparkling wine *cuveé* before and after KHT stabilization by several methods.

Stabilization method	Alcohol (vol/vol)	Sugar-free extract (g/L)	pH	Total acid (g/L)	Tartaric acid (g/L)	K^+ (mg/L)
Untreated	9.37	21.26	3.31	7.55	2.50	720
Chilling	9.43	21.74	3.23	7.25	2.50	715
Contact	9.46	20.34	3.20	7.15	1.95	565
Ion exchange	9.35	20.74	3.26	7.45	2.40	360

SOURCE: (Rhein and Neradt 1979).

ubilization of KHT crystals. This step must be performed at the same temperature as the stabilization procedure. Table 15–5 compares several wine parameters after cold stabilization by chilling, contact seeding, and ion exchange.

Ribereau-Gayon and Sudraud (1981) reported a comparison of KHT stabilization techniques in 16 white wines and 11 reds. Each lot was prefiltered and evaluated after 14 and 21 days of storage at −4°C (24.5°F) and results compared with contact seeding. CP values were determined for each wine after respective treatments. Eighty percent of the wines treated by contact seeding had lower final CP values than did their conventionally cold-stabilized counterparts. In all cases, stability values for contact seeded wines were at least equivalent to conventionally cold-stabilized wines.

KHT crystals can be reused after removal from the treated wine. After repeated use, however, crystal size increases, resulting in decreased surface area available for growth (Rhein and Neradt 1979). Regrinding of crystals eventually is necessary to optimize performance, although most wineries will not find it economically feasible to regrind. Winemakers should also be aware of the potential for microbiological contamination arising from the reuse of tartrate crystals.

Mixtures of calcium carbonate and calcium tartrate have been studied for their ability to increase the rate at which tartrate stability is achieved in wines. Upon addition of this mixture, added calcium carbonate reacts with the tartaric acid in wine to form insoluble calcium tartrate. The precipitation is aided by the seeding effects of the calcium tartrate crystals present. In laboratory trials, the tartaric acid concentration was reduced to a low enough level that cold (bitartrate) stability was achieved (Clark et al. 1988). One potential advantage of this technique is that these materials function at room temperature.

Ion Exchange

Ion exchange may be used to bring about KHT stabilization, but is not regarded as a desirable practice in premium wine production. It is occasionally used in conjunction with refrigeration to avoid deterioration in wine palatability.

With ion exchange, the cation concentration of a wine can be reduced by exchanging it with either H^+ or Na^+. Reduction in the concentration of the cation component (K^+) directly results in a decrease in the solubility coefficient and hence decreases the likelihood of salt precipitation at low temperatures. Unlike chill-proofing or seeding, the concentration of tartaric acid (tartrates) does not change. Hence, there are not the changes in TA and pH that may accompany stability by storage of the wine at low temperatures.

CALCIUM TARTRATE STABILITY

Calcium tartrate (CaT) precipitation is a relatively infrequent occurrence in table wines, but is more common in sparkling and fortified products. CaT precipitation is of particular concern due to post-bottling crystal formation and the difficulty in predicting crystal formation.

The mechanism of CaT precipitation in model solutions involves a two-step process, the formation of soluble unionized CaT followed by the nucleation of soluble CaT to form a precipitate (McKinnon et al. 1993).

CaT precipitation can occur in sherry with calcium levels as low as 50 mg/L (Crawford 1951). Approximately 30% of the precipitate is present as calcium oxalate. McKinnon (1993) reported that all untreated wines in Australia are supersaturated with respect to CaT and that wines containing 80 mg/L or more calcium are at risk of precipitation. Precipitation of CaT is thought to be influenced by agitation, pH, alcohol, and complexing factors (McKinnon 1993).

Agitation decreases the time required for precipitation by affecting nucleation. This may explain the formation of CaT in sparkling wines shortly after disgorgement. McKinnon (1993) suggested that filtration to remove particles that promote nucleation may indeed increase the likelihood of CaT precipitation due to agitation.

CaT precipitation is highly pH dependent. The higher the pH the lower the degree of solubility. Any activity that increases pH, including malolactic fermentation (MLF) and blending, will significantly increase the likelihood of precipitation.

As the alcohol content increases, the CaT solubility decreases and, therefore, precipitation increases. Using model solutions, McKinnon (1993) demonstrated that fortification may decrease the time needed for CaT precipitation by a factor of 12, illustrating the potential problem in fortified wines.

The presence of complexing factors may also be highly significant in determining CaT precipitation. As with KHT, CaT inhibitors work by preventing nucleation or by reducing the rate of crystal growth. The complexing factors that may help to inhibit CaT precipitation include malic, citric, and lactic acids, amino acids, and sugars. The influence of malic and citric acids on CaT is significant, further illustrating the relationship between MLF and precipitation (McKinnon 1993).

Methodology for Estimating Calcium Tartrate Stability

As stated, it is unlikely that CaT stabilization can be brought about by cooling. Therefore, chill tests are not effective stability predictors. Traditionally, CaT stability was determined by CP values. This method has been found to be of little value due to lack of solubility product data (Muller et al. 1990) although Curvelo-Garcia (1987) developed solubility product tables for wine pHs previously unavailable.

Concentration products for CaT have traditionally been calculated from the equation:

$$CP = [Ca^{2+}]\ [T^{2-}]$$

where Ca^{2+} and T^{2-} represent the free or ionized calcium and tartrate concentration. In determining CP values, the total calcium concentration has frequently been used (as measured by atomic absorption), whereas the free Ca^{2+} concentration should be measured (by specific ion electrode). Since a large percentage of total calcium may be bound, analysis of total (rather than free) leads to high CP values.

Muller et al. (1990) reported the use of a minicontact conductivity measurement to predict CaT stability. In this procedure, wines are stirred in the presence of CaT crystals. A decreased conductivity indicates crystal growth with regard to CaT. Success of this procedure is influenced by the quality of the seed crystals.

ANALYSIS

The following is a cross-reference of analytes and procedures found in Chapter 20. Tartaric acid by HPLC, paper chromatography, and spectrometry; potassium bitartrate stability by freeze test, conductivity test, Concentration Product values; potassium by atomic absorption, flame emission, and ion selective electrode; calcium by atomic absorption, and ion selective electrode; sodium by atomic absorption and flame emission; pectins and glucans by a rapid diagnostic screen; presence of suspected tartrate deposits—KHT, CaT, and calcium oxalate by rapid diagnostic screen.

CHAPTER 16

FINING AND FINING AGENTS

PRINCIPLES OF FINING

Fining is the addition of a reactive or adsorptive substance to remove or reduce the concentration of one or more undesirable constituents. Fining agents are added to juices, wines, and sparkling wine cuveés for the purposes of enhancing clarity, color, aroma, flavor, and/or stability modification. Fining agents can be grouped according to their general nature.

1. Earths: bentonite, kaolin
2. Proteins: gelatin, isinglass, caseins, pasteurized milk, albumens, yeasts
3. Polysaccharides: alginates, gum arabic
4. Carbons
5. Synthetic polymers: PVPP, nylon
6. Silica gel (silicon dioxide)
7. Tannins
8. Others: including metal chelators, blue fining, and enzymes

The mechanisms of action of fining agents may be electrical (charge) interaction, bond formation, and/or absorption and adsorption. In the case of electrical interaction, particles of opposite charge to the fining agent are induced to coalesce with the agent, forming larger particles. Due to its greater density, the complex eventually settles from solution.

Turbidity in a juice and wine may be due to grape tissue, yeast and bacteria, colloids derived from the grape, or from changes occurring during aging or storage. These particles may be in the form of proteins, pectins and gums, metallocolloids (formed by flocculation of insoluble oxidized salts), and degradation products of polyphenols. Absorption/adsorption of colloidally suspended material by fining agents may enhance filterability.

Common fining agents are compared in Table 16.1 in decreasing order of activity and effectiveness. Effectiveness of fining agents is dependent on the agent, method of preparation and addition, concentration, wine or juice pH, metal content, temperature, age, and previous treatments.

Table 16.1. Comparison of selected fining agents with respect to desired effects and potential problems in decreasing order of activity or effectiveness.

Color reduction	Tannin reduction	Volume of lees formed	Clarity and stability	Potential for overfining	Quality impairment
carbon	gelatin	bentonite	bentonite	gelatin	carbon
gelatin	albumen	gelatin	ferrocyanide[a]	albumen	bentonite
casein	isinglass	casein	carbon	isinglass	casein
albumen	casein	albumen	isinglass	casein	gelatin
isinglass	bentonite	isinglass	casein	ferrocyanide[a]	albumen
bentonite	carbon	ferrocyanide	gelatin		isinglass
ferrocyanide[a]	ferrocyanide[a]	carbon	albumen		ferrocyanide[a]

[a]*Not permitted in the United States.*

SOURCE: Berg 1981.

Fining is a surface reaction, and therefore the method of hydration and addition is important. To duplicate laboratory trials in the cellar, the same lot of fining agent must be prepared and used in the same way. Carefully controlled laboratory fining trials must be performed before cellar addition because the effects of fining agents can be unpredictable. The winemaker must note and record how each fining trial alters clarity, lees formation and compaction, stability, aroma character and intensity, color, body (front, middle, and finish), astringency, bitterness, finish, aging potential, and overall wine palatability. Table 16–2 equates laboratory fining trials to required cellar additions in g/hL or lb/1,000 gal.

Methods of Addition

It is essential that fining agents be properly added in the cellar if winery trials are to duplicate laboratory results. Common methods of fining agent

Table 16–2. Laboratory fining trial equivalents to g/hL or lb/1,000 gal. Fining trial levels using 1 L of juice or wine (soln. vols. in mL)[a]

Level of treatment		Tannic acid soln.[b]	Gelatin soln.	Sparkolloid stand. soln.	Bentonite	Casein[c]	Any solid
g/hL	lb/1,000 gal	(1%)	(1%)	(0.4%)	(5%)	(2%)	(g)
3	0.25	3	3	1.3	0.6	1.6	0.03
6	0.5	6	6	2.6	1.1	3	0.06
9	0.75	9	9	3.8	1.9	4.6	0.09
12	1	12	12	5	2.4	6	0.12
18	1.5	18	18	7.5	3.6	9	0.19
24	2	24	24	10	4.9	12	0.24
36	3	36	36	15	7.1	18	0.36
48	4	48	48	20	9.6	24	0.49
60	5	60	60	25	12	30	0.6
72	6	72	72	30	14.3	36	0.71
84	7	84	84	35	16.9	42	0.84
96	8	96	96	40	19.1	48	0.96
108	9	108	108	45	21.6	54	1.10

[a]Protein fining agents should be prepared fresh prior to use, but can be preserved for several days by the addition of 500 mg/L benzoic acid.
[b]Prepared in 70% ethanol.
[c]Potassium caseinate or milk casein, are prepared by dispersing casein in alkaline solution (5 g sodium carbonate then 20 g/L casein).

additions include (1) uniformly and slowly through a "y" on the suction side of a pump while transferring or mixing; (2) uniformly and slowly through a proportioning "in-line" pump; (3) uniformly and slowly through a "T" into a Guth-type tank mixer; (4) added slowly as a slurry to a barrel using a dowel stirrod in a "Figure-8" motion through the bung hole.

FINING AND WINE STABILITY

Wine fining can significantly affect wine stability by removing complexing factors formed between polyphenols and tartaric acid in red wines and proteins and tartaric acid in whites (see Chapters 8 and 15). Therefore, there is a direct relationship between wine fining and potassium bitartrate and protein stabilization. Protein stability should be determined after pH adjustment because any variation in pH influences the charge density and hence stability. Other considerations include potential changes in cold stability and malolactic fermentation.

The fining abilities of filtered vs. unfiltered wines often differ because of protective colloids remaining in unfiltered wines. Gelatin is particularly sensitive to this phenomenon, albumin less sensitive, and casein and isinglass are least affected (Ribereau-Gayon et al. 1972). Thus, if initial fining does not give sufficient clarification, it may be repeated after racking.

SUMMARY OF IMPORTANT CONSIDERATIONS IN FINING

1. The smallest quantity of agent necessary should be used.
2. Only fining agents of the highest purity, free from undesirable odors and flavors, should be used.
3. The contact time should be limited to only that necessary for complete reaction and formation of lees; centrifugation may shorten exposure time.
4. Careful laboratory trials must be carried out before cellar treatment. Laboratory fining agents should be prepared by the same method intended for cellar treatment. For example, the sheer force exerted by laboratory blenders cannot be duplicated in the cellar.
5. To ensure maximum utilization and reactivity of a fining agent, it is essential that thorough mixing of the material and juice or wine be achieved.
6. Wines to be fined should be low in dissolved CO_2. Evolution of gas will serve to maintain particulates in solution and impede settling.
7. Young wines are more "forgiving" than older wines to the action of protein fining agents.
8. Lower pH wines will require less fining agent for clarification than the corresponding wine of higher pH.
9. A high metal content in either the fining slurry or the wine may adversely affect flocculation and activity of the agent. Although the action of some agents such as gelatin depends on the presence of metal ions (iron), many winemakers choose to use deionized water for hydration.
10. Temperature is important in reaction time; therefore, laboratory fining trials must be carried out at the same temperature as the juice or wine to be treated. Protein fining agents are more effective in removing phenolic compounds at lower vs. higher temperatures. The difference between 10°C (50°F) and 25°C (77°F) relative to phenol absorption is significant.
11. The effectiveness of a fining agent depends on the agent, method of preparation and addition, levels of addition, pH, metal content, temperature, presence of CO_2, and prior wine treatments.
12. Table I–10 of Appendix I summarizes juice and wine additives approved for use by both the U.S. Bureau of Alcohol, Tobacco and Fire Arms and the Office International de la Vigne et du Vin (OIV).

BENTONITE

Bentonites are mined from several areas of the world and come in different levels of purity, particle size, adsorption and swelling capacity. The utilization of bentonite generally increases the aluminum content by at least two-fold (McKinnon et al. 1992) and increases in the iron content can also occur as a result of bentonite fining (Pocock 1983). Possible heavy metal pickup suggests the need for careful laboratory trials to determine minimal dosages.

The type and source of bentonite may affect protein removal due to variations in swelling capacity. In the refining process bentonites from several sources are stockpiled and tested for purity, adsorption and polymerization capacity, and lead and iron content. They are then dried, ground, and packaged. Due to these variations, there may be differences between lots. Lab trials must be conducted using the same lot as will be used in the cellar. The cation-exchange capacity of bentonite may be determined (American Colloid Co, 1983).

American bentonite, a volcanic clay like material, is a complex hydrated aluminum silicate with exchangeable cationic components ($Al_2O_3 \cdot 4SiO_2 - H_2O$). Also known as Montmorillonite clay, after the French town where it was first mined, American bentonite is mined in Wyoming and is also known as "Wyoming clay." Although calcium, sodium, or magnesium forms are available, the most commonly used form in the United States is sodium bentonite, due to its superior protein binding ability. Bentonite finds its principal application in removal of proteins from white wine and juice, and removal of enzymes, such as tyrosinases (polyphenol oxidases), which may catalyze oxidation and browning in juice (see Chapter 3). The mechanism of removal is by adsorptive interaction between the flat negatively-charged surfaces of the bentonite platelets and the positively-charged proteins. In that the platelet edges are positively-charged, some limited binding of negatively-charged species may occur.

The prevention of protein haze or deposit in bottled white wines is a universal concern. Protein precipitation is usually not a problem in bottled red wines because the relatively high concentration of phenols bind with and precipate labile protein well before the wine is ready for bottling. Wines low in phenols, such as rosé, light reds, and whites, should be checked for protein stability before bottling. The total protein nitrogen content of a wine does not correlate well with the potential for protein instability. Protein haze is a complex of protein, polysaccharides, polyphenols and metals (See Chapter 8).

Some cultivars present more problems than others. Sauvignon blanc and Muscat often have large concentrations of unstable proteins and may require more extensive treatment. Although there is considerable variation in protein content among grape cultivars, wines of lower pH generally require less bentonite for stabilization.

Because of variations in the fundamental chemistry of wine proteins, types of bentonites, as well as addition and mixing protocols, protein stability can be confusing and a major production problem. Proteins which are most sensitive to heat denaturation have molecular weights between 12,000–30,000 daltons and isoelectric points between 4.1 and 5.8 (Hsu and Heatherbell, 1987; Hsu, et al., 1987; Heatherbell and Flores, 1988).

The greater the difference between wine pH and the isoelectric point of the protein fraction, the greater is the net charge on that fraction, and therefore, the greater is the potential for reacting with fining materials of the opposite charge. Therefore, protein instability is difficult to correct with fining when wine pH is near the isoelectric point of the various unstable fractions. The point of maximum instability and hence, tendency toward natural precipitation, is reached when positive and negative charges are equal. (See Chapter 8).

Bentonite may indirectly bind phenols that have complexed with proteins. However, as seen in Table 16.1, its activity toward phenolics is relatively low. Bentonite affects red wine color by binding with positively-charged anthocyanin monomers and may result in color decreases depending upon the age of the wine. Bentonite may also remove more color in younger wines because of the greater action on colloidally colored material found in younger wines (Bergeret 1963). Addition of bentonite to red wines at levels of 6 to 12g/hL (0.5 to 1 lb/1000 gal) improves membrane filterability due to reduction in colloidally suspended particles.

Bentonite fining of juice may remove peptides and some amino acids, potentially affecting rate and completion of fermentation. In that bubble retention and quality are related to the concentration of protein and peptides, sparkling wines made from heavily bentonited cuveés may be lacking in these areas. Winemakers facing large bentonite additions may wish to utilize multiple additions rather than a single large dose. This may reduce the overall bentonite requirement, especially if the wine in question is low in suspended solids.

Bentonite fining is known to indirectly prevent or impede formation of copper, and possibly iron casse in wines where metal levels may be a problem. In the case of copper casse, this is probably due to removal or reduction in the levels of proteins and peptides known to be involved in the formation of haze and precipitate (see Chapter 12).

Preparation of Bentonite

Preparation of bentonite greatly impacts its activity toward proteins. In solution, bentonite swells to many times its dehydrated dimensions. Its activity is much like that of a multiplateted, long-chain, linear, negatively-charged molecule (Singleton 1967). During the hydration phase, charged platelets repel each other and begin to separate. Water molecules partially

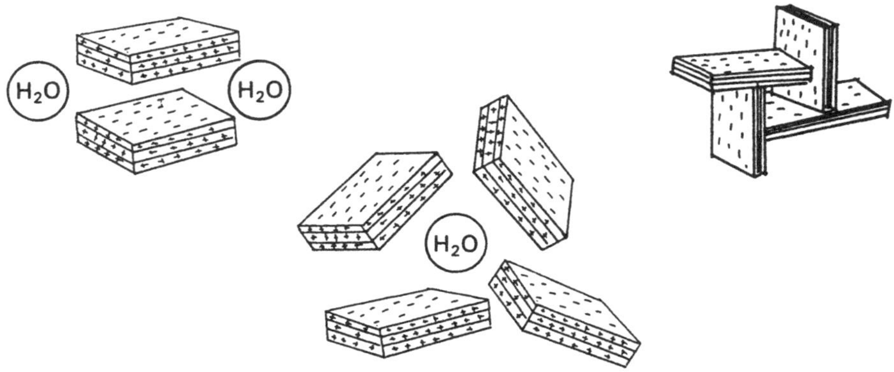

Fig. 16–1. Fully hydrated bentonite molecule depicting "house of cards" configuration.

neutralize and separate exposed surfaces, exposing a large matrix of reactive surface (each silicaceous platelet has calculated dimensions of 1 nm × 500 nm). When fully hydrated, 1 gram of sodium bentonite has an estimated adsorption surface of 5 square meters (Peynaud, 1984). As seen in Fig. 16.1, properly dispersed, bentonite exists as a network resembling a "house of cards." The presence of water molecules within the network prevents flocculation and precipitation.

The winemaker should refer to the supplier's technical literature when preparing bentonite. The water used in the hydration phase should have a low mineral content. Dissolved metal cations present in slurry water preferentially replace sodium ions on the clay surface and detrimentally affect the hydration, viscosity, and binding capacity of the bentonite (American Colloid Co.). In the USA, the total quantity of water must not exceed 1% of the wine volume treated.

It is common in the USA to use agglomerated bentonites which can be hydrated in cool water. These bentonites disperse approximately 5 times faster than non-agglomerated forms and exhibit fairly low viscosities. Non-agglomerated bentonites should be slowly hydrated in hot water to avoid clumping. Typically, the bentonite-to-water ratio for slurries is 5–6% (w/v). Heating non-agglomerated bentonite allows the platelets to fully separate and the slurry resembles a gel.

Sodium vs. Calcium Bentonite

Sodium bentonite finds widest application in the United States. Lees compaction is better with the sodium form (Ferenczy 1966). Expected sodium pickup from sodium bentonite may range from 1.7–3.5 grams/100 grams of the bentonite (Amerine and Joslyn 1970).

Calcium bentonite platelets tend not to separate as well as the sodium form and hence have less exposed surface area for protein binding. Calcium bentonite finds its widest acceptance in Europe (where sodium levels are restricted) as a riddling aid in methode champenoise production.

Bentonite and Wine Lees

A problem with sodium bentonite is excessive and loose compaction of lees. Typically, bentonite lees volumes range from 5–10% of the total volume of juice or wine treated. Several techniques may be employed to minimize these problems.

1. Bentonite requires only minutes to react with and precipitate protein. Seventy five percent of the total protein removed by bentonite occurs in the first minute after contact (McLaren et. al. 1958). Since adsorbed protein may "slough off" bentonite platelets upon standing, prolonged contact may result in less efficient removal of protein. Therefore, in-line centrifugation or filtration to remove bentonite rather than traditional gravity clarification may be desired.

2. Bentonite may be hydrated in the wine to be treated rather than in water. Although this may significantly reduce the binding capacity due to premature flocculation, the lees volume should be about half that obtained by conventional hydration protocol.

3. To enhance lees compaction, bentonite fine during conventional cold stabilization. Potassium bitartrate crystal formation compacts and reduces the lees volume. Cold stabilization and protein fining may be effectively linked especially if the wine pH is below the pK_1 value for tartaric acid (3.65). At pH values of less than 3.65, formation of potassium bitartrate releases a free hydrogen ion into solution and thereby maintains or decreases pH. Since lower pH favors increased positive charges on proteins, the activity of bentonite is also enhanced. Additionally, cold stabilization may reduce the concentration of proteins in wine.

4. Counterfining of bentonite with sparkalloid, silicon dioxide, gelatin, etc., often aids bentonite precipitation and compaction.

Fermentation with Bentonite

Bentonite additions, especially those exceeding 48 g/hL (4 lbs/1,000 gal), may strip wine flavor, body, and in the case of young red wines, significant color. Further, it may impart an earthy character to the wine. Some winemakers choose to ferment settled juice in contact with bentonite to aid protein stability and to eliminate or reduce the amount of bentonite needed to stabilize the wine. The procedure for fermentation of white juice in contact with bentonite used in the USA is as follows:

1. Settle juice to remove non-soluble solids by refrigeration and/or fining agents (A high solids level could foul the bentonite utilized during fermentation and reduce overall efficiency). Add the desired quantity of bentonite in-line while racking into the fermentor.
2. Add yeast nutrient and any needed sugar and/or acid where allowed.
3. Add yeast inoculum to juice surface.

A yeast nutrient addition is preferred when fermenting in the presence of bentonite. Bentonite may deplete must assimilable nitrogen due to electrostatic binding and adsorption resulting in fermentation sticking and/or H_2S production. Addition of an exogenous source of nitrogen helps to eliminate these potential problems (See Chapter 9).

The quantity of bentonite required is determined empirically or analytically. Many winemakers fermenting in contact with bentonite simply add about 24 g/hL (2 lbs/1,000 gal). Methods for predicting specific bentonite levels needed in the juice for subsequent wine stabilization are available (See Chapter 20).

Kaolin is also a silicaceous clay refined from the mineral kaolinite. Kaolinite has less than 10% of the hydrative and adsorptive properties of bentonite. Therefore, while permitted by both BATF and OIV, it is not frequently used.

The difficulties encountered with the use of bentonite to obtain protein stability have lead to examination of other techniques and materials including ultrafiltration and proteases.

POLYSACCHARIDES

Alginates

The most commonly used polysaccharides for fining are alginates. Alginic acid or algin is a structural polymer in the cell wall of algae. Commercially, it is extracted from marine brown algae. Alginic acid exists as a high-molecular weight, long-chained polymeric salt of *beta*-1,4-D-manuronic acid, and L-guluronic acid. The polymer usually consists of three components: (1) L-guluronic acid, (2) D-manuronic acid, and (3) alternating segments of monomers of both 1 and 2.

Alginic acids are positively charged and are usually bound to some inert carrier such as diatomaceous earth to facilitate settling. Reactivity and clarification is best accomplished if the juice or wine pH is less than 3.5. Clarification may be accelerated with small additions of counterfining agents such as gelatin or bentonite. Some proprietary products include 5 to 10% (wt/wt) gelatin.

Sparkalloid and Klearmor are polysaccharides used in the United States. In solution, both are present as positively charged alginates on a diatomaceous earth carrier. Sparkalloid is sold as a cream-colored powder that forms a viscous colloidal solution. Sparkalloid and Klearmor find their principal application in enhancing clarity and filterability. These compounds have little absorptive ability, and do not affect color, odor, or flavor. Both are usually prepared by hydration in hot water at 60 g/L (0.5 lb/gal water at 82°C/180°F) and added to juice or wine while still hot. Addition levels are generally those of bentonite, but as with any other fining agent, laboratory trials should precede cellar addition.

As compared with bentonite, polysaccharides produce relatively compact lees. They are occasionally used as counter fining or "top dressing" agents after bentonite fining to aid in lees compaction.

Acaci (Gum Arabic)

Acaci is a polysaccharide, primarily arabinose. It delays bitartrate precipitation by interrupting crystal growth. This delay is rather short term and not as effective a method of preventing bitartrate casse formation as cold stabilization (see Chapter 15). Acaci is added after the last fining or filtration, just before bottling (see Table I–10 in Appendix I).

CARBONS

Activated carbon adsorbents are used as a means of modifying the sensory character of juices, wines, and spirits. The activation process develops pores of molecular dimensions within the carbon particle. The vast number of pores in each particle gives the carbon extremely high internal porosity and surface area. Typically, activated carbons contain surface areas from 500 to 2,000 m^2/g.

Activated carbon is a relatively nonspecific adsorptive agent that tends to bind with weakly polar molecules, especially those containing benzene rings or their derivatives. Thus, some phenolic compounds in juices and wines (or their derivatives), are effectively removed by carbon. Because of the nature of the carbon particle surface, smaller phenolics are preferentially bound. Singleton (1967) pointed out that the active adsorptive surface of most carbons is confined to micropores, so compounds larger than flavonoid dimers are excluded.

Carbon-Catalyzed Oxidation

Carbon-catalyzed oxidation of phenols to quinones has been reported (Singleton and Draper 1962); however, quinones are strongly adsorbed and removed from solution. Because activated carbons contain a great deal of air (oxygen) within the carbon particles, quick and thorough removal from the wine may aid in reducing carbon-catalyzed oxidation. Activated carbons may also induce oxidation of alcohol, resulting in increased levels of acetaldehyde. Because carbon particles tend to remain suspended in solution, co-fining may aid settling. Some winemakers add carbon with the diatomaceous earth body feed during filtration. Such "in-line" additions, and removal, minimize the time of carbon-wine contact and reduces the potential for carbon-induced oxidation. The oxidative properties of carbon in wine may also be reduced by adding ascorbic acid (Singleton and Draper 1962). Lowering the pH and elevating the temperature also increases the adsorptive activity of the carbon (Amerine et al. 1980). The addition of carbon to juice rather than wine is perhaps the most desirable method of minimizing carbon-induced oxidation.

Activated carbons used in the wine industry are of two types: (1) decolorizing carbon, which removes brown coloration, and (2) deodorizing carbon, which removes odors and flavors. Decolorizing carbon can effectively "strip" color and at excessive levels, impart a "carbon" taste to the product. In addition, wines treated with decolorizing carbon may undergo significant browning from carbon catalyzed oxidation. In the case of decolorizing carbon, it is often desirable to artificially age the treated sample in the laboratory and measure its subsequent browning rate against an untreated, heated control. Such aging may be accomplished by holding the samples in an incubator at 49°C (120°F) for 1 to 3 days.

Deodorizing carbon is often added to wine after addition of wine spirits, as well as to the spirits themselves to mask the hot or harsh character. Because much of the vinous character may be removed by charcoal addition, care must be taken to establish acceptable fining levels by preliminary laboratory trials. Use levels for deodorizing carbon seldom exceed 48 g/hL (4 lb/1,000 gal).

SILICA DIOXIDE

Silica gel and kieselsol are generic names for aqueous suspensions of silica dioxide. Silica dioxide was first used in Germany as a substitute for tannic acid in gelatin fining. Tannins, like silica dioxide, electrostatically bind with positively charged proteins (such as gelatin) and initiate flocculation and settling. Silica dioxide electrostatically binds and adsorbs compounds onto its particle surfaces. The extent of this interaction depends on particle size, shape and nature of the surface (the most important parameter), size distribution within the suspension, and particle charge density. Charge density is determined by the number of hydroxy groups that are available for binding on the surface of the particle. This is determined, in part, by the pH of the gel suspension.

The most frequent use of silica dioxide is in clarification as a replacement for tannic aid during protein fining. The effectiveness of the silica dioxide-gelatin complex is dependent on the above-mentioned parameters, and, additionally, juice or wine chemistry, mixing procedures, and on the "bloom number" of the gelatin used. Some of the advantages of silica dioxide vs. tannin in counterfining include decreased lees volume, faster precipitation, superior clarity, and less "stripping" of wine character. In finings with proteinaceous agents to reduce phenolics, silica dioxide plays a role not only in precipitating protein but also in helping to bring down protein-tannate complexes. They play another role in enhancing clarification as well as compaction of lees.

Practical Considerations Regarding Silica Gels

1. As with other fining agents, laboratory trials must precede cellar additions. (Winemakers should review the manufacturer's recommendations.)
2. Silica dioxides have a limited shelf life (less than 2 years). Avoid freezing, which may detrimentally affect particle suspension.
3. The order of addition may be important. It is usually recommended that gelatin be added first. However, some manufacturers recommend that their silica dioxide be added before the protein fining agents. In such cases, care must be taken to avoid overfining.

With few exceptions, use levels of silica dioxide should not exceed 120 g/hL (10 lb/1,000 gal). According to the BATF, silicon dioxide must be completely removed by filtration.

PROTEIN FINING AGENTS

Protein and protein-like fining agents have selected affinity for wine polyphenols. The mechanism of interaction is by hydrogen bonding between the phenolic hydroxyl and the carbonyl oxygen of the peptide bond (Fig. 16–2).

Hydrogen bonds are rather weak; therefore, the capacity of a protein fining agent is partially a function of the number of potential hydrogen bonding sites per unit weight. Thus, the techniques used for hydration or swelling, and addition and mixing are important in achieving desired effect.

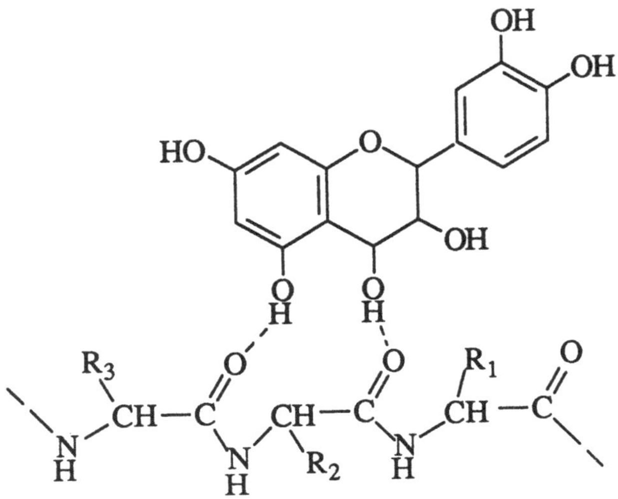

Fig. 16–2. Hydrogen bond interaction between phenolic compound and protein.

Selectivity of protein fining agents is, in part, based on bond strength (number of potential bonds formed) between the fining agent and phenols. The combination of phenol and fining agent that produces the strongest total hydrogen bonding usually occurs preferentially. Therefore, larger phenolics with more available hydroxyls and, hence, more potential hydrogen bonding sites, are preferentially bound.

As the wine ages, some monomeric phenols polymerize to form larger molecules. Protein fining agents such as gelatin, casein, and isinglass preferentially remove condensed tannins of molecular weight over 5,000 (Rossi and Singleton 1966). Therefore, the strongest complexes are usually formed between dimeric and larger tannins whereas only weak bond formation occurs with monomeric species. Polymeric phenols are largely responsible for the perception of astringency, lower molecular weight forms for bitterness. As a result, use of protein fining agents may, by selective removal of polymeric group, unmask bitterness.

Another consideration in selection of a protein fining agent is its solubility and flexibility. To be effective, the fining agent must align itself with phenolic hydroxyls. Insoluble agents (such as PVPP) generally have a much lower capacity for phenol removal than soluble species due to reduced binding sites available for bond formation. In addition, overfining is occasionally a problem with protein fining agents, resulting in stripping body and possibly proteins remaining in the wine. Juices and young wines are much more "forgiving" of the action of protein fining agents than aged wines with a greater degree of phenol polymerization.

Gelatin

Gelatin is prepared from collagen, the major structural protein in skin and bones. Hydrolysis of the multistranded polypeptide in solutions of acid and base causes strand separation, producing gelatin. The peptide chain ranges in size from 15,000 to over 140,000 daltons, and contains high levels of the amino acids glycine, proline, and hyroxyproline compared with most proteins (Singleton 1966).

The isoelectric point of gelatin is pH 4.7. Therefore, it occurs in juice or wine as a positively charged entity capable of reaction with negatively charged species such as tannins via hydrogen bond formation. It finds principal application in clarification as well as in modification of astringency. It is also employed to reduce harshness (astringency) and improve clarity in juice before fermentation.

Gelatin preferentially binds with larger molecules having more phenolic groups and potentially more hydrogen bonding sites (Singleton 1967). Thus gelatin has a less dramatic effect on color and tannin reduction in younger wines than in older products. The latter generally have a greater percentage of larger polymeric phenolics. Gelatin additions may result in color shifts in red wines from tawny (brown) to a more ruby red, perceived visually as a shift in hue (the ratio of absorbances at 420 and 520 nm, see Chapter 7). Gelatin helps reduce the phenol level and brown color in press juice before fermentation. Such applications usually occur in conjunction with silica dioxide fining.

Gelatin may strip wine of its character; therefore laboratory trials should precede any cellar addition. Only good quality gelatin, free from undesirable flavors and odors, should be used. Commercial gelatin is available in several forms and grades, and is usually rated according to purity as well as "bloom." Bloom refers to gelatins' ability to absorb water; usually 6 to 10 times its weight. Determination of bloom number is achieved by allowing a 6.66% solution of gelatin to age for 18 hours at 10°C (50°F). The weight (in grams) required to force a 0.5 inch stamp into the gel to a depth of 4 mm determines the bloom number (Hahn and Possman, 1977). Thus the higher the bloom rating, the greater is its adsorbing capability. Gelatin recommended for wine treatment ranges from 80 to 150 bloom. Using a higher bloom gelatin in conjunction with silica dioxide may result in unreacted gelatin being left in solution.

In that the number of potential bonding sites determines its effectiveness, the size of the gelatin molecule is also an important consideration.

Lower molecular weight gelatin reduces the rate of precipitation but enhances clarification and lees compaction.

The quantities of gelatin needed to achieve clarification may reduce wine astringency to undesirably low levels. Most white wines have such a low precipitable phenolic content that an exogenous source of tannic acid or silica gel is needed for reaction with the excess gelatin. Tannic acid is often added 24 hours before gelatin fining. The ratio of tannin to gelatin is usually 1:1 (wt/wt), but will vary depending on the individual wine.

Most winemakers prefer to counterfine with silica dioxide rather than tannin. The replacement of tannin with silica dioxide moderates the activity of gelatin on wine flavor (see section on Silica Dioxide). Gelatin is occasionally used as a counterfining agent after bentonite additions to remove residual haze. In this case, the negatively charged planar surface of the bentonite platelets reacts with positively charged gelatin and the two precipitate from solution. This technique may additionally help in compaction of troublesome bentonite lees. Overfining with gelatin, as with most protein fining agents, may render the wine unstable with respect to heat-labile proteins as well as potential biological activity. Additionally, where copper levels exceed "safe limits," residual gelatin may increase the possibility of casse formation (see Chapter 12).

Gelatin is commonly available as a powder but also may be purchased as a liquid concentrate at 30 to 46% or in sheets. Liquid gelatins are produced by hydrolysis which lowers the molecular weight and prevents geling at high concentration. Liquid gelatin concentrates are typically stabilized with benzoates and/or sulfur dioxide.

Before use, dry gelatin must be hydrated in warm water (44°C/112°F), at 60 g/L (1 lb gelatin/2 gal water), and added to juice or wine by the methods outlined earlier. Prolonged or excessive heating may result in denaturation of the protein and reduced activity. Due to potential biological deterioration, gelatin solutions should not be stored over extended periods.

Addition levels generally range upward from 0.75 g/hL (1/16 lb/1,000 gal). In red wines, levels often range from 4.8 to 10 g/hL (0.4–0.8 lb/1,000 gal). Much larger doses are used in juice fining, especially press juice. Heavy press juice may require up to 48 g/hL (4 lb/1,000 gal) to reduce astringency and oxidized color. Counterfining with silica dioxide or bentonite is recommended.

Casein

The principal protein in milk, casein occurs in solution as a positively charged macromolecule with a molecular weight of approximately

375,000. It is presently available in dry form as (1) purified milk casein which is insoluble in acid but soluble in alkali solution, and (2) sodium or potassium caseinate which is soluble in water. Some proprietary caseinates contain potassium bicarbonate to enhance solubility of the casein and its salts in water. Caseinate can simply be hydrated with water before use, but milk casein must first be dissolved in water at a pH greater than 8.0. Regardless of the particular casein or casein preparation, dry forms should be fully hydrated in water, never juice or wine, before use. At wine pH, casein flocculates, and the resulting precipitate adsorbs and mechanically removes suspended material as it settles.

In general, casein is used in white wines and sherries to reduce oxidized (brown) color and character. Its use is not uncommon in Burgundy, particularly in white wines that have spent a second winter in the cellar.

Casein has been used to impede or prevent pinking in susceptible wines such as Chardonnary or Pinot Blanc. It is also used as a substitute for carbon in color modification in juice and white wines, and in reducing or removing dark color and cooked flavor from sherry. Although not as effective as carbon, casein does not catalyze the oxidative deterioration associated with the use of carbon in wine. Casein reduces the concentration of both copper (by up to 45%) and iron (by 60%).

As is the case in gelatin fining, casein finings of white wine are often preceded by tannic acid or silica dioxide additions. In this case, however, the ratio of tannic acid to casein should be 0.5:1 (wt/wt). Such additions are usually made approximately 24 hours before planned casein addition.

Use of casein for clarification may not yield the desired results. However, when used for this purpose, it is usually prepared in more concentrated form. The use of working solutions prepared at concentrations of at least 25 g/L appears to improve clarification.

Only the purest available grade of casein should be used so as not to impart off flavors and characters to the wine. Addition levels to wines usually range from 1.25 to 24 g/hL (1/8 to 2 lb/1,000 gal) depending on the purpose and the form used. In laboratory trials casein is frequently pipetted to the floor of the fining vessel, then mixed by shaking. This avoids the formation of clots on the wine surface. Neither BATF nor OIV regulate the use of casein.

Pasteurized Milk

Pasteurized milk is approved for addition into white wines and sherry by BATF. The purposes are the same as those listed under casein, although use is limited to no more than 2.0 L of whole or skim milk per 1,000 L (0.2% vol/vol).

Egg Albumen

Egg albumen is commonly used in red wines. Fresh egg whites contain approximately 12.5% (wt/wt) protein, corresponding to 3 to 4 grams of the active product per white (Peynaud 1984). The principal proteins in egg white are albumen and globular proteins. Albumens are soluble in water, whereas globular fractions are soluble in neutral dilute salt solutions. Egg whites contain 7 to 8 g/L salts. If whites are diluted with water before addition to wine, the salt concentration is lowered and the globular fraction becomes partially insoluble, producing a turbid solution. In order to solubilize the entire protein content and, hence, maximize fining efficiency, a pinch of potassium chloride is often added to the diluted white. Sodium chloride is not permitted by BATF.

In preparation for fining, fresh or frozen whites should be gently whipped with a whisk before addition to the main volume of wine. Excessive mixing, although effective in reducing the troublesome gelatinous character of the white, yields a foam, some of which will remain on the wine surface. As compared with frozen whites, fresh egg whites appear to have a larger phenol adsorption capability. This may amount to as much as a twofold difference. Egg whites are added to barrelled wines and mixed with a dowel through the bung hole rotated in a "Figure-8" motion or by barrel mixers.

BATF has published the following guidelines for use of albumen. A working solution is prepared by dissolving 1 oz (28.35 g) of potassium chloride and 2 lb (907 g) of egg white in 1 gal (3.8 L) of water. Use levels of this solution may not exceed 0.25 L/hL (2.5 gal/1,000 gal) of wine.

Addition levels range up to eight whites per 225 L barrel. Neither BATF nor OIV regulates the quantity of egg whites used. However, BATF does limit the volume of water to 0.15 L/hL or approximately 330 mL per barrel. The OIV expects no water to be used except that necessary to dissolve dried albumen.

Egg white is seldom used in white wine fining because of the need for counterfining. As compared with gelatin, egg white appears to remove less of the fruit character (See Table 16–1). For this reason egg white fining has its principal application in barrel fining aged reds to round and reduce astringency before bottling. As a rule-of-thumb, one gram of albumen precipitates two grams of tannin.

Blood Albumen

Blood albumen has a long history as a fining agent. Blood albumen is available in powder form, and is permitted by the OIV but not BATF. It requires slightly less tannin to precipitate than egg white (Recht 1993a,b).

Isinglass

Isinglass is a protein fining agent produced from sturgeon collagen. It is available in two forms: a prehydrolyzed form that hydrates in 20 to 30 minutes and a fibrous form of flocced isinglass. Hydration should be carried out in cool water (≤ 15°C/60°F). If prepared in hot water, isinglass undergoes partial hydrolysis resulting in the formation of smaller molecules. The reduction in molecular weight from 140,000 to 15–58,000, results in differences in fining characteristics, and a product that is more gelatin-like in its activity (Rankine 1984).

A typical hydration protocol for flocced isinglass calls for adding 6 g of the agent to 1 L of pH-adjusted, cool water (< 15°C/60°F). The pH is adjusted to 2.4 to 2.9 with either citric or tartaric acid. The isinglass is mixed over a 24 hr period until uniformly dispersed. An addition of 0.1 mL of this stock solution to 80 mL of wine is equivalent to 0.75 g/hL (1 oz/1,000 gal).

Isinglass is principally used in white wine fining to bring out or unmask the fruit character without significant changes in tannin levels. Isinglass is less active toward condensed tannins than either gelatin or casein (Rankine 1984). Because condensed phenolics are principally responsible for astringency, isinglass has a less dramatic effect on the reduction of both wine astringency and body than most other protein fining agents. It has the added benefit of not requiring extensive counterfining as compared with other proteinaceous fining agents. Ribereau-Gayon (1972) reported that most wines contain sufficient quantities of endogenous tannin needed to glean residual protein potentially present after fining. Isinglass may be used at several stages in the winemaking process. Many vintners fine with the agent after aging (particularly barrel aging) and before bottling to "round out" background astringency and produce a brilliantly clear white wine without the stripping effect seen by other protein fining agents. Isinglass is also used as a riddling aid in methode champenoise production at levels of 1.5 to 4.0 g/hL (1/8–1/3 lb/1,000 gal).

Isinglass has several advantages over gelatin in fining of white wines. The agent is active at lower concentrations, produces enhanced clarification and a more brilliant wine, and is much less temperature dependent than gelatin, which shows enhanced properties at low temperature (Ribereau-Gayon et al. 1972).

Isinglass does, however, have several significant drawbacks. The low density of flakes forming after addition to the wine can result in voluminous lees formation (>2%), and particulates tend to hang on the sides of barrels and casks. This problem may be muted by use of counterfining agents such as bentonite. Isinglass can degrade with time, particularly if stored warm, imparting an unpleasant, fishy odor to the slurry. For this reason, only isinglass of the highest quality should be used. No limit on the use of isinglass is imposed by either BATF or the OIV.

YEAST FINING

Yeast fining involves the addition of fresh yeast (as much as 10%) to a wine with subsequent removal by centrifugation or filtration. Consisting of approximately 30% protein, the yeast cell wall may play an important role in complexation of wine and must polyphenols and metals. Singleton and Esau (1969) report that 10 g of yeast may adsorb up to 1 g of leucoanthocyanidin. Yeast production during fermentation is on the order of 10 g/L and at this level, up to 1,000 mg/L of tannin may be removed due to absorption by yeasts (Singleton and Esau 1969).

Yeasts are effective in reducing metals during fermentation. Their use in post-fermentation metal reduction may be important, especially in the absence of previously available agents. Langhans and Schlotter (1987) as well as our own trials have shown reductions from 50 to 85% upon addition of recently rehydrated wine active dry yeast. Screening procedures for identification of metal casses are given in Chapter 20.

In addition to their adsorbent activity, actively fermenting yeasts have been used to revitalize oxidized wines. As such, yeast fining has been found to have application in reduction of excessive browning and other oxidative characters in problem wines. This procedure has also been employed for the removal of herbaceous and off odors.

POLYVINYLPOLYPYROLIDONE

Polyvinylpolypyrolidone (PVPP) is a synthetic, high-molecular-weight fining agent composed of cross-linked monomers of polyvinylpyrolidone (PVP). PVPP is a "proteinlike" fining agent with affinity for low-molecular-weight phenolics. The mechanism of action is hydrogen bond formation between carbonyl groups on the polyamide (PVPP) and the phenolic hydrogens. As a selective phenol adsorbent, PVPP is available in several particle sizes. Unlike the soluble protein fining agents that remove larger polyphenols by conforming with the molecule and interacting with many hydroxyl groups, insoluble PVPP contacts relatively few reactive groups. Therefore, PVPP finds its major application in binding with and removing smaller phenolic species such as catechins and anthocyanins, which conform to the PVPP molecule. These compounds are precursors to browning in white wines and browning and bitterness in red.

PVPP may be effective in "toning down" bitterness or the potential for this problem in wines. PVPP is most beneficial as a post-fermentation treatment, but is occasionally added to juice and removed by settling before fermentation. Versini (1982) reported that PVP treatment of Traminer must reduced the formation of 4-vinylguaiacol, an important character impact compound for the variety. It may be a benefit in prefermentation fining of juice from moldy grapes and for removing excess color for blush wine production. Ough (1960) found that PVPP removed more tannin and anthocyanin than gelatin. This represents a potential problem in use of this agent in red wine fining. In addition, PVPP is used in removal of browning or pinking precursors in young white wines. When used for browning removal, it is often used in conjunction with decolorizing carbon. In cofining with carbon, PVPP can aid precipitation. Recommended use levels are from 12 to 72 g/hL (1–6 lb/1,000 gal). Addition slurries are prepared at 5 to 10% (w/v) with wine or must and mixed for a minimum of 1 hour to ensure swelling (O'Reilly 1993). PVPP has the ability to strip wine complexity. It is therefore desirable to fine juice or young wines before the development of aged bouquet. Spacek and Jelinkova (1991) reported that Amidap (co-polymers of polyamide 6-polyoxyethylene), which has been used in beer, was effective in removing wine flavonoids.

Regeneration of PVPP is more common in Europe than America. The method is based on the fact that the absorbed polyphenols may be easily removed from the PVPP by treatment with a weak, warm (50°C/122°F) caustic solution. According to the BATF, PVPP must be removed from wine by filtration before bottling.

TANNIN

Tannin or tannic acid (as oak bark tannin) and enological tannins (extracted from nutgalls or grape seeds) are occasionally used as a fining agent. In solution, tannin is negatively charged and is used in conjunction with gelatin to enhance clarification. Tannin is also used to increase astringency or "backbone" in juices and wines deficient in grape tannins. Tannic acid additions to wines are usually at levels of less than 3 g/hL (0.25 lb/1,000 gal). Laboratory solutions of tannic acid are usually prepared as a 1% (wt/vol) solution in 70% ethanol.

BATF regulations state that the total tannin shall not be increased by more than 150 mg/L by addition of tannic acid. Only tannin that does not impart color may be used.

METAL REMOVAL

Historically, German winemakers found that excess iron and copper could be removed from wine by the addition of ferrocyanide. Also known as *Blauschonung* or "blue fining," the practice has been used in Europe under very strict governmental regulation for treatment of white and rosé, and in some cases, red wines. One milligram iron (III) requires 5.65 mg of potassium ferrocyanide (Peynaud 1984). In practice, this may vary, ranging from 6 to 9 mg ferrocyanide/mg iron. Laboratory trials for each wine must be conducted to optimize fining additions. Because the major decomposition product of the reaction is cyanide, the practice is not permitted in the United States.

The reaction of ferrocyanide with iron (III) is slow, requiring up to 7 days (Castino 1965). Some workers recommend the use of ascorbic acid, at 50 mg/L, to reduce the iron present to the Fe(II) state, thus increasing the rate and completeness of the reaction. Properly used, ferrocyanide removes not only excess iron, but also most of the copper without adversely affecting the wine's flavor or bouquet.

The proprietary compounds Cufex and Metafine have been used in the United States for metal removal, but currently neither product is being marketed. In the absence of chemical agents, yeast fining may be the best method to lower metal levels by absorption onto the yeasts' cell membrane.

Reaction of Metals and Blue Fining Agents

Removal or reduction in the concentration of copper and iron by blue fining takes advantage of the fact that cyanide ion forms very stable complex ions with transition metal ions. However, excess ferrocyanide remaining in solution may degrade, producing traces of free hydrocyanic or "Prussic" acid (HCN):

$$Fe(CN)_6^{4-} + H_2O \rightarrow Fe(CN)_5^{3-} + OH^- + HCN$$

Wines with residual cyanide in excess of 1 mg/L are unacceptable under both state and federal standards. Thus, where potential cyanide may exist, it is necessary to test for its presence (see Cyanide in Chapter 20).

RIDDLING AIDS

Riddling (remuage) is the process of conveying the sediment to the neck of the bottle in methode champenoise production. Proper riddling causes the heavy particles to ride over and bring down the lighter more flocculent particles.

To enhance riddling, disgorgement, and possibly wine palatability, some add riddling aids at the time of *cuveé* bottling. Such aids may enhance the riddler's ability to convey the yeast to the neck of the bottle. The most common riddling aids are sodium and calcium bentonite; Clarifying Agent C; Adjuvants H and 84; isinglass; Colvite; tannin; Boltane; gelatin; and diatomaceous earth.

Clarifying agent C is a proprietary bentonite preparation used with phosphate mazure; Adjuvants H and 84 are proprietary bentonite-based agents sometimes used with tannin; Colvite is a proprietary isinglass; and Boltane is a proprietary tannin formulation; all are of European origin.

Bentonite is the most popular riddling aid in the United States. It is added at levels rarely exceeding 6 g/hL (1/2 pound/1,000 gal) sometimes with alginates. In Europe, calcium bentonite at levels of 3 g/hL (1/4 lb/1,000 gal) is frequently used. Calcium bentonite produces more compact lees than sodium bentonite (Ferenczy 1966). The decision as to which riddling aid to use should also be based on the expected time *sur lie.* Clays are often preferred for young wines, gelatins for aged or older wines.

The major disadvantage of riddling aids is that their effects on both riddling ease and sparkling wine palatability are not predictable. Because each cuveé is different, the winemaker must wait until riddling and disgorgement to review the merits or deficiency of the riddling aid(s) employed. This is a primary reason why bentonite is the most frequently used riddling aid. It seldom has a detrimental effect on product palatability at the levels employed.

Utilization of Enzymes in Juice and Wine Production

Katherine G. Haight, Research Associate
Viticulture and Enology Research Center, Fresno, CA

Several issues need to be considered when choosing a commercial enzyme preparation. These include the wine type, processing conditions, desired effect, and the cost/benefit relationship. Enzyme activity is a function of pH, temperature, and contact time. Knowing the pH and temperature profile of the particular enzyme preparation determines the enzyme dosage and the cost to the winery. Decreases in pH, temperature, and/or contact time, along with increases in other chemical characteristics of the must/wine (e.g., SO_2), increases enzyme usage (and conversely). An example of a processing decision is the timing of bentonite additions. Bentonite will dramatically inhibit the activity of an enzyme: therefore, it is very important not to add bentonite until enzyme reactions are complete.

For optimum stability, enzyme preparations should be stored at approximately 4°C (39°F). To prevent activity loss, the products should not be diluted until they are needed. When stored properly, minimal activity loss should occur over a 1-year period. Before addition, the enzyme preparation should be diluted to a 10% solution with cool clean water to facilitate dispersion. Recommended dosage rates are available from the manufacturer, and may vary depending on the activities of the product. Due to the specificity of today's enzyme preparations, it is important to start with laboratory trials.

GLUCANASES

White wines produced from botrytized grapes often present serious problems in clarification and filtration. The origin of this problem is usually *beta*-glucan, a polymer of glucose synthesized by *Botrytis cinerea*. *Beta*-glucan impedes settling of particulates (including yeast and bacteria) resulting in very turbid wines. The presence of *beta*-glucan in wine can be established by an alcohol test, during which a threadlike white precipitate is formed (see Chapter 20, Diagnostics). After degradation of *beta*-glucan by the use of a specific *beta*-glucanase, it is possible to clarify wine with considerably reduced filtration costs.

Pectinases have not proven to be successful for this application. The use of beer *beta*-glucanase has also not proven to be satisfactory for improved clarification and filtration of wines (Janda 1983). Wine *beta*-glucan has a structure consisting of *beta*-1,6-linked side chains attached to a *beta*-1,3-linked backbone. Barley *beta*-glucan, which causes filtration problems in beer, is different from the *Botrytis* glucan as the latter is comprised mainly of *beta*-1,4-linked glucose units.

When botrytized grapes are used, *beta*-glucanases can be added at either juice clarification or post-fermentation. When added to chilled juice or wine the dosage rate needs to be increased due to the unfavorable conditions of temperature and shorter contact times. Janser (1994) suggested that optimal results are obtained when the enzyme preparation is added post primary fermentation after the racking stage. Tannins can irreversibly inactivate enzymes in the same manner as bentonite (Villettaz et al. 1984). Therefore, red wines require a higher dosage than white wines. Favorable results have also been reported by staggering the dosages at intervals of about six days (Urlaub 1989). Some glucanases are prohibited by BATF because they are of non-*Aspergillus* origin (see Table I–10 in Appendix I).

PECTINASES

Pectinases are used to reduce viscosity of grape must through the hydrolysis of pectin. The primary benefits of pectinase use include improved free run yields, improved juice settling rates and clarification, and enhanced filterability. Berg (1959b) reported an average increase of 4% in white juice yield. Subsequently, Ough and Berg (1974) reported an average increase in white juice yield of 15%, and an average increase in red juice yield of 5.6% in laboratory trials.

Most commercial pectinase preparations are derived from strains of *Aspergillus*, primarily *Aspergillus niger* (Canal-Llauberes, 1993). This species is accepted as GRAS (generally recognized as safe) by the USFDA and accepted by the OIV and BATF for the production of enzyme preparations.

Pectinases work on pectic substances that occur as structural polysaccharides in the middle lamella and primary plant cell wall. Pectins, and pectic substances, play an important role in juice extraction and subsequent downstream processing. Colloidal substances such as pectin may suspend grape particlulates, causing turbidity, and therefore slowing juice clarification. Pectin exists as a polymer of galacturonic acid linked via *alpha*-1,4-glycosidic bonds with varying degrees of methylation.

Commercial pectinase preparations typically contain varying ratios of pectin methyl esterase (PME), pectin lyase or transeliminase (PL), and polygalacturonase (PG), along with other side activities (see Fig. 16–3).

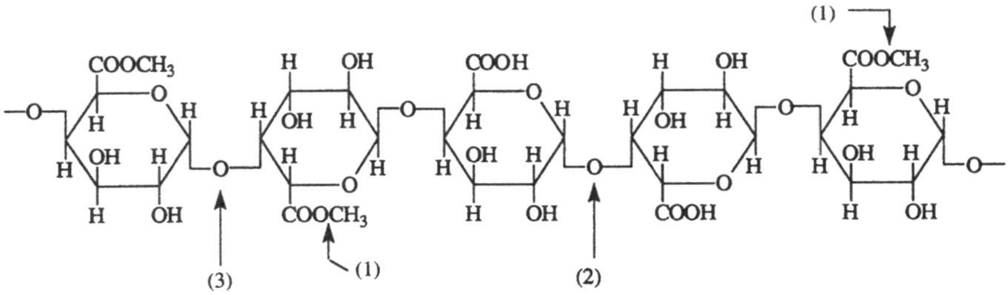

Fig. 16–3. Portion of a pectin polymer showing sites of pectinase activity: (1) PME activity, (2) PG activity, and (3) PL activity.

Each of these individual pectolytic enzymes attack the galacturonic acid-derived backbone of the pectin molecule differently. Methyl galacturonic acid is de-esterified by PME, producing a low methoxyl pectin, pectic acid, and methanol. Pectate and low methoxyl pectin are the best substrates for PG. PG hydrolyzes the 1,4-glycosidic linkages in the galacturonan chain. Highly esterified pectins are the most suitable substrates for PL. This endoenzyme randomly splits the glycosidic linkages in the methylated galacturonan chain (see Fig. 16–3).

MACERATING ENZYMES

In the mid-1980s macerating enzymes for winemaking were developed. Like pectinases, this was due to earlier success in the apple juice industry. These formulations typically contain pectinase, cellulase, hemicellulase, and other carbohydrase activities. They improve juice yields (Haight and Gump 1994) by degrading structural polysaccharides that interfere with juice extraction, clarification, and filtration. Cellulases and hemicellulases degrade cell wall polysaccharides, breaking down the entire structure and causing solubilization of the middle lamella. Due to this more complete degradation of cell wall and middle lamella components, macerating enzymes may improve press yields (especially with hard-to-press varieties) and/or lees settling/clarification rates when compared with pectinases alone (Sims et al. 1988). Other benefits may include improved color extraction in some red grape varietals (Shoseyov et al. 1990), increased aroma and flavor, and improved body and structure (10–30% increase in tannins, procanthocyanidins) (Plank and Zent 1993; Zent and Inama 1992).

Throughout most of the world, macerating enzymes are rapidly replacing traditional pectinases used in winemaking (Plank and Zent 1993).

Macerating enzymes are not always solely derived from *Aspergillus* sp. and therefore are not approved for winemaking in the U.S. However they have GRAS in U.S. fruit juice production.

Color extraction

Not only is the initial red juice color significant, but the stability of this color is of great importance. Increases in desired color are due to improved extraction of the colored pigments brought about by the breakdown of pectin (Schwimmer 1981). However, not all enzyme formulations give identical results. Their suitability for color extraction depends on the ability to "loosen" the structure of the middle lamella. Protopectin is part of the three-dimensional hemicellulose network rich in galacturonic acid residues. Protopectinase activity results in the release of highly methyl-esterified, unmodified pectin directly from the solids and further breakdown of this substrate (Felix and Villettaz 1983). Besides increased color extraction, degradation of the protopectin allows juice to flow more freely from crushed grapes resulting in increased juice yield.

Flavor and Aroma Extraction

Grape flavor components consist of free volatiles and sugar-conjugated flavor precursors. Various free volatile terpenes play an important role in the varietal character of Muscat and several non-Muscat varieties, including Johannisberg Riesling and Gewürztraminer. Monoterpenes are present in the grape either in a free state or bound to sugars in the form of glycosides. Free monoterpenes include the volatile aroma compounds, known as terpenols, and the odorless terpenoid hydroxylated linalool compounds (polyols). The monoterpene precursors of terpenols are the bound terpenes and the polyols. Bound terpenes are a mixture of disaccharide glycosides of various monoterpene alcohols, and possess flavor potential that can be realized by using a suitable enzyme (Williams et al. 1982, 1985; Wilson et al. 1984). Monoterpene glycoside precursors are nonvolatile and therefore without significant aroma. The quantity of these precursors can be higher than the amount of aromatic terpenols, indicating increased flavor potential (Dimitriadis and Williams 1984). Linalool and geraniol are two of the most abundant bound terpenols (Gunata et al. 1985; Ribereau-Gayon 1975). These two terpenols are also the most aromatic, meaning they have a very low olfactive threshold (Gunata et al. 1985, 1988; Marais 1983; Ribereau-Gayon 1975).

In aromatic grape varieties, the terpenols in various states of oxidation form the major part of the aroma (Gunata 1985; Williams et al. 1980, 1981). Terpenols are known to interact synergistically with one component increasing the aroma of another component (Marais 1983). Terpenols originate from bound terpenes, either by direct pathways or by way of polyols. Bound terpenes can undergo both acid or enzyme-catalyzed hydrolysis to provide volatile aroma compounds. Alternatively these bound terpenes can form odorless polyols, which themselves undergo acid hydrolysis to produce volatile aroma compounds (Williams 1990; Williams et al. 1980, 1985).

Monoterpenes are bound to disaccharide glycosides, namely *alpha*-L-arabinofuranosyl-*beta*-D-glucopyranosides or *alpha*-L-rhamnofuranosyl-*beta*-D-glucopyranosides, and a nonsugar portion (the aglycone), which is often a phenolic derivative. Hydrolysis of the terpenol precursor can be accomplished by *beta*-glucosidase, *alpha*-arabinosidase, *alpha*-rhamnosidase, *beta*-xylanosidase, and *beta*-apiosidase activity (Cummings 1994; Plank and Zent 1993; Williams 1990). For instance, enzymatic hydrolysis of the major grape

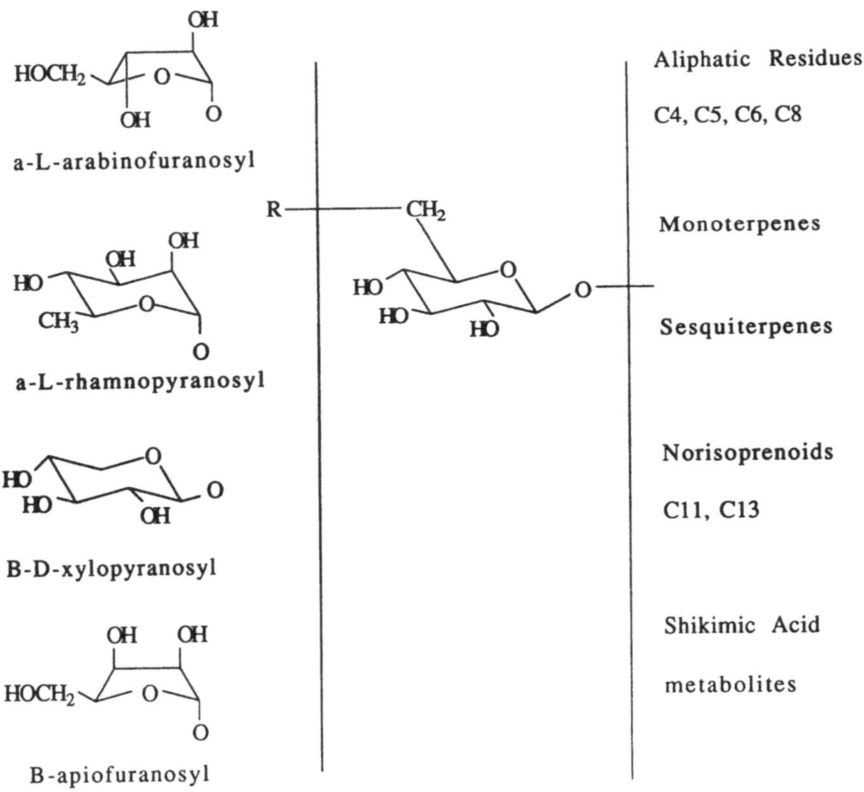

Fig. 16–4. Flavor precursors/products in wine flavor development.

monoterpenyl diglycosides involves first the removal of the terminal nonreducing unit by cleavage of the *alpha*-1–6 linkage by either *alpha*-L-arabinosidase or *alpha*-L-rhamnosidase with accompanying production of monoterpenyl glucosides. Terpenols are then liberated by the action of *beta*-D-glucosidase (see Fig. 16–4). Grape *beta*-glycosidases are known to lack or to have low levels of *alpha*-L-arabinosidase or *alpha*-L-rhamnosidase activities. Thus expression of these latent terpenoids can be enhanced by addition of specific enzymes.

The *beta*-glucosidase inhibition by glucose is of a competitive type. Thus it is recommended that these products be added to the must toward the end of fermentation. Conjugated aroma precursors may be important sources of flavor in non-terpene varieties (see Fig. 16–4 and Ch 3).

ULTRAFILTRATION

Recently certain enzyme preparations have been identified for both increased product flux rates and cleansing of ultrafiltration (UF) membranes. Pectinases with hemicelluolytic side activities, have been reported to be successful in increasing flux rates on apple juice (Stutz 1993). He also reported that relative flux was increased by using a purified rhamnogalacturonase, indicating that this particular activity is of great importance. Depectinized grape juice has been treated with cellulase for increased filtration flux rates (Haight and Gump, 1991, unpublished data).

By causing a slimming effect, soluble colloidal substances, such as pectin residues and neutral polysaccharides, can interfere with UF flux rates. Enzymatic cleaning, along with a proper chemical cleaning, is an excellent alternative when chemical cleaning alone is not working. Results can usually be achieved using an enzyme dosage rate of 500 ppm (Haight and Gump, 1991, unpublished data; Stutz 1993), however, contact time and temperature need to be considered. Usually several hours to an overnight enzyme contact time is sufficient. It is important to check enzyme and membrane upper temperature limits before determining a cleaning protocol.

FUTURE DEVELOPMENTS

Wine proteases

The use of an acid protease to hydrolyze the grape proteins involved in wine haze is eagerly awaited. Proteases hydrolyze peptide linkages between amino acids in proteins. Presently, acid proteases are strictly in the research and development stage (Bisson 1993a). Without heating, commercially available enzymes have not been effective in producing protein stability of white wines (Bakalinsky and Boulton 1985; Bisson 1993b; Modra and Williams 1988; Waters et al. 1990)

Laccases and Tannases

Laccases oxidize a wide range of phenolic compounds. Recently, the use of laccase has been investigated in an effort to induce phenolic browning associated with white wine production (Plank and Zent 1993). Browning and precipitation of juice phenols allows for lighter and color-stable finished wines. Cantarelli et al. (1989) demonstrated that laccase treatment can enhance the effect of traditional fining treatments. Laccases are similar to tyrosinase (polyphenoloxidase); however, the two enzymes have different substrate specificities and reaction mechanisms (see Chapters 3, 7, and 14).

In theory, tannases could be used to decrease tannin levels in wine. However, the color stability of treated red wines is in question. Use of these enzymes are still very much in the research stage and there is, as yet, no viable commercial use.

Glucose Oxidase and Catalase

Villettaz (1986) proposed using glucose oxidase and catalase in juice to convert glucose into gluconic acid that cannot be metabolized by the yeast. Thus, the subsequent wine has reduced alcohol and increased acidity. Glucose oxidase reactions may lead to oxidation of other components, including flavor components. Therefore, sufficient amounts of catalase are necessary to convert hydrogen peroxide produced into water. This may be an alternative way of producing low alcohol wine, without the use of expensive equipment, such as reverse osmosis filters.

SUMMARY

During the past four decades, there have been dramatic changes in enzyme preparations used in fruit processing. These beneficial changes include both increased specificity and purity, therefore allowing the winemaker to use enzyme products for a particular processing need. Genetic engineering technologies are expected to be of great importance to the next generation of food grade enzyme products. One can also look forward to more and better options in juice/wine processing through selective enzyme use.

PROCEDURES

The following is a cross-reference of procedures found in Chapter 20. Rapid determination of amorphous materials (protein and protein-complexes); pectins and glucans.

CHAPTER 17

WINERY SANITATION

Uncontrolled proliferation of microorganisms eventually leads to product deterioration and spoilage. Traditionally, SO_2 has represented one important tool for control of microbiological growth. However, application levels have dropped dramatically in the last few years, leading to spread of microorganisms. Given the likelihood of further reductions, and potentially, elimination of the compound altogether, we are faced with identifying alternative strategies for dealing with microbial problems. The single most important of these is sanitation.

Unlike sterilization where success is measured as 100% kill, sanitation is usually defined as reduction of viable cell number to acceptably low numbers. Additionally, sanitation accomplishes a second important goal: elimination of hospitable environments for growth.

Typically, monitoring involves microbiological sampling of surfaces before and after sanitation operations. Careless use of detergents and sanitizers poses a health and safety threat; sanitation personnel must be trained in their proper use and supplied with appropriate safety clothing (gloves, glasses, boots, etc.). Further, U.S. safety regulations require that Material

Safety Data Sheets (MSDS) be made readily available to all employees involved in these operations.

WATER QUALITY

Water is basic to most detergents/sanitizers and consideration should be given to its chemical as well as sensory properties, if any. Water supplies contain varying amounts of calcium and magnesium (and other alkali metals) that contribute to "hardness." Hard water interferes with the effectiveness of detergents (particularly bicarbonates) contributing to precipitate formation. Such precipitates serve as sites for accumulation of organic debris and microorganisms, thus making subsequent sanitation more difficult. United States Geological Survey (U.S.G.S.) definitions for hardness levels are presented in Table 17–1.

Table 17–1. U.S.G.S. definitions for water hardness levels.

Hardness	Parts per million (mg/L)	Grains per gallon (gpg)[a]
Very hard	>180	>10.5
Hard	120–180	7.0–10.5
Moderately hard	60–120	3.5–7.0
Soft	0–60	0–3.5

[a]gpg × 17.1 = mg/L (ppm).

If one has hard water, a water softener should be installed in the delivery system. Although detergent formulations may include chelating agents that help mitigate the problem, their extensive use is more expensive than initially providing soft water systems.

Water temperature is an important concern with respect to "scale" formation. With few exceptions, hot water is superior to cold. However, where hard water is used, maximum scale formation occurs at 82°C (180°F). Where soft water is used, this is not an issue.

PRELIMINARY CLEANING

In any cleaning/sanitation process, it is necessary to remove as much of the first level (visible) debris as possible before use of detergents. This is accomplished either manually or by automated cleaning systems (spray balls, etc.).

CLEANERS (DETERGENTS)

Once visible debris and film have been removed, detergents are used for solubilization of adhering deposits. Each detergent has unique properties of action as well as formulation for most effective application. Generally, increasing the concentration beyond recommended levels provides little additional benefit and is not cost effective. However, increasing water temperature to 38°C to 43°C (100°–110°F) and delivery pressure improves cleaning operation while decreasing both the amount of water used and time required for this operation. It is best that spray be directed at an angle to the surface being cleaned.

Several different components may be present in a particular detergent formulation. In varying proportions, these include alkalies, polyphosphates, various surfactants, chelators, and acids.

Alkalies

Strong alkalies, including NaOH (caustic soda or lye) or KOH (caustic potash), and sodium carbonate (Na_2CO_3) are the most commonly used detergents. Both NaOH and KOH have excellent detergent properties and are strongly antimicrobial; they are active against viable cells as well as spores and bacteriophage. Unfortunately, they may also be corrosive, even to stainless steel, if recommended application levels are exceeded. Handling strong alkalies requires use of protective gloves and eyeware.

Sodium *ortho*- and metasilicates (Na_2SiO_3) are less caustic than NaOH and have better detergency properties. They are also less corrosive. Where the organic load is not heavy, mild alkalies such as sodium carbonate (soda ash), or trisodium phosphate (TSP) find application. Sodium carbonate is an inexpensive, frequently used detergent. Unfortunately, in hard water, Na_2CO_3 contributes to precipitate formation.

Sequestering agents

Due their abilities to chelate calcium and magnesium and prevent precipitation, polyphosphates are often included in detergent formulations. Examples include sodium hexametaphosphate (Calgon) and sodium tetraphosphate (Quadrofos). The amount included depends on hardness of water. With the exception of TSP, the group is noncorrosive (Brown 1973).

Organic chelating compounds such as EDTA may also be included. Although more expensive than polyphosphates, they have the advantage of being relatively heat stable.

Surfactants

Having both water and oil-miscible properties, surfactants are generally used as wetting agents to reduce surface tension and facilitate contact between detergent and the surface being cleaned. Various anionic, nonionic, and cationic surfactants are available.

Acids

Acids are used in specialized detergent formulations (at approximately 0.5%) to reduce mineral deposits and soften water. Maximum effectiveness occurs at pH 2.5. At low pH, acid solutions are very corrosive toward stainless steel (and other metals). Phosphoric acid is preferred because of its relatively low corrosiveness and compatibility with nonionic wetting agents.

SANITIZERS

Once deposits are removed and the surface is visibly clean, it is sanitized. Two general categories of chemical sanitizing agents are currently in use: the halogens, including chlorine and iodine, and Quarternary Ammonium Compounds (QUATS). Some also use sulfur dioxide as a sanitizer agent (see Table 17–2 for comparative properties).

Chlorine-Based Sanitizers

Chlorine in its active form, hypochlorous acid (HOCl), is a powerful oxidant and antimicrobial agent. Molecular hypochlorous acid is present in highest concentration at near pH 4, decreasing rapidly with increased pH. At pH greater than 5, hypochlorite (OCl^-) increases whereas at pH less than 4, chlorine gas (Cl_2) increases. Neither chlorine gas nor hypochlorite have been shown to be active against microorganisms (Mercer and Somers 1957); however, both are very corrosive. Formation of $Cl_2(g)$ is a safety issue. Because there are still substantial amounts of HOCl present at pH greater than 6.5, sanitizing operations are typically carried out in the range pH 6.5 to 7.0.

Because chlorine is an oxidant, activity will prematurely degrade if organic residues (reflecting inadequate cleaning) are present. Reaction time is also temperature dependent. Up to 52°C (125°F), the reaction rate (and corrosive properties) doubles for each 18°F increase in temperature. Although hypochlorites are relatively stable, solubility of Cl_2 decreases rapidly at temperatures above 50°C (122°F).

Table 17–2. Characteristics of Sanitizers.

Properties	Hypochlorites (HOCl)	Iodophores (I_2)	'QUATS'	SO_2
Activity towards microbes	All microbes, bacteriophage, spores[a]	All microbes except phage and spores	Many microbes except phage and spores	See Chapter 10
Corrosiveness (at recommended levels)	Some metals	Not	Not	Not
Effect of organics	Decreased	Decreased	Effective	Decreased
pH	Ineffective at pH>8.5	Slow at pH >7.0	Wide range	Increases with lower pH
Temperature	Volatilizes 50°C/122°F	Volatilizes 49°C/120°F	Stable	Volatilizes with higher T°C
Use and levels	Stainless, wood tanks, 200 mg/L	Rubber fittings, tile: 25 mg/L Walls: 200 mg/L Concrete: 500–800 mg/L	200 mg/L	>200 mg/L

[a]*Bacterial endospores, mold spores.*

Sanitizing surfaces require active chlorine concentrations of 100 to 200 mg/L. Although chlorine is compatible with stainless steel surfaces at recommended levels, severe oxidation (pitting) may result from use of larger amounts. Upon completion of the operation, thorough rinsing is required to remove remaining sanitizer. The effectiveness of this operation may be monitored by use of "test kits" or the procedures outlined in Chapter 20 (Chlorine-Residual).

There are several forms of chlorine available. They are shown in Table 17–3.

Iodine

Formulations including iodine and nonionic wetting agents are called iodophors. Iodine (I_2) is the active principal and thus iodophors are most effective in the range pH 4 to 5 where the concentration of I_2 is maximum. To ensure activity, formulations typically include phosphoric acid. The sanitizer has the advantage of lower use levels: 25 mg/L of iodophore is equivalent to 200 mg/L chlorine (Jennings 1965). Iodophores are frequently used for bottling sanitation, followed by a cold-water rinse.

Compared with HOCl, iodophores are not as readily degraded by organics and are nonirritating (at recommended levels). I_2 becomes volatile

Table 17–3. Forms of Chlorine available for use as sanitizers (expressed as % active chlorine).

Chlorine gas	100%
Pure calcium hypochlorite	99%
Formulated proprietary $Ca(OCl)_2$	70–75%
Chlorine dioxide	263%
Chlorine dioxide decahydrate	17%
Household bleach (sodium hypochlorite)	5.25%
Chlorinated TSP	3.5 %
Chlorinated orthophosphate	3.3 %

SOURCE: York 1986.

at temperatures greater than 49°C (120°F). Formulations containing iodophores may stain polyvinylchloride and other surfaces.

Quaternary Ammonium Compounds

QUATS function by disrupting microbial cell membranes. Historically this group was reported to have differential activity toward microbes with Gram-positive bacteria (LAB) most affected, gram-negatives less affected, and no activity against bacteriophage. Modern formulations using QUATS are considerably stronger than their predecessors and have extended activity over a broad pH range. They have additional advantages of being heat stable and noncorrosive. In wineries QUATS find application in controlling mold growth on walls and tanks. The formulation is sprayed on the surface and left without rinsing. Depending on environmental conditions and extent of mold growth, a single application may last for several weeks (Hall 1994).

Detergent-Sanitizer Formulations

Suppliers currently market formulations employing both detergents and sanitizer. These typically include a surfactant ("wetting agent") and other necessary adjuncts discussed previously.

Sulfur Dioxide

In some cases, winemakers use SO_2 as a sanitizing agent. The effectiveness of SO_2 against microbes is pH dependent (see Chapters 10 and 18). Depending on the physical properties of the surface and level of organic debris, circulating a solution of 10 g/hL SO_2 (or 20 g/hL potassium metabisulfite) and 300 g/hL citric acid at 60°C (140°F) may be effective.

Physical Sterilants

Hot (>82°C/180°F) water or steam is an ideal sterilant: it has penetrative properties, works against all wine/juice microorganisms, is noncorrosive, leaves no residues, and is relatively inexpensive. Bottling line sterilization can be accomplished with steam or hot water. Where hot water is employed to sanitize lines, the authors recommend temperatures greater than 82°C (180°F) for more than 20 minutes. The temperature should be monitored at the farthest point from the steam source (i.e., the end of the line, fill spouts, etc.). When steam is used to sterilize tanks, the recommendation is to continue until condensate from valves reaches temperatures greater than 82°C for 20 minutes. Dismantling valves, racking arms, etc., while desirable for cleaning, may not yield the time and temperature relationships necessary for santitation. Ultraviolet (UV) light is directly effective against microbes. Unfortunately it has low penetrative capabilities and even a thin film will serve as an effective barrier between radiation and microbes. Thus, its use is generally restricted to laboratory applications for surface sterilization. Skin and eyes must be shielded (by glass) from continued exposure to UV-light. Ozone (O_3) is being used in water treatment. Ozone degrades rapidly in warm (>35°C/95°F) water. Thus, at present, its primary application is cold water recirculation systems.

CLEANING AND SANITATION MONITORING

The cleaning process will significantly lower microbial populations. Bacterial endospores are most resistant to cleaners, followed by gram-positive non-spore-forming bacilli, micrococci, and gram-negative rods (Maxcy 1969)

Following sanitation, all surfaces must be thoroughly rinsed to flush residual sanitizer and solubilized debris. Tank/hose sanitation is typically followed by citric acid rinse to neutralize residual alkali. The final rinse water should be tested for residual oxidants (see Chapter 20). This may be done in the laboratory or cellar by use of "kits" designed for this purpose.

The most frequently encountered method for evaluating cleaning operations is sensory. Visually, does the surface appear clean, by touch, does it feel clean, and equally important, does it smell clean? A slippery surface or the presence of "off" odors is indicative of inadequate cleaning or rinsing of detergent. Although a quick sensory review may be adequate for fermenters and storage tanks, other areas (e.g., bottling lines) require further examination. In these cases, follow-up microbiological examination should be conducted to evaluate the effectiveness of sanitation. As discussed below, a variety of tests have been proposed, and are being used. These generally involve sampling a defined area with sterile cotton swabs, agar surfaces, or special adhesive strips.

Each technique has common problems that make quantitation difficult. These include (1) the nature of the surface (smooth vs. pitted, flat vs. irregular); (2) definition of area to be sampled; (3) amount of pressure applied to surface; and (4) the time of application. Further, one cannot be certain of complete recovery of microbes from cotton swab. By standardizing the sampling procedure, one can improve success and make general observations; that is, the operation may be rated as "good," "fair," or "poor."

Swab Tests

Swab testing involves application of a sterile cotton swab over a defined surface area for a defined period of time. The swab is then transferred to a sterile diluent (e.g., peptone) and shaken thoroughly to separate adhering viable cells/spores. The diluent is then membrane-filtered and the membrane transferred to appropriate agar media for growth. In some instances, the swab is directly rolled or streaked across the agar surface.

Direct Contact Tests

Where surfaces are flat and smooth, agar plates filled with the appropriate media can be pressed directly against the sanitized surface. In theory, viable cells are transferred directly to the agar plate. Variables affecting success include contact time and pressure. Examples of commercially available direct contact "kits" include the Rodac plate (Sobolesky 1968) and the Monoflex plate (Howard 1970).

Various "tapes" have been used in a manner similar to agar plates. In this case, "tape" is applied to surface and subsequently reapplied to agar surface.

ANALYSIS

The following is a cross-reference of procedures/analytes covered in Chapter 20. Chlorine (residual); iodine (residual); SO_2; as well as cross-reference in Chapter 18.

CHAPTER 18

Microbiology of Winemaking

This chapter is intended to highlight microorganisms of importance to the winemaker.

MOLDS

Molds are multicellular, filamentous fungi. Most are saprophytes, requiring a readily available source of exogenous nutrient. Macroscopically, mold growth appears as a mass of threadlike elements, the mycelium. Typically, a mold mycelium is composed of branched elements called hyphae. Depending on the organism in question, these may or may not be septate.

The usual mode of reproduction among the molds is asexual, that is, without union of sex cells. In most species, the asexual elements, or spores, are borne on specialized aerial hyphae termed sporangiophores or conidiophores, depending on the mold in question. Sporangiospores and conidiophores may be differentiated in that the former are normally protected within the body of a sporangium until dehiscence, whereas the latter are borne unprotected at the apices of conidiophores.

Molds are important in winemaking because, when conditions permit, they may cause damage to the fruit in the vineyard, and ultimately wine palatability. Common organisms involved in vineyard spoilage include *Penicillium, Aspergillus, Mucor, Rhizopus,* and *Botrytis.* Because vineyard molds are discussed in Chapter 3, they will not be considered further here.

Mold Growth in the Winery

Molds may also be present in the winery, where they may grow on the surface of cooperage as well as on walls and other porous surfaces. Aside from being aesthetically unpleasing, mold growth on the outside of cooperage may potentially result in leaching of their metabolites, thus imparting moldy (musty) odors and flavor to the wine.

Mold Growth on Packaging Materials

In recent years, winemakers have become increasingly aware of mold growth on/in wine cork. Potent metabolites, perceivable in the low parts-per-trillion (ppt) range, may result from mold growth on certain chlorinated phenols. The generally accepted culprit is 2,4,6-trichloroanisol; however, there are increasing reports of "corked-like" compounds that may be produced during the winemaking process (Muller and Fugelsang 1994a). This issue is covered further in Chapter 19.

Molds are aerobic organisms and are generally rather intolerant of alcohol. One exception to this is *Penicillium.* This mold has been reported to grow on cork at alcohol levels as high as 16% (Marais and Kruger 1975). Further, its growth has been implicated in corkiness. In other cases, mold fragments in aging wine are most likely a carryover of earlier stages of processing or from contact with contaminated equipment or storage containers. Thus microorganisms that can grow directly in wine are limited to alcohol tolerant lactic and acetic acid bacteria and several species of yeast.

YEASTS

Yeast Morphology

Microscopically, most yeasts appear oval to ellipsoidal in shape. Like molds, they usually reproduce asexually, in this case, by budding. In the laboratory, and presumably in nature, certain yeasts may undergo a sexual cycle of replication. In this case, the mother cell produces ascospores that subsequently germinate, producing new vegetative cells.

Yeast taxonomy is complicated somewhat by the fact that not all species have an identified sexual phase. In isolations where ascospores are not observed, the yeast is given the "imperfect" designation (e.g., *Brettanomyces*), whereas another isolate may appear identical except that it produces ascospores on special media. In this case, the microbiologist would then use the "perfect" taxon (e.g., *Dekkera*). Thus, in laboratory identification of yeast, the mode of reproduction (both sexual and asexual) varies and is an important criterion in identification. To the winemaker, this discussion is largely academic in that both "species" may behave similarly in wine.

In addition to cell morphology (shape), which varies with age of the culture and type of media on which it is grown, the position of buds on the mother cell provides initial information of value in identification. With the exception of a few genera, buds normally arise on the shoulders and axial areas of vegetative cells. In some genera where the vegetative cell has no apparent axis, buds may arise at any place on the surface. This type of budding is called multilateral budding.

Other species exhibit restricted bud formation. For example, budding limited to the axial areas of the cell is referred to as polar budding. This type of reproduction is characteristic of the apiculate or lemon-shaped yeast such as *Kloeckera/Hanseniaspora.*

As compared with budding, asexual reproduction in *Schizosaccharomyces* occurs via fission in a manner that is visually similar to bacteria. In this case, cell division occurs by formation of cross walls or septa without constriction of the original cell wall. When septum formation is complete, the newly formed cells separate.

Several species of yeast typically grow as spreading films at the surface of wines exposed to air. These films, originating from the budding of single cells, proliferate rapidly when growing in wines stored under oxidative conditions. However, film formation results from incomplete separation upon completion of budding. Although the mother and daughter cells are functionally separate, they tend to adhere and bud further. The result of this type of budding is a spreading type of growth called a "pseudomycelium." The latter differs from a true mycelium in that there is no cytoplasmic continuity between cells comprising the pseudomycelium. This type of growth is seen among in the several species of *Candida* and *Pichia.*

Yeasts of Importance to the Winemaker

Film Forming Yeasts

Several species of yeast, including strains of *Saccharomyces,* may grow oxidatively as a film on the surface of improperly stored wine exposed to the air(see previous section). Film-forming yeasts include those native vineyard yeasts *Pichia, Candida, Hansenula,* and *Metschnikowia* that survive the fermentative phase. Under oxidative conditions, ethanol as well as glycerol and organic acids (primarily malate) serve as substrate in formation of acetaldehyde, acetic acid, and ethyl acetate (van Zyl 1962). Uncontrolled growth results in decreased alcohol levels and TA/pH changes.

In that film yeasts grow oxidatively, one method of control is to maintain wine in properly topped tanks and sealed barrels, thereby depriving them of the surface needed for growth. Reliance on SO_2 for control is generally not effective once film formation is noted. As reported by Thomas and Davenport (1985), native yeasts such as *P. membranaefaciens* and *C. krusei* are resistant to free molecular SO_2 of more than 3 mg/L. Depending upon pH levels (pH 3.2–3.8) this corresponds to total SO_2 additions of 75 mg/L (at pH 3.2) and 273 mg/L (at pH 3.8). Growth of film yeasts is slowed and largely suppressed at lower cellar temperatures. Dittrich (1987) reported no growth in wines of 10 to 12% alcohol at temperatures ranging from 8°C (47°F) to 12°C (54°F), whereas growth was observed at 14% alcohol at higher temperatures. Survival and numerical dominance during alcoholic fermentation also appear to be temperature-linked (Heard and Fleet 1988b; Sharf and Margalith 1983). They concluded that low temperatures enhance alcohol tolerance of native yeast species and thus extend the time in which they may be active during fermentation.

***Candida* sp.** The asporogeneous ("imperfect") genus *Candida* includes numerous species. Those most common in wine include *C. vini, C. stellata,* and *C. pulcherimmia.* Aside from similar morphology, the group is linked by the absence of observed ascospore formation. In wine, *Candida* typically grows as a surface yeast, producing a chalky white film on the surface of low alcohol wines. In the presence of oxygen, *Candida* sp. are fast growers, using ethyl alcohol in addition to wine acids as carbon sources producing a mixture of oxidized end products. Acetic acid levels resulting from *C. krusei* and *C. stellata* were up to 1.3 g/L (Shimazu and Watanabe 1981). In the case of *C. krusei,* ethyl acetate levels ranging from 220 to 730 mg/L have also been reported. As a group, *Candida* sp. are weak fermenters.

Microscopically, these yeasts appear as long cylindrical cells of approximate dimensions 3 to 10 × 2 to 4 μm. Asexual reproduction is by budding. Incomplete separation of mother and daughter cells leads to formation of an extensive pseudomycelium as well as a true mycelium.

***Pichia* sp.** *Pichia* may also be found as part of the film-yeast community. Important species of this ascospore-producing yeast include *P. membranaefaciens, P. vini,* and *P. farinosa.* Although most species are inhibited by alcohol levels of near 10%, depending on cellar temperature, growth may be found in wines of up to 13% alcohol. Growth on the wine surface appears as a very heavy, "balloonlike," chalky film. Unrestricted growth of this organism imparts an aldehydic character to the wine.

In addition to SO_2 resistance, *P. membranaefaciens* has been isolated from spoiled mango juice (pH 3.5) stabilized with sodium benzoate at concentrations ranging up to 1,500 mg/L (Ethiraj and Suresh 1988).

Actively growing cells appear as short ellipsoids to cylindrical-shaped rods. Reproduction is usually by multilateral budding, which leads to development of an extensive pseudomycelium. Ascospores, when present, usually appear as round to hat-shaped, ranging from 1 to 4 per ascus.

***Hansenula* sp.** *Hansenula* sp. are widely distributed and may be present in significant numbers in fermentations originating from native grape flora. *H. anomala* has both fermentative and oxidative metabolism. The yeast is capable of very limited fermentation of juice (producing 0.2–4.5% alcohol). *Hansenula* sp. form large amounts of acetic acid (ranging from 1 to 2 g/L) as well as the volatile esters, particularly ethyl acetate (up to 2,150 mg/L) but also isoamyl acetate (Shimazu and Watanabe 1981; Sponholz and Dittrich 1974). When active during the early stages of fermentation, formation of these volatile esters has been reported to add limited flavor and bouquet to the wine (Saller 1957; Wahab et al. 1949). Additionally, *H. anomala* is able to completely oxidize malic acid thereby bringing about decreases in TA and corresponding increases in pH.

Microscopically, *H. anomala* appears as an oval to oblong-shaped cell of approximate dimensions 2.5 × 5 to 10 μm. Members of the genus reproduce by asexual budding. Frequently, cells do not separate, resulting in formation of pseudomycelium. Sexual reproduction yields 2 to 4 hat-shaped (Saturn-shaped) ascospores per ascus.

Fermentative Yeasts

There are several genera of yeasts occurring in juice and wine that typically do not grow as a film. Rather, their activity is largely fermentative.

***Brettanomyces* sp.** *Brettanomyces* and its sporulating counterpart, *Dekkera*, have been isolated in wines worldwide. Microscopically, these yeasts resemble *Saccharomyces cerevisiae* although usually somewhat smaller. Classically vegetative cells are described as ogival in shape. Reminiscent of gothic arches, the ogival cell shape results from repeated polar budding characteristic of the yeast. It is more apparent in older populations but, even in these cases, may represent less than 10% of the population (Smith 1993). In older cultures maintained on laboratory media, most cells appear elongated. Occasionally, incomplete separation of daughter cells may result in chains being present.

Brettanomyces has historically been viewed as principally a problem in red wines. However, it has also been reported from white table wine (Wright and Parle 1974) and even in German sparkling wine *cuveés* (Schanderl 1959). Although it is usually viewed as a problem in barrel-aging wines, both *Brettanomyces* and *Dekkera* are fermentative, capable of producing 10 to 11% alcohol (Fugelsang et al. 1992). Thus, in the case of wineries using extensive barrel fermentations and/or where sanitation is neglected, it is possible that substantial populations of these yeasts may develop over the course of the season.

Brettanomyces is particularly difficult to control because its presence may go unnoticed until the wine is permanently tainted. It appears to spread from winery to winery through contaminated wine and/or equipment. Once established, insects, particularly fruit flies, are important vectors. Formation of the volatile phenol, 4-ethyl phenol, from *p*-coumaric acid is attributed directly to presence and growth of *Brettanomyces* and its sporogenous counterpart *Dekkera* sp. Formation requires the activity of two enzymes; Cinnamate Decarboxylase and Vinylphenol Reductase. Neither acetic acid bacteria, LAB, nor other genera or species of wine yeasts produce the compound (Chatonnet et al. 1992). Thus, the presence of 4-ethyl phenol may be used as a marker for present/past *Brettanomyces/Dekkera* activity (see Chapter 20).

Wineries planning to buy wine are advised to quarantine such lots until laboratory tests confirm there is no contamination. We also encourage sterile filtration of all wines entering the premises.

Kloeckera/Hanseniaspora. The apiculate (lemon) shape cells characteristic of this group arise from repeated budding at both poles. These fermentative yeast occur in abundance during the early stages of native fermentations where they frequently represent the dominant species in unsulfited must/juice (Osborn et al. 1991). In this regard, *K. apiculata* has been reported to grow in the presence of up to 150 mg/L SO_2 (Heard and Fleet 1988a). Both *Kloeckera* and *Hanseniaspora* are capable of producing substantial (and potentially inhibitory) levels of acetic acid and ethyl acetate (see Chapters 6 and 11). Further, killer strains of *Hanseniaspora uvarum* have been reported and thus may be of concern when sensitive strains of *S. cerevisiae* are used for fermentation (Radler et al. 1985).

Zygosaccharomyces. In the case of wines sweetened with concentrate, the osmophilic yeast *Zygosaccharomyces bailii* may be troublesome. It is resistant to SO_2 (>3 mg/L, molecular), sorbic and benzoic acids (>800 and 1,000 mg/L, respectively) and to the sterilant DMDC (see Chapter 13) at levels greater than 500 mg/L. Further, it survives at pH levels below 2.0 and in alcohol levels greater than 15% (Thomas and Davenport 1985). This organism must be controlled by sterile packaging.

Saccharomyces sp.

Morphologically, this group of yeasts appears spherical to ellipsoidal in shape with approximate dimensions of 8 × 7 μm, depending on the organism and the growth medium. Asexual reproduction is by multilateral budding. Most strains of *S. cerevisiae* are capable of producing alcohol levels of up to 16%. Use of supplemented "syruped" fermentations (addition of sugar at intervals) may enable the yeast to produce as much as 18%, or more, alcohol. Although these yeasts are usually thought of as fermentative, they may also grow oxidatively as part of the surface film community. *S. beticus*, for example, is a commercially available "flor" sherry yeast.

Attributes of Wine Yeasts

The advantages of selected pure yeast cultures over fermentations by natural strains include (1) rapid onset of active fermentation and predictable rate of sugar-to-alcohol conversion; (2) complete utilization of fermentable sugars; and (3) improved alcohol tolerance. Additionally, winemakers may consider other properties such as (4) SO_2 and H_2S production; (5) reduced formation of acetic acid, acetaldehyde, and pyruvate; (6) reduced tendency to foam; and upon completion of fermentation, (7) clarification (e.g., flocculation). Many winemakers are currently utilizing native yeast fermentation properties (see Native Yeast Fermentations).

Yeast Starter Preparation

Commercially, wine yeasts are available in dehydrated form and in already-expanded liquid culture. Prior to addition to must, yeast must be expanded such that final viable cell numbers, upon addition, are on the order of 2 to 5×10^6 cells/mL. This corresponds to an active starter inoculum of 1 to 3% vol/vol. Potentially troublesome fermentations, such as late harvest and/or *Botrytis*-infected grapes should receive slightly higher addition levels. In addition to ensuring relatively high cell numbers compared with competitive native species, proper preparation of yeast starters ensures that upon inoculation, the majority of yeasts in the must are in, or entering, the log phase of growth.

Before use, active dry yeast must be rehydrated. This should be done in warm (40°C/104°F) water (Kraus et al. 1981). Physically, rehydration in warm water or must disperses the yeast to a much greater extent than is seen when added to cold must. In the latter case, pellets tend to remain intact (clump), resulting in reduced nutrient and oxygen incorporation.

Viability is greatly reduced at low rehydration temperatures. Hydration at 15°C (60°F) may result in up to 50 to 60% cell death (Cone 1987). By comparison, activation at 40°C quickly reestablishes the membrane barrier and function before essential soluble cytoplasmic components escape. When yeasts are rehydrated in water, they should not be allowed to remain there for more than 20 minutes before transfer to must. Longer hydration periods tend to reduce viability.

Prolonged growth under semi-anaerobic conditions reduces the steroid (ergosterol) content of the cell membrane, thereby making the yeast more sensitive to the effects of alcohol. Because synthesis of yeast cell membrane components (fatty acids and steroids) requires molecular oxygen, conventional starter preparation may be modified to include incorporation. Simple mechanical agitation or pumping over may not increase the oxygen concentration to levels needed; it is suggested that sterile-filtered compressed air be bubbled directly into the tank. Use of pure oxygen should be avoided because of its toxicity (Fugelsang 1987).

Actively growing yeast should not be directly transferred to chilled must. Cold shock may reduce the viable cell count by up to 60%, and in general results in slow growth and increases the potential for problem fermentations. Further, sudden drops in temperature may result in hydrogen sulfide production (Monk 1986). When fermentation at lower temperatures is desired, starters should be acclimated to growth at the lower temperature before inoculation.

Monitoring Starter Viability and Cell Number

After rehydration and transfer of yeast to starter tanks, 24 to 72 hours may be required before cell numbers have reached the level at which the starter can be added to must. Growth should be followed microscopically, noting viability as well as percentage of budding cells. Cell viability is usually determined using dyes such as methylene blue, Ponceau-S, or Walford's stain. Procedures for use of these dyes and interpretation of results are presented in Chapter 20. In actively growing starters, budding cells, upon addition to the must, should comprise 60 to 80% of the total cell number.

Transfer of starter cultures should occur before sugar levels are fully depleted. Nutrient exhaustion may force growing yeast into a secondary lag phase, significantly reducing cell viability and delaying activity.

The practice of holding back 5 to 10% of the starter to serve as inoculum for fresh sterile juice, although convenient, may introduce contamination from other yeast and bacteria. Further, prolonged use of this technique may result in depletion of critical cell membrane components, thereby increasing the potential for troublesome fermentations. Such activities may also increase the production of ethyl carbamate (see Chapter 8).

When purchasing active dry yeast, the vintner should plan to order only enough to fulfill needs for that season. Dehydrated yeast, even when stored under ideal conditions, loses viability. For example, at 4°C (39.5°F) activity is reduced by 5% per year, and at 20°C (68°F) activity drops by 20% per year.

Native Yeast Fermentations

There is interest among some U.S. winemakers in native yeast fermentations; the perceived benefits include added complexity and intensity as well as a fuller, rounder palate structure. The latter may be the result of near threshold levels of sugar remaining after fermentation. This also reflects the fact that native fermentations are frequently not as efficient in sugar-to-alcohol conversions as are fermentations with cultured strains. A wide range of yeasts have been found on grapes and in grape juice and wine. Some of these native species have positive sensory properties. For example, the fusel oil phenylethanol has both textural properties and a distinct "roselike" nose that may contribute to wine character.

Population densities of native species vary from less than 160 to more than 10^5 CFU/berry (Reed and Nagodawithana 1991). The most frequently isolated native species are the apiculates *Hanseniaspora uvarum* and its asexual counterpart *Kloeckera apiculata,* typically accounting for over 50% of the total. Other frequently encountered species include *Metschnikowia pulcherrima* and its asexual counterpart *Candida pulcherrima.* Less frequent isolates include *Pichia membranaefaciens, Hansenula anomala, Candida stellata,* as well as *Cryptococcus* and *Rhodotorula* sp. *Saccharomyces* sp. are infrequently isolated from vineyards where winery wastes are not reincorporated as soil amendments. In instances where such practices are used, resident populations may be high (Pardo et al. 1989).

By comparison to vineyards flora, *Saccharomyces cerevisiae* of many strains represents the dominant flora in grape juice. Although numerically less prevalent, other indigenous winery yeasts (e.g., *Brettanomyces/Dekkera* sp. and non-*cerevisieae* strains of *Saccharomyces*) may, in the absence of sulfite, also begin to multiply in the juice. Studies by Fugelsang et al. (1993) have shown that the presence of *Brettanomyces/Dekkera* sp. in co-culture with *Saccharomyces cerevisiae* at the start of fermentation, and inoculated midway through fermentation, significantly, and negatively, impacted *Saccharomyces* populations compared with pure culture *Saccharomyces* fermentations. The causative factors are high levels of acetic as well as octanoic and decanoic acids. In this regard the practice of "cold soaking" red musts for several days before initiation of fermentation should be considered. The "cold soak" before inoculation of yeast starter encourages oxidative polymerization of phenols, which may be important in wine color stability (Chapter 7). However, it may also permit growth of native yeast flora.

The qualitative and quantitative distribution of native yeast species potentially present on the fruit and during fermentation is substantial. Thus, routine reliance on spontaneous or native yeast fermentation must be approached with caution. Spontaneous fermentations (in unsulfited juice/must) occur as a succession of yeast populations (arising from the vineyard and winery environs) beginning with relatively weak, although numerically superior, species present on the fruit. However, these strains are susceptible to increasing alcohol levels and are not as metabolically predictable as wine strains of *S. cerevisiae.* In a review of the subject, several winemakers report no unusual problems associated with native yeast fermentation and, in fact, state that the added complexity results from such fermentation warrant the potential risks involved (Goldfarb, 1994).

Over a period of a couple of days, depending on temperature, the activity of native non-*Saccharomyces* species declines and indigenous populations of *Saccharomyces cerevisiae* establish themselves and carry on the fermentation.

Killer Yeasts

Species and strains possessing "killer" property kill sensitive members of their own species and, frequently, those of other species and genera. Killer toxins have been isolated in several strains of *S. cerevisiae* used in wine. The killer characteristic is also found in yeast from other genera, including *Candida, Pichia, Hansenula,* and *Torulopsis.*

Killing results from production of an extracellular protein toxin or glycoprotein. After binding to the receptor site of the sensitive cell, the toxin interacts directly with protein components of the cell membrane, disrupting the normal state of electrochemical ion gradients.

Several commercially available strains of killer yeast are currently available.

Immobilized Yeasts

Studies have shown the potential for using immobilized yeast in table and sparkling wine fermentations (Fumi et al. 1988). Yeast can be immobilized within calcium alginate carrier beads. The beads may be packed into a column of suitable dimensions and a volume of juice pumped over the immobilized yeast. In methode champenoise, the yeast-impregnated beads are placed directly into the bottle with the *cuveé* facilitating riddling and disgorging.

The utility of individual strains may become more important as the field of genetic engineering develops. There is potential for development of strains for specific applications as well as inclusion of fully expressed gene(s) for the malolactic fermentation within selected yeast strains. Initial work in this area has been successful in implanting the appropriate genes but the level of expression is far less than is needed for practical importance to the winemaker.

WINE BACTERIA

Acetobacter and *Gluconobacter*

Bacteria of these genera use ethanol (and glucose) aerobically in the formation of acetic acid. Thus growth of acetic acid bacteria in wine as well as in musts and on deteriorating grapes may significantly increase the volatile acid content. For details regarding the analysis of volatile acidity in juice and wines, the reader is referred to Chapters 3, 5, and 20. The taxonomy of acetic acid bacteria has undergone considerable revision in the last 35 years. In Bergey's *Manual of Determinative Microbiology*, 9th edition, both *Acetobacter* and *Gluconobacter* are placed in the family Acetobacteriaceae. The genus *Gluconobacter* is comprised of a single species, *G. oxydans*, whereas *Acetobacter* consists of four species: *A. aceti*, *A. pasteurianus*, *A. liquefaciens*, and *A. hansenii*. *Acetobacter* and *Gluconobacter* are Gram-negative (becoming Gram variable in older culture) rods frequently occurring in pairs or chains. Although size varies (depending on growth media) cell dimensions range from 0.6 to 0.9 μm by 1 to 3 μm.

Vaughn (1955) reported *A. aceti* and *A. oxydans* (currently *G. oxydans*) to be the two most commonly encountered species of acetic acid bacteria in California wines. *A. aceti* also was found to be the most commonly isolated species in Bordeaux wines (Joyeux et al. 1984a). By comparison Drysdale and Fleet (1985) reported *A. pasteurianus* to be the most common isolate among aging Australian red wines.

Optimal growth temperatures for *Acetobacter* and *Gluconobacter* also vary. Reported growth optima for *A. aceti* range from 30°C to 35°C (86–95°F), whereas *G. oxydans* grows best at around 20°C (68°F). Joyeux et al. (1984a) reported growth of *A. acetic* in wine at 10°C (50°F).

Both *Acetobacter* and *Gluconobacter* have historically been viewed as having totally respiratory (aerobic) metabolisms. Thus their growth generally occurs on the surface of wine and is seen as a translucent, adhesive film. This film may separate, resulting in a patchy appearance. Formation of surface films or pellicles may cause the wine to appear hazy or cloudy. It is currently thought that these bacteria may be able to survive at very low (semianaerobic) oxygen concentrations and may be found in stored wines (Drysdale and Fleet 1989).

The principal physiological similarity of interest to the winemaker between *Acetobacter* and *Gluconobacter* is the ability of both to carry out oxidation of ethanol (and glucose) to acetic acid. However, the extent to which this oxidation occurs varies with the organism (DeLey and Schell 1959). In the case of glucose metabolism, both genera use the pentose phosphate pathway in the formation of acetic and lactic acids. An important taxonomic distinction between the two is that *Acetobacter* may carry the oxidation a step further, converting acetate and lactate to CO_2 and H_2O via the TCA cycle. *Gluconobacter,* by comparison, does not demonstrate this overoxidation. The reasons for the inability of *Gluconobacter* to oxidize acetic acid further lie in the fact that the organism lacks functional key enzymes of the TCA cycle. Specifically, *alpha*-ketoglutarate dehydrogenase (catalyzing formation of succinyl-CoA from *alpha*-ketoglutaric acid) and succinate dehydrogenase (catalyzing the formation of fumaric from succinic acid) are not operational (Greenfield and Claus 1972).

In slowly fermenting or stuck fermentations, the growth of *Acetobacter* and *Gluconobacter* may result in formation of gluconic acid from oxidation of glucose (DeLey 1958; Vaughn 1938). Depending on must chemistry (pH and sugar concentration), gluconic acid may not be further metabolized and thus accumulates. Levels of up to 70 g/L have been reported in grape musts where growth of *G. oxydans* has occurred (Joyeux et al. 1984b). Production of gluconic and ketogluconic acids also has been reported in grapes where the growth of *Gluconobacter* and *Acetobacter* has taken place. Some species of *Acetobacter* and *Gluconobacter* do not oxidize fructose or do so to a limited extent (Joyeux et al. 1984b). As a result of the relative sweetness of fructose (compared with glucose), disproportionate utilization of glucose by these species may result in a wine (or stuck fermentation) with a "sweet-sour" character. In this case the sweetness is attributed to unoxidized fructose (Vaughn 1938, 1955).

Under oxidative conditions *G. oxydans* is capable of oxidizing glycerol to dihydroxyacetone (Eschenbruch and Dittrich 1986; Hauge et al. 1955), which may play a role in the sensory properties of wine. Because the reaction is inhibited by alcohol concentrations of greater than 5% (Yamada et al. 1979), its formation in wine is questionable. However, it is likely that glycerol oxidation occurs in infected grapes and musts. The growth of acetic acid species may produce significant quantities of dihydroxyacetone, which may end up in the fermented wine. In one study, dihydroxyacetone levels of 260 mg/L were reported in must infected with *G. oxydans*. Dihydroxyacetone was detected in wine produced from this must at 133 mg/L (Sponholz and Dittrich 1985).

Bacillus sp. and Other Bacteria

Normally, *Bacillus* sp. are soil-borne organisms that are secondarily found in water supplies. However, there are reports of *Bacillus*-associated wine and brandy (Murrell and Rankine 1979) spoilage. Gini and Vaughn (1962) were able to grow isolates of *B. subtilis, B. circulans* and *B. coagulans* obtained from spoiled wines, and subsequently reinoculate cultures into wine where further growth was noted. Lee et al. (1984) isolated *B. coagulans* from cork and upon reinoculation into wine obtained growth. The extent to which *Bacillus* sp. may represent a problem in winemaking is uncertain.

Streptomyces, also a soil-borne bacteria, has been occasionally isolated from cork and wooden cooperage. The organism is capable of producing guaiacol using naturally occurring lignin as substrate.

LACTIC ACID BACTERIA

The malolactic fermentation (MLF) is a catabolic pathway in which L-malic acid is enzymatically oxidized to L-lactic acid and carbon dioxide. Depending on the strain(s) of lactic acid bacteria (LAB) involved, several byproducts may be produced that impact the sensory properties of the wine.

Chemically, the most significant changes observed during the course of MLF are increases in pH and corresponding decreases in titratable acidity. Depending on the concentration of malic acid and the extent of microbial growth, increases in pH of 0.3 have been documented (Pilone et al. 1966; Rankine 1977). Decreases in TA are generally on the order of 1 to 3 g/L.

Therefore, successful induction of MLF in high acid, low pH wines is, potentially, a useful technique for acid and pH adjustment. However, in the case of a high pH wine, the MLF may significantly (and negatively) impact the sensory properties of the wine. Additionally, instability, both chemical and microbial, may occur as a result of increased pH. High pH wines tend to be susceptible to subsequent growth of microorganisms. In cases where MLF may occur in low acid wines, follow-up adjustments in acidity may be desirable.

The occurrence of MLF is common to all wine-producing areas of the world. Studies have shown that LAB probably originate on the grape, where they may be isolated from the berry surface and grape leaves. Their numbers, however, are rather low—in most instances, less than 10^2 CFU/mL (Lafon-Lafourcade et al. 1983). Other studies suggest that winery equipment is an important source of infection. Contamination, in these cases, is secondary, resulting from improperly sanitized sites where LAB have been allowed to accumulate and proliferate.

Taxonomy of Lactic Acid Bacteria

Taxonomically, LAB are in the families Lactobacillaceae (*Lactobacillus* sp.), encompassing Gram-positive rod-shaped species, and Streptococcaceae (*Leuconostoc* sp.), which includes spheroid to lenticular ("coccobacilloid") shaped cells. In some cases, individual cells may adhere, resulting in chains (*Leuc. oenos*) or filamentation. In the case of *Lactobacillus trichoides,* filament formation may be extensive, leading to the appearance of mycelial-like growth. This characteristic has led to its descriptive name "cottony bacillus."

The amount of lactic acid (and other metabolites) formed from glucose and the pathway of its formation serve to separate LAB into two groups, the hetero- and homofermenters. Homofermenters produce primarily lactic acid as the end product of glucose metabolism. LAB in this group use the Embden-Meyerhoff Parnas (EMP) pathway. Pyruvate is reduced to lactic acid yielding 2 moles of lactic acid and 2 moles of ATP per mole of glucose. Heterofermenters, however, lack the aldolase enzyme, which mediates cleavage of fructose-1,6-diphosphate. This group, instead, uses the oxidative pentose-phosphate pathway yielding 1 mole of ATP per mole of glucose. Once glyceraldehyde-3-phosphate is produced by cleavage of ribulose-5-phosphate, the pathway leading to lactic acid is the same as that used by homofermenters. The second product of the cleavage (acetyl-phosphate) is either reduced to ethanol via two successive reduction steps involving the coenzyme NADH, or oxidized to produce acetic acid.

Parameters Affecting Growth of LAB

The lactics are microaerophilic to facultatively anaerobic, requiring reducing (low oxygen) conditions for normal growth. Further, lactics are nutritionally fastidious microorganisms, requiring complex organic media for growth. Having lost their ability to synthesize many specific compounds required for activity, LAB require preformed compounds such as vitamins and amino acids for growth, in addition to other organic compounds. This is reflected in laboratory cultivation where it is necessary to augment standard culture media of yeast extract and protein hydrolysates with fruit or vegetable juices.

Table wines contain sufficient carbohydrates to serve as energy sources for growth of LAB. LAB have been found to use seemingly minute concentrations of sugars in "dry wines" at less than 0.1% (Amerine and Kunkee 1968). Melamed (1962) identified the most commonly used sugars as glucose and arabinose. Glucose levels as low as 0.5 µM/mL have been found sufficient to support the growth of lactic bacteria (Pilone and Kunkee 1972).

Thus variations in susceptibility of wines to LAB are partly due to differences in available nutrients and metabolic intermediates, as well as variations in particular LAB strains. If other factors are not limiting, the addition of small amounts of yeast autolysate may stimulate growth of these bacteria in wine (Carter 1950). The B-complex vitamins produced by yeasts are especially important to the growth of lactic bacteria, and it is believed that addition of lactic starter cultures during the course of alcoholic fermentation takes advantage of increased supply of nutrient provided by the yeast.

Processing Considerations

Processing protocol is known to play a role in predisposing a wine to MLF. Prefermentation processing such as cold clarification and fining may not only reduce native populations of lactic bacteria but may additionally reduce nutrient levels to a point where their growth may be impeded. Skin contact enhances the growth of lactics but it is unclear whether this is due to increases in pH and/or extraction of nutrients.

Certain yeast strains may inhibit successful growth of LAB when grown in co-culture. This antagonism may result from competition for nutrients and/or from production of soluble antimicrobial agents (King and Beelman 1986). Demand for and accumulation of amino acids by yeast is reported to deplete available pools needed for bacterial growth (Beelman et al. 1982). Certain strains of wine yeast are known to produce SO_2 at levels sufficient to inhibit lactic bacteria (Fornachon 1963). Further, Labatut et al. (1984) reported the importance of octanoic and decanoic acid production by yeast in LAB inhibition.

LAB Starters and Inoculation

Rapid onset and successful completion of MLF requires preparation of LAB starters of high cell density ($>10^6$ CFU/mL) and vigor. Unlike wine yeast addition, however, selection is usually limited to strains of *L. oenos.* Two commonly available strains, ML-34 and PSU-1, differ largely in the area of pH tolerance. PSU-1 is generally recommended for use in must/wine in the pH range 3.1 to 3.3, whereas ML-34 is recommended for higher pH applications. Less frequently, *Lactobacillus sp. (L. brevis)* have been used.

The methodology of propagation and expansion of lactic starters will have an impact on final activity of the bacteria upon addition to wine or must. Although rather elaborate preinoculation growth media have been reported, most techniques call for use of unsulfited grape or apple juice (5–10% sugar). Hydrogen ion concentration is adjusted to 4.0 to 4.5 using

$CaCO_3$. Yeast extract is added at 3 to 5%. Propagation of starters may be carried out in co-culture with *Saccharomyces cerevisiae* or as a pure LAB culture. The former has the advantage that the yeast provides essential growth factors and intermediates for the lactics. However, aside from direct microscopic examination of the culture, growth of the bacterial component is difficult to assess. This problem is overcome by pure culture propagation. The culture is then expanded to greater than 1% of the final wine.

It is recommended that stationary phase LAB be used for inocula. In addition to bringing about the most rapid MLF, cells at this stage of development are also more resistant to lytic bacteriophage than are younger cultures (Henick-Kling 1988).

Over the past several years, a number of commercially available high titer lyophilizates and cultures have been developed. Their advantage is that the lag time needed to prepare sufficient volume of active starter is reduced significantly from that needed to bring up cultures stored on laboratory media.

Although commercial concentrated cultures are available, direct transfer to wine will result in high rate of mortality. Therefore, before use it is necessary to create a high titer inoculum and expand the final volume of starter to more than 1% of the final volume of wine. One supplier markets a freeze-dried culture of *Leuconostoc oenos* reported to survive direct inoculation into wine without rehydration/reactivation (Chr. Hansen Laboratory 1993).

Timing of LAB Inoculation

There is no unanimous opinion as to timing the addition of LAB starters. They may be added along with the yeast at the crush/clarification stage, during the course of, or upon completion of alcoholic fermentation.

Concern regarding addition of LAB starters at the beginning of fermentation centers around the potential for heterolactic LAB species producing acetic acid in the presence of fermentable sugars. In one report, the onset of alcoholic fermentation was delayed by addition of *L. oenos* at a cell density of 10^7 CFU/mL (Lafon-Lafourcade et al. 1983). During the extended lag phase, unimpeded growth of lactics on sugars produced higher than normal levels of acetic and lactic acids bringing about inhibition of fermentative yeasts (Lucramet 1981). Beelman and Kunkee (1985) did not report increased acetic acid levels when MLF and alcoholic fermentations occurred concomitantly. Under such conditions, acetic acid attributed to LAB increased by only 0.2 g/L (Davis et al. 1985). However, growth of LAB on sugars either before MLF or upon completion of MLF resulted in elevated production of acetic acid (Lucramet 1981).

Many winemakers use high titer (>10^6 CFU/mL) bacterial starter additions during the course of or shortly after completion of alcoholic fermentation. At this point, potentially inhibitory levels of SO_2 have been reduced and yeast growth has proceeded to the point where bacteria have little impact on their activity. Further, there is ample nutrient availability in the form of yeast autolysate, which serves as an important source of B-complex vitamins to stimulate bacterial growth and activity (Weiller and Radler 1972).

In the case of red wine fermentations, many vintners prefer to add lactic cultures after pressing or before the fermentation has finished. Still others add starters to freshly fermented wines or rely on endogenous populations present in barrels to bring about the conversion. Successful completion of the MLF at this stage may be difficult because of alcohol levels and nutrient depletion. To overcome the problem of nutritional status, winemakers may use a period of lees contact to make available the necessary nutrients for growth of the bacteria. Due to the sensitivity of LAB to SO_2, its use post-fermentation is usually delayed until completion of MLF.

Monitoring Progress and Completion of MLF

The occurrence and/or progress of MLF may be easily monitored by paper chromatographic separation, with the absence of malic acid on the developed chromatograph being a *somewhat reliable indication* of the progress of the bacterial fermentation. Reduction in the intensity (or disappearance) of malic acid spot is not always indicative of a bacterial MLF. Malic acid is used to a varying degrees by *Saccharomyces*. Some strains are capable of using up to 50% (Shimazu and Watanabe 1981). Additionally, the fermentative activities of the yeast *Schizosaccharomyces* sp. as well as *Zygosaccharomyces* sp. may result in complete oxidation of the malic acid to ethanol and carbon dioxide.

Kunkee (1968) reported visual resolution for malic acid at approximately 100 mg/L. As pointed out by Gump et al. (1985) this is not sensitive enough to ensure stability with regard to further LAB activity. In the survey of U.S. winemakers by Fugelsang and Zoecklein (1993), 34% considered wine as "safe" at malate concentrations of 30 mg/L whereas 66% required 15 mg/L (well below the resolution limits of paper chromatography). As a result, it is recommended that more sensitive procedures for malic acid analysis be pursued. Two methods are presented, an enzymatic procedure and high-performance liquid chromatography (see Chapter 20).

Sensory Aspects of Malolactic Fermentation

In addition to its importance in acid balance, byproducts of MLF may play important sensory roles, potentially contributing to complexity. These products include principally diacetyl, and secondarily acetoin and 2,3-butanediol, in addition to acetic acid and its esters.

Diacetyl is perceived as buttery in character and is normally produced by yeasts during alcoholic fermentation at very low concentrations (<1 mg/L). Bacterial production during MLF represents the primary source for diacetyl. Its formation during MLF varies with the strain of LAB involved. At concentrations greater than 5 mg/L, its presence may be regarded as objectionable (Rankine 1977). At lower levels, and in combination with other components, diacetyl may add to the complexity, and thus may be used as a stylistic tool.

Timing of inoculation may also play an important role in levels of diacetyl present in wine. In that yeast are capable of using the compound, inoculation towards the end or at the completion of alcoholic fermentation and before first racking may result in lower concentration. Conversely, inoculation after first racking (when yeast titer is low) results in higher levels of diacetyl remaining in the wine. The presence and activity of spoilage lactics (*Pediococcus* or *Lactobacillus* sp.) may result in objectionably high levels of diacetyl. Because of its sensory importance, diacetyl concentration is monitored by some wineries (see Chapter 20). Henick-Kling (1991) reported the presence of a compound other than diacetyl in MLF wines that has a strong butterlike character.

Ethyl lactate may play a sensory role in wines having undergone MLF. Post-MLF ethyl lactate levels range upwards to 110 mg/L (Dittrich 1987), which is thought to contribute to enhanced "mouth feel."

Lactic populations in wine may reach levels equivalent to yeast populations (10^6–10^8 CFU/mL). Aside from their primary transformation, it might be expected that byproducts of their metabolism would have a continuing influence on the wine long after the bacterial cells are gone. It is believed that the activity of bacterially produced esterases, lipases, and proteases may play an important sensory role in wines undergoing MLF. However, the nature of enzymatic activity is still unclear (Davis et al. 1988).

Henick-Kling (1991) reported that MLF significantly affected the fruity aromas of Chardonnay, not by reduction but frequently by enhancement. In the case of wines not undergoing MLF, sensory descriptors from taste panel responses were "green apple, citrus, fruity, melon rind." By comparison, parallel lots having undergone MLF were described as "apple, nutty, smokey, fruity, buttery, melon and sweaty."

Grape variety and vinification techniques also influence the sensory properties of wines having undergone MLF. In aromatic varieties, the contribution of MLF to flavor may be overshadowed. *Sur lie* contributes to wine-enhanced complexity in the form of yeasty and nutty aromas, whereas barrel fermentation and aging imparts smoky and spicy odor notes (Henick-Kling 1991).

Spoilage Resulting from LAB Growth

Polysaccharide Formation

In low acid wines, growth of some LAB, particularly *Pediococcus* sp., may result in formation of extracellular dextrins. Termed "ropiness," this defect is visually manifest as an increased viscosity and oily character. Ropiness usually begins in the bottom of cooperage, corresponding to onset of LAB growth. Eventually, growth envelopes the container. LAB involved in this defect include *Pediococcus,* as well as, in some instances, *Leuconostoc* sp. (Mayer 1974).

Mannitol Taint

In high pH, sweet wines, heterolactic LAB may produce mannitol from reduction of fructose. In the process, NADH is reoxidized. Energetically, this is important to the LAB in that reoxidation of NADH during formation of mannitol "permits" the bacteria to produce acetate from acetyl-phosphate (rather than regenerate oxidized coenzyme by reduction to ethanol) and thus, produce an additional ATP. Other than acetic acid and mannitol formation, diacetyl is also typically present in objectionable concentrations.

Acetic Acid

Compared with acetic acid levels from other sources, that attributed by MLF is not excessive, being on the order of 0.1 to 0.2 g/L (Peynaud 1984). However, both hetero- and homolactic LAB are capable of producing acetic acid from growth on sugars using pathways already discussed. Prefermentation production of acetic acid by LAB may be sufficient to inhibit *Saccharomyces* resulting in potentially stuck fermentation.

Acrolein Formation

Acrolein is produced by dehydration of glycerol to the intermediate 3-hydroxypropionaldehyde (Sliniger et al. 1983). With time or exposure to elevated temperatures, the aldehyde is further dehydrated to acrolein. Rentschler and Tanner (1951) reported that acrolein induced bitterness upon reaction with wine pigments.

CONTROLLING MICROBIAL GROWTH IN WINE (A SUMMARY)

Growth of microorganisms in juice and wine depends on several physical and chemical properties considered below. For more detail, the reader is referred to appropriately referenced chapters.

Hydrogen Ion Concentration (pH)

Hydrogen ion concentration establishes whether microbes will grow in wine, which species (or strains) will grow and their growth rate, as well as the concentration of metabolites produced during the growth cycle. For example, *Lactobacillus* and *Pediococcus* sp. are generally not reported growing in wines of pH less than 3.5. Bousbouras and Kunkee (1971) reported completion of malolactic fermentation using *L. oenos* in 14 days at pH 3.83. The same wine with pH adjusted to 3.15 required 164 days for completion. A similar pH-dependent trend is seen in wild lactic fermentations.

Conversion of malic acid is greatest at pH 3.0, whereas utilization of glucose increased with increasing pH (Hennick-Kling et al. 1991). Also, preculture pH is important in subsequent vitality of starters. Cultures activated and expanded at wine pH 3.5 show improved rates of malic acid conversion compared with cultures prepared at higher pH.

Acetic acid bacteria are generally pH tolerant, requiring low pH for optimal growth. However, this is closely tied to alcohol levels. Dupuy and Maugenet (1962, 1963) found little growth in wines of pH less than 3.2 and/or alcohol levels greater than 13%. Subsequently they reported *A. pasteurianus* grew at pH 3.4 in 12.5% alcohol whereas in 8% alcohol its survival was noted at pH 3.0.

The other major role of pH is related to the nature of antimicrobial agents used in winemaking. As discussed in Chapter 10, the percentage of molecular SO_2 increases with decreased pH. The same relationship exists for sorbic and benzoic acids (Chapter 13).

Alcohol

Table 18–1 compares relative alcohol tolerance of the major groups of wine microbes. The table represents a group summary; each group may have members with unique levels of tolerance. Also, alcohol tolerance is closely tied to storage temperature and wine pH.

Sulfur Dioxide

Sulfur dioxide is an effective inhibitor of microbial growth. However, many studies attempting to address inhibition fail to distinguish between molecular and other forms. Generally, molecular SO_2 levels of 0.8 mg/L are sufficient to prevent growth of wine bacteria (Beech et al. 1979). Levels of bound SO_2 approaching 50 mg/L may also be inhibitory to LAB (Hood 1984).

Wine strains of *S. cerevisiae* are less sensitive to the effects of molecular SO_2 than are most spoilage yeasts. However, as noted earlier, some native

Table 18–1. Alcohol tolerance of wine microorganisms at 20°C.

Group	Max. alcohol level for growth (% vol/vol)
Wine yeasts	16[a]
Non-*Saccharomyces* sp.	10–13
Acetobacter sp.	10–15[b]
Gluconobacter oxydans	<5
LAB (except *L. trichoides*)	10–16[c]

[a]Harrison and Graham 1970.
[b]Drysdale and Fleet 1988.
[c]Davis et al. 1988.

film-forming yeast species can survive levels of more than 3 mg/L molecular SO_2.

During fermentation, yeasts produce large amounts of acetaldehyde, which rapidly binds added SO_2. This is the principal reason why it is difficult to stop yeast fermentations with the use of SO_2. Reactivity is slower at lower temperatures, so reducing storage temperatures slows formation of bound SO_2. Reportedly, the combination of SO_2 use at levels of 200 mg/L and storage at 4°C (39°F) was effective in storing juice (Splittstoesser 1981). Successful application of this technique requires well-clarified juice produced from clean, sound fruit of low pH as well as reducing initial yeast populations by cold settling or centrifugation.

If the winemaker's goal is to control microbial growth, a single large dose of SO_2 is more effective than several smaller doses. This is particularly important in high pH wines because of the relatively low percentage of free SO_2 in the active molecular form. Because formation of bound SO_2 is favored by higher pH, acidulation (if planned) before addition improves the antimicrobial activity.

Sorbic Acid

Sorbic acid exhibits inhibitory activity toward fermentative yeasts (Chapter 13). Although the preservative inhibits *Saccharomyces*, it exhibits little activity toward *Brettanomyces/Dekkera* or *Zygosaccharomyces* sp. Additionally it has little inhibitory activity toward acetic acid bacteria or LAB, the latter group may use it as substrate in production of the volatile compound 2-ethoxyhexa-3,5-diene ("geranium tone"). Thus, sorbic acid must be used in combination with SO_2 additions at bottling to prevent further bacterial growth (Chapters 10 and 13).

Fumaric Acid

At levels of 300 to 400 mg/L, fumaric acid may be useful in controlling the growth of LAB; however, in one study levels of 1,500 to 2,000 mg/L were required (Pilone et al. 1974). Fumaric acid may be utilized by yeast, so it is usually added to wine *after* the first racking (Ough and Kunkee 1974).

Use of fumaric acid presents three problems. It is difficult to dissolve and, in some cases, effective concentrations may approach solubility limits of the acid. Secondly, fumaric acid has unique sensory properties, and its contribution to the overall wine profile should be considered. Lastly, the concentration of fumaric acid used appears to be critical. When the concentration of the acid is inadequate for bacterial inhibition, it may be used (in addition to malic acid) as a carbon source.

The effectiveness of fumaric acid is enhanced by low pH, adequate levels of molecular SO_2, and low microbial titer. Fumaric acid is a relatively strong organic acid so its additions may result in drops in pH and correspondingly increased TA (Cofran and Meyer 1970). Fumaric acid additions of 1 g/L are equivalent to an addition of 1.29 g/L tartaric acid.

Carbon Dioxide and Pressure

At pressures above 1 atmosphere, CO_2 is inhibitory toward yeasts. Table 18–2 illustrates the relationships between pressure and yeast growth. This may be used in the production of *mutés*. Although replication is inhibited, fermentative activity may proceed at higher pressures; it is therefore important that only still (nonfermenting) juice be used. Many wineries use small pressurized vessels held at 70 to 90 psi (4.7–6 atm) with limited amounts of SO_2 (<200 mg/L) for *muté* storage. The success of the method is predicated upon low initial yeast populations, adequate pressures in addition to low storage temperatures, and adequate levels of SO_2.

Although growth of most yeast is significantly affected by increased pressures, *Brettanomyces* sp. have been isolated in sparkling wine fermentations

Table 18–2. Effect of pressure on yeast cell growth.

Pressure (atm)	Yeast cell titer (cells/mL)
0	104
2	15
3	11
4	6
5	3
6	<0

SOURCE: Schmitthenner 1950.

(Schanderl 1952), and carbonated soft drinks (van Esch 1992). Additionally, some lactics may grow at pressures exceeding 7 atm; therefore the technique may not be effective in controlling their growth. Because of their requirements for oxygen, growth of acetics is inhibited in the presence of CO_2.

Nitrogen Availability

The presence of assimilable nitrogen sources has a significant impact on the potential for microbial growth. Except for certain conditions of excessive amelioration and/or clarification, low must nitrogen is generally not a limiting factor during primary fermentation. In sparkling wine fermentations, nitrogen additions may be necessary, particularly in the case of aged *cuveés*.

Several post-fermentation processing decisions may affect the potential for bacterial activity. These include extended lees contact (*sur lie*) as well as protein and tartrate stability. Therefore, if the vintner wishes to prevent MLF, early and continued racking is needed. Bentonite fining depletes nitrogen sources important for growth of the organisms. Cold stabilization also is known to impede the activity of lactics, possibly by reducing the levels of critical nutrients and potentially lowering pH.

Biological Control

Bacteriophages, specific for *L. oenos,* have been isolated by a number of workers worldwide (Davis et al. 1985; Gnaegi and Sozzi 1983; Lee 1978; Nel et al. 1987), and may play a key role in the control of LAB in wine. Young (pre-stationary phase) LAB cultures appear to be most prone to activity of lytic phage (Henick-Kling 1988), whereas stationary phase LAB are most resistant. The enzyme lysozyme is widely distributed in nature. It brings about hydrolysis and, eventual destabilization of bacterial cell wall peptidoglycan. Its role in preventing MLF has been reported (Amati et al. 1992).

PROCEDURES

The following is a cross-reference of procedures found in Chapter 20. Isolation and identification of acetic acid bacteria, lactic acid bacteria and yeasts; catalase test; viable and total cell count; monitoring methods of MLF, paper chromatography, enzymatic, HPLC and gasometric, microscopic cell counting; staining techniques (Gram, methylene blue, Nigrosine); 4-ethylphenol.

CHAPTER 19

CORK

It is estimated that 12 to 13 billion bottles of wine annually are packaged with cork stoppers (Hagen and Lemble 1990). Cork is the product of the suberized outer tissues of the cork oak, *Quercus suber*. Commercial stands are limited to the western Mediterranean, where Potugal and Spain produce 55 and 28% of the total, respectively. The balance is produced in France, Italy (Sardinia), Tunisia, Morocco, and Algeria.

Cork is not harvested until the tree reaches about 30 years of age; it is stripped at 9-year intervals thereafter. Upon harvest, the corkwood is stacked to air dry for varying amounts of time (generally 6 months to 1 year before processing). Upon arrival at the processing facility, cork slabs are first boiled and then restacked to dry before punching into cylinders. They are then cut into appropriately sized strips and the cylinders punched perpendicular to the cross-diameter of the strip. After punching, corks are washed and usually bleached in either hypochlorite or hydrogen peroxide baths. However, increasing numbers of unbleached ("natural") cork are being marketed. Following this, cork is washed, dried, and depending on the destination, stamped with the winery's logo and treated with one or a combination of surfacing agents.

CORK MICROBIOLOGY

The microflora of unprocessed cork consist primarily of molds, including *Penicillium* of several species, *Mucor, Aspergillus, Monilia, Trichoderma,* and *Cladosporium* (Lacey 1973). Of the penicillia found on unprocessed cork, *P. granulatum* and *P. glabrum* are reported as the most common (Lee et al. 1984). Processing removes surface mold; however, spores lodged in lenticels and cracks escape early processing. Thus microbiological examination of processed cork reveals that it contains variable and potentially high numbers of mold spores representing the groups already identified. Reflecting this, Davis et al. (1981) reported the level of mold contamination to range from 0 to 10^8 CFU/cork.

Yeast and bacteria are less frequent isolates from cork (<100/cork). Several genera of yeast have been identified from cork. These include *Candida, Cryptococcus, Rhodotorula, Sporobolomyces,* and *Saccharomyces* sp. Among bacterial isolates, *Bacillus* sp. and *Streptomyces* represent the most frequent isolates (Fumi and Colagrande 1988; Lefebvre et al. 1983). On occasion, blue-green bacteria have been isolated (Naes et al. 1988).

In cases of wines exhibiting corkiness, *Penicillium* is frequently identified. Unlike most molds, some species of *Penicillium* are relatively resistant to alcohol. Marais and Kruger (1975) report that *Penicillium* is only partially inhibited at the alcohol levels of table wine and concluded that sporulation and growth may occur, potentially leading to cork-tainted wines. Where taint is identified in bottled wine, *Penicillium roqueforti* is frequently reported (Lefebvre et al. 1983). This species has not been identified in earlier processing.

IDENTITY AND PROPERTIES OF ODOR-ACTIVE METABOLITES

Over 50 naturally occuring volatile compounds have been identified in defect-free cork (Rigaud et al. 1984). These include phenolic aldehydes, such as vanillin, phenols, and fatty acid esters and furans such as furfural. Cork taint however, results from the contributions of a relatively small group of microbially produced volatile metabolites linked by their musty odor and flavor-active properties. These include chloroanisoles, 1-octen-3-one, and 1-octen-3-ol, 2-methylisoborneol, and guaiacol.

In the U.S., the incidence of cork taint is reported to range from 0.5% to more than 2%. Whether or not cork exhibits taint upon bottling may depend on its intrinsic structural integrity. Imperfections, such as cracks, increase available surface area for extraction, and therefore, the potential for development of off odors.

Cork grade appears to make little difference in the incidence of cork taint. Smith (1992) reported that more expensive corks show greater incidences of taint, suggesting that these may be stored for special use and hence potentially subject to storage-related mold growth.

Chloroanisoles and Their Derivatives

Corkwood, punched cylinders, and wood surfaces, in general, contain substantial amounts of simple phenols, including phenol, that arise from degradation of lignin. These serve as the precursors for chlorophenol (and, subsequently, chloroanisole) and related compounds.

The primary compound implicated in cork taint is 2,4,6-trichloroanisole (TCA). Its extraordinarily low sensory threshold in wine (reported to be 1.4 ng/L, Duerr 1985) makes TCA particularly problematic. Other chlorophenols such as pentachlorophenol (PCP), used as wood preservative, and 2,3,4,6-tetrachlorophenol (present as an impurity in PCP formulations) may also be methylated yielding pentachloroanisol and 2,3,4,6-tetrachloroanisole. These latter compounds also have very low sensory thresholds.

The frequently ascribed origin of 2,4,6-TCA is hypochlorite bleaching of the cork cylinders. Under the alkaline conditions present in the bleaching bath, chlorine reacts with simple phenols forming a mixture of chlorophenols, including 2,4,6-trichlorophenol (see Fig. 19–1). Under appropriate environmental conditions, and at sublethal levels, mold growth (arising from germination of spores or recontamination of cork) produces the less toxic (but odor and flavor-active) methyl ether, 2,4,6-TCA (Crosby 1981).

Chloroanisoles and their precursors may also be acquired by cork during transport and storage. Storage of bagged or baled cork in mold- and/or chlorophenol/chloroanisole-contaminated containers has been shown to be an important source. This is particularly important when containers are exposed to high temperature and humidity (Tinsdale 1987).

Chloroanisoles have also been identified in wines before bottling. Where chlorine or chlorinated cleaners and wood preservatives (pentachlorophenol) are (or have been) used in wineries, resident molds have been shown to produce 2,4,6-TCA (Maujean et al. 1985). Tanner and Zanier (1981) reported that 2,3,4,6-tetrachloroanisole-contaminated bentonite and fining agents led to taint in treated wines.

Other Compounds

A variety of compounds, other than chloroanisoles, have been reported to elicit musty odors in wine. These include guaiacol, geosmin, and 2-methylisoborneol.

Guaiacol is an infrequently encountered metabolite produced by *Strep-*

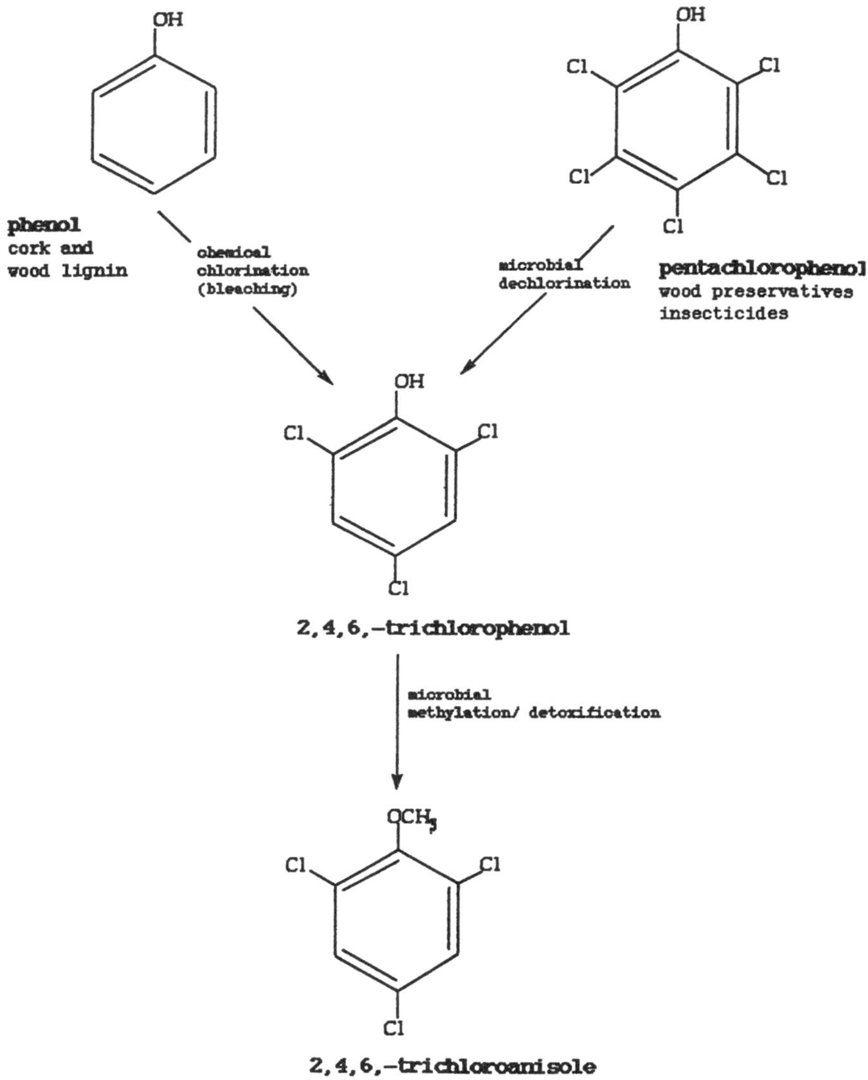

Fig. 19–1.

tomyces growing on lignin in stockpiled unprocessed corkwood (Riboulet 1991). In that growth of the microbe is visually detectable as a yellow discoloration ('yellowspot') on incoming raw material, contaminated wood is usually eliminated at early stages in processing. Guaiacol is normally present at concentrations well below the threshold level of 20 μg/L (ppb)

(Simpson et al. 1986). These workers speculate that, where it has been identified in cork, acquisition probably resulted from careless screening of raw material or environmental contamination during shipment/storage.

Geosmin is also a metabolite of molds (Simpson and Lee 1990) as well as blue-green bacteria growing on (or in the immediate proximity of) cork. Algal formation of geosmin results as a byproduct of the formation of carotenoid pigments (Naes et al. 1988) whereas bacterial (*Streptomyces*) formation uses a sesquiterpene precursor (Bentley and Meganathan 1981). Although geosmin has a reported sensory threshold of 25 ppt (Amon et al. 1989), it is unstable at wine pH (Amon et al. 1989) and the products of its breakdown reportedly lack sensory properties (Gerber 1979).

2-Methylisoborneol is produced by molds as well as Cyanobacteria and *Streptomyces* (Simpson and Lee 1990). In the case of *Streptomyces,* it is reported to be formed by methylation of a monoterpene (Bentley and Meganathan 1981). Its sensory threshold in wine is reported to be 30 ppt (Amon et al. 1989).

Microbial utilization of saturated- and unsaturated fatty acids may produce volatile aldehydes such as hexanal and its higher molecular weight analogs (Rocha et al. 1993), which are of sensory importance. Other compounds produced by degradation of lipids include 1-Octen-3-one and the corresponding alcohol, 1-Octen-3-ol. Some flavor and odor-active compounds may be synthesized chemically without intervention of microorganisms. One recently reported compound, 2,4,6-trimethyl-1,3,5-trithiane, produced from H_2S and acetaldehyde precursors, has musty/mushroom-like properties (Muller and Fugelsang 1994a).

PREPARATION OF CORK FOR SHIPMENT

Sterilization of cork is usually done after bagging and before sealing. Various methods have been suggested for cork sterilization, including ethylene oxide (Borges 1985), as well as UV-light and gamma-irradiation. Although ethylene oxide is an effective sterilant, it is potentially flamable and, thus, requires special handling. Marais and Kruger (1975) reported that gamma-irradiation provided the most effective means of sterilization. Sulfur dioxide gas is currently used for commercial sterilization. Injected into the bag at sealing, SO_2 is reported to reduce bacterial and mold populations by 80 to 100% (Davis et al. 1982). Although effective against growing cells, SO_2 has little or no effect on metabolites already present from earlier processing. In that SO_2 doesn't immediately break down, it has the additional advantage of providing residual activity. Use levels must be controlled;

overdosing may induce brittleness as well as mercaptan and pyrazine formation in cork (Schanderl 1971).

Because it is currently not economically feasible to produce cork entirely free of mold spores, effort must be directed toward preventing spore germination and subsequent growth. The most important parameter in control of mold growth is water activity (A_w) during processing and subsequent shipment and storage of cork. Unfortunately, in the early stages of processing, excess moisture is pervasive. Upon harvest, cork slabs are stacked in open fields while they air-dry and 'cure.' However, the most important period for microbial activity occurs in the processing facility. Before punching the cork cylinders, the cork wood must be remoisturized. This is accomplished by immersing the slabs in a vat of near-boiling water. In traditional processing, cork bark is then restacked, and allowed to air dry under ambient conditions before punching. During this period, substantial mold growth (and potentially TCA formation) occurs, due largely to recontamination. In this regard, the boiling step itself may play an important role. It has been reported that low levels of chlorine may enter the process from chlorinated plant water (Amon and Simpson 1986), and it is not uncommon for a processor to operate for several days without changing the water. However, the authors have raised the issue of chlorination with several cork processors and each claims to use only unchlorinated well water.

If TCA and other compounds implicated in cork taint are already present in the cork bark (before boiling), as some contend, it would seem that, given their extraordinary volatility, immersion in a hot water bath would serve to volatilize them. In this regard, Rocha et al. (1993) did not identify chloroanisols in cork wood before processing. Chlorophenol precursors may be present in some lots of cork bark and, during the period between boiling and punching, molds growth converts these to their anisol end products. Given the odor and dust present in cork processing facilities, sensory detection of TCA is likely impossible.

Thus, one important control is to lower the moisture content of cork as soon as possible after boiling to a point where mold germination is not promoted. Some facilitites have installed an accelerated drying component to the process. In other cases, cork slabs are autoclaved instead of boiled. This shortens the time before punching to hours rather than days. Thereafter, moisture content must be kept below 8% to ensure that spores do not germinate. None of the mold isolates of Daly et al. (1984) germinated at A_w of 0.75 or less (corresponding to 8% moisture).

ANALYSIS

The following is a cross-reference of the analytes and procedure found in Chapter 20. These include physical measurements (length, width, moisture content) as well as microbiological methods (cork microbiology) that are within the capabilities of most winery laboratories. Additionally, there is increasing interest in evaluating the compressability and recoil properties of cork as well as extraction force. However, these involve use of available, but rather unique equipment. For details of these analyses and equipment, contact the International Organization for Standardization (ISO), in Geneva, Switzerland, or the Cork Quality Council in Napa, California.

The authors recommend use of Military-Standard, version 105E (MIL-STD-105E) to determine the number of corks to collect for each test. The subject is covered by Fugelsang and Callaway (1995).

CHAPTER 20

LABORATORY PROCEDURES

I. GENERAL LABORATORY INFORMATION

Successful analytical measurements in the laboratory are generally the result of well-developed and practiced skills in handling laboratory glassware and instruments, care and attention in the preparation of reagents and samples, closely following procedural instructions, and an overall understanding of what a procedure is attempting to measure and the key steps and potential pitfalls involved.

The authors assume that practitioners will have some basic background in quantitative chemistry, and access to a general quantitative analysis text. The latter should serve as a general source of information regarding analytical measurements and as a source of basic information on typical instrumental methods widely used in the industry.

Readers should be aware that the analytical process involves several steps: *sampling, choice of method, separating the analyte from matrix material, performing the quantitative measurement, and evaluating the results.* Too often a bad analytical result is due to failure on the part of the analyst to consider all the factors involved. A laboratory sample that is representative of a load of grapes, a tank, or a barrel, etc., is necessary for quality work. It is *usually* not simple to obtain such a sample; one must typically sample a tank at several depths and mix to get a representative tank sample.

Single "best" analytical methods for a particular element or component do not exist. Methods are selected based on a number of factors such as ease of performance, accuracy, sensitivity, selectivity, speed, and cost. Some methods require a preliminary separation of the component of interest from the matrix. Incomplete recovery of the analyte or contamination of a presumed "clean" sample will affect results. As mentioned above, performing the analysis well requires a degree of care and experience. Finally, we frequently fail to ask ourselves, "is this value reasonable?". Looking at one's analytical results from a perspective of past values and common sense expectation of possible ranges or magnitudes of values, can save some embarrassment.

Accuracy and Precision

Accuracy and precision are important concepts when making analytical measurements. Accuracy refers to the "correctness" of your answers. It is obtained by using correct methods, quality equipment, and quality reagents. Throughout the following procedures the authors have specified the use of standard solutions of analytes (solutions of known, exact concentration), and careful calibration practices. It is always good practice to include known samples (reference materials) or samples with known amounts of standard analyte added to them in order to be assured that one's results are accurate.

Precision involves how reproducibly you can repeat a test or measurement. It is often reported in terms of the range of analytical values obtained, or as the "standard deviation" of the set of analytical values. Precision is directly related to how often you have practiced an analysis. Although it is expected that two or more runs on the same sample will produce different analytical values, the more often you perform an analysis, the smaller the spread of these values will be.

Sampling and Storage Procedures

Methods for vineyard sampling and sample processing are given in Chapter 3. Analysis of juice should occur fairly quickly on settled samples. Refrigeration of juice may cause potassium bitartrate precipitation, influencing the analysis of pH, titratable acidity, and potassium.

One may freeze grapes or must from vineyard experiments for later evaluation. Spayd et al. (1987) reported that freezing musts and wines results in lower acidity, higher pH, and with little change in soluble solids.

Wine produced from frozen grapes have higher potassium values, lower total phenols and caffeoyl tartrate concentrations, and similar color compared with wines made from fresh grapes. Freezing grapes to −7°C does not influence the sensory response of Chardonnay, Chenin blanc, or white Riesling wines.

Sampling methods are crucial to ensuring that the sample collected is representative of the tank or barrel. Top samples may not be representative of a large tank due to stratification, microbial growth, and/or oxidative degradation. Furthermore, bottom or racking valve samples may also not be representative unless an adequate volume of wine has run through the valve before collection. Large tanks can be difficult to mix to provide uniform distribution of fining agents, etc. The use of sparging gas in addition to pump mixing may help to overcome this problem (see Chapter 14).

II. NEW VS. TRADITIONAL METHODS

Traditionally, wet chemical-titrametric, pH-meter, simple physical, colorimetry (visible spectrometry), and flame photometry methods have been used in the winery laboratory. Today these are being supplemented with ultraviolet (UV) spectrometry, atomic absorption, gas and liquid chromatography, enzymatic, and, occasionally, gas chromatography–mass spectrometry techniques. Part of the driving force in changing to new methods is the selectivity toward specific analytes that they provide. Another reason is that they require smaller amounts of samples and reagents and therefore produce less "hazardous waste," which is becoming a significant expense for all laboratory operations.

Spectrometric Analysis

Light (or electromagnetic radiation) is a carrier of energy. The energy varies in an inverse manner with wavelength; short wavelengths (ultraviolet radiation, 195–380 nm) carry more energy than the longer [visible (VIS), 380–800 nm] wavelengths. The energy carried by light can interact in very specific ways with atoms, molecules, and ions; various species of these chemicals will absorb only specific wavelengths from a light beam.

The basis of spectrometric measurements is to pass a quantity of light of a selected wavelength through a sample. The molecules (or atoms or ions) in the sample will absorb the light as a function of their concentration. The light not absorbed passes on to a detector system where its amount is measured. Colored solutions absorb specific wavelengths of light in the VIS region of the spectrum; colorless solutions do not absorb visible light, but may absorb in the UV region.

Spectrometers are instruments that have a light source (tungsten lamp for VIS, hydrogen or deuterium lamp for UV), a monochrometer (device for selecting the wavelength you wish to use for your analysis), a container for the sample (cuvette, and in simple instruments a test tube), and a detector system (phototube or photomultiplier tube or light-sensitive diode). In quantitative spectrometry the amount of light absorbed by the sample (called the absorbance) is measured. Beer's law (also Beer-Lambert law) states that absorbance (A) is directly related to concentration (C). A graph of absorbance vs. concentration using known standards should produce a straight line. The absorbance measurement from an unknown sample can be related directly to its concentration.

The natural colors of juices and wines are used for some spectrometric measurements. In other cases a chemical reaction is used to convert some component, such as phenols, into colored compounds. Some components in wine can be measured directly with a UV spectrometer *or* be chemically converted into a colored species and measured with a VIS spectrometer.

Enzymatic Methods

One category of spectrometric methods are the enzymatic analysis procedures. These have been developed for a number of analytes. Kits of reagents for conducting these procedures are now available, as well as instruction sheets that provide instructions for the specific analyses purchased. The kits generally list several wavelengths that can be used for analysis, so these analyses can be completed with a VIS spectrometer, as well as with a UV spectrometer. Advantages of these analyses include (1) being specific for the analyte, (2) being performed with an inexpensive spectrometer, and (3) producing a minimum amount of "waste" requiring proper disposal. *However:* In the authors' experience that these are difficult procedures to carry out successfully, due to potential interferences in juice/wine samples and the requirement to pipette very small volumes of samples and reagents. When using these kits, readers are advised to run several known standards to validate their abilities to use the kit and to validate the constants provided in the kit procedure. An example of an enzymatic procedure that works well is the determination of L-malic acid. The procedure relies on the enzymatically catalyzed reaction between the acid and NAD^+ to produce NADH:

$$\text{L-malate} + NAD^+ \overset{\text{enzyme}}{\leftrightarrow} \text{oxaloacetate} + NADH + H^+$$

The change in concentration of NADH is measured spectrophotometrically at 340 nm (the absorption peak for NADH). The reagents are purchased in the form of a test kit of sufficient material for approximately 20 determinations. The general procedure calls for mixing the sample with reagents (except the enzyme), and measuring the absorbance. The enzyme is added, and the absorbance is read again. The difference in absorbance values taken before and after addition of enzyme is used to calculate the concentration of L-malic acid.

Atomic Absorption and Emission Spectrometry

Atomic emission techniques measure the radiation (either UV or VIS) emitted by atomic species excited by the flame during their return to ground state. Sample elements of interest in liquid form are aspirated into the flame, desolvated, and converted into a cloud of ground state atoms. These atoms are available to participate in the atomic emission process. The energy emitted at specific wavelengths is monitored using a typical monochrometer and photodetector. In emission techniques the intensity of the emitted light is measured in terms of transmittance (%T) and is directly related to concentration. In principle, atomic absorption (AA) measures the absorption of UV or VIS light by neutral (ground state) atoms present in the gaseous phase. As such, it is similar to solution spectroscopy except that the absorption lines for atoms are very narrow compared with relatively broader bands characteristic of absorption spectra of molecules. As a result, in atomic absorption, the light source is an interchangeable hollow cathode lamp that emits radiation of the same wavelength as the element to be measured. Because absorption of light is restricted to those species absorbing at the same wavelength as the source, the method is specific and rather free of interferences. Accuracy of ±2% and sensitivity in the range of 0.1 mg/L can be attained. At present, some 68 elements may be determined using AA spectrophotometry.

Chromatographic Techniques

Chromatography is a technique for separating various solute molecules or compounds in a mixture. It is a very powerful technique in its modern variations, capable of performing separations of quite similar molecules. Separations are achieved because different molecules interact differently with their environment due to differences in size, functional group, geometry, charge and charge distribution, and solubility. A chromatographic system consists of a packed bed of some solid (or liquid coated solid) material with a mobile phase (either a liquid or a gas) percolating through it. Solutes placed into this system distribute between these two phases and therefore move at different rates, effecting a separation.

There are a number of different mechanisms used in chromatographic systems to cause separations. The most common mechanisms are partition, adsorption, ion exchange, and size exclusion.

The common modes of chromatography are paper chromatography (PC), thin-layer chromatography (TLC), gas chromatography (GC), and high-performance liquid chromatography (HPLC), listed in order of the complexity of the mechanical hardware required to perform them.

Paper chromatography is performed using a sheet of filter paper placed inside a jar or chamber. Samples (which, for example, could be a mixture of organic acids) are placed at one end of the paper in a process called "spotting," and the paper is dipped into a water-organic solvent mobile phase in the chromatographic chamber. The phases climb the paper by capillary action, and the solute molecules are carried upward as a function of their individual partition coefficients. The chromatogram (record of what happened in the chromatographic process) is allowed to develop until the solvent front has climbed a distance sufficient to allow the various solute components to separate. The solutes (spots) are not eluted or removed from the chromatographic system; instead one measures the distance a spot travels to calculate its R_f value. One generally runs mixtures of standards along with unknowns so that direct comparisons of unknowns and standards may be made on the same paper chromatogram.

Thin-layer chromatography is quite similar to paper chromatography. The sorbent material is spread as a thin layer on some support material such as glass, aluminum, or plastic, producing what are called "thin-layer plates." Typical sorbents used to make these plates are silica gel, alumina, and crystalline cellulose. The plates are spotted in the same way as above, placed in chambers, and the developed spots visualized by use of chemical sprays.

In terms of the hardware necessary to perform them, GC and HPLC techniques are considerably more sophisticated than are those of paper and thin-layer chromatography. GC and HPLC systems use columns packed with stationary phase, through which the mobile phase percolates driven by a pressurized mobile phase source. Sample placement at the inlet of the chromatographic column requires special injection devices. Because sample components are eluted off the column, some type of detector or sensor is required to "see" these components.

The common detectors used in gas chromatography are the (1) thermal conductivity detector, TCD; (2) flame ionization detector, FID; (3) electron capture detector, ECD; (4) flame photometric detector, FPD; and (5) mass selective detector, MSD. The common detectors used in HPLC are the (1) refractive index, (2) UV (fixed and variable wavelength), and (3) conductivity detectors.

III. STANDARD ACID-BASE SOLUTIONS

Standard Sodium Hydroxide Solutions are prepared as follows: Stock solution (1 + 1): In a 1-L Erlenmeyer flask, carefully mix one part CP grade NaOH and one part boiled and cooled deionized water. Upon dissolution and after reaction has cooled, transfer to polyethylene container for storage. After Na_2CO_3 precipitation is complete (several days),and the solid has settled out, the following solutions may be prepared by dilution of the concentrated solution (see Table 20–1). *Note:* Do not shake this solution and remix the solid Na_2CO_3.

Table 20–1. Dilution procedures for preparation of working solutions from stock sodium hydroxide.

Desired approximate normality	mL of 1 + 1 stock solution per 1.0 L
0.01	0.54
0.02	1.08
0.10	5.40
0.50	27.00
1.00	54.00

Standardization: Due to the indeterminate effect of carbonate, the above solutions represent only an approximate concentration. It is necessary to standardize them relative to a primary standard acid such as potassium hydrogen phthalate (KHP). This reagent can be purchased in prestandardized liquid form or as a powder of defined purity. In either case an accurately determined quantity is titrated with base to the phenolphthalein end point. In cases where prestandardized liquid KHP is used, the normality of base may be calculated according to the following relationship:

$$\text{Normality of base} = \frac{(\text{Normality KHP})\ (\text{mL KHP used})}{(\text{mL of base used})}$$

In the case of powdered KHP, the following relationship applies:

$$\text{Normality of base} = \frac{(\text{g KHP})\ 1{,}000}{(\text{mL base used})\ (204.229)}$$

Standard Hydrochloric Acid Solutions

The concentration of stock HCl ranges from 35 to 37%. The dilutions from stock required to achieve the approximate concentrations desired are presented in Table 20–2.

Table 20–2. Dilution procedures for preparation of working solutions from stock hydrochloric acid

Desired approximate normality	mL of stock HCl per 1.0 L
0.01	0.89
0.02	1.78
0.10	8.90
0.50	44.50
1.00	89.00
2.00	178.00

Standardization: For routine laboratory uses, anhydrous sodium carbonate (ACS grade) may be used for standardizing HCl solutions. Dissolve 1 to 3 g (accurately weighed) of ACS grade anhydrous sodium carbonate (Na_2CO_3) in 40 mL deionized water. Titrate with HCl using 4 drops of bromocresol green indicator until solution begins to change color slightly. At this point, transfer to an electric burner and boil gently for 2 min. Cool, titrate until color of reference solution. The reference solution consists of 80 mL boiled deionized water and 4 drops of bromocresol green indicator.

The normality of HCl may be calculated using the following relationship:

$$N\,\text{HCl} = \frac{(\text{g Na}_2\text{CO}_2)\ 1{,}000}{(\text{mL HCl})\ (52.994)}$$

Standard Sulfuric Acid Solutions:

Table 20–3 presents the volume of stock sulfuric acid that must be used per 1.0 L of final solution. Standardization: Acid solutions can also be standardized against previously standardized sodium hydroxide solutions. Using a volumetric pipette carefully transfer an aliquot of standard NaOH to an Erlenmeyer flask. Add several drops of indicator (phenolphthalein or methyl red) and titrate to an end point. Calculate normality of acid using the following relationship:

$$N_{\text{Acid}} = \frac{(\text{mL NaOH})\ (\text{Normality NaOH})}{\text{mL HCl}}$$

Table 20–3. Dilution procedures for preparation of working solutions from stock sulfuric acid.

Desired approximate normality	mL of stock H_2SO_4 per 1.0 L
0.01	0.28
0.02	0.57
0.10	2.84
0.50	14.18
1.00	28.35

IV. DIAGNOSTIC PROCEDURES

Identification of Hazes and Precipitates

On occasion sediments and/or hazes develop in bottled wines or juices. These hazes, clouds, and/or deposits encountered in juices and wines are due to metals, tartrates and oxalates, oxidized coloring matter and tannins, excess fining agent, protein, polysaccharides, filtration media, bacteria, and yeast.

Yeasts appear as a film, fine haze, and/or precipitate and wine may be gassy. Acetic acid bacteria appear as a graphite colored film or precipitate. Lactic acid bacterial growth appears as an amorphous sediment or silky haze, sometimes as a streaming cloud. Wines are usually gassy. Microbiological hazes and precipitates are identified in the section on Yeast and Bacteria Identification.

This section provides screening tests for the isolation and identification of nonbiological hazes or precipitates. Preliminary sample examination should be made using a high-intensity light source to determine the nature of the haze or sediment, categorized as crystalline, amorphous, or fibrous (Quinsland 1978). The history of the product before bottling should be carefully reviewed.

The haze or precipitate should be isolated by either centrifugation, membrane filtration, or aspiration directly from the floor of the vessel. Material can then be examined microscopically and screening tests performed. Tanner and Vetsch (1956) recommended washing the collected material in 5 mL of 95% ethanol and recentrifuging.

The following list summarizes the nonbiological categories of juice and/or wine hazes and sediments:

Crystalline Deposits	Amorphous Materials
Potassium bitartrate	Protein
Calcium tartrate	Tannin
Calcium oxalate	Pectin
Cork	Glucan

Fibrous Materials
- Cellulose
- Case lint
- Asbestos

Starch
Diatomaceous earth
Copper and iron compounds

CRYSTALLINE DEPOSITS

Tartrate Deposits

Tartrate instability is discussed in Chapter 15. The characteristics of these crystalline deposits are as follows:

Potassium bitartrate

a) appearance—coarse sediment, sediment shows color of the beverage
b) microscopic appearance—prisms
c) solubility—0.492 g/100 mL H_2O (at 20°C).

Calcium tartrate

a) appearance—crystalline, medium coarse, sediment shows color of beverage
b) microscopic appearance—prisms
c) solubility—0.0322 g/100 mL H_2O(at 20°C)

Calcium oxalate

a) appearance—sediment, mostly clear beverage, sediment shows color of beverage
b) microscopic appearance—small cubic crystals
c) solubility—0.00067 g CaC_2O_4/100 mL (at 13°C)

Tartrate Test

I. Equipment
1. Bright field microscope, slides, and cover slips.
2. Membrane filters and appropriate housing. Most laboratory filtration units use 47-mm membranes, although a 10 mL syringe fitted with a membrane housing will be adequate. For general purposes, 1–5 μm cellulose acetate filters are useful.
3. Clinical centrifuge.
4. Magnesium oxide rods.
5. Cobalt filter.

II. Reagents
1. Sulfuric acid, (1 + 3): Carefully add 1 volume concentrated H_2SO_4 to 3 volumes deionized water.

2. Sodium metavanadate solution (3% wt/vol): In a 100-mL volumetric flask dissolve 3 g $NaVO_3$ in hot (<70°C) deionized water. Cool to room temperature and bring to volume. *Note:* Sodium metavanadate does not completely dissolve and must be filtered through Whatman #2 paper before use.

III. Procedure:
   1. Using filter apparatus and appropriate membrane or by aspiration or centrifugation, collect a portion of the sample containing the sediment.
   2. Rinse sediment/crystals with a small volume of deionized water and remove water by filtration, aspiration, or decanting.
   3. Transfer membrane and/or precipitate to a watch glass. Place a drop of dilute sulfuric acid on precipitate.
   4. Add a drop of metavanadate solution.

Interpretation: Tartrate present in sediment turns yellow-orange in color.

Test for Potassium

1. "Load" a magnesium oxide ("magnesia") rod with sediment using the following technique:

(a) using the flame from a Bunsen burner, heat the end of rod and insert into collected sediment.

(b) reheat rod and again insert into sediment.

(c) repeat (a) and (b) if necessary.

2. When sediment is concentrated onto magnesia rod, hold loaded end in the outer portion of the flame.

Interpretation: Using a cobalt blue filter disk, the presence of potassium is indicated by a rose-color flame.

pH Test for Potassium Bitartrate and Calcium Tartrate

Potassium bitartrate and calcium tartrate can be differentiated based on the ability to form crystals at defined pHs.

I. Reagents
   1. Concentrated hydrochloric acid (HCl)
   2. 50% sodium hydroxide (NaOH)

II. Procedure:
   1. Add acid or 50% sodium hydroxide to adjust the pH of a centrifuged sample, then chill.

Interpretation: Crystal formation at pH 3.6 indicates potassium bitartrate; crystal formation at pH 6.0 indicates calcium tartrate.

Test for Calcium

I. Reagents
   1. Oxalic acid
   2. Concentrated sulfuric acid
   3. Methyl alcohol
II. Procedure:
   1. To centrifuged wine sample (approximately 10 mL) add oxalic acid (spatula tip).
   2. Crystal formation is indicative of the presence of calcium. To confirm add several drops of concentrated sulfuric acid to dissolve precipitate. The addition of excess methyl alcohol and heat will then cause precipitate to reappear.

Alternative Test for Tartaric Acid and Tannins (Silver Mirror Test)

I. Reagents
   1. Silver mirror reagent. Mix 50 mL of 0.1 *N* silver nitrate (17.0 g/L) solution with 5 mL of 10% sodium hydroxide. Add concentrated ammonium hydroxide drop by drop until silver hydroxide precipitate has dissolved. The reagent should be stored in a brown bottle and kept from drying as there is a danger of explosion if it is dry.
   2. Cation exchange resin (regenerated with 5% hydrochloric acid and rinsed with deionized water to remove all traces of the acid).
II. Procedure:
   1. Dissolve some of the precipitate in hot water in a test tube.
   2. Add three "kitchen knife tips" of cation exchange resin and shake tube after each addition.
   3. Filter into a test tube and add 5 mL silver mirror reagent.
   4. Heat tube gently for 5 min over an open flame.

Interpretation: Colloidal silver produced will form a mirror on the walls of the tube and is a positive test for tartaric acid and tannin.

Cork Dust

Cork dust (lignin) appears "crystallike" microscopically, and may be confused with bitartrate or tartrate unless further examination is performed. (For a discussion of cork, see Chapter 19.) The following procedure uses a stain that reacts with lignin, a structural macromolecule in cork.

I. Equipment
   1. Bright field microscope, several slides, and cover slips.
   2. Membrane filters and appropriate housing. For general purposes, 1–8 µm cellulose acetate filters are useful.
   3. Clinical centrifuge.

II. Reagents

1. Phloroglucinol stain: Mix 2 g of phloroglucinol in 100 mL of 10% HCl. This is a near-saturated mixture and supernatant must be decanted from any crystals that do not dissolve.

   This solution should be made fresh daily. Therefore, unless many samples are to be run, the analyst may wish to reduce proportionately the volume made at one time.

III. Procedure:

1. Collect a portion of wine containing the debris by filtering through an 8-μm membrane filter.
2. Wet the filter and sediment with phloroglucinol stain. Hold stain in contact with sediment for 5 min.
3. Apply vacuum to filter to remove stain. Rinse the filter with deionized water.

Interpretation:

1. Examine the sediment microscopically. Cork debris appears as red crystal-like aggregations of cells (see photomicrograph 20-1).
2. Case lint also stains red using this technique. However, it is fibrous in appearance (see photomicrograph 20-2).

FIBROUS MATERIALS (CELLULOSE, CASE LINT, AND ASBESTOS)

Fibrous materials found in finished wines are usually cellulosic in nature, originating from the filter pad matrix or from case lint present in the bottles before filling. Although asbestos is not used in the United States it is used in wine processing in some countries.

I. Equipment

1. Bright field microscope and several slides and cover slips.
2. Membrane filters and appropriate housing. Most laboratory filtration units use 47-mm diameter membranes, although a 10-mL syringe fitted with a membrane housing is adequate. For general purposes, 1–5 μm cellulose acetate filters are useful.
3. Clinical centrifuge.

II. Reagents

1. Phloroglucinol stain: see cork dust.
2. Cellulose stain: Dissolve 200 g zinc chloride in 100 mL of deionized water. Add 20 mL of iodine solution made by dissolving 10 g KI and 4 g I_2 in deionized water and bringing to 100 mL final volume.

III. Procedure:

1. Collect by membrane filtration a portion of the wine containing the sediment onto two separate membranes.

2. Stain the first membrane using phloroglucinol technique listed under "Cork Dust."
3. Treat second membrane with cellulose stain.
   (a) Flood the filter and particulates with stain. Hold 5 min.
   (b) Filter to remove stain. Rinse with several milliliters of deionized water.

Interpretation: Microscopically, all cellulosic material appears light blue. *Note:* Best results are obtained when the preparation is examined fresh. Color intensity diminishes after 30 min.

IV. Supplemental Note
   1. Asbestos fibers are not stained using either of these techniques and therefore can be differentiated from cellulose and cork lignin.

AMORPHOUS MATERIALS

Protein/Phenolics

Particulates lacking defined shape include protein and phenolics. Protein haze and precipitates are mainly complexes of protein, polysaccharides, and polyphenols (see Chapters 5, 7, 8 and 16).

Protein, Protein-tannate characteristics
   a) appearance—haze or fine amorphous sediment showing the color of the beverage.
   b) microscopic appearance -amorphous

I. Equipment
   1. Polycarbonate membrane filters (47-mm) and housing.

II. Reagents
   1. Amido black 10-B protein stain. Dissolve 2 g amido black 10-B (syn: Naphthol blue black) in 100 mL of methanol:acetic acid solvent. Solvent is prepared by mixing 90 mL methanol and 10 mL acetic acid.
   2. Eosin Y protein stain: available as a commercial preparation.
   3. Folin-Ciocalteu reagent: available as a commercial preparation (also see Phenols, Spectrometric Analysis).
   4. Concentrated sulfuric acid.

III. Procedure for Protein:
   1. Filter sample containing sediment through an 8 μm polycarbonate membrane. (*Note:* Cellulose acetate membranes are unstable in presence of protein stain).
   2. Wet membrane with stain and hold 10 min.
   3. Remove stain by applying vacuum. Rinse with methanol-acetic acid solvent until filter is white.

Interpretation: Proteinaceous materials will stain blue-black.

4. Using the above procedure, Eosin Y stain may also be used instead of amido black 10-b.

IV. Procedure for Phenolics:

1. Collect sediment by filtering sample through an 8 μm membrane filter.
2. Rinse with several milliliters deionized water and, with a spatula, transfer sediment onto a watch glass.
3. Dilute Folin-Ciocalteu reagent 1:10 with deionized water, then add a drop to sediment.

Interpretation: Phenolic complexes dissolve to yield a slate-grey to blue turbid solution.

V. Alternative Procedure for Protein and Phenols:

1. Collect a portion of sediment into another test tube. Add 1 mL of concentrated H_2SO_4 and heat gently.

Interpretation:

(a) Phenolics (pigments and tannins) present in sample turn dark red.

(b) Carbonization is suggestive of protein.

Alternative Test for Residual Protein Fining Agents

Protein fining agents can on occasion remain suspended in a wine forming a slight haze or cloud (see Chapter 16).

I. Reagents

1. Tannic acid solution, 1% (wt/vol) in 70% ethanol.

II. Procedure:

1. Add 5 mL of tannic acid solution to 100 mL of suspected wine.

Interpretation: The formation of a much more pronounced haze or precipitate is indicative of excess protein in the wine.

Polysaccharides Pectins/Glucans:

Polysaccharides can form protective colloids in juices and wines inhibiting clarification, fining, and filtration. In grape juices and wines polysaccharides may be in the form of pectins and/or glucans, each forming gelatinous aggregates in an alcohol solution (see Chapters 3, 5, and 16).

Pectin Instability

Pectins are structural components of plant cell walls (see Chapter 16). A simple test for the presence of pectins is given. If pectins are present, the addition of pectolytic enzymes to a laboratory sample and subsequent pectin precipitation test is recommended.

Procedure:

1. To a 25-mL aliquot of the wine containing unidentified haze, add 50 mL of a 95% ethanol: 1% HCl or alternatively, isopropanol: 1% HCl reagent.

Interpretation: Formation of gel after several minutes is indicative of pectin.

Glucan Instability

Dubourdieu et al. (1981) developed two precipitation tests for glucans. The first procedure given is for the presence of glucans in concentrations greater than 15 mg/L, the second for levels as low as 3 mg/L. Even at low concentrations, glucans can cause filtration problems. A positive test for the presence of glucans should be followed by a laboratory fining trial using glucanases and retesting (see Chapter 16).

I. Procedure for Glucans > 15 mg/L:

1. Add 5 mL of 96% ethanol (vol/vol) acidulated with 1% HCl to a tube containing 10 mL of wine.

Interpretation: The formation of a white filament is indicative of the presence of glucans at levels greater than 15 mg/L. Because much lower levels can cause problems, an additional test that will detect glucans at concentrations above 3 mg/L may be warranted.

II. Procedure for Glucans > 3 mg/L:

1. 5 mL of wine is mixed with 5 mL of 96% ethanol (vol/vol) acidulated with 1% HCl.
2. After 30 minutes at room temperature the mixture is centrifuged at 3,000 g for 20 min.
3. The supernatant is carefully removed and the precipitate redissolved in 1 mL water. The precipitate is then mixed with 0.5 mL acidulated ethanol.

Interpretation: The formation of filaments is indicative of glucans.

Starch Test

Starch can cause hazes in apple juices and ciders affecting clarity and filtration. The following is a simple screening test for starch.

I. Reagents

1. Iodine reagent: Dissolve 1.0 g iodine and 20 g potassium iodide in 1 L deionized water.

II. Procedure:

1. To 10 mL of juice add 1.0 mL of iodine reagent.

Interpretation: A dark blue or violet color indicates the presence of starch. In some cases the blue color is visible only for a short moment following reagent addition. This is also a positive confirmation.

Diatomaceous Earth

Diatomaceous earth, also known as diatomite or DE, is a common filtration media composed of the fossilized remains of marine plants called diatoms. Occasionally DE may bleed through the final filtration and appear as an amorphous deposit. Confirmation may be made by microscopic examination (see photomicrograph).

Metal Instabilities

Metal instabilities are discussed in Chapter 13 and include copper casse, iron (or "white") ferric phosphate casse, and rarely ferric tannate or "blue casse."

Copper Casse Characteristics

a) appearance—a white haze and later a red-brown precipitate in white wines stored in the absence of air.
b) microscopic appearance—amorphous

Iron or Ferric Phosphate, White Casse Characteristics

a) appearance—a milky white cloud appearing in white wines eventually forming a white precipitate.
b) microscopic appearance—amorphous

Ferric Tannate or Blue Casse Characteristics

a) appearance—a cloud in a white wine that has had a tannin addition, and an inky blue cloud, later a blue deposit, in red wines.

Preliminary acidification of suspect sample using 10% HCl is useful in separation of metal-containing complexes from complexes of protein and phenolics.

I. Reagents

Potassium ferrocyanide, 0.5%: Dissolve 0.5 g $K_4Fe(CN)_6 \cdot 3H_2O$ in 100 mL deionized water
Hydrochloric acid, 10%
Hydrogen peroxide, 30%

II. Procedure:

1. Collect 15–20 mL of suspect wine.
2. Add 3–5 mL of hydrochloric acid (10%) and note whether haze dissipates or remains.
3. If haze solubilizes, it is likely copper sulfide, copper proteinate or ferric phosphate; proceed using diagnostic scheme presented below.

4. If haze remains, the instability is probably due to protein or complexes of protein, protein-tannins, or phenolics-phenolics (i.e., pigment-tannin). For presumptive identification of the later group, see Protein/Phenolics.

(a)

1. Collect 15–20 mL of suspect wine.
2. Add 5 drops of hydrogen peroxide.
3. If the haze dissipates, Cu^{2+} is suspected. If haze remains, see options under (b).
4. If the haze can be concentrated by centrifugation or if sediment is present, collect sufficient amount on a stainless steel laboratory spatula. *Slowly* dry the sediment over a bunsen burner and, when completely dry, attempt to ignite by more intensive exposure to the flame.

Interpretation: If the haze consists primarily of complexes of copper and organics, sediment is partially burnt. However, copper sulfide and ferric phosphate casse will not burn.

(b)

1. Collect 20 mL of turbid wine into two test tubes.
2. To Tube 1 add 5 mL of 0.5% potassium ferrocyanide.

Interpretation: Formation of red coloration is a positive presumptive test for copper and its complexes.

3. To Tube 2 add 5 mL of potassium ferrocyanide (0.5%) and 5 mL HCl (10%).

Interpretation: Formation of blue coloration is a positive presumptive test for iron.

Potential Metal Instabilities (prebottling)

The following tests are included to assist the winemaker in making judgments regarding the potential for instability in "borderline" wines.

Procedure for Iron:

(a)

1. Pipette 10 mL of filtered wine into two test tubes. To one test tube add citric acid equivalent to 0.7 g/L.
2. Add several drops of 3% H_2O_2 to each.
3. Thoroughly aerate each sample by mixing.
4. Examine the next day.

Interpretation: Haze and/or sediment suggests the likelihood of future instability. If the test tube receiving citric acid and H_2O_2 shows no sign of haze, use of the acid should be considered.

Procedure for Copper:
(b)
1. Collect 10 mL of filtered wine into a test tube.
2. Add several drops of 100 ppm sodium sulfide.

Interpretation: Haze formation suggests copper levels >0.5 mg/L.

V. STANDARD LABORATORY PROCEDURES

ACETALDEHYDE: AROMA SCREEN

Acetaldehyde is the principal aldehyde present in wine. It is produced as an intermediate during alcoholic fermentation and as a result of oxidation of ethanol during wine storage. The levels normally encountered in newly fermented table wines is <75 mg/L with sensory thresholds ranging from 100 to 125 mg/L. Above this range, acetaldehyde can impart an odor to the wine described as overripe, bruised apples, sherrylike, and/or nutty. Bisulfite rapidly reacts with acetaldehyde producing an addition product (hydroxysulfonate), which has limited volatility and therefore odor. The following screening procedure for acetaldehyde uses sulfur dioxide binding as a means of helping to mask excessive acetaldehyde present in wine.

Procedure:

1. To 100 mL wine samples add 5, 10, 15, 20, and 25 mg/L sulfur dioxide, as per Table 20–4.

Table 20–4. Volumes of stock 1,000 mg/L SO_2 solution for 100 mL volumes of wine.

Volume (mL) of stock SO_2/100 mL wine	Approximate SO_2 concentration in wine (mg/L)
1	10
1.5	15
2	20
2.5	25

2. Let wine stand for approximately 10 min, pour samples into wine glass, and evaluate wine odor compared to an untreated control.

Interpretation: Reduction of overripe, bruised apple/sherry, or nutty like odor is indicative of acetaldehyde. Specific analysis procedures for acetaldehyde are provided below in this chapter.

II. Supplemental Note

1. 44 mg/L acetaldehyde binds with 64 mg/L sulfur dioxide or 1 to 1.45 ratio.
2. High acetaldehyde levels are responsible, in part, for high bound and low free sulfur dioxide (see Chapters 10 and 14).

ACETALDEHYDE: ENZYMATIC/SPECTROPHOTOMETRIC (VIS)

The presence of acetaldehyde may be used as an indicator of spoilage in sparkling wine cuveés as well as table wines. The enzymatic assay for acetaldehyde uses the enzyme aldehyde dehydrogenase to catalyze the quantitative oxidation of acetaldehyde to acetic acid. During the course of reaction, the coenzyme nicotinamide-adenine dinucleotide (NAD) is reduced to NADH; the amount of reduced NADH formed is measured spectrophotometrically. Kits containing the necessary components for the analysis are commercially available. The reader is referred to the product sheet that accompanies the acetaldehyde test kit for details.

I. Equipment

The acetaldehyde procedure will require the following equipment:
Spectrophotometer capable of reading absorbance at 334, 340, or 365 nm.
Single or multiple-range micropipettes (0.05–3.00 mL).

II. Reagents

Acetaldehyde Enzymatic Assay Kit (Boehringer-Mannheim).
Deionized water.

III. Procedure

1. Follow directions on product information sheet for proper dilution of sample.
2. Mix an aliquot of the sample in a cuvette with all the reagents provided *except* the aldehyde dehydrogenase solution. Read the absorbance of the mixture.
3. Add the aldehyde dehydrogenase solution to cuvette and mix. After the reaction is complete, again read the absorbance of the mixture.
4. The acetaldehyde concentration in the sample is calculated from the difference in the two measured absorbances.

IV. Supplemental Notes

1. Suppliers of enzyme kits will include their various cautions and suggestions on the detailed product sheet. Read this carefully before beginning the analysis.
2. Because standard acetaldehyde solutions are not used for the calibration, accuracy in this method is dependent on the care with which one delivers the precise volumes of reagents and samples specified.

ACETALDEHYDE: GAS CHROMATOGRAPHIC DETERMINATION (SEE FUSEL OILS)

ACETALDEHYDE: TITRAMETRIC METHOD USING SULFITE BINDING

Acetaldehyde can be determined by taking advantage of its strong binding affinity with sulfur dioxide. A wine sample is distilled and the distillate is titrated at both pH 2.0 and 9.0 with iodine. At pH 2.0 the acetaldehyde-bisulfite complex is not dissociated (Jaulmes and Espezel 1935). The oxidative changes occurring during the alkaline titration may be minimized by addition of a chelating agent (EDTA) and the addition of a small amount of isopropyl alcohol (Burroughs and Sparks 1961).

I. Equipment
   Distillation equipment (See alcohol distillation procedure)
   750 mL Erlenmeyer flasks
   1-L volumetric flasks
   Volumetric pipettes 10, 50

II. Reagents
   Potassium metabisulfite ($K_2S_2O_5$): Dissolve 15 g potassium metabisulfite in deionized water, add 70 mL of concentrated hydrochloric acid, and dilute to 1L.
   EDTA: Dissolve 4.0 g of EDTA, disodium salt, and 200 g of $Na_3PO_4 \cdot 12H_2O$ in deionized water. Dilute to 1 L.
   Sodium Borate: Dissolve 100 g of boric acid and 170 g sodium hydroxide in deionized water and dilute to 1 L.
   Starch Solution, 0.2%.
   Iodine Solution, 0.1 *N* and 0.02 *N*
   Hydrochloric acid (HCl) 3 *N*

III. Procedure
   1. Pipette 50 mL of wine containing less than 30 mg/L acetaldehyde into distilling flask.
   2. Add 50 mL of sodium borate solution.
   3. Distill 50 mL into a 750 mL Erlenmeyer flask containing 300 mL of water, 10 mL of potassium metabisulfite solution, and 10 mL of phosphate-EDTA solution. The pH of the solution in the receiving flask should be between 7.0 and 7.2 before distillation.
   4. After distillation add 10 mL of 3 *N* hydrochloric acid and 10 mL of 0.2% starch solution to the receiving flask.
   5. Mix and immediately titrate with 0.1 *N* iodine solution to a faint blue end point.

6. Immediately add 10 mL of sodium borate solution and rapidly titrate with 0.02 *N* iodine solution to the same blue end point which should be in the pH range of 8.8–9.5.
7. The acetaldehyde content is determined according to the following:

$$\text{Acetaldehyde mg/L} = \frac{(N)\ (V_1)\ (22.0)\ (1{,}000)}{V_2}$$

where: N = Normality of iodine solution
V_1 = Volume of iodine used in step 6
22.0 = Equivalent weight of acetaldehyde
V_2 = Volume of wine sample distilled
1,000 = Factor to convert to mg/L

III. Supplemental Notes

1. Wine sample should be diluted so that the acetaldehyde content is less than 30 mg/L. Table wines immediately after fermentation contain < 75 mg/L. The sensory threshold is 100–125 mg/L. The first titration oxidizes excess SO_2 not used to bind acetaldehyde. If the first titration takes only a few drops, the wine contained excessive acetaldehyde or the bisulfite solution was not made properly.
2. If a yellow color is prevalent in the second titration or if excessive amounts of iodine are consumed the buffer solution was too alkaline.

ACETIC ACID GAS CHROMATOGRAPHIC ANALYSIS

Acetic acid has been determined successfully using gas chromatography. Carbon-based packing materials are used to avoid the problem of acetic acid adsorbing on the column. Quantitative measurement of the acetic acid peak is accomplished by using standards of known concentration.

I. Equipment

Gas chromatograph equipped with a flame ionization detector
Electronic integrator or strip-chart recorder
Hypodermic syringe (10 μL)

II. Reagents

Acetic acid-water stock standard (100 g/L): Dilute 10 mL glacial acetic acid to 100 mL in a volumetric flask with deionized water.

n-Pentyl alcohol internal standard solution: Dilute 1.3 g *n*-pentyl alcohol to 1 L in a volumetric flask with deionized water.

Table 20–5. Operational parameters for gas chromatographic separation of acetic acid.

Carrier gas, N_2	Flow rate, 30 mL/min
Oven temp, 160°C	Injector temp, 200°C
Detector temp, 250°C	

Ethanol (10% vol/vol): Dilute 100 mL absolute ethanol to 1 L with deionized water.

III. Procedure

1. Install one of the following columns in the GC oven.
   a. 6 ft × 2 mm i.d. glass, packed with 6.6% Carbowax 20 M on 80–100 mesh Carbopack B.
   b. 6 ft × 2 mm i.d. glass, packed with 5% Carbowax 20 M on 60–80 mesh Carbopack B.
2. Set the operating conditions for the gas chromatograph as per Table 20–5.
3. Adjust the air and H_2 gas flows to values specified in the instrument's operation directions (approximately 300 and 30 mL/min, respectively). These may have to be altered somewhat to optimize conditions for the particular instrument and column used.
4. Prepare acetic acid calibration standards in 100-mL volumetric flasks as indicated in Table 20–6. Store the standard solutions in a refrigerator.
5. Dilute each standard and wine sample 1 + 1 with internal standard.
6. Inject 1.0 μL of each of the standards into the gas chromatograph. Record the chromatogram, and note the retention times of the various components. Determine the peak area ratios (integrator) or peak height ratios (recorder) for the acetic acid peak relative to the *n*-pentyl alcohol peak. Calculate the respective response ratios (RR').
7. Inject 1.0 μL of prepared wine sample into the gas chromatograph. Record the chromatogram and note the retention times of the various components. Identify acetic acid by comparison of its retention time with that of the standard. Determine the appropriate response ratio for the acetic acid peak to internal standard peak (RR).
8. Calculate the percent acetic acid in the wine sample:

$$\%\ \text{Acetic acid} = \frac{RR}{RR'} \times \%\ \text{acetic acid in std.}$$

IV. Supplemental Notes

1. Carbopack columns are capable of resolving complex mixtures of compounds in a variety of alcoholic beverages such as beer and

Table 20–6. Preparation of calibration standards from 100 g/L acetic acid-water stock solution.[a]

Volume acetic acid stock (mL)	Final conc. (g/L) in 100 mL volume	Volume acetic acid stock (mL)	Final conc. (g/L) in 100 mL volume
1.0	0.10	7.0	0.70
2.0	0.20	10.0	1.00
4.0	0.40	15.0	1.50

[a]Final dilution made with 10% (vol/vol) ethanol.

wine. They can also be used to detect fusel oils and other low molecular-weight carboxylic acids that are usually adsorbed on other columns.

2. Carbopack columns work best with small or dilute samples. If too large or concentrated samples are used, tailing and poor separation efficiency will result. This is due to the low percent stationary phase used on the packing.
3. Columns should be purchased with the ends packed with phosphoric acid-treated glass wool. Untreated or silanized glass wool will adsorb acetic acid and inhibit its elution from the column.
4. Some workers recommend saturating the carrier gas with formic acid vapor to reduce tailing and ghosting of the acetic acid peak. Others have not found this step to be necessary.
5. GC supply houses are constantly improving and upgrading chromatographic packing materials. If the above listed columns are not available, check with your supplier for their current recommendations.

ACETIC ACID: HPLC ANALYSIS (INCLUDES GLYCEROL AND ETHANOL) IN GRAPE JUICE

HPLC can be used to quantify certain mold, yeast, and bacterial metabolites present in a juice sample. This procedure was developed to measure the amounts of glycerol, ethanol, and acetic acid in fruit. Filtered juice samples are injected into a special HPLC column that separates the three components from each other and from any other matrix compounds. Quantification is accomplished by comparing component peak areas to those from standard solutions chromatographed in the same way.

I. Equipment

HPLC unit with column oven (>40°C capability) and refractive index detector

Bio-Rad fruit quality column (or equivalent)
Bio-Rad Cation H guard column (or equivalent)
0.45-μm membrane filter
50-μL syringe

II. Reagents

Glycerol, acetic acid, ethanol standard solution: In approximately 800 mL of deionized water, dissolve 500 mg glycerol, 200 mg acetic acid, 1.0 mL ethanol, 115 g glucose, and 115 g fructose. Mix thoroughly. Adjust to standard temperature and bring to final volume of 1 L with deionized water.

Mobile phase—Sulfuric acid (approximately 0.002 N): Prepare from chromatography grade water by adjusting to pH 2.7 with sulfuric acid.

III. Procedure

1. Install the column and an appropriate guard column in the HPLC mobile phase flow stream.
2. Begin pumping the mobile phase. Adjust the flow rate to 1.2 mL/min.
3. Set the column oven temperature at 55°C, and allow the system to stabilize.
4. Turn on the refractive index detector, and allow it to warm up and stabilize.
5. Fruit samples are crushed and the juice passed through cheesecloth for primary filtration. An analysis sample is prepared by filtration through a 0.45-μm disposable membrane filter.
6. Inject a 10-μL sample of the standard mixture. The peaks should be eluted in the order sugars, glycerol, acetic acid, and ethanol. There may be a "water dip" occurring between the sugar and glycerol peaks. Record the peak heights or areas.
7. Inject 10 μL of the juice sample. Identify peaks of interest by comparison of retention times to those of the standards. Again record the peak heights or areas. (See sample chromatogram.)
8. Concentrations of the components of interest may be calculated directly from the electronic integrator readouts or by use of the following equation:

$$\text{Conc (Unk)} = \frac{\text{Peak area (Unk)} \times \text{Conc (Std)}}{\text{(Peak area std)}}$$

IV. Supplemental Notes

1. Temperature variation of the analytical column does not appear to be a critical factor in this analysis. Ambient and higher temperatures

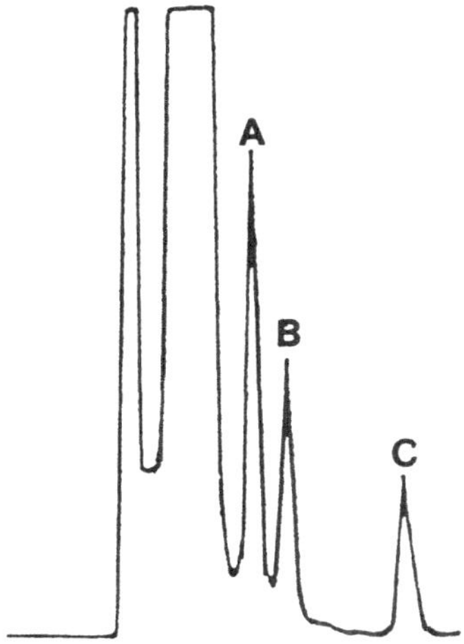

Fig. 20–1. Chromatogram of sample prepared from infected grapes on HPLC fruit quality column. Peaks for glycerol (A), acetic acid (B), and ethanol (C) are clearly present.

can be used with only minimal changes in the retention times. For consistency of operation, it is suggested that the column be operated at a constant temperature of 40°C or higher.

2. Variable sugar levels affect the glycerol response. As sugar levels increase, the fructose peak expands into the glycerol peak, resulting in a lower peak response. Thus, one would prefer to prepare calibration standards with sugar levels similar to those of the samples being analyzed.
3. The purpose of the guard column is to trap any compounds in the filtered juice sample that might damage the analytical column. As the guard column becomes used, resolution of the sample peaks (and possibly the peak responses) decreases. Recalibration with standards is necessary, as is eventual replacement of the guard column.

ACETIC ACID: ENZYMATIC/SPECTROPHOTOMETRIC

Quantitative recovery of acetic acid from wine using steam distillation technique is claimed to be relatively poor (see Chapter 11). Further, other steam-distillable acids may serve as major interferences in the subsequent acid-base titration, a problem that has been overcome with the development of an enzymatic analysis (McCloskey 1976).

Rather than present a detailed procedure here, the reader is referred to the product sheet accompanying the acetic acid kit for the most recent information and a procedure specifically designed for the kit obtained.

In general the acetic acid procedure will require the following:

I. Equipment
   A spectrophotometer capable of reading absorbance at 340, 334, or 365 nm.
   Micropipettes (10, 200 μL)
   Volumetric pipettes (1.0, 2.0 mL)

II. Reagents
   The reagents used in this analysis are those of Boehringer-Mannheim and are sold in the form of test kits of sufficient volume for approximately 25 analyses.

III. Procedure
   The general thrust of the procedure is as follows:
   1. Samples either are used without any preparation (red wines with acetic acid levels of more than 0.1 g/L and white wines) or are treated with PVPP, filtered, pH-adjusted, and diluted two-fold.
   2. The samples are mixed with reagents (except the malate dehydrogenase and acetyl-CoA-synthetase) in a cuvette and the absorbance values measured. Malate dehydrogenase is added and the absorbance again read.
   3. Finally the acetyl-CoA-synthetase is added and the final absorbance measurement taken. A blank is run with the samples.
   4. Because reduction of $NADH^+$ in this assay is not linear, it is necessary to calculate corrected absorbance measurements for sample and blank. The corrected absorbance values are used to calculate the concentration of acetic acid.

ACETIC ACID: STEAM DISTILLATION OF VOLATILE ACID USING CASH OR MARKHAM STILL

In this procedure, steam distillation of the sample is followed by titration (of the distillate) with standardized sodium hydroxide. Results are reported as acetic acid (g/L).

I. Equipment
   Cash volatile acid still assembly (or Markham still)
   10-mL volumetric pipettes
   250-mL Erlenmeyer flask graduated in 50-mL increments
   50-mL burette

II. Reagents

Standardized sodium hydroxide of 0.1 *N* or less
1% phenolphthalein indicator
0.02 *N* iodine or other dilute standardized iodine reagent
1 + 3 sulfuric acid
1% starch indicator
Deionized water

III. Procedure (see Fig. 20–2)

(a) Distillation:

1. Turn on condenser cooling water
2. By releasing clamp *B*, fill boiling chamber *A* with deionized water to the approximate level indicated in Fig. 20–2. When filled, make certain that clamps *B* and *C* are secure!
3. With stopcock *G* positioned so that sample will be delivered to the inner chamber through channel *D*, volumetrically transfer 10 mL of wine to funnel *E*. Rinse the sample into inner chamber *F* with deionized water.
4. Return stopcock to closed position.
5. Plug heater unit *H* into a regulated power supply (Variac). Bring water in chamber *A* to moderate boiling.
6. Collect 100 mL of distillate into receiving flask.
7. Immediately upon completion, *turn the heater unit off.*
8. With stopcock positioned to deliver to the inner chamber, add approximately 15 mL of deionized water to funnel for "self-cleaning operation." Repeat at least two times.
9. Release clamp *C*, allowing water to drain from boiling chamber.
10. Refill boiling chamber with deionized water as previously described.

(b) Titration of distillate:

11. Add 2–3 drops of phenolphthalein indicator to distillate and titrate, using previously standardized NaOH, to a pink end point lasting 15–20 sec.
12. Record the volume of NaOH used in titration and calculate the volatile acidity (VA in g/L) according to the following equation:

$$\text{VA (g/L)} = \frac{(\text{mL NaOH})\ (\text{N NaOH})\ (0.060)\ (1{,}000)}{\text{mL wine}}$$

(c) Sulfur dioxide correction:

Governmentally regulated limits for the volatile acid content of a wine are listed as *exclusive of* SO_2. Therefore, in cases where the volatile acidity approaches legal limits, it is necessary to correct results for the contribution of SO_2.

Free Sulfur Dioxide:

13. Immediately upon completion of VA titration, cool the sample.
14. Add approximately 1 mL of starch indicator and 1 mL of 1 + 3 sulfuric acid.
15. Titrate with standardized iodine solution to a faint blue-green end point.
16. Calculate the *free sulfurous acid equivalent* according to the following equation:

$$\text{F.S.A.E. (g acetic acid/L equivalent to } SO_2 \text{ present)} = \frac{(\text{mL } I_2)\ (N_2)\ (32)\ (2)\ (60)\ (1{,}000)}{(1{,}000)\ (\text{sample vol})\ (64)}$$

IV. Supplemental Notes

1. The Cash still should *never* be operated without water in the boiling chamber! During operation, water levels in the boiling chamber should reflect the approximate level depicted in Fig. 20–2.
2. Carbon dioxide may be present in water used to fill the boiling chamber. To remove, vent steam through funnel *E* for 10–15 sec before closure of stopcock *G*.
3. Direct line current to heating coil may result in exceptionally vigorous boiling. If this problem is noted, a transformer (Variac) may be necessary in line from the 110-volt outlet to the heating coil.
4. The connection between the condenser and the distillation unit should be routinely checked to ensure proper seating.
5. Where the unit is run continuously throughout the day, funnel *E* may become warm, introducing the possibility of premature volatilization of sample. It is recommended that the funnel be filled with deionized water during operation.
6. Several interferences may be encountered in using this procedure, as summarized below.
   a. Carbon dioxide, titrated in the distillate as carbonic acid, would constitute a major source of error. It may be removed from samples with vacuum agitation for several minutes, or by heating an aliquot to incipient boiling under a condenser and cooling immediately. Purging the distillate for 30 seconds with low-pressure nitrogen gas, diffused through an air stone, can also be used to drive off carbon dioxide.
   b. Sulfur dioxide can be corrected for if the volatile acidity approaches legal limits, or if the levels of sulfur dioxide in the sample have been found to be extraordinarily high.

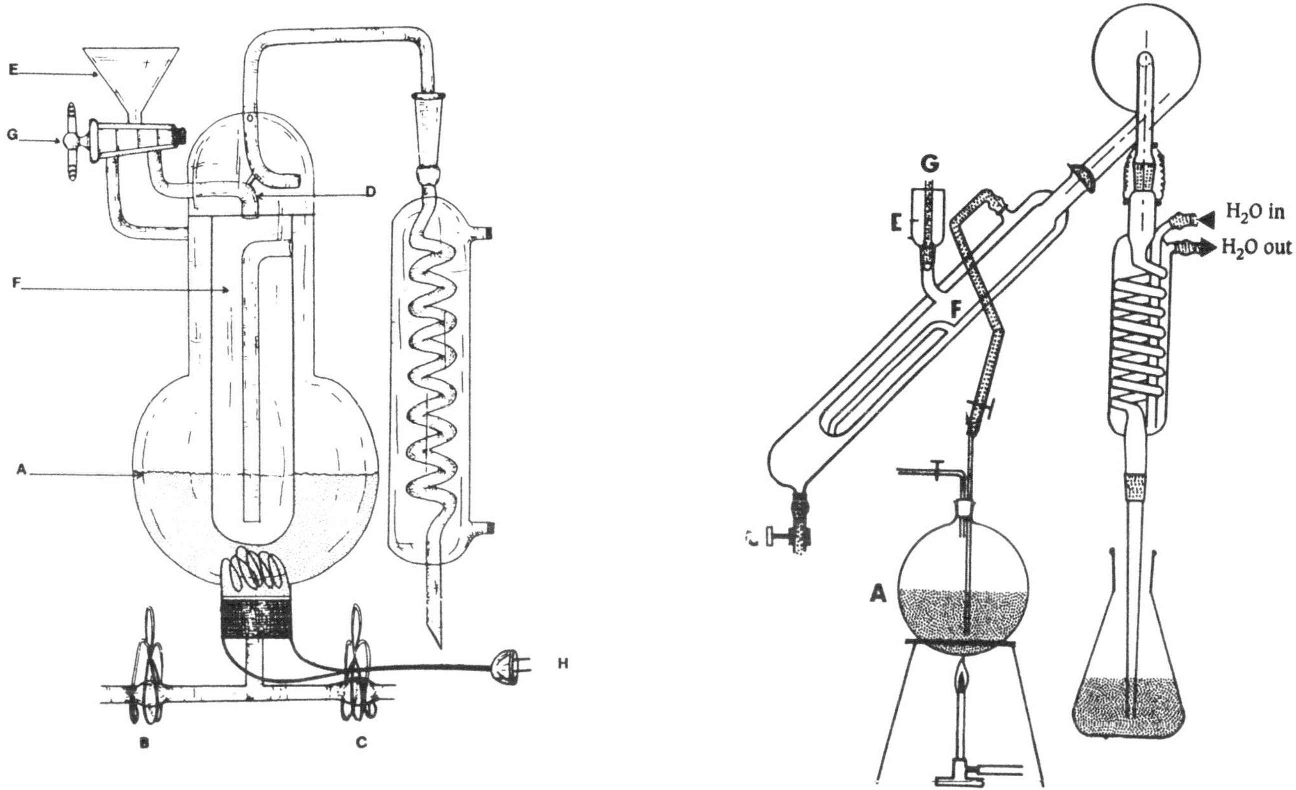

Fig. 20–2. Cash and Markham Volatile Still Assemblies for Volatile Acid Determinations. Lettered Components Are Described in the Procedure.

c. Sorbic acid is almost completely volatilized by steam distillation. Where sorbic acid is known (or thought) to be an additive in a wine, and uncorrected volatile acidities approach legal maxima, correction for its presence may be essential: 1 g sorbic acid = 0.535 g acetic acid

7. Results should be reported as ± 0.05 g/L.

ACTIVE AMYL ALCOHOL: GAS CHROMATOGRAPHIC PROCEDURE (SEE FUSEL OILS)

ALCOHOL: SEE ETHANOL AND FUSEL OILS

ALPHA AMINO NITROGEN (FAN); SPECTROMETRIC DETERMINATION

Six major *alpha*-amino acids are found in musts and wine: arginine, alanine, threonine, serine, glutamic, and aspartic acids. Quantitatively, this group represents the major amino acid constituents of must and wine; so its measurement provides an indication of the potential for problem fermentations as well as the need for nitrogen supplementation. Unfortunately, published analytical results appear to be inconsistent, reflecting problems with the method and interferences associated with it.

After removal of proteins and large peptides from a must sample, 2,4,6-trinitrobenzene sulfonic acid (TNBS) will react with the primary amino groups of amino acids to form trinitrophenylated amino complexes absorbing at 420 nm (Crowell et al. 1985). The reaction initially is carried out in pH 9.5 borate buffer; the pH then is lowered to 7 with phosphate to avoid interference from hydroxyl ion. Sulfite ion concentrations of between 0.5 and 2 mM are optimum for full color development.

I. Equipment
   Spectrometer capable of measurements at 420 nm
   Vortex mixer
   Various pipettes
   Centrifuge

II. Reagents
   Primary Amine Standard (L-arginine monohydrochloride):
   In a 1-L volumetric flask, dissolve 2.11 g of the amino acid in 500 mL of deionized water and bring to volume. Prepare a 200-μM dilution of amine standard by diluting 1 mL of stock to a final volume of 50 mL using deionized water.

Sodium sulfite (0.1 M): In a 100-mL volumetric flask, dissolve 1.26 g sodium sulfite in deionized water and bring to volume.
Sodium Dihydrogen Phosphate (0.1 M): In a 100-mL volumetric flask, dissolve 1.38 g of reagent in deionized water and bring the solution to volume.
Sodium Sulfite-Sodium Dihydrogen Phosphate Solution: Dilute 1.5 mL of 0.1 M sodium sulfite with 98.5 mL sodium dihydrogen phosphate solution. This solution should be prepared daily.
Sodium Borate Solution (0.1 M). In a 100-mL volumetric flask, dissolve 2.01 g of sodium borate in 0.1 *N* NaOH. Bring to volume using 0.1N NaOH.
2,4,6-trinitrobenzene sulfonic acid (TNBS, also known as Picrylsulfonic Acid): In a glass test tube, dissolve 645 mg TNBS in 2.0 mL of deionized water.
Trichloroacetic acid (TCA): Dissolve 60g of TCA in deionized water, and bring to a final volume of 100 mL.

III. Procedure

1. Prepare must samples by centrifugation at 2,000 rpm for 20 min. Filter the supernatant through Whatman No. 1 filter paper and then through a 0.45-μm membrane filter. Add 2 mL of 60% (wt/vol) TCA to 1-mL sample, mix the solution thoroughly, and refilter it through a 0.45-μm membrane filter. Dilute 1.5 mL of this now clarified must sample to 50 mL with deionized water.
2. Dilute 0.25, 0.50, 0.75, and 1.00 mL of the 200-μM stock arginine standard to 1.00 mL with deionized water. These calibration standard solutions are equivalent to 0.7, 1.4, 2.1, and 2.8 mg primary amino nitrogen/L.
3. To 0.5 mL of diluted sample or working standard in a small glass test tube add 0.5 mL borate buffer. Mix the sample, add 20 μL TNBS, and mix it again. Set the test tube aside for exactly 5 min. After 5 min, add 2.0 mL of the sodium sulfate/sodium dihydrogen phosphate solution to stop the reaction.
4. Measure absorbance of the solution within 10 min at 420 nm, against a reagent blank prepared by substituting 0.5 mL water for the sample.
5. Calculate the concentration of free amino nitrogen using the following equation:

$$\text{Free amino nitrogen (mg/L)} = \frac{A_{420\text{ nm}}}{\text{Slope}} \times \text{Dilution Factor}$$

where the dilution factor = 100.

IV. Supplemental Notes

1. The results obtained from this test are expressed as arginine equivalents. Alanine, glycine, serine, and threonine give comparable responses to TNBS.
2. Proline does not react with TNBS, as it lacks a primary amino group. Proline is the major amino acid present in wine grapes. However, it usually is not used by yeast.
3. Although this method has been tested on wines, Crowell et al. (1985) have found that it gives falsely high results and thus is of limited utility.
4. In the authors' experience solid TNBS is required in this procedure, Puchased solutions of TNBS have not proven to be viable.

AMMONIA: (AMMONIUM ION) BY ION SELECTIVE ELECTRODE

The ammonia selective ion electrode has been successfully used for analysis of ammonia (NH_3) and ammonium (NH_4^+) ion concentrations in must and wine (McWilliams and Ough 1974). The electrode is constructed with an interior pH-sensing combination electrode separated from the outside by an ammonia-permeable membrane. When the electrode is placed in a sample, ammonia diffuses across the membrane and into the electrode body. The ammonia then reacts with the filling solution of the electrode, producing a shift in pH that is sensed (and measured) by the internal electrode. In an alkaline sample, ammonium ion is converted to ammonia and measured in similar fashion.

I. Equipment

Expanded-scale pH/mV meter or specific ion meter
Ammonia selective ion electrode
Magnetic stirrer and stir-bar

II. Reagents

Ammonia stock solution (1,000 mg/L): In a 1-L volumetric flask, dissolve 3.141 g of reagent grade ammonium chloride (NH_4Cl) in deionized water. Add 0.1 mL of 1N HCl. Mix and bring to volume with deionized water. The concentration of this solution is 5.9×10^{-2} M expressed as NH_3.

Sodium Hydroxide (10 M): In a 1-L volumetric flask dissolve 400 g of NaOH in approximately 500 mL deionized water. Mix, and bring to volume with deionized water.

III. Procedure

1. Prepare calibration standards of 1.0, 10.0, and 100.0 mg/L from the ammonia stock solution according to Table 20-7. Bring solutions to final volume with deionized water.

Table 20–7. Preparation of ammonia calibration standards from 1,000 mg/L stock solution.

Volume of stock (mL)	Final NH_3 concentration (mg/L) in 1-L volume
1.0	1.0
10.0	10.0
100.0	100.0

2. Transfer 100 mL of the 1.0 mg/L standard to a 150-mL beaker. Place the electrode in the standard, add 1 mL NaOH, and begin stirring. Allow the mV reading to stabilize, and record value.
3. Repeat the process with 10.0 and 100.0 mg/L standards. Between readings, rinse the electrode and blot dry.
4. Construct a calibration curve of mV readings vs. log concentration. This can best be done using semilog graph paper.
5. Treat the wine sample as in step 2. Record the reading, and determine the ammonia concentration by reference to the calibration curve.
6. If the reading is off the scale, measure the ammonia concentration in a diluted sample as follows: Place 20 mL sample in beaker, add 80 mL deionized water, and proceed as instructed.
7. Multiply the answer obtained from the calibration curve by 5 to account for the sample dilution.

IV. Supplemental Notes

1. The calibration curve should be checked periodically by rerunning one or more of the standards. Experience will determine how often one must do this.
2. Most electrode manuals will contain instructions for use with different ion meters, as well as for known addition techniques. These meters and techniques should produce answers equivalent to those obtained using the above procedure.

AMORPHOUS MATERIALS: VISUAL RAPID DIAGNOSIS OF PROTEIN, AND PHENOLS, PECTINS, GLUCANS, STARCH, DIATOMACEOUS EARTH, COPPER AND IRON COMPOUNDS (SEE DIAGNOSTICS)

ANTHOCYANINS: SPECTROMETRIC (VIS) ESTIMATION (SEE PHENOLS: SPECTROMETRIC ANALYSIS OF JUICE/WINE) (SEE PIGMENTS: HPLC DETERMINATION)

ARGININE (FAN): SPECTROMETRIC (VIS) (SEE *ALPHA-AMINO NITROGEN*)

AROMA: SENSORY EVALUATION

Aroma assessment is not a simple, straight forward task. Juice separation from a grape sample for aroma evaluation must be done with care. Rapid enzymatic oxidation can occur if berries are warm, or if juicing has occurred. In addition to phenolic oxides, oxidation products include aldehydes, as a result of enzymatic oxidative cleavage of linoleic and linolenic acids to hexenal (Drawert et al. 1973). These aldehydes can produce grassy aromas that mask fruit characteristics and make aroma assessment difficult. The following procedure is that of Jordan and Croser (1983).

I. Equipment
   Cone-in-cone juicer, or potato ricer, or hand press
   Sieve

II. Reagents
   Pectolytic enzymes
   Ascorbic acid
   Sulfur dioxide (potassium metabisulfite or equivalent)
   N_2 and CO_2 gases

III. Procedures
   1. Chill grapes sample to <2°C (<28°F).
   2. Press the chilled grape sample.
   3. Estimate the quantity of juice that the sample will yield. Mix and then add 10 mg/L ascorbic acid and 30 mg/L sulfur dioxide. Add pectolytic enzyme at the supplier's recommended level.
   4. Lightly sparge the juice with nitrogen, and pour through a sieve into CO_2-filled sample bottles. Seal bottles and cold-settle at <2°C.
   5. Decant the cold-clarified juice into CO_2-filled bottles, and carry out aroma/flavor evaluation and any chemical analyses desired.

IV. Supplemental Notes
   1. Ascorbic acid is an antioxidizing agent that along with sulfur dioxide will help minimize degradation of aroma components.
   2. The sample preparation method (i.e., pressing, degree of pressing vs. crushing) affects the titratable acidity and pH.
   3. Under optimum conditions, juices prepared using this methodology and stored at 0°C will remain viable for aroma/flavor assessment for up to several days.
   4. As fruit matures, the aroma develops from "green" underripe tones to the floral complexes characteristic of the particular cultivar. Nat-

urally, the more neutral the grape variety, the more difficult it is to use aroma as a maturity indicator. Several of the commonly used descriptors noted during maturation of Cabernet Sauvignon, Pinot noir, Sauvignon blanc, and Chardonnay are listed in the following table.

Table 20–8. Commonly used descriptions for maturation in four premium grape varieties.

Cabernet Sauvignon	*Chardonnay*
Green, unripe, spinach	Green, unripe, citrus
Slightly herbaceous	Cucumber
Herbaceous	Melon, ripe figs
Minty, black currant, blackberry	Honey
Pinot Noir	*Sauvignon Blanc*
Green, unripe	Green, unripe, grassy
Lightly herbaceous	Lightly herbaceous
Slightly brambly	Stalky/leafy
Spicy, roselike, violets	Tobacco

AROMA: SENSORY EVALUATION (ALTERNATIVE PROCEDURE)

Most of the aroma components of grapes are located in the skins. This procedure involves separation of the pulp and seeds from the skins and the solubilization of skin components in ethanol.

I. Equipment
   Colander, potato ricer, hand press, or simply a plastic bag
   pH meter
   Gram scale (optional)
   Airtight flasks or jars

II. Reagents
   15% (vol/vol) ethanol
   Tartaric acid

III. Procedure
   1. Approximately 200 berries are lightly crushed and pressed.
   2. The skins are separated from the pulp and seeds and placed in 200 mL of 15% (vol/vol) ethanol, which has been adjusted to pH 3.0 with tartaric acid.
   3. The skin alcohol mixture is kept in an airtight flask or jar for several days to 1 week. The alcohol solution is decanted and the aroma evaluated for character and intensity.

IV. Supplemental Notes
   1. Aroma can be evaluated using the aroma wheel.
   2. For comparative purposes it is suggested that a fixed weight of fruit be evaluated during maturation.

ASBESTOS: MICROSCOPIC PROCEDURE USING METHYLENE BLUE (SEE DIAGNOSTICS)

BACTERIA: ISOLATION AND CULTIVATION OF WINE MICROORGANISMS (INCLUDING YEAST)

Examination of the Wine

Microbial instabilities in wine may be due to bacterial or yeast growth. However, the analyst should not overlook the potential for haze and/or sediment resulting from nonmicrobiological sources. It is suggested that personnel dealing with instability problems become aware of the appearance of typical chemical instabilities as well as confirmatory tests for them (see Diagnostics for rapid identification of hazes and precipitates).

Whether the problem is chemical or biological, a thorough review of processing records for the wine in question is often useful. Also, sensory evaluation of the product may provide some insight as to the nature of problems, with the analyst paying particular attention to the nature of precipitates (e.g., crystalline or refractile vs. amorphous), as well as turbidity and gas formation. Peculiarities in color, nose, and taste may also provide important clues (e.g., geranium tone, hydrogen sulfide, yeastiness).

Once preliminary opinions are formed, the analyst may wish to conduct appropriate laboratory tests. These typically include VA, free and total sulfur dioxide, pH, and possibly malic acid, as well as metals.

Where the instability is present as a light haze, it may be necessary to concentrate the suspension such that it may be examined microscopically. This can be accomplished by centrifugation or membrane filtration. In the latter event, the membrane, or pieces of it, may be examined directly under the microscope. In some cases, allowing the bottle to stand upright overnight will yield sufficient sediment to collect the necessary amount for examination.

Routinely, isolation (and enumeration) techniques employ collection (concentration) of microorganisms on sterile filter membranes. The advantage of this technique is that because of membrane porosity and the free flow of nutrient, cells subsequently can be cultured by placing the membrane directly on the appropriate growth media. In the case of pre-bottling evaluation and samples from the bottling line, potential microbial contamination can be determined by filtering a known volume (750+ mL) using membranes of appropriate pore diameter (usually 0.45 μm). Membranes for this purpose usually have a grid-marked surface to facilitate counting, and they may be purchased with white or dark backgrounds. The latter are particularly convenient for visualization of light-colored colonies.

Collection of Sample and Isolation of Microbes

I. Equipment
   Autoclavable filter holder and funnel of approx. 300 mL capacity
   Vacuum flask and line
   Prepackaged sterile Petri plates and 47-mm presterilized membrane filters (0.45 μm)
   Pads (media-impregnated)
   Forceps
   Bunsen burner

II. Reagents
   Ethanol (70% vol/vol) for sterilizing foceps

III. Procedure
   1. Dip forceps in alcohol, flame, and cool slightly.
   2. Carefully open the filter package and, using sterile technique, place a sterile filter membrane (grid side up) in the sterile filter housing.
   3. Connect the vacuum source.
   4. In the case of bottling-line samples, aseptically open the bottle. It is recommended that the neck be thoroughly swabbed with alcohol before removal of the cork, which should then be removed so as to minimize intrusion of outside air. Once the cork has been removed, flame the neck and transfer contents to the funnel.
   5. Apply vacuum.
   6. Once the wine has been filtered, allow the pressure across the filter to equilibrate and transfer the membrane to the appropriate growth medium (grid side up!). The filter must be placed so that it is in complete contact with the substrate. Voids between filter and growth medium prevent the free flow of nutrient through the membrane.
   7. For wines in which bacteria and/or yeasts are known to be present, preliminary dilution (or reduction in sample volume) may be re-

quired. Because of the potential for uneven distribution of cells on the membrane surface, one should avoid using initial sample volumes of less than 10 mL. A better distribution of colonies can be achieved by first pouring 20–30 mL of sterile water into the funnel and then the measured amount of wine. *Note:* Unless you are transferring less than 1 mL volumes, this technique does not create a dilution in final calculations of cell number.

8. If potassium sorbate or large quantities of other inhibitors have been added to the wine, it may be necessary to follow the wine with a sterile water rinse (20–50 mL).
9. Incubate in an inverted position. Both bacterial and yeast cultures may be incubated at room temperature [25°C (77°F)]. *Saccharomyces* generally develop in 3–4 days. *Brettanomyces/Dekkera* and *Zygosaccharomyces* (ascospore confirmation) take 6–7 days to develop. Lactic acid bacteria take 5–10 days, depending on the organism, to develop. Growth of lactics is somewhat enhanced in a reduced-oxygen atmosphere. Such conditions may be achieved by placing plates in a 1-gal jar (or desiccator). Position a lighted candle on the top plate and seal the container. The candle will burn until the majority of the oxygen is depleted.
10. Once colonies have formed, count and express results as cells/mL, taking into account any preliminary dilution.
11. Microscopically examine the colonies to identify the nature of microorganisms. Examination is facilitated by use of phase microscopy. Although bright-field scopes are adequate, it generally is necessary to use stains to enhance the contrast between the material to be observed and the background. Such stains or dyes include nigrosin, methylene blue, and Ponceau-S. Where bacteria and/or yeast are noted, make preliminary identification based on cell shape. Prepare appropriate media for re-isolation and identification. Depending on the organism (yeast vs. bacteria) and the source of isolate (wine suspension vs. sediment in bottom of barrel) from one to several subsequent transfers may be necessary to obtain a pure culture. Yeasts may be particularly troublesome in this regard because they often secrete a rather extensive capsule when recently isolated from natural sources, making pure culture isolation difficult. When isolation has been achieved, note size and shape of colony, whether it has a shiny or opaque surface, as well as any coloration that might be seen.

Once the organism is isolated and in pure culture, it should be stored on appropriate media pending further identification. In the case of aerobes, agar slants work well. Where the organism is thought to be an acetic, calcium carbonate-supplemented media is useful to neutralize acid as formed. Calcium carbonate-impregnated slants should also be used if *Brettanomyces/Dekkera* is suspected. If LAB are found, one should plan to use agar deeps into which the culture is inoculated by stab technique.

Media and Reagents for Culture and Identification of Yeasts

For analysts not wishing to prepare lab media, prepackaged sterile media are commercially available. These usually consist of sterile Petri plates containing media-impregnated absorbant disks (pads). When ready for use, sterile water is added to rehydrate the medium.

General Yeast Media

For yeast growth, nutrient, grape juice, or wort agar is routinely used. Media may be prepared in liquid form or solidified with inclusion of 2% agar. Wine yeasts differ in their sensitivity to actidione (cycloheximide), and this somewhat diagnostic feature may be evaluated early in the identification procedure by transfering suspect yeast to plates with and without actidione. Use levels of actidione in media vary, but 50 mg/L appears to be the most commonly used. Also, biphenyl (10 mg/L) may be added to culture media to inhibit mold growth.

Selective utilization of media serves as a valuable tool in identification of troublesome microbes that may be found in wines. Among strains of ascomycetous yeasts, this may include growing isolates on selective substrates that promote sporulation. In most cases, speciation requires establishing nitrogen and carbon utilization profiles, which is accomplished by growing the isolate on carbohydrate- and nitrogen-free media supplemented with the appropriate substrates. Media may also be prepared to include dyes or other components that react with metabolites characteristic of certain strains so that color development in and around colonies provides a means of rapid screening. In both yeast and bacterial identification, fermentation broths are routinely used for identification of fermentative utilization of sugar.

The following section includes examples of media (both solid and liquid) of use to the wine microbiologist in maintenance of cultures as well as diagnostic work.

Grape Juice, Wort, Malt, and Yeast Extract-Malt Extract (YM) Agars

Grape juice agar is used as a general, nonselective medium for isolation and growth of wine/juice microorganisms. It is prepared by mixing 20 g of agar in approximately 800 mL of grape juice diluted 1 + 3. Fully dissolve agar by holding in a boiling water bath and bring to 1 L volume with reserve juice. Sterilize by autoclaving.

Wort, malt, and YM agars are commercially available general growth media for yeasts and mold.

Note: For maintenance of acid-producing microbes (i.e., *Brettanomyces*) on malt agar, it is recommended that calcium carbonate (2% wt/vol) be included in the above procedure. Even with the inclusion of carbonate, cultures should be transferred regularly to minimize toxicity due to production of acetic acid.

Zygosaccharomyces-Selective Media

This yeast is unique among wine/juice species in that it grows in the presence of 1% acetic acid. Colonies are then transfered to either YM or wort agar. Formation of club-shaped asci after 6 days is considered presumptive identification.

In approximately 600 mL deionized water, add the following:

100 g glucose
10 g tryptone
10 g yeast extract
25 g agar

Bring to 1 L with deionized water. Once in solution, sterilize by autoclaving.

When autoclave cycle is finished, cool media. Add 10 mL glacial acetic acid with thorough mixing. Pour plates.

Brettanomyces/Dekkera-Selective Media

In approximately 600 mL deionized water, add the following:

2.4 mL Bactoglycerol (or equivalent)
48.3 g wort agar
50.0 mg actidione (cycloheximide)

Actidione may be incorporated directly into water used in making the agar. Expected reduction in "activity" after autoclaving is approximately 50%. At the recommended addition level, this should not present a problem in that *Saccharomyces* are generally inhibited by <20 mg/L. Alternatively, it may be added to autoclaved media by preparing a sterile stock solution (500 mg/100 mL distilled water). From stock, 1 mL/100 mL of agar is equivalent to 50 mg/L. *Note:* actidione must be thoroughly distributed throughout the agar and plates poured before solidification.

WL and WL-Differential Media

Both media contain complete growth supplements for yeast and bacteria. In addition, both include the pH indicator Bromcresol Green, which allows detection of those isolates producing acid. WL-differential includes actidione to inhibit yeasts (*Saccharomyces*)/mold while allowing *Brettanomyces* and bacteria to grow. Both are commercially available and are typically used in tandem.

Yeast Sporulation Media

Yeast identification relies on demonstration of the presence or absence of a sexual cycle, that is, the presence or absence of ascospores. Such demonstration is often difficult. The fact that ascospores are not observed does not necessarily mean that the yeast is an "imperfect" form. It may mean that it does not sporulate on the particular media used or that some other requisite conditions were not met. Further, the isolated strain may be heterothallic, that is, conjugation with a compatible mating type must precede formation of ascospores.

A variety of media have been used to demonstrate sporulation. Details regarding makeup of these can be found in Lodder (1970).

Media for Observation of Formation of Pseudomycelia

Demonstration of mycelial or pseudomycelial formation in yeasts is often of diagnostic importance. The usual media recommended are either corn meal or potato dextrose agars. Both agars are available commercially.

Identification of Microbial Isolates

This section is designed to tentatively identify wine spoilage bacteria. The reader is referred to appropriate references (Lodder 1970, and a current edition of Bergey's *Manual of Determinative Microbiology*) for more definitive testing, if necessary.

I. Lactic Acid Bacteria (Table 20–9)

1. Morphology. In the case of wine lactics, typical cell shapes are bacilloid (rods) characteristic of the genus *Lactobacillus* and coccoid, ranging to lenticular or cocco-bacilloid. The homofermentative genus *Pediococcus* is typically coccoid whereas its heterofermentative counterpart *Leuconostoc* tends toward cocco-bacilloid.
2. Generally, the first test of taxonomic value in the identification of bacteria is the Gram stain reaction. In the case of lactics, the Gram stain is positive (blue), whereas acetics are Gram negative (pink). Staining should always be done on young cultures of approximately the same age.

Table 20–9. Physiological characteristics of wine bacteria.

Organism	Gram reaction	Catalase	Oxygen reqs.	Maj. Endproduct	Sporulation
Gluconobacter	neg	+	Aer	acetic	neg
Acetobacter	neg	±	Aer	acetic	neg
Lactobacillus	pos	−	Aer/Ana	lactic	neg
Leuconostoc	pos	−	Fac/Ana	lactic	neg
Pediococcus	pos	−	Aer/Ana	lactic	neg
Bacillus	pos	+	Aer	several	pos

3. At the same time, the analyst may wish to screen for oxygen requirements. This is most easily done by the catalase test. A drop of 3% (vol/vol) hydrogen peroxide is placed on a microscope slide and a loopful of suspect colony added. Bubbling (foam formation) is positive for catalase activity characteristic of aerobic organisms such as *Acetobacter* and *Gluconobacter*. No foaming suggests LAB. For comparison and as a positive control, a loopful of *Saccharomyces* may be used.

$$2H_2O_2 \xrightarrow{\text{Catalase}} O_2 + 2H_2O$$

4a. Initial separation of *Pediococcus* from *Leuconostoc* can be accomplished by testing for CO_2 resulting from glucose fermentation. The medium used is Rogosa except that apple juice is replaced by tomato juice at the same dilution. Because this is a test for CO_2 formation, expected decarboxylation of malic acid present in apple juice would serve to only create positive results.

Inoculate 0.1 mL of actively growing culture into approximately 5 mL of sterile medium in culture tubes. Chill and overlay with liquid (hot) sterile vaspar. For best results a tight seal must be effected. Incubate at 20°C for 1 week and examine for gas formation.

Production of CO_2 from glucose in the absence of malic acid is a general test for heterofermentative species (i.e., *Leuconostoc* sp. and some lactobacilli), whereas the absence of gas production is indicative of homofermenters (*Pediococcus* sp).

4b. Another diagnostic test separating wine hetero- from homolactic species is formation of mannitol crystals from fructose. Heterofermenters such as *Lactobacillus brevis* use fructose-6-phosphate in production of mannitol. In the process, the coenzyme NADH is reoxidized.

Prepare apple Rogosa broth as described below. Include in the preparation 2% fructose in place of glucose.

Inoculate 5 mL of sterile broth with 0.1 mL of active bacterial culture.
Incubate at 20°C for 1 week.
Transfer to evaporation dish and hold until liquid phase has evaporated.
The presence of mannitol crystals is indicative of heterofermentative LAB.

Media, Reagents, and Supplies for Identification of Lactic Acid Bacteria

The lactic acid bacteria are fastidious, microaerophilic facultatively anaerobic organisms, which may be difficult to culture on synthetic media. Thus, most media include a combination of vegetable juice and other digests.

For best results, lactic bacteria should be cultured under conditions of low oxygen tension, such as in a candle jar or in specially designed incubators. Preparation of pour-plates is also useful in this regard.

Although commercial media for growth of LAB (i.e., Difco tomato juice agar) are available, the authors have found the best growth to occur using the medium presented below:

Modified Tanner-Vetsch Medium

Dilute 355 mL of tomato juice to 1,500 mL with deionized water and filter.

To 1 L of the above solution, add:

20 g agar
20 g tryptone
5 g peptone
5 g yeast extract
3 g glucose
2 g lactose
1 g liver extract
1 mL of 5% aqueous Tween 80

Heat in a boiling water bath until dissolution is complete. Adjust acidity of the final *cooled* media to pH 5.5. Sterilize by autoclaving.

For cultivation of *Lactobacillus trichoides,* the above media is recommended *without the inclusion of agar.* After the broth has cooled, 10% (vol/vol) NSFG is added.

Apple Rogosa Medium

Dilute 200 mL of apple juice (use either frozen or juice without preservatives) to 1 L using deionized water. Transfer 700–800 mL of diluted juice to a separate 1-L flask.

Dissolve the following into this volume:

20 g tryptone
5 g peptone
5 g glucose
5 g yeast extract

Bring to volume with diluted apple juice. Adjust pH to 5.5. Sterilize by autoclaving.

Note: for demonstration of CO_2 production and separation of homo- vs. heterofermentative LAB, tomato juice (at the same dilution)is used in place of apple juice.

Vaspar Plugs

Vaspar is prepared by melting 1 part paraffin with 6 parts vaseline (wt/wt) and autoclaving. Before use, tubes containing Vaspar should be held in a water bath at a temperature sufficient to keep the product liquified.

Media for Isolation of Acetic Acid Bacteria

Acid Plates

It may be difficult to cultivate some acetic acid bacteria on routine laboratory media where pH levels exceed 5.0. To overcome this problem, acid plates are often employed. Media preparation is as follows:

In an appropriate volume of water, dissolve the following:

5 g yeast extract
20 g dextrose (glucose)
5 g KH_2PO_4
20 g agar

Hold in boiling water bath until components are fully dissolved. Bring to a final volume of 1 L with water. Sterilize by autoclaving. Once media has cooled and before solidification, carefully adjust acidity of media to a final pH of 4.0.

Calcium Carbonate-Ethanol Medium

In wine samples with probable contamination due to acetics, isolations should be made on calcium carbonate-supplemented media. In alcoholic media, some acetic acid bacteria produce acetic acid that neutralizes $CaCO_3$ present, resulting in a zone of clearing around the colonies.

In an appropriate volume of water, dissolve the following:

20 g $CaCO_3$
5 g yeast extract
20 g agar

Bring contents into solution by holding in a boiling water bath. Autoclave according to defined procedure. After sterile media has cooled (not solidified), add 3% (vol/vol) NSFG. *NOTE!* $CaCO_3$ should be uniformly suspended throughout prior to pouring plates.

Modified Carr Agar

Carr medium includes a pH indicator to facilitate identification of acid-producing bacteria.

In approximately 600 mL of deionized water, add the following:

20 g yeast extract
0.02 g bromocresol green
Adjust pH to 5.5.
20 g agar

Sterilize by autoclaving. Allow agar to cool. While warm (and before solidification), add NSFG to a final concentration of 2% (vol/vol). Mix thoroughly and pour plates.

BENZOIC ACID: HPLC DETERMINATION (INCLUDES SORBIC ACID)

Benzoic acid, as well as sorbic acid, can be determined using HPLC. A commercial organic acid column with acetonitrile-modified sulfuric acid mobile phase will resolve these two acids. Detection is best accomplished at 233 nm.

I. Equipment

HPLC unit with variable-wavelength detector
Organic acid HPLC column and appropriate guard column
Millipore filters (0.45 μm) or equivalent
2 mL pipette
50 mL volumetric flasks

II. Reagents

Benzoic acid stock (100 mg/L): Weigh 118 mg of reagent grade sodium benzoate and bring to 1 L in a volumetric flask with mobile phase.

Sorbic acid stock (100 mg/L): Weigh 134 mg potassium sorbate and bring to 1 L in a volumetric flask with mobile phase.

Mobile phase (85% 0.01 $N\,H_2SO_4$: 15% acetonitrile): Prepare approximately 0.01 N H_2SO_4 by adding 2.8 mL concentrated acid to 10 L chromatography grade deionized water. To the 10 L of dilute acid add

Table 20–10. Preparation of benzoic and sorbic acid calibration standards from 100 mg/L stock solutions.

Volume of benzoic acid or sorbic acid stock (mL)	Final concentration (mg/L) in 100 mL volume[a]
2.0	2.0
5.0	5.0
10.0	10.0
20.0	20.0

[a]Dilute to volume with prepared mobile phase.

1.76 L chromatography grade acetonitrile and mix. Store in large sealed containers.

III. Procedure

1. Prepare calibration standards of benzoic and sorbic acids at concentrations of 2, 5, 10, and 20 mg/L from benzoic and sorbic acid stock standards (see Table 20–10). Use mobile phase as the solvent for these standards.
2. Transfer 2.0 mL sample to 50-mL volumetric flask. Dilute to mark with mobile phase.
3. Set up chromatograph with 1.0 mL/min flow rate of freshly sparged (air-free) mobile phase. Bring column temperature to 65°C and set detector to 233 nm. Allow unit to stablize.
4. Before injection of standards and samples, filter through 0.45-μm membrane filter disks.
5. Inject 10 μL of each standard into chromatograph. Note retention times and response factors (Figure 20–3). If desired, construct calibration curve.
6. Inject 10 μL each diluted sample. Record values provided by data processing unit, or by reference to prepared calibration curve. If sample peak response is greater than that of the most concentrated standard, dilute sample by appropriate amount and chromatograph again.
7. Calculate concentrations of original samples by multiplying chromatographic result by appropriate reciprocal dilution factor (e.g., 50 mL/2 mL).

IV. Supplemental Notes

1. An alternative procedure for the HPLC analysis of benzoic acid can be found in *Official Methods* (AOAC) section 12.018–12.021 (Fifteenth Edition, 1990). This procedure uses a C_{18} reverse phase column and a fixed wavelength (254 nm) UV detector.

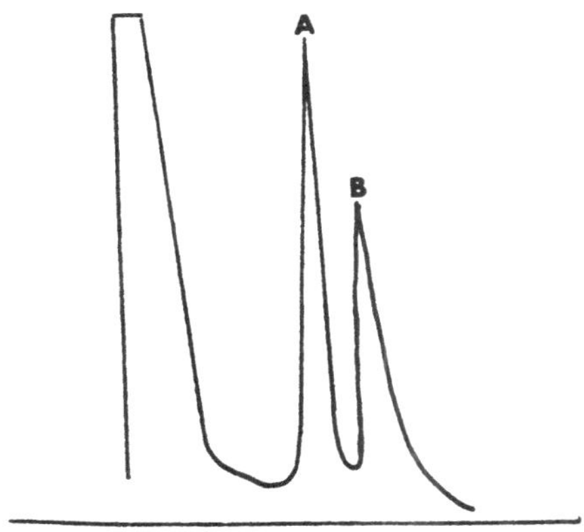

Fig. 20–3. HPLC analysis of sorbic and benzoic acids using "fast acid analysis" column (Bio-Rad) and a mobile phase consisting of 85% (0.01 *N*) H_2SO_4:15% acetonitrile. The column is operated at 65°C with a mobile phase flow rate of 1.0 mL/min. Detection is by UV absorbance at 233 nm.

BENZOIC ACID: SPECTROMETRIC (UV) DETERMINATION

Benzoic acid can be determined by measuring its absorbance at 272 nm. The acid is first extracted into ether to separate it from other absorbing substances in the sample. Then the absorbance spectrum of the acid is scanned between 265 and 280 nm and a baseline constructed from the two absorbance minimums. Net absorbance is measured from this baseline to the peak of the absorbance maximum (at approximately 272 nm) and related to concentration of benzoic acid. As presented, the method is applicable to samples containing from 200 to 1,000 mg/L benzoic acid.

I. Equipment
   UV spectrometer
   Stoppered silica cuvets
   500-mL separatory funnel
   10-mL pipette

II. Reagents
   Benzoic acid stock solution (1,000 mg/L): Weigh 1.00 g reagent grade benzoic acid, dissolve in diethyl ether, and bring to volume in a 1-L volumetric flask. Keep tightly closed during storage.
   Saturated NaCl solution: Add enough sodium chloride to 1-L deionized water in a beaker to saturate. Decant saturated solution and store in a bottle.

Table 20–11. Preparation of benzoic acid calibration standards from 1,000 mg/L stock solution[a]

Volume of benzoic acid stock (mL)	Final concentration (mg/L) in 100 mL volume Acid
2	20
5	50
7.5	75
10	100

[a]Dilute to volume with diethyl ether.

HCl (1 + 999): Dilute 1 mL concentrated HCl to 1 L with deionized water.

Diethyl ether; used without further preparation.

III. Procedure

1. Prepare calibration standards of benzoic acid in ether from the 1,000 mg/L stock solution as per Table 20–11. Keep standards tightly stoppered to prevent evaporation.
2. Scan the absorbance spectrum of the 50 mg/L standard over the range of 265–280 nm. Note wavelength of maximum absorbance (around 272 nm) and wavelengths of adjacent absorbance minimums (around 267.5 and 276.5 nm).
3. Measure absorbance values for remaining standards at maximum and minimum wavelengths determined in step 2.
4. For each standard, determine the average *minimum* absorbance value. Subtract this calculated average from the maximum absorbance value and plot this corrected absorbance against concentration.
5. Transfer 20 mL wine sample to separatory funnel, dilute to 200 mL with saturated NaCl solution, and acidify using litmus paper, as an indicator, with HCl.
6. Extract acidified sample sequentially with 70, 50, 40, and 30 mL portions ether, mixing well to ensure complete extraction. Drain and discard aqueous phase.
7. Wash combined ether extracts sequentially with 50-, 40-, and 30-mL HCl and discard aqueous washings.
8. Dilute combined ether extracts to 200 mL with additional ether. Determine the absorbance of the sample in a stoppered cuvette at the above determined minimum and maximum wavelengths.
9. Average two minimum absorbance values as before and subtract from maximum absorbance value. Determine concentration of benzoic acid from standard curve.
10. Multiply concentration value by 200 mL/20 mL (reciprocal dilu-

tion factor) to obtain benzoic acid concentration in original sample. Multiply this value by 1.18 (MW sodium benzoate/MW benzoic acid) to obtain equivalent concentration of sodium benzoate.

IV. Supplemental Notes

1. In some instances an additional purification step may be necessary. The washed ether extracts from step 7 may require further extraction sequentially with 50, 40, 30, and 20 mL of 0.1% NH_4OH, and the ether phase discarded. Neutralize the combined NH_4OH extracts with HCl and add 1 mL excess. Extract this acidified solution sequentially with 70, 50, 40, and 30 mL ether. Combine ether extracts and proceed with procedural step 8.

BITARTRATE STABILITY: CONCENTRATION PRODUCT CALCULATION USING RESULTS FROM ANALYSES OF ETHANOL, POTASSIUM, AND TARTARIC ACID SEE CHAPTER 15 FOR AN EXAMPLE OF CONCENTRATION PRODUCT CALCULATION

BITARTRATE STABILITY: THE CONDUCTIVITY TEST

This determination involves saturation of a juice or wine sample with potassium bitartrate powder (seed crystals) at a defined and controlled temperature. Because K^+ is the major conducting species in wine, bitartrate precipitation can be measured by changes in conductance from the beginning to the end of the test. Changes in conductance of <5% are considered to be indicative of a stable product, although some producers use smaller changes. Results are only valid at, or above, the temperature at which the test was run; the wine may be unstable at lower temperature (see Chapter 15). In the United States most stabilize whites at 0°C and red wines at +5°C.

I. Equipment

Conductivity meter: Unit should be capable of measuring in the range of 100–1,000 micro-Siemens. A Siemens unit is 0.94073 International Ohms. Meters are also available that combine conductivity, pH, and mV capabilities.

Controlled temperature water bath ± 1.0°C.

Magnetic stirrer

Thermometer

II. Reagents

Potassium bitartrate powder (40–70 μm). See Supplemental Note 1 for further details.

III. Procedure

1. Using a 100-mL volume, equilibrate wine to the desired stability temperature before proceeding. Measure conductivity of original sample.
2. Add 1.0 g of potassium bitartrate powder. Place conductivity probe in sample.
3. While constantly mixing the sample, take readings (conductivity and temperature) at 5-min intervals. Continue mixing at defined temperature until conductance readings stabilize—usually 20–30 min—or until several successive readings show no change in conductance.

IV. Interpretation

The final conductivity value corresponds to that of the stable product. It can be used for a comparison with conductivity values of samples taken from the winery during full-scale stabilization treatment to determine if and when stability has been reached. The difference between the conductivity value before addition of powdered potassium bitartrate and the final value is then a measure of potential KHT instability.

1. Conductance readings >5% between beginning (unseeded sample) and end of test (stable conductivity reading) suggest the wine is unstable with respect to bitartrate precipitation.
2. Conductance readings <5% are taken to mean that the sample is stable (some producers use lower values).

V. Supplemental Notes

1. Bitartrate seed used in laboratory evaluations should be the same as will be used in the cellar operations. Optimally, crystal size should be 40–70 μm.
2. Conductivity values are affected by temperature. The sample temperature must be constant throughout the test. As the temperature rises, conductivity readings will increase due to temperature effects.

BITARTRATE STABILITY: THE FREEZE TEST FOR POTASSIUM BITARTRATE STABILITY AND CHILL HAZE

The freeze or slush test involves freezing or chilling a wine sample at a defined low temperature for a prescribed period of time. Upon thawing, the sample is examined for the presence of crystals. If present, the wine is judged to be "unstable."

Although it has been common practice to filter the sample to remove noncrystalline amorphous material, this procedure has been called into question. The freeze test using a filtered sample is essentially a precipitation rate test (see Chapter 15).

In addition to conventional freeze tests, many wineries routinely evaluate stability at refrigeration temperatures. In this case, however, the instability, seen as a haze, is usually due to protein-tannate and polysaccharide complexes that become insoluble at lower temperatures.

I. Equipment

Refrigerator or freezer for constant temperature storage.

II. Procedure

1. Place wine or juice sample into 4-oz vial.
2. Place sample in freezer for predefined time period determined by experience. Retain a similarly treated "control" at room temperature.
3. Compare sample upon thawing with control for the presence of precipitated crystals. Note the crystal volume, making sure to distinguish between crystals and precipitated amorphous material (see Supplemental Note 4). The sample is reevaluated at room temperature to note the extent of crystal resolubilization.
4. If all the crystalline precipitates formed redissolve upon warming the sample to room temperature, some winemakers consider the original wine to be stable. However, subsequent exposure to low temperatures may bring about recrystallization.

IV. Supplemental Notes

1. If membrane filtration is used to clarify the sample, nuclei needed for crystallization are removed. In this instance the procedure is not a measure of bitartrate stability, but rather the capacity of the sample to form nuclei and their subsequent rate of precipitation.
2. The above procedure is subjective. Various wineries use different times for the test, different temperatures, and different interpretations of stability based on the amount of crystal formation.
3. In addition to bitartrate crystals, other solid materials may fall out of solution during the freeze test. These materials consist of proteins, phenols, pectins, and other colloidals and may be indicators of wine instability.
4. In order to distinguish between bitartrate crystals and other solid materials, one should use a high-intensity light source held at right angles to the line of vision. Check for refraction of light from the bitartrate crystals. (See Diagnostics for further identification tests.)

BOTRYTIS (MOLD): SPECTROMETRIC ESTIMATION VIA LACCASE ASSAY

Botrytis cinerea produces the oxidative enzyme laccase, which is partially responsible for oxidative deterioration of grapes. Dubourdieu et al. (1984)

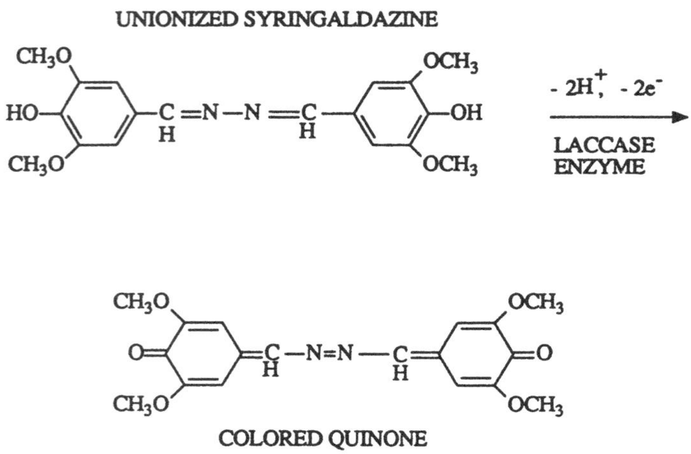

Fig. 20–4. Reaction of syringaldazine with laccase.

developed an assay for this enzyme that can be used to help establish fruit quality and as a means for grower compensation. Lacasse values in grapes range from 0.1 to 7.8 nanomoles/mL (nmol/mL). Values above 4.6 may suggest oxidative problems with the must.

The procedure involves the reaction of syringaldazine in the presence of laccase to produce a colored quinone. Lacasse activity is measured at 20°C in a 0.1 M sodium acetate buffer solution at pH 5. The oxidized quinone formed has an absorption maximum at 530 nm (Fig. 20–4).

I. Equipment

Visible range spectrophotometer

1.2-μm membrane filter or a syringe fitted with glass wool plug

II. Reagents

0.1% syringaldazine: Weigh 0.1 g syringaldazine and add to 100 mL absolute ethanol. Place container in sonic bath to completely dissolve solid. Store in capped bottle.

Sodium acetate (0.1 M): Dissolve 0.82 g reagent grade sodium acetate in deionized water. Bring to volume in a 100 mL volumetric flask.

Activated PVPP: Activate commercial PVPP at 100°C in 6 M HCl for 1 hour. Follow by thorough rinsing with deionized water until neutral.

III. Procedure

1. Transfer 5 mL of juice to an Erlenmeyer flask.
2. Add 0.8 g of *activated* PVPP.
3. After 10 min of contact, filter the sample through a 1.2-μm membrane filter or a syringe packed with glass wool. (For heavily colored wines, it may be necessary to repeat this treatment.)

4. In a spectrophotometer cell, mix the following:
   a. 0.6 mL syringaldazine-ethanol solution
   b. 1.0 mL juice sample (diluted if necessary)
   c. 1.4 mL buffer
5. Read the change in absorbance at 530 nm over several minutes. The response should be linear for at least 10 min.
6. Determine ΔAm, corresponding to the change in absorbance per minute.
7. Calculation:
   a. One laccase unit (U) is equal to the quantity of enzyme catalyzing the oxidation of 1 nmol of syringaldazine per min.
   b. ΔAm is the recorded change in absorbance per min at 530 nm.
   c. Laccase activity in laccase units (U) per mL (M/mL= nmol/mL) is defined as:

$$U(\text{nmol/mL}) = \frac{\Delta \text{Am} \times 300 \text{ nmol}}{6.5 \text{ mL}}$$

Rearranging the above equation:

$$U = \frac{\Delta \text{Am} \times 3 \text{ mmol}}{65{,}000 \text{ mL}}$$

IV. Supplemental Notes

1. Polyphenols interfere with the final reading of laccase activity. Best results are obtained by eliminating or reducing the levels of polyphenols present in juice by prior treatment with *activated* PVPP.
2. The above analysis has been developed in "kit" form for field assay of laccase.
3. It has been reported that laccase activity can also be measured in a closed system using an oxygen probe (J. Pineau, personal communication).

BRETTANOMYCES: (SEE BACTERIA-YEAST)

BRETTANOMYCES: GAS CHROMATOGRAPHIC ANALYSIS OF 4-ETHYLPHENOL

Brettanomyces in wine has been shown to produce 4-ethylphenol. The 4-ethylphenol residues can be determined using conventional extraction and capillary gas chromatographic techniques. The following procedure is adapted from one for volatile phenols in wines by Chatonnet and Boidron (1988).

I. Equipment
   Gas chromatograph equipped with a flame ionization detector
   Capillary column: 50 meters × 0.33 mm i.d., coated with Carbowax 20M (polyethylene glycol), film thickness 0.5 μm
   Hamilton syringe, 10 μL
   Centrifuge
   Volumetric flasks, 1-L and 100-mL

II. Reagents
   Stock 4-ethylphenol (4EP) solution (1,000 mg/L): Dissolve 100 mg 4-ethylphenol in matrix solution in a 100-mL volumetric flask. Mix to dissolve and bring to volume with matrix solution.
   Working standard I of 4-ethylphenol (10 mg/L): Transfer 1.0 mL stock 4EP to a 100-mL volumetric flask. Bring to volume with matrix solution.
   Working standard II of 4-ethylphenol (1.0 mg/L): Transfer 10.0 mL Working standard I 4EP to a 100-mL volumetric flask. Bring to volume with matrix solution.
   Internal standard, dimethyl-3,4-phenol (100 mg/L): Dissolve 25 mg dimethyl-3,4-phenol and 7.5 g ammonium sulfate in absolute ethanol in a 250-mL volumetric flask. Mix to dissolve and bring to volume with absolute ethanol.
   Matrix solution: Dissolve 10 g glycerol, 5 g tartaric acid, and 2 g glucose in 120 mL absolute ethanol and deionized water. Adjust pH to 3.50 using 1 N NaOH. Bring to 1 liter with deionized water. Store in a plastic bottle.
   Sodium bicarbonate 5% (wt/vol): Dissolve 50 g reagent grade sodium bicarbonate in 1 liter of deionized water. Mix to dissolve and store in a plastic bottle.
   Sodium hydroxide 0.5% (wt/vol): Dissolve 5 g reagent grade sodium hydroxide pellets in 1 liter deionized water. Mix to dissolve and store in a plastic bottle.
   Dichloromethane
   Absolute ethanol

III. Procedures
   i) Sample preparation
   1. Prepare calibration standards of 4-ethylphenol of 100 and 10 μg/L (0.1 and 0.01 mg/L, respectively) in the matrix solution from the Working standard II:

Table 20–12. Preparation of 4-ethylphenol calibration standards from 1.0 mg/L (1,000 μg/L) working standard II.

Volume 4EP Working Standard II (mL)	Final concentration (μg/L) in 100 mL Volume*
1.0	10
0.10	1.0

*Dilute to volume with matrix solution

2. Clarify a 100 mL wine sample by centrifugation. Transfer to a 200-mL flask and add 1 mL of the internal standard solution.
3. Extract sample three times with 20, 10, and 5 mL, respectively, dichloromethane. Use a magnetic stirrer for 5 minutes at ~ 600 rpm for mixing the phases.
4. Decant the respective organic phases and collect in a 100-mL flask. If emulsions are a problem, the extracts can be centrifuged at 4°C to assist in separation of the phases.
5. Wash the dichloromethane extracts two times with 50 mL of a 5% sodium bicarbonate solution to remove carboxylic acids present. Stir 5 minutes at ~ 600 rpm.
6. Isolate the volatile phenol fraction by extracting the combined dichloromethane extracts two times at pH 13 with 50 and 25 mL, respectively of 0.5% sodium hydroxide. Stir 5 minutes at ~ 600 rpm. The phenolic fraction will transfer into the aqueous, alkaline phase. Discard the dichloromethane phase.
7. Acidify the aqueous extracts to pH 1 using 1+1 HCl and extract with 15, 5, and 5 mL of diethyl ether (stabilized with 2% ethanol).
8. Transfer the organic phase to a graduated, conical bottom centrifuge tube. Direct a stream of nitrogen gas into the tube and concentrate to a final volume of 1 mL. Rinse the walls of the tube several times with diethyl ether-hexane (1:1) during the evaporation process. Remove the water layer that forms with a pipette.
9. Store the concentrated extract at 4°C until analysis.
10. Extract aliquots of calibration standards using the same procedure as described for the wine samples.

    ii) Chromatography
11. Following manufacturer's guidelines set up the following instrumental conditions: Splitless injector @ 230°C and a split time of 15 sec; flame ionization detector @ 260°C with 300 mL/min purified air and 20 mL/min hydrogen; carrier (helium) pressure set to maintain column flow rate @ 1.7 mL/min, a septum purge of ~ 0.5

mL/min, and vent flow of ~ 130 mL/min. The oven temperature is programmed from 45 to 230°C at 3°C/min, with a final hold time of 25 min.

12. Inject 1 µL of sample or Calibration standard extract into the gas chromatograph.
13. Determine peak area ratios for 4EC/Internal standard. Using ratios from injected Calibration standards, prepare a standard curve of 4EC/Internal standard ratios vs 4EC concentrations (in µg/L). Compare the peak area ratios for wine sample extracts to the standard curve to determine their 4EC concentrations.

°BRIX: (SEE SOLUBLE SOLIDS BY HYDROMETRY AND REFRACTIVE INDEX METHODS)

CALCIUM: ATOMIC ABSORPTION SPECTROMETRIC ANALYSIS

Analyses of calcium in juice and wine often are performed using spectroscopic techniques. Both flame absorption and plasma emission techniques can be used.

I. Equipment
   Atomic absorption spectrometer
   Calcium hollow cathode lamp

II. Reagents
   Stock calcium solution (1,000 mg/L): Dissolve 2.500 g of predried $CaCO_3$ in approximately 100 mL of deionized distilled water. Add 5–7 mL stock HCl and warm on a hotplate to drive off CO_2. *Quantitatively* transfer contents to a 1-L volumetric flask and bring to volume with deionized water at defined temperature.
   Note: the deionized water used in makeup of working and calibration standards should be boiled (and cooled to defined temperature) before use to eliminate entrapped CO_2. Commercially available calcium standards may be conveniently substituted for the above reagent preparation.
   Lanthanum Oxide/HCl: Dissolve 10 g La_2O_3 in 10 mL of concentrated HCl. When dissolution is complete, dilute to 100 mL volume with deionized water.

III. Procedure
   1. Prepare a 100 mg/L working calcium standard by diluting 10 mL of calcium stock solution to 100 mL in a volumetric flask.

Table 20–13. Preparation of calibration standards from 100 mg/L working calcium standard.

Volume of working standard (mL)	Final Ca^{2+} concentration (mg/L) in 100-mL volume[a]
1	1.0
2	2.0
5	5.0
10	10.0

[a]Dilute to volume in a solution of 1% La_2O_3/HCl.

2. Prepare 1, 2, 5, and 10 mg/L calibration standards in 100-mL volumetric flasks (see Table 20–13).
3. Install the calcium lamp. Set appropriate wavelength, slit width, and power supply current as listed in instrument operating instructions.
4. Light and adjust burner flame, optimize analytical conditions, and prepare calibration curve (as per instrument operating instructions).
5. Dilute 1 mL wine to 10 mL with the lanthanum oxide/HCl solution. Aspirate into the flame and read absorbance value (or concentration). If necessary, prepare second sample at more appropriate dilution level.
6. Read the calcium concentration from the calibration curve and multiply by 10 (or other dilution factor, if used) to obtain calcium concentration in original wine sample.

IV. Supplemental Notes

1. Phosphorus in wine will react with calcium in the air-acetylene flame resulting in a depression of the measured calcium absorbance. This interference is overcome by adding lanthanum at a level of about 10,000 mg/L to all standards and samples. Phosphorus preferentially reacts with lanthanum so that the former cannot bind with calcium in the flame.
2. One can also avoid the phosphorus interference by using the hotter nitrous oxide-acetylene flame. In this instance, lanthanum is not needed as an additive. At higher flame temperatures, an ionization interference may occur due to a certain amount of ionization of sodium and potassium atoms in the nitrous oxide-acetylene flame. To eliminate the effects of this interference, all standards and samples should be prepared to contain 2,000–5,000 mg/L potassium.
3. Analyses conducted at the low mg/L concentration level require careful attention to cleanliness and careful pipetting and diluting of standards and samples.

CALCIUM: ANALYSIS OF IONIZED CALCIUM IN WHITE WINE BY ION SELECTIVE ELECTRODE

An alternative technique for calcium in white wines has been developed by Cardwell et al. (1991). This method involves the use of potentiometry and a calcium ion selective electrode that responds to ionized calcium ion rather than to the total calcium concentration of a white wine. The method was developed to see if ionized calcium levels correlate better with the potential of a white wine to throw a calcium tartrate precipitate.

I. Equipment

Expanded scale pH/mV meter or specific ion meter
Calcium selective ion electrode
Double-junction reference electrode with the outer junction filled with 10% NH_4NO_3
Beakers 100-mL
Pipettes 10 and 1 mL
Magnetic stirrer and stir-bar

II. Reagents

Calcium stock solution (40,000 mg/L): Dissolve 110.98 g oven-dried (3–4 hours at 105°C) calcium chloride in deionized water. Make up to 1 liter in a volumetric flask.

Calcium working standards I (4,000 mg/L) and II (400 mg/L): Transfer 10.0 mL and 1.0 mL, respectively, of calcium stock to two 100-mL volumetric flasks. Dilute to volume with deionized water and label as calcium working standards.

Ethanol (aqueous): Prepared from absolute (or 95%) ethanol and deionized water. The concentration prepared should match that of the wine sample being analyzed.

Potassium chloride (3 M): Dissolve 22.3 g potassium chloride in deionized water. Bring to 100 mL volume.

Sulfuric Acid (0.1 M): See Table 20–3 for preparation.

III. Procedure

1. Prepare separate calcium calibration standards in range of 0.4 to 4,000 mg/L (10^{-5} to 10^{-1} M) by transferring appropriate volumes of calcium Working Standards I and II to respective 100-mL volumetric flasks (see Table 20-14). Bring to approximately 90 mL with aqueous ethanol (ethanol concentration matching that of the wine samples). Adjust pH with 0.1 M H_2SO_4 to that of the wine samples and adjust the ionic strength with 3 M potassium chloride (to the potassium level determined by atomic absorption spectroscopy). Mix and bring to volume with aqueous ethanol. *Note:* the calibration standards are prepared fresh for each wine sample at the time of analysis.

Table 20-14. Preparation of calcium working standards I and II and calcium calibration standards

Volume of 40,000 mg/L calcium Stock	Volume of 4,000 mg/L calcium Working Standard I	Volume of 400 mg/L calcium Working Standard II	*Concentration of Calibration Standards in 100 mL volume
10.0			4,000
1.0			400
	1.0		40
		10.0	4.0
		1.0	0.40

*Prepared for each wine sample; diluted to volume with aqueous ethanol

2. Place electrodes in one of the calibration standards. Stir and allow meter reading to stabilize as per manufacturer's recommendation. Record the final mV reading.
3. Rinse electrodes, dry, and place in other calibration standards in sequence. Again, stir and allow meter reading to stabilize. Record the final mV reading.
4. Prepare a calibration curve by graphing millivolt readings vs. log concentration. This also may be accomplished by using semi-log paper (graph concentration values on the log axis). *Note:* The slope of the calibration curve should match that specified by the manufacturer.
5. Prepare wine samples the same as standards. Place clean, dry electrodes into the sample, stir, allow the meter to stabilize, and record the MV reading. Refer to the calibration curve and read the ionized calcium concentration off the graph.

IV. Supplemental Notes
1. The electrodes should be recalibrated for each sample, unless one has several with similar pH, alcohol, and potassium levels.
2. Cardwell, et al (1991) demonstrated that with a 10-fold dilution (using 0.1 M KCl), the mean value for the ionized calcium fraction approximates 75% of the total calcium level in the wine sample. Therefore, the measured concentration of the diluted sample, relative to similarly diluted standards, is multiplied by 1.33 to give an estimate of the total calcium level.

CALCIUM OXALATE: (SEE DIAGNOSTICS)

CALCIUM TARTRATE: (SEE DIAGNOSTICS)

CARBOHYDRATES: HPLC ANALYSIS OF GRAPE AND WINE SUGARS

Sugars can be separated and quantitated by using an appropriate chromatographic column and mobile phase. The method allows the analyst to determine all the sugars present in the sample, both fermentable and nonfermentable. In the following procedure, a preliminary separation of the sugars and fixed organic acids in the sample is used. A typical HPLC chromatogram is shown in Fig. 20–5.

I. Equipment
   HPLC with refractive index or variable wavelength (UV) detector
   Bio-Rad HPX-87H carbohydrate column (or equivalent) plus appropriate guard column
   Disposable polypropylene column
   Pipettes (1 mL) and volumetric flasks (10 mL)

II. Reagents
   Mobile phase (0.01 $N\,H_2SO_4$): See Table 20–3 for preparation
   Bio-Rex 5 resin: 100–200 mesh (Cl− form)
   Standard solution (for use in must or juice): In an appropriate volume of deionized water, dissolve 1.0 g glucose, 1.0 g fructose, 0.5 g sucrose (and other sugars of interest). Bring to 1 L with deionized water.
   Standard solution (for use in wine): In an appropriate volume of deionized water, dissolve 0.1 g glucose, 0.10 g fructose, and appropriate amounts of other sugars of interest. Bring to 1 L with deionized water.

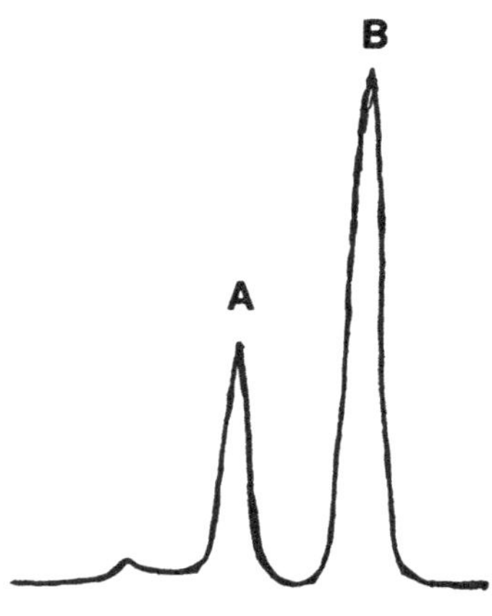

Fig. 20–5. A typical HPLC chromatogram of (A) glucose and (B) fructose in sweet white wine.

III. Procedure

1. Install a Bio-Rad HPX-87H carbohydrate column (or equivalent) in the HPLC. This column should be preceded by an appropriate guard column in the mobile phase flow stream.
2. Begin pumping 0.01 $N\,H_2SO_4$ mobile phase through the column at a flow rate of 0.6 mL/min.
3. Set the column oven temperature to 25°C, and allow the system to stabilize.
4. Turn on the variable-wavelength UV detector and allow it to warm up. Set the wavelength to 210 nm. Alternatively, turn on the refractive wavelength detector and allow it to warm up.
5. Prepare and chromatograph the sample as follows:
   a. Pipette 2.00 mL of wine or grape must and approximately 0.2 mL of concentrated ammonium hydroxide into a capped vial. Mix the contents of the vial and hold until ready to place contents on the anion exchange resin.
   b. Prepare the ion exchange column by slurrying 1 g Bio-Rex 5, 100–200 mesh (Cl^- form) resin in 3 mL deionized water, and pouring it into a disposable polypropylene sample preparation column. Drain off the water and wash column with additional 5 mL deionized water. Do not allow the column to dry out before use.
   c. Add 1 mL sample to the column and allow to drain through. Wash the sugars off the column with deionized water (9 mL) and collect in a 10 mL volumetric flask. Bring to the mark with deionized water and mix well before injection.
   d. If it is desired that the wine acids also be analyzed, carefully add 2 mL of 20% sulfuric acid and enough deionized water to collect 10 mL of eluent containing the acids. Again mix this sample well before injecting onto the column.
6. Qualitative analysis may be accomplished by injecting standard solutions of the various sugars expected in the sample. Compare observed retention times of peaks in the sample chromatogram with those in the standards chromatogram.
7. Standard solutions of various sugars at known concentrations can be injected and chromatographed. Peak heights (recorder) or peak areas (integrator) are compared to those from samples, and the unknown concentrations calculated using the equation:

$$\text{Sugar conc. (g/L)} = \frac{A}{A'}\,\text{(sugar conc in std, g/L)}$$

where:

A = response of unknown sugar

A' = response of standard sugar

IV. Supplemental Notes

1. There are a number of resin columns currently available for carbohydrate analysis. One should consult column manufacturers' literature for columns suitable for the specific analysis desired.
2. The column listed in the above procedure can also be used for the determination of organic wine acids.

CARBOHYDRATES: (SEE REDUCING SUGARS)

CARBON DIOXIDE: CARBODOSER TECHNIQUE FOR APPROXIMATE VALUES

The Carbodoser (marketed by Scott Laboratories, Petaluma, CA) is a device for measuring the volume of carbon dioxide contained in a wine sample. By controlling the temperature of the sample and careful agitation a rapid approximate value can be obtained. The device comes with tables for converting the readings obtained directly into mg/L CO_2.

CARBON DIOXIDE: TITRIMETRIC DETERMINATION USING CARBONIC ANHYDRASE

The procedure of Caputi et al. (1987) calls for first converting all CO_2 in the sample to the carbonate ion (CO_3^{-2}) by adjusting the pH of the sample to 11–12 using a known quantity of 0.1 N NaOH and adding carbonic anhydrase. In earlier work, without the use of anhydrase, it was not possible to recover theoretical amounts of CO_2 present in standardized samples. Accuracy was, at least partially, dependent on the speed at which the analysis was performed. This suggested a slow hydration of localized CO_2:

$$CO_2 + H_2O \rightarrow CO_2{\cdot}H_2O \rightarrow H_2CO_3 \text{ (slow)}$$

The solution to this problem was to add the enzyme, carbonic anhydrase, immediately after the addition of NaOH. Carbonic anhydrase catalyzes the rapid hydration of CO_2 (to H_2CO_3) so that it can be quantitatively converted into carbonate ion upon addition of strong base.

$$CO_2 + H_2O + \xrightarrow{\text{carbonic anhydrase}} H_2CO_3 \text{ (fast)}$$

$$H_2CO_3 + 2OH^- \rightarrow CO_3^{-2} + 2H_2O \text{ (instantaneous)}$$

This latter reaction is quite rapid. The sample is then titrated with acid to bring the pH down to 8.45, which neutralizes the excess hydroxide ions and converts the carbonate, CO_3^{2-}, present to bicarbonate, HCO_3^-.

$$Na^+ + OH^- + H^+ \rightarrow H_2O + Na^+ + Cl^-$$

$$CO_3^{2-} + H^+ \rightarrow HCO_3^-$$

Because anions other than bicarbonate are present in wine samples, a degassed (CO_2-free) separate blank is titrated with standard base for each sample analyzed. The difference in volumes of standardized base minus the amount of standard acid used to titrate the sample to pH 8.45 is directly related to the quantity of CO_2 (in mg/L) present in the sample.

I. Equipment
   Burette (25 mL)
   pH meter (pH 4.00 and 10.00 standard buffers)
   Magnetic mixer and stir-bar
   Vacuum flask and source (28 torr)
   Volumetric Pipette

II. Reagents
   NaOH (0.1 N): See Table 20–1 for preparation. Store in plastic bottle.
   Carbonic anhydrase (0.1 mg/mL): Dissolve 10 mg of carbonic anhydrase enzyme in 100 mL deionized water. Store under refrigeration until use.
   Sulfuric acid (≤ 0.100 *N*): See Table 20–3 for instructions.

III. Procedure
   (a) Sample Titration
   1. To a 150-mL beaker add 20 mL standardized 0.1 *N* NaOH.
   2. Cool sample to <0°C. Carefully open sample bottle, avoiding unnecessary agitation.
   3. Transfer a 20-mL aliquot of wine to the beaker with the pipette tip below the surface of the NaOH.
   4. Add 3 drops of carbonic anhydrase, and immerse electrodes.
   5. Titrate to pH 8.45 using standardized 0.1 *N* H_2SO_4.
   (b) Blank Titration (to compensate for other wine acids)
   6. Transfer a 50-mL aliquot of wine sample to a vacuum flask (with stir-bar). Stopper, place on stirrer plate, and degas under a vacuum (≥28 inches) for 5 min with maximum mixing.
   7. Transfer 20 mL of degassed sample to an appropriate sized beaker. Add 40 mL of recently boiled (and cooled) deionized water. Insert pH and reference electrodes.

8. Using standardized 0.1 *N* NaOH, titrate to a pH of 7.75. Record buret volume.

(c) Calculations

9. Calculate the carbon dioxide content of the wine sample in units of mg/L:

$$CO_2\ (\text{mg/L}) = \frac{(A-B) \times N(\text{NaOH}) - \text{mL } H_2SO_4 \times N\,(H_2SO_4) \times (44) \times 100}{\text{sample volume}}$$

A = milliliters NaOH added to wine sample

B = milliliters NaOH used to titrate degassed "blank" to pH 7.75

IV. Supplemental Note

1. A *separate* blank must be determined for *each* wine sample run.

CATALASE: DIFFERENTIATION BETWEEN ACETIC ACID AND LACTIC ACID BACTERIA (SEE BACTERIA: ISOLATION AND CULTIVATION)

CELL COUNTING: CELL COUNTING OF YEAST

Monitoring and enumerating yeast and bacterial populations during the winemaking process provides valuable information about changes occurring in the wine. In the case of yeast, total and viable cell numbers as well as the percentage of budding provides vital information to the winemaker in preparation and maintenance of yeast starters.

Classically, procedures for monitoring cell number call for collecting representative samples (potentially containing yeast and/or bacteria) and plating on appropriate media that will support growth. Plate counting presumes that each colony arose from a single cell; but in the case of wine yeasts that tend to aggregate this may not be a valid assumption.

In most cases, however, the winemaker requires a quick decision about the presence of contaminants and/or physiological status. Thus, the delay associated with colony formation is frequently unacceptable.

To shorten the time needed and to improve the accuracy of cell counting, a variety of other techniques have been evaluated, including direct count procedures using fluorescence microscopy and dye reduction techniques. Fluorescence microscopy makes use of the fact that some compounds absorb UV light and reemit a portion of the absorbed energy as light in the visible spectrum; that is, they fluoresce. Because glass lenses block UV light, a quartz condenser lens is required to focus radiation from the UV-light source upon the sample. Light in the visible spectrum is reemitted from the sample, so conventional optics are used to collect and magnify the sample image.

Operationally, samples containing microbes are collected on membrane filters and stained with a fluorescent dye such as acridine orange. With this stain, viable yeast/bacteria fluoresce green, whereas dead cells appear orange. However, cells exposed to preservatives such as sorbate fluoresce intermediate shades (Meidell 1987) and this makes interpretation difficult.

The most commonly used method of direct counting in the wine industry is dye reduction. The basis for this viable staining technique is that there is a change in the color of a reducible dye (usually methylene blue) when it is used as a hydrogen acceptor in lieu of oxygen. In order to quantitate cell numbers, it is necessary to use a cell-counting slide. The hemocytometer (see Fig. 20–6) is specially ruled so that the volume present in the counting area is constant.

In counting wine yeast, the initial problem is the tendency of the yeast to aggregate in groups that are not only difficult to count but also significantly bias the final result. Thus, mixing to break aggregation is critical to good correlation.

I. Equipment
   Cell-counting slide (Neubauer-Lowenthal or Levy-Hausser) and cover glass
   Pipettes
   Dilution bottles
   Test tubes
   Vortex mixer

II. Reagents (for preparation see end of this procedure)
   Methylene blue
   Peptone dilution blanks (1:10, 1:100)

III. Procedure
   1. Depending on the stage of yeast growth, prepare the appropriate dilution from the collected sample in peptone dilution blanks. For example in monitoring starter tanks, initial $>10^{-4}$ dilutions may be needed in order to obtain countable slides, whereas post-fermentation counts may be made directly without prior dilution.
   2. Vortex each diluted sample before going to the next dilution in the series.
   3. After vortexing the final dilution in the series, transfer a small volume from the center of the dilution tube (bottle) to the counting slide.

4. Using the cover glass provided with the kit, carefully position it over the counting grid. Using a pipette, transfer well-mixed sample to counting slide. Allow diluent and yeast to flow under the cover glass and into the counting area via the groove provided
5. Count all the cells within the bounded area of the grid (400 squares). By using methylene blue in the final diluent one may also score viable/nonviable cells as part of the total cell count. In this case, viable cells will be colorless (i.e., they reduce the blue dye to its colorless or leuco-form). Nonviable cells, by comparison, appear blue or have concentrated areas of blue stain within the boundaries of the cell wall (see Supplemental Notes 1 and 2).

   Tally nonviable and viable as well as percent budding cells present in each field. Each counting slide is provided with two fields for duplication purposes. Count the second and average the results.
6. When one counts all the cells present within the grid network the expression of results (in cells/mL) may be calculated using the following equation:

$$\text{Cells/mL} = (\text{Number cells counted})(10^4) \times \text{Dilution factor}$$

IV. Supplemental Notes

1. Although statistically approved procedures exist for counting portions of the grid field, they apply to measurement of blood cells. In cases where uniform distribution of cells is difficult to achieve (e.g., wine yeasts) it is recommended that the entire field of 400 squares be counted. For reproducible results it is necessary that each laboratory establish whether cells lying on grid boundaries be included in the count or rejected. Depending on the initial dilution created, a difference of five or more cells may have a significant effect on the final calculation of cell number.
2. Methylene blue, itself, eventually will poison living cells. Thus it is imperative that once the dye is mixed with the final dilution, that cell numbers, and especially viability, be determined as soon as possible.
3. When bottling-line samples are screened microscopically it is necessary to concentrate (by membrane filtration) the microbes potentially present in the samples. The membrane may then be stained and examined directly.

Ponceau-S and Ponceau-S/Methylene Blue Viable Staining

The procedure for use of Ponceau-S in conjunction with methylene blue (or by itself) is presented below (Kunkee and Neradt 1974):

DAY ____________

TIME ____________

PERSON COUNTING CELLS ____________________

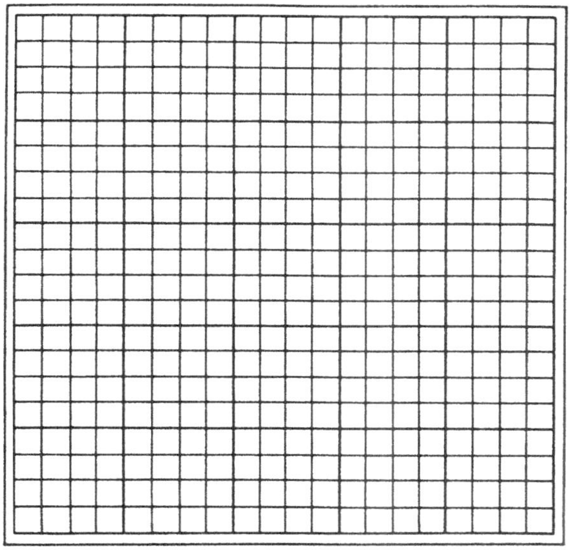

(1) Number yeast side 1 __________

Number yeast side 2 __________

$\bar{x}$ cell number __________

Fig. 20–6. Hemocytometer.

1. Using an appropriate laboratory filter, collect a known volume of suspect wine on a 1-μm membrane.
2. While still assembled, cover membrane with a small volume of methylene blue (0.01%) prepared in citrate buffer (pH 4.6).
3. Hold 20–30 seconds and filter stain through. Transfer filter to microscope slide and examine.
4. **Interpretation:** clear cells are viable and blue-colored cells are dead.
5. Collect a fresh sample of suspect wine or transfer previously stained membrane to an absorbant pad (or several thicknesses of laboratory filter paper) soaked with Ponceau-S stain.
6. Once stained, transfer to similar pads soaked with 5% acetic acid (decolorizer). Transfer between pads at 3–4 minute intervals (total 5–6 transfers) until membrane appears white to light pink.

7. Transfer membrane to microscope slide, quickly pass over flame of bunsen burner, and examine under the microscope.
8. Subtracting results of methylene blue staining from results of Ponceau-S stain, yields viable cells. Many laboratories simply use Ponceau-S as a general staining tool without consideration of the above technique.

Stains/Dyes

Nigrosin

Commonly used as a background or counterstain, nigrosin is very effective in differentiating gross morphological differences between microorganisms (viz., separation of cocci from bacilli). In practice, any nonselective counterstain, such as India ink, may be substituted for nigrosin.

Preparation:

Dissolve 10 g water-soluble powder in 100 mL deionized water. Heat the solution in a boiling water bath for 20 min. Cooling to room temperature, membrane filter the solution before use.

Methylene Blue

Preparation:

Dissolve 0.3 g methylene blue (methylthionine chloride) in 30 mL of 95% ethyl alcohol and 100 mL of citrate buffer (pH 4.6). Before use, the dye should be membrane-filtered and examined for biological activity.

Citrate Buffer: In a 125-mL flask, dissolve 2.1 g NaH_2-citrate and 2.4 g NaH-citrate. Bring to a final volume of 100 mL with deionized water.

Ponceau-S Stain

Ponceau-S is prepared by dissolved the following in approximately 70 mL of deionized water:

0.9 g Ponceau-S
13.4 g trichloroacetic acid
13.4 g sulfosalicylic acid

When dissolution is effected, bring to a final volume of 100 mL using distilled water. Already prepared Ponceau-S staining kits may be purchased from Millipore Corporation.

Peptone

Peptone, as a dehydrated medium, is available from commercial suppliers. For makeup, follow supplier's recommended procedures.

CELLULOSE: SEE DIAGNOSTICS

CHILL HAZE: SEE BITARTRATE STABILITY-FREEZE TEST

CHLORINE (RESIDUAL): VISUAL QUALITATIVE TEST USING IODIDE/STARCH

At pH <8, chlorine liberates free iodine from potassium iodide in solution. The iodine may then be titrated with a standard solution of sodium thiosulfate using starch as the indicator. Because of undefined oxidation of thiosulfate to sulfate at near neutral pH, the reaction must be run under acidic conditions (pH 3–4).

I. Equipment
   Erlenmeyer flasks
   Stir-bar/mixing table
   Burette

II. Reagents
   Glacial acetic acid
   Standardized sodium thiosulfate (0.01 *N* or 0.025 *N*): commercially available
   Standardized iodine (0.0282 *N*): Commercially available
   Starch indicator solution (1%): Mix 10 g soluble starch and 1 L deionized water. Heat the solution to incipient boiling. Cool, and store in a refrigerator.
   Note: Discard any starch indicator that appears turbid.

III. Procedure
   Sample Titration:
   1. Collect 500 mL rinse water in an appropriate-sized Erlenmeyer flask.
   2. Add 5 mL acetic acid (or sufficient volume to reduce pH to 3–4.0) and approximately 1 g KI.
   3. Place stir-bar in flask and mix.
   4. Titrate with standardized thiosulfate until light straw color. Add 1 mL starch indicator until blue disappears.

   Blank Titration:

   The volume of titrant should be corrected for presence of oxidizing/reducing agents in reagents as well as compensation for iodine bound to starch at the endpoint. This leads to either consumption of titrant (resulting in a value that will be subtracted from sample titrant volume) or I_2 present in excess. In the latter case, sample is back titrated with $Na_2S_2O_3$ resulting in a blank value that must be added to sample titrant volume:

1. Using deionized water at the same volume used in sample titration, follow steps 2 and 3 above. If blue color develops, titrate with $Na_2S_2O_3$ per step 4 above. In this case, the blank will be subtracted from sample volume.
2. If blue color does not develop upon addition of deionized water to KI and acetic acid, titrate with standardized I_2 until blue color develops. Back titrate with standardized $Na_2S_2O_3$, recording volume of titrant used. In this case, the blank will be added to volume of titrant used for sample titration.
3. Results are calculated as:

$$Cl_2\ (\text{mg/L}) = \frac{(A + B) \times N \times 35.45 \times 1{,}000}{\text{sample vol (mL)}}$$

where:

A = vol. titrant
B = vol. blank (±)
N = normality of $Na_2S_2O_3$
35.45 = atomic weight of Cl
1,000 = conversion to mg/L (ppm)

IV. Supplemental Notes
1. Titration should be conducted under subdued light (not in direct sunlight).
2. This method can analytically measure chlorine to 1 mg/L.

CHLORINE (RESIDUAL): VISUAL QUALITATIVE TEST USING SILVER NITRATE

Chlorine-based sanitizers are widely used in the wine industry. Because chlorine is such a strong oxidizing agent, residual levels in the final tank rinse water should be measured before fill. This can be done with commercial swimming pool test kits or by using the following procedure:

I. Reagent
0.2 M silver nitrate solution ($AgNO_3$): Dissolve 34 g of silver nitrate in deionized water. Bring to 1-L volume. Store in a dark bottle.

II. Procedure
To 10 drops of tank rinse water add 2 drops of silver nitrate solution. The formation of a white precipitate (silver chloride, AgCl) is indicative of the presence of residual chloride ions.

CITRIC ACID: HPLC DETERMINATION (SEE ORGANIC ACIDS)

CITRIC ACID: PAPER CHROMATOGRAPHIC METHOD (SEE MLF—PAPER)

CITRIC ACID: ENZYMATIC SPECTROMETRIC (VIS) ANALYSIS

Enzymatic analysis procedures for the determination of citric acid have been developed, as reported in the OIV (1990). This procedure relies on the coupled enzymatic reactions where first citrate is converted into oxaloacetate and acetate, and then oxaloacetate and its decarboxylation derivative, pyruvate are reduced to malate and lactate, respectively:

$$\text{citrate} \overset{\text{CL}}{\leftrightarrow} \text{oxaloacetate} + \text{acetate}$$

$$\text{oxaloacetate} + \text{NADH} + \text{H}^+ \overset{\text{MDH}}{\leftrightarrow} \text{L-malate} + \text{NAD}^+$$

$$\text{pyruvate} + \text{NADH} + \text{H}^+ \overset{\text{LDH}}{\leftrightarrow} \text{L-lactate} + \text{NAD}^+$$

CL: citrate lyase
MDH: malate dehydrogenase
LDH: lactate dehydrogenase
NADH: reduced nicotinamide-adenine dinucleotide

The formation of NAD^+ is stoichiometrically proportional to the quantity of citric acid, and thus the concentration of the latter may be determined by measurement of the decrease in amount of NADH present. Kits of reagents (e.g., see Boehringer-Mannheim catalogue) for conducting this procedure are now available.

Refer to the product sheet that accompanies the citric acid kit of reagents, for up-to-date information and procedural details specific for the analysis of citric acid. In general the citric acid procedure will require the following:

I. Equipment
   A spectrometer capable of reading absorbance at 340, 334, or 365 nm plus suitable 1 cm cuvettes
   Adjustable micropipettes (20-μL to 2.0 mL)

II. Reagents
   As mentioned, the reagents used may be obtained from Boehringer Mannheim or other supplier. They are sold in the form of a test kit of sufficient material for a number of citric acid determinations.
   Polyvinylpolypyrrolidone (PVPP)

III. General procedure

1. The determination is generally carried out on undiluted wine, provided that the citric acid content is less than 400 mg/L. When necessary, dilute a wine sample with deionized water so that the citric acid concentration is between 20 and 400 mg/L.
2. Heavily pigmented red wine samples should be treated with PVPP before analysis. Place 0.2 g of pre-moistened PVPP into 10 mL of wine in a 50-mL Erlenmeyer flask. Agitate for 2 to 3 minutes, and filter.
3. The sample (or treated/diluted sample) is then mixed with reagents (except citrate lyase, CL), and the absorbance values are measured.
4. The citrate lyase, CL, is then added, and the absorbance again is read after the reaction is complete (about 5 min).
5. A blank is run along with the samples, and the absorbance values are corrected for the blank reading.
6. The difference in the corrected absorbance values taken before and after the addition of citrate lyase, CL, is then used to calculate the concentration of citric acid using the equation given in the procedure.

COLOR: SPECTROMETRIC DETERMINATION OF HUE (TINT) AND INTENSITY (DENSITY)

Values for a juice or wine color can be assigned by measuring the absorbance at different wavelengths. For white wines or juices, absorbance is usually determined at 420 nm. In the range of 400–440 nm, increases in the brown coloration of white wines are readily detected. For red wines or juices, samples frequently are diluted and the absorbances measured at 420 and 520 nm. When a dilution procedure is employed, the sample must be diluted with deionized water adjusted to the pH of the original wine. The extent of dilution is a function of the original intensity of red color, and may range from 1 + 3 to 1 + 9. Microcuvettes can be used to avoid dilution. All particulate matter must be removed from samples before analysis. This is usually accomplished by filtration through a 0.45–1.2 μm membrane filter.

A measure of color density or intensity can be achieved by summation of absorbance readings at 420 and 520 nm (summation of A_{420}, A_{520}, and A_{620} in the OIV procedure). By comparison, hue or tint is measured as the ratio of absorbance readings at 420–520 nm.

I. Equipment

Single beam visible spectrometer capable of measurements at 420 and 520 nm

Matched cuvettes

pH meter

II. Reagents
   Concentrated sulfuric acid
   Deionized water
III. Procedure
   1. Switch on the instrument, and allow appropriate time for instrument to warm up.
   2. Before placing cuvette into sample chamber, adjust zero transmittance.
   3. Set instrument to desired wavelength.
   4. Adjust meter to read 100% transmittance (zero absorbance) using reference cuvette.
   5. Remove the reference cuvette, and replace it with sample in cuvette S. *Note:* Red wines should be quantitatively diluted with water adjusted to wine pH (or use microcuvettes). Record absorbance (or % transmittance).
   6. For each sample, record absorbance at 520 and 420 nm. Where necessary one may record these data in $\%T$ and convert to absorbance using the relationship:

$$A = \log \frac{100}{\%T}$$

   7. A measure of the color intensity or density can be determined by summation:

$$\text{Density or intensity} = (A_{420} + A_{520})$$

$$\textit{Note: } I = (A_{420} + A_{520} + A_{620}) \text{ for OIV}$$

   8. The ratio of absorbance at 420 and 520 nm is a measure of hue or tint:

$$H = (A_{420})/(A_{520})$$

IV. Supplemental Notes
   1. For intensely colored red wines, Ribereau-Gayon (1974) recommends the use of microcell cuvettes (0.1 cm thickness) rather than dilution procedures to obtain absorbance readings within the range of the spectrophotometer (see Table 20–15).

COLOR: THE 10-ORDINATE METHOD FOR COLOR SPECIFICATION

Although 30 selected ordinates are generally used in food products that exhibit greater spectral diversity, a 10-selected-ordinate system will suffice in products such as wine that have relatively simple absorption spectra (see

Table 20–15. Comparison of wine color data from diluted vs. nondiluted wine samples in regular vs. microcuvettes.

Wine	Diluted sample (1 + 9 using 1-cm cuvettes)			Undiluted sample (using 0.1-cm microcell cuvettes)		
	$A_{420\ nm}$	$A_{520\ nm}$	Tint or hue	$A_{420\ nm}$	$A_{520\ nm}$	Tint or hue
Red wine A	0.147	0.250	0.588	1.295	2.076	0.624
Red wine B	0.315	0.377	0.836	1.775	1.900	0.934
Red wine C	0.248	0.270	0.919	1.675	1.670	1.000

Figs. 7–11 and 7–12). The three series of 10 wavelengths at which transmittance measurements are taken using this 10-ordinate system are defined in Table 20–16. Transmittance data are then summed and totals multiplied by their corresponding factors to yield tristimulus values.

I. Equipment

Recording spectrophotometer covering the range of wavelengths from 400 to 660 nm
Matched cuvettes
pH meter

II. Reagents

Concentrated sulfuric acid
Deionized water

III. Procedure

1. Using membrane-filtered (0.45 μm) wine sample, collect transmittance data at the three sets of ten wavelengths as per following table:

Table 20–16. Measurement of CIE color values using the ten-selected ordinate procedure and iluminant A (Tungsten Filament lamp at 2854 K).

Ordinate number	X_{nm} (Red)	Y_{nm} (Green)	Z_{nm} (Violet)
1	516.9	507.7	424.9
2	561=.4	529.8	436.0
3	576.3	543.7	443.7
4	587.2	555.4	450.5
5	596.5	566.3	456.8
6	605.2	576.9	462.9
7	613.8	587.9	469.2
7	613.8	587.9	469.2
8	623.3	600.1	476.8
9	635.3	615.2	487.5
10	655.9	639.7	508.4
Normalizing	0.10984	0.10000	0.03555

See steps 1 and 2 of Procedure for directions for the use of this table.

(Note: red wines should be quantitatively diluted with deionized distilled water that has been adjusted to wine pH.)

2. Using the following format, total the collected %T values Totals by their corresponding factors:

X = Total $\%T$ (for X) (0.10984)
Y = Total $\%T$ (for Y) (0.10000)
Z = Total $\%T$ (for Z) (0.03555)
Total $X + Y + Z$

$$x = \frac{X}{X+Y+Z}$$

$$y = \frac{Y}{X+Y+Z}$$

$$z = 1 - (x + y)$$

IV. Supplemental Notes
1. The number of selected ordinates chosen will depend on the complexity of the spectral curve measured. In the case of wine, the goal of color specification can be accomplished by use of ten selected ordinates. This reflects the relatively simple nature of white and red wine spectra.
2. Transmittance data are collected at different wavelengths for X, Y, and Z tristimulus values. This procedure reflects directly on the standard observer curves which report human visual perception of color, *not* objective monitoring.

COPPER: ATOMIC ABSORPTION ANALYSIS

Atomic absorption (AA) is used in many laboratories as a rapid alternative to wet chemical procedures. Caputi and Ueda (1967) were among the first to report its value in heavy metals analysis in wine.

I. Equipment
Atomic absorption spectrophometer equipped with a copper hollow cathode lamp (or a multielement lamp containing copper as one of its constituents).
Volumetric pipettes

II. Reagents
Ethanol (200° proof)
Glucose

Table 20–17. Preparation of copper calibration standards from 10.0 mg/L working standard.

Volume of working standard (mL)	Final Cu^{2+} concentration in 100 mL volume[a] (mg/L)
2	0.2
4	0.4
5	0.5
8	0.8
10	1.0

[a]Dilution is made with 12% (vol/vol) ethanol.

Copper stock solution (100 mg/L): In a 1-L volumetric flask dissolve 0.1000 g of copper metal (available as either foil, powder, shot, or turnings) in several milliliters of dilute nitric acid. When dissolution is complete, bring to volume with deionized water at defined temperature. *Note:* The copper used should be of accurately defined purity. Alternatively, many laboratories use copper salts of defined purity as well as prestandardized commercially available solutions of copper.

III. Procedure

1. Prepare a 10.0-mg/L working copper standard by diluting 10 mL of the copper stock solution to 100 mL in a volumetric flask.
2. Prepare 0.2–1.0 mg/L copper calibration standards in 100-mL volumetric flasks according to Table 20–17. Final dilution is made with 12% (vol/vol) ethanol.
3. Consult the operator's manual for instrument setup and calibration.
4. Aspirate the blank [12% (vol/vol) ethanol] and calibration standards through the "sample inlet," recording absorbance values after readings stabilize. Between samples, aspirate a sufficient volume of the blank solution to flush the system.
5. Aspirate filtered wine samples, recording absorbance readings. Graph absorbance of standards vs. their respective concentrations. Compare absorbance of unknown(s) to the standard curve, expressing concentration in mg/L.

IV. Supplemental Note

1. Caputi and Ueda (1967) call for making up standard solutions in 12% (vol/vol) ethanol and 5% (wt/vol) glucose for sweet wines. Values are reported to be within ±10% of the actual copper concentration. In dry table wine analyses, 12% ethanol alone is generally sufficient.

COPPER: SPECTROMETRIC (VIS) ANALYSIS USING DIETHYLDITHIOCARBAMATE

Copper will react with diethyldithiocarbamate to form a colored complex measured spectrometrically at 450 nm. Extraction of this complex into an amyl acetate-methanol solvent mixture isolates it from other colored compounds in the wine sample. Amyl acetate is very volatile; it is good practice to conduct the extractions in a hood.

I. Equipment
   Pyrex test tubes (of more than 25 mL capacity)
   Spectrometer (Bausch and Lomb Spectronic 20 or equivalent)
   Matched cuvettes
   Volumetric flask (100 mL)
   Volumetric pipettes (1.0, 2.0 mL)
   Whatman #40 filter paper
   Glass funnels
   Watch glasses (2.5–3 in. in diameter)

II. Reagents
   Stock copper solution (1,000 mg/L): In a 1-L volumetric flask, dissolve 1.000 g of copper metal (available as either foil, powder, shot, or turnings) in several milliliters of dilute nitric acid. When dissolution is complete, bring to volume with deionized water at defined temperature. *Note:* As previously described, the copper used should be suitable for standardization.
   Ammonium Hydroxide (5 *N*): Dilute 666 mL of stock NH_4OH to 2 L final volume with deionized water.
   HCL-Citric Acid Solution: In a 1-L volumetric flask, dissolve 94 g of reagent grade citric acid and 62 mL concentrated HCl. Bring to volume with deionized water.
   Amyl Acetate-Methanol Solvent (2:1): In a convenient-sized flask, mix 2 volumes of amyl acetate with 1 volume of methanol. *Note:* The fumes are very volatile. Hold the solvent in a ventilated area (under fume hood) until needed.
   Sodium Diethyldithiocarbamate (1%): Transfer 1 g sodium diethyldithiocarbamate into a 100-mL volumetric flask, dissolve, and bring to volume with deionized water. This solution should be prepared on a daily basis.

III. Procedure
   1. Prepare a 100-mg/L working copper standard by diluting 10 mL of the copper stock solution to 100 mL in a volumetric flask.
   2. Prepare 0.5, 1.0, and 2.0-mg/L calibration standards in 100-mL volumetric flasks according to Table 20–18.

Table 20–18. Preparation of copper calibration standards from 100-mg/L working standard solution.

Volume of working standard solution (mL)	Final Cu^{2+} concentration in 100 mL volume (mg/L)
0.5	0.50
1.0	1.0
2.0	2.0

3. Volumetrically transfer 10-mL portions of the calibration standards into separate test tubes. Blank preparation consists of 10 mL of deionized water.
4. To each tube, add with thorough mixing:
   a. 1 mL HCl-citric acid solution.
   b. 1 mL 5 *N* NH_4OH.
   c. 1 mL sodium diethyldithiocarbamate. Mix the contents well, and allow tubes to stand for approximately 1 min.
5. Add 15 mL amyl acetate-methanol solvent to each and mix each tube's contents thoroughly.
6. Allow phase separation to occur.
7. With a 10-mL graduated pipette, draw off the upper organic phase and filter it through Whatman #40 filter paper directly into a cuvette. Cover the funnel with a watch glass during filtration.
8. Set the spectrophotometer at 450 nm. Using a deionized water reagent blank (following steps 3–6 above), determine the absorbance or transmittance of standards.
9. Plot the measured or calculated absorbance against the appropriate concentration (mg/L) of the copper standards.
10. Using 10 mL of wine sample, proceed as in the preparation of standards, comparing absorbance measurements to the standard curve.

IV. Supplemental Notes

1. All glassware must be exceptionally clean. The laboratory water and reagents should be checked for copper contamination. Failures using the above procedure can generally be attributed to contamination from these sources. It is recommended that laboratory personnel maintain separate glassware exclusively for copper analyses. Sloppy technique and/or reagent contamination may be detected as color development in the blank.
2. Post-filtration sample opacity is attributed to incomplete separation of water into the aqueous phase. Such samples may be "cleared" by the addition of anhydrous sodium sulfate powder.

3. In this procedure, HCl-citric acid is used to chelate any iron in solution, since iron will interfere with color development.
4. To avoid interference with wine pigments, the copper-carbamate complex is extracted with amyl acetate-methanol solution. Methanol acts to prevent or reduce the tendency for emulsion formation between wine and amyl acetate.
5. Standard solutions should be made fresh each time a standard curve is prepared. Metal ions, in low concentrations, are adsorbed by glass upon storage, producing anomalous results.

COPPER: VISUAL QUALITATIVE TEST FOR DETERMINATION OF CASSE (SEE DIAGNOSTICS)

CORK: MICROBIOLOGICAL, PHYSICAL, CHEMICAL, AND SENSORY PROPERTIES

Cork Sterility Testing

Although sterility tests will identify the presence of viable microbes, they will not detect already produced metabolites. The following procedure is that of the International Organization for Standardization (ISO-1993).

I. Equipment
   250 mL Erlenmeyer flasks
   Petri plates (sterile)
   0.45 μm membrane filter and sterilized housing
II. Media/Reagents
   WL Nutrient (WLN)-agar (available commercially).
   Alcoholic-Malt Extract Broth:
   1. Malt extract is available commercially. Prepare according to manufacturer's instructions.
   2. Adjust pH to 4.0 using tartaric acid.
   3. Sterilize by autoclaving.
   4. Once broth is cooled, add alcohol (as NSFG) to a level of 8% (vol/vol).
III. Procedure
   1. Transfer 100-mL aliquots of alcoholic broth to previously sterilized 250-mL Erlenmeyer flasks.
   2. Aseptically, transfer 4 corks to each flask. Make certain that each is fully immersed in growth medium.
   3. Incubate at 25°C (77°F) for 24 hours.

4. Remove cork from each flask and filter broth through 0.45 μm membrane.
5. Transfer membrane to WLN-agar plates. Incubate at 25°C for 5 days.
6. Count colonies and tabulate as to type: bacteria, yeast, mold.
7. Express results as CFU/cork.
8. The ISO recommends running a media blank, consisting of incubated uninoculated broth, as a check on media and technique.

Sensory Defects

I. Equipment
   1. 500-mL Erlenmeyer flasks (screw-cap)
   2. Wine glasses
   3. Watch glasses

II. Supplies
   1. Neutral white wine.

III. Procedure
   1. Collect lots of 5 corks each into a 500-mL screw cap Erlenmeyer flask.
   2. Fill to top with neutral dry white wine. Some wineries elect to use the actual wine that is planned for bottling.
   3. Allow to stand at room temperature for 24 hours.
   4. Decant wine into glasses and cover with appropriately-sized watch glass.
   5. Evaluate vs. a control.

IV. Supplemental Notes
   1. In that this procedure provides for enormous surface area for extraction, natural woody characters may be picked up. Thus the panel used for evaluation should be trained in perception of cork taint.
   2. 2,4,6-trichloroanisol and other reference compounds can be purchased from biochemical supply houses.

Dimensional Analysis

Length and diameter measurements are most easily made using calipers. Various models may be purchased from suppliers of scientific equipment. Models should be selected to read to 0.01 mm. Length measurements are reported ± 1mm whereas diameter are reported to ± 0.5 mm (FP Portocork 1993 Personal Communication)

Cork Moisture

Germination and growth of mold is not observed at A_W <0.75 (Daly et al. 1984). This corresponds to the recommended maximum moisture content of 8% for microbiological stability (Lee et al. 1984). At moisture levels <5% brittleness and potential damage to the cork results when compressed. The recommended range is 6–8%. The following method is that of the ISO.

I. Equipment
   Analytical balance
   Laboratory drying oven

II. Procedure
   1. Weigh individual corks. Cut each into four pieces.
   2. Transfer to drying oven at 103°C for 3 hours.
   3. Remove, transfer to desiccator, hold for 30 min and reweigh.
   4. Express results (moisture % wt/wt) as:

$$= \frac{(\text{wt before drying}) - (\text{wt after drying})}{(\text{wt before drying})}$$

III. Supplemental Notes
   1. Many laboratories use moisture meters that provide immediate response.

Cork Dust: Microscopic Determination in Bottled Wine

Cork dust appears "crystallike" microscopically, and may be confused with bitartrate unless further examination is performed. The recommended procedure is to use a stain that reacts with lignin, a structural macromolecule in cork.

I. Equipment
   1. Bright field microscope and several slides and coverslips.
   2. Membrane filters and appropriate housing. Most laboratory filtration units use 47 mm membranes. For general purposes, 1–5 μm cellulose acetate filters are useful.
   3. Clinical centrifuge.

II. Reagents
   1. Phloroglucinol Stain: Mix 2 g of phloroglucinol in 100 mL of 10% HCl. This is a near-saturated mixture and supernatant must be decanted from any crystals that do not dissolve. This solution should be made fresh daily. Therefore, unless many samples are to be run, the analyst may wish to reduce proportionately the volume made at one time.

III. Procedure
   1. Collect a portion of wine containing the debris by filtering through an 8-μm membrane filter.
   2. Wet the filter and sediment with phloroglucinol stain. Hold stain in contact with sediment for 5 minutes.

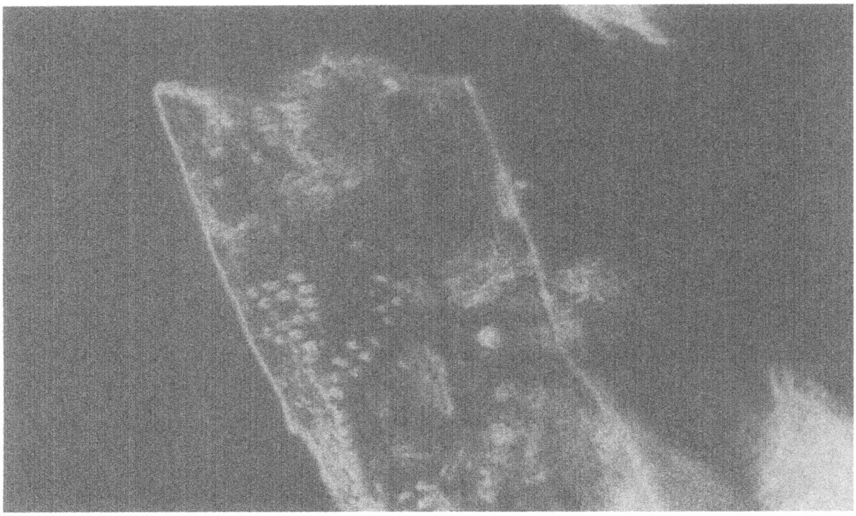

Photomicrograph 20–1. Cork debris.

3. Apply vacuum to filter to remove stain. Rinse the filter with deionized water.

IV. **Interpretation**
1. Examine the sediment microscopically. Cork debris appears as red crystal-like aggregations of cells (see Photomicrograph 20–1).
2. Case lint also stains red using this technique. However, it is fibrous in appearance (see Photomicrograph 20–2).

Cork Dust

Fine particulates (dust) are inevitable in processing cork. During several steps in processing, corks are washed and tumble-dried in an effort to remove adhering particulates. However, dust may remain trapped in lenticels and subsequently leach into bottled wine. Where cork dust is the suspected cause of wine sediment, it may be collected on a membrane filter and stained (see previous procedure). The procedure presented below is designed to detect the problem on incoming cork.

I. Equipment
1. Orbital variable speed shaker or magnetic stirring plate.
2. 500-mL Erlenmeyer flasks
3. Analytical balance

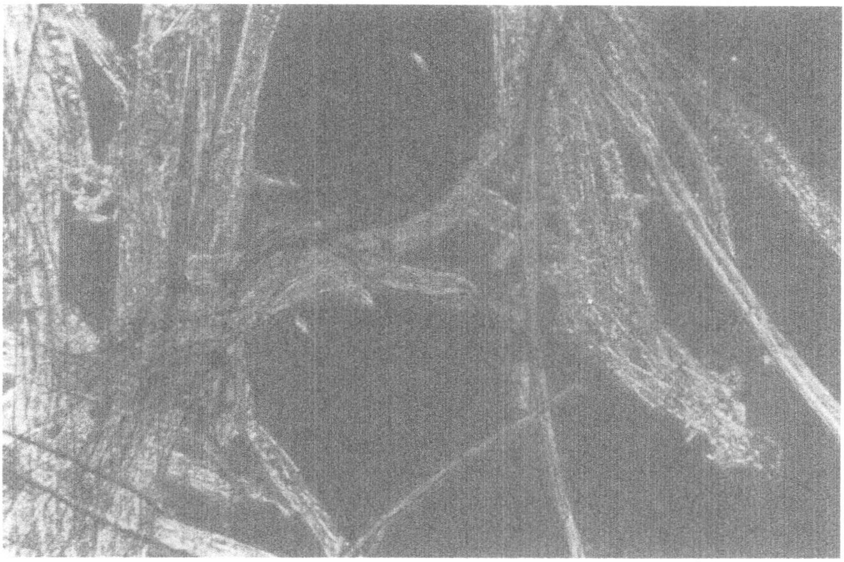

Photomicrograph 20–2. Cellulosic debris.

4. Filter paper (Whatman #40 or equivalent)
5. Vacuum filter assembly
6. Drying oven

II. Reagents

1. Wine Simultant: In a 1-L Erlenmeyer flask mix 100 mL of isopropyl alcohol and 900 mL of deionized water.

III. Procedure

1. Transfer 3–5 randomly selected corks to a 1-L Erlenmeyer flask containing 500 mL of wine simulant.
2. Place flask on variable speed mixing table and agitate for 15 minutes.
3. Preweigh filter paper and place in filter housing.
4. Filter wine simulant.
5. Dry to constant weight at 60°C (140°F).
6. Determine the weight of dust/cork as:

$$\frac{(\text{wt of filter} + \text{debris}) - (\text{wt of filter})}{\text{number of corks tested}}$$

IV. Supplemental Note

Use of Whatman #40 filter paper facilitates gravity filtration, thus obviating the need for a vacuum source.

Cork Capillarity

I. Equipment
   Petri plates
   Filter paper
II. Procedure:
   1. Transfer 25 mL of red wine to Petri plate.
   2. Collect 4 corks to be tested. Stand each on end in the wine. Measure liquid level on corks at start.
   3. Hold at room temperature for 24 hours.
   4. Remove cork and allow excess wine to run off.
   5. Using a ruler, measure distance travelled by wine up the cork, making certain to subtract initial liquid level. Express results to nearest 0.5 mm.
III. Supplemental Notes:
   1. The ISO is developing a model wine (10% vol/vol alcohol/water + colored dye) for use in this test. As of this writing, the nature of the dye has not been specified.

Coatings (Stability)

Cork coating agents, applied as part of finishing operations, accomplish three important goals. First, they reduce frictional forces and, thus, facilitate high-speed bottling. Second, coating agents help in restoring much of the cork's natural impermeability lost during boiling and bleaching. Third, proper selection and application of agent(s) facilitates cork extraction. Several coating agents have been used. These include paraffins (C_{21}–C_{40} hydrocarbons) and silicons, as well as mixtures of the two. Polymeric coatings are also in use. The ratio of paraffin to silicon oils as well as selection of lower vs. higher molecular weight fractions will impact sealing properties as well as tenacity of cork-glass contact. Further, the amount and distribution of coating agent is crucial. Too much and/or unequal distribution of paraffin may result in abrasion during corking with fragments potentially ending up in the wine. Amon and Simpson (1986) report that some polymeric coatings are unstable at alcohol levels common to dessert wines. The following procedure is described by Amon and Simpson (1986).

I. Equipment
   1. 1-L Erlenmeyer flask
II. Reagents
   1. Alcohol simulant: Using NSFG and deionized water, prepare an alcohol-water solution corresponding to the alcohol level of wine to be bottled.

2. Immerse several corks in the solution and hold for 2–4 days.
3. Cloud formation or flaking suggests coating instability. A sample should be examined microscopically to verify the nature of haze.

Defect Appraisal

A variety of physical criteria, in addition to those discussed, are used in selection and ordering of cork. These include potential problems stemming from climatic and management practices in the cork forests as well as insect and processing defects. These issues are considered in a series of articles by Fugelsang and Callaway, 1995.

CORK DUST: MICROSCOPIC DETERMINATION OF LIGNIN SEDIMENT (SEE DIAGNOSTICS)

CRYSTAL-LIKE DEPOSITS: MICROSCOPIC RAPID DIAGNOSIS OF SEDIMENTS (SEE DIAGNOSTICS)

CRYSTALLINE DEPOSITS: VISUAL RAPID DIAGNOSIS OF BITARTRATE AND TARTRATE

CYANIDE: HUBACH DISTILLATION PROCEDURE WITH VISUAL INSPECTION OF TEST PAPERS

The current procedure for determination of residual cyanide in wine is Hubach's modification of an earlier analysis developed by Gettler and Goldbaum (1947) for the medical field. The procedure involves reduced pressure distillation of a dealcoholized aliquot of wine through specially prepared, alkaline ferrous sulfate-impregnated, filter paper. A positive reaction is demonstrated by formation of a characteristic blue spot corresponding to Prussian blue. The reaction between ferrocyanide and iron (III) is specific with no known interferences:

$$\underset{\text{in wine}}{K_4Fe(CN)_6} + \underset{\text{on paper}}{Fe^{3+}} \leftrightarrow \underset{\text{Prussian blue}}{KFe[Fe(CN)_6]} + 3K^+$$

The equilibrium in the above reaction is displaced to the right-hand side. During analysis of a wine sample, reduced pressure distillation in the presence of a catalyst, hydrocyanic acid, is quantitatively liberated from any remaining ferrocyanide. The HCN thus formed then reacts with iron on the test papers, forming Prussian blue. Upon completion of the distillation step, the Hubach paper is acid-treated to dissolve iron hydroxides, which tend to mask the development of Prussian blue (Bonastre 1959).

$$[Fe(CN)_6]^{4-} + 4Fe(OH)_3 + 6H_2SO_4 \leftrightarrow Fe_4[Fe(CN)_6]_3^{4-}$$

The intensity of color formation is proportional to the concentration of HCN and may be determined by comparison to color development of standards.

I. Equipment
   Hubach apparatus (see Fig. 20–7)
   1-L Pyrex beaker
   Hot plate
   Thermometer, range 0°–110°C
   5-inch evaporating dish
   Vacuum source

II. Reagents
   Potassium ferrocyanide stock solution (100 mg/L as cyanide): Dissolve 270.6 mg $K_4Fe(CN)_6 \cdot 3H_2O$ in 1-L of deionized water in a volumetric flask. Mix well.
   Sulfuric Acid Solution (1 + 1): to 1 volume of deionized water, carefully add 1 volume of concentrated sulfuric acid. Cool to room temperature before use.
   Hydrochloric Acid Solution (1 + 3): to 3 volumes of deionized water, add 1 volume of concentrated HCl. Cool to room temperature before use.
   Sodium Hydroxide Solution (6 *N*): Dissolve 200 g of reagent-grade NaOH in approximately 800 mL of deionized water. When cooled to room temperature, bring to 1 L volume with deionized water.
   Cuprous Chloride: Because CuCl acts as a catalyst and is regenerated during the distillation, the quantity used is not crucial. Dissolve 1.5 g of reagent grade CuCl in 1 *N* H_2SO_4 decanting the blue supernatant. The process is repeated two times or until only a light blue color remains in the acid solution. Cuprous chloride should be stored in an amber bottle, taking care to avoid exposure to air.
   Hubach Test Papers: Test papers are most conveniently purchased at nominal cost through Scott Laboratories (Petaluma, California). However, for those wishing to prepare their own, the following guide is presented:
   1. Ferrous Sulfate Solution: Dissolve 5 g of $FeSO_4 \cdot 7H_2O$ in 50 mL of deionized water, adding a drop or two of concentrated sulfuric acid to remove turbidity. This solution should be prepared shortly before

Table 20–19. Preparation of standard solutions from stock (cyanide = 100 mg/L) solution.

Volume of stock (mL)	Final concentration (mg/L) in 100-mL volume
10	10
1.0	1.0

use. Immediately before use, add 1 drop of concentrated sulfuric acid.

2. Alcoholic Sodium Hydroxide: From stock 50% (wt/vol) NaOH, transfer 11 g of solution to 100-mL volumetric flask, bringing to volume with 95% ethanol.
3. Using Whatman #50 filter paper of convenient size, immerse in ferrous sulfate solution for 5 min. Remove, and air dry at room temperature. When completely dry, immerse paper in alcoholic sodium hydroxide solution. Allow 5 min for soaking. Remove and air dry at room temperature. When completely dry, paper should be light green to tan in color. At this point, the paper may be cut into appropriate-sized sections for analysis. It is recomended that test papers be stored in a desiccator jar away from light

III. Procedure

(a) Standard Preparation

1. Prepare 1.0- and 10.0-mg/L cyanide working standards from the 100 mg/L stock solution (see Table 20–19).
2. Thoroughly clean the Hubach apparatus, using deionized water, to remove traces of residual cyanide. Allow it to air dry.
3. Preheat the water in beaker to 80°–90°C. The water depth should be above wine level in the aeration tube as indicated in the reference figure. *Note!* To ensure complete hydrolysis of ferrocyanide, it is essential that the temperature of the water bath remain between 80°C and 90°C during the distillation step.
4. Immerse the apparatus in the water bath, using utility clamps for support.
5. Turn on the condenser water.
6. Transfer 20 mL of standard solution to receiving port B.
7. Add 1 mL of CuCl, being sure to mix the suspension thoroughly. Add 1 mL 1 + 1 H_2SO_4.
8. With a pair of forceps, insert Hubach paper at the union of the aeration tube and the condenser (A). Moisten the paper with one or two drops of deionized water.
9. Connect the aeration tube and the condenser, using a Thomas clamp, and turn on the vacuum slowly so as to draw air through the system at a rate just short of forming a steady bubble stream.

10. Aspirate the sample for 10 min.
11. Remove the paper from the unit using forceps, and place it in an evaporating dish containing 1 + 3 HCL.
12. Allow the paper to soak until it is white.
13. A blue stain is a positive result for cyanide. Rerun the test using a deionized water blank. When the unit is clean (as demonstrated by the absence of a blue spot in the water blank distillation), continue with the next standard.

(b) Sample Preparation and Analysis

14. Before each analysis, a 20-mL deionized water blank should be run to demonstrate that the apparatus is free of residual cyanide.
15. Wine samples are first neutralized with 1 + 1 NaOH to prevent excessive foaming and then dealcoholized by evaporation over a steam bath until they occupy approximately 5 mL volume. *Note!* Samples must be brought back to volume with deionized water before distillation.
16. Substituting the dealcoholized wine sample for the standard, continue according to steps 6 through 13 above.

IV. Supplemental Notes

1. It has been reported (Roberts 1988) that the aeration oxidation glassware can be adapted to perform Hubach analyses.
2. It has been the authors' observation that the Hubach test paper must be wetted with a drop of deionized water, prior to placement in the Hubach apparatus.
3. To ensure the integrity of the unit and test paper, it is essential that standards be run before each day's work. After standard testing, the unit should be thoroughly cleaned, and a blank, consisting of reagents and deionized water, run to make certain that the apparatus is free of residual HCN. Thorough cleaning between trials in any day's operation is also essential.
4. Hubach test papers must be kept in a desiccator and are viable only for a limited period of time.
5. Failure to produce a definite blue stain on Hubach test papers *with standards* may result from:
   a. Improperly prepared test papers.
   b. Improper addition of reagents.
   c. Incorrect water bath temperature. The test sample must be kept between 80°C and 90°C to ensure complete hydrolysis of the ferricyanide.
   d. Condenser water that is too warm.
   e. An insufficient aeration period.
   f. Leakage around flanges.

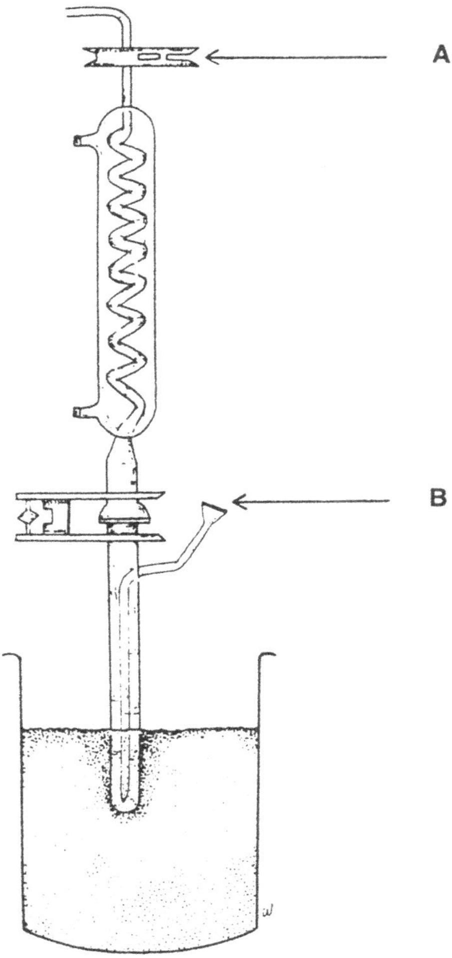

Fig. 20–7. Hubach apparatus. (A) Flange for holding Hubach paper. (B) Receiving port for addition of samples and standards.

DEKKERA: (SEE BACTERIA-YEAST) (SEE 4-ETHYL PHENOL-GAS CHROMATOGRAPHIC PROCEDURE)

DIACETYL: SPECTROMETRIC (VIS)

Diacetyl is a byproduct of the metabolism of microorganisms. The presence of diacetyl as a butterlike aroma in concentrations above 2–4 mg/L (depending on the wine) is an indication that a malolactic fermentation has taken place. In this procedure, diacetyl is first distilled out of a wine sample. The absorbance (520 nm) of a red-colored complex of diacetyl and ferrous ion, formed in the presence of tartrate, is used to quantitate the amount of diacetyl present. The method works well at low mg/L concentrations.

I. Equipment

Distillation unit: consisting of a 500-mL 3-neck flask with a distillation column connected to first a connecting adapter, 75° bend, and then a condenser followed by a tube adapter, 105° bend. A second neck of the flask is fitted with a thermometer adapter holding a Pasteur pipette. The position of the pipette is set so that its tip extends to just above the surface of the sample. The third neck of the flask is used for admitting the sample and is otherwise fitted with a stopper.

Heating mantle with rheostat
Steam bath
Visible spectrometer with 1-cm cuvettes
CO_2 tank with regulator
Pipettes
Assorted beakers

II. Reagents

Diacetyl stock solution (800 mg/L): Dissolve 0.800 g fresh or redistilled diacetyl in 100 mL alcohol. Transfer to a 1-L volumetric flask and dilute to volume with deionized water.

Hydroxyalmine-acetate reagent:

(a) Hydroxylamine hydrochloride—Dissolve 43.75 g reagent grade hydroxylamine hydrochloride in approximately 500 mL deionized water. Transfer to a 1-L volumetric flask and dilute to volume with deionized water. Prepare fresh every 2 weeks.

(b) Sodium acetate—Dissolve 52.75 g anhydrous sodium acetate in approximately 200 mL deionized water. Transfer to a 250-mL volumetric flask and dilute to volume with deionized water. Prepare fresh every 2 weeks.

(c) Mixed reagent—Mix 200 mL of hydroxylamine hydrochloride solution with 50 mL of sodium acetate solution. Filter through Whatman No. 1. This reagent is stable for 2 weeks.

Potassium phosphate–acetone reagent: In a 1-L volumetric flask dissolve 144 g potassium hydrogen phosphate ($K_2HPO_4 \cdot 3H_2O$) in approximately 500 mL deionized water. Add 200 mL reagent grade acetone and dilute to volume with deionized water. Mix well, and filter through Whatman No. 1 paper. Prepare fresh every 2 weeks.

Sodium potassium tartrate–ammonium hydroxide solution: Dis-

Table 20–20. Preparation of diacetyl calibration standards from 40 mg/L working standard.

Volume working standard (mL)	Final diacetyl concentration in 100-mL volumetric flask (mg/L)
12.5	5
25	10
50	20
100	40

solve 900 g Rochelle salt (sodium potassium tartrate) in approximately 1,200 mL deionized water with gentle warming. Cool to room temperature, add 387 mL concentrated ammonium hydroxide, and mix. Dilute to 2 L with deionized water and filter through Whatman No. 1 paper. Prepare fresh every 2 weeks.

Ferrous sulfate (3.5% wt/vol)–sulfuric acid (1%) solution: Add 1.45 mL concentrated sulfuric acid to approximately 200 mL deionized water. Dissolve 8.75 g ferrous sulfate in this solution, mix, and dilute to 250 mL with deionized water. Filter through Whatman No. 1 paper. Prepare fresh daily.

III. Procedure

(a) Preparation of Standards

1. Prepare a diacetyl working standard solution (40 mg/L diacetyl) by diluting 50 mL of diacetyl stock to 1,000 mL with deionized water.
2. Prepare diacetyl calibration standards of 5, 10, 20, and 40 mg/L in 100-mL volumetric flasks (see Table 20–20).

(b) Distillation of Wine Sample

3. Transfer 1.5 mL hydroxylamine–acetate reagent and 10- to 15-mL deionized water to a 100-mL graduated beaker.
4. Place the beaker under the delivery tip from the condenser so that the tip is beneath the surface of the liquid in the beaker. Position an ice bath such that the lower part of the beaker is surrounded by ice.
5. Connect a tube from the CO_2 cylinder to the Pasteur pipette in the three-neck flask. Start the flow of CO_2 through the system at a rate of approximately one bubble per second.
6. Add sample (125 mL) of wine to the distillation flask. Rinse the contents into the flask with deionized water. Place stopper into neck of flask.
7. Begin heating the distillation flask. Once distillation begins, continue for approximately 40 min or until 60 mL of distillate have been collected in the receiver beaker.

8. Place the receiver beaker on a steam bath and concentrate the volume of distillate to approximately 20 mL. Allow the concentrated distillate to cool.

(c) Spectrophotometric Analysis of Diacetyl

9. Transfer the concentrated distillate with several small deionized water rinsings into a 50-mL volumetric flask.
10. Transfer 25 mL of each diacetyl calibration standard plus 1.5 mL hydroxylamine-acetate solution into separate 50-mL volumetric flasks.
11. Prepare a blank by transferring 25 mL deionized water plus 1.5 mL hydroxylamine-acetate solution into another 50-mL volumetric flask.
12. To each flask add 4 mL potassium phosphate-acetone solution, mix, and let sit 5 min.
13. To each flask add 12.4 mL potassium tartrate–ammonium hydroxide solution, 0.8 mL ferrous sulfate (3.5%)–sulfuric acid solution, and enough deionized water to bring contents to volume. Mix well.
14. Set instrument at zero (or 100% T) with the prepared blank. Read absorbance of each flask at 520 nm using 1-cm cuvettes.
15. Prepare a calibration curve from the absorbance readings of the standards. Read the concentration of diacetyl in the wine sample distillate directly from the calibration curve.
16. The concentration of diacetyl in the original wine sample is obtained by multiplying the value obtained from the calibration curve by a factor of 5:

$$\text{Diacetyl (mg/L)} = \text{value from graph} \times \frac{125}{50} \times \frac{50}{25}$$

IV. Supplemental Notes

1. The first correction factor (125/50) accounts for the fact that 125 mL of wine produced 50 mL of distillate plus reagents. The second correction factor (50/25) accounts for the dilution of standards (plus reagents) into 50-mL volumetric flasks.
2. There is some evidence that diacetyl production during MLF may be related to grape variety (Steinschrieber 1984).

DIMETHYLDICARBONATE: GAS CHROMATOGRAPHIC ANALYSIS

Dimethyldicarbonate in wine reacts to form breakdown products of methanol, carbon dioxide, and ethyl methyl carbonate. The ethyl methyl car-

Table 20–21. Preparation of ethyl methyl carbonate calibration standards from 500 mg/L EMC stock standard.

Volume EMC stock standard (mL)	Final concentration (mg/L) in 1,000 mL volume[a]
2.0	1.0
5.0	2.5
10.0	5.0
15.0	7.5

[a]Dilute to volume with 12% (vol/vol) ethanol

bonate residues in wine can be determined using conventional extraction and gas chromatographic techniques. An estimate of the original addition level of dimethyldicarbonate can be calculated from the ethyl methyl carbonate and percent ethanol analytical results (Stafford and Ough, 1976).

I. Equipment

Gas chromatograph equipped with a flame ionization detector and a stainless steel column: 10 feet × 1/8 inch o.d. packed with 10% Carbowax 20M on 100/120 mesh Chromosorb G. Instrument conditions include: helium carrier flow rate = 20 mL/min, oven temperature = 160°C, injector temperature = 150°C, and detector temperature = 200°C.

Hamilton syringe, 10 µL

Strip chart recorder or electronic integrator

1-L and 250-mL volumetric flasks

II. Reagents

Carbon disulfide: In a separatory funnel wash 200 mL carbon disulfide with 20 mL fuming nitric acid. Draw off excess acid and wash organic layer with 20 mL portions of deionized water until neutral.

Ethyl methyl carbonate (EMC) standard: Place 50 mg of EMC in a 100-mL volumetric flask. Bring to volume with 95% (vol/vol) ethanol.

Diethylcarbonate (DEC) internal standard: Place 50 mg DEC in a 100 mL volumetric flask. Bring to volume with 95% vol/vol ethanol.

III. Procedures

1. Prepare calibration standards of ethyl methyl carbonate (EMC) in 12% (vol/vol) ethanol at concentrations of 1, 2.5, 5, and 7.5 mg/L from EMC standard solution (see Table 20–21)
2. Add 100 mL wine sample *or* working standard plus 1-mL DEC internal standard solution plus 1-mL absolute ethanol to a 250 mL separatory funnel.
3. Extract sample with 20 mL carbon disulfide for 10 min using a gentle rolling motion. Allow layers to separate.

4. Transfer 10 mL of organic phase to a 15-mL conical bottom centrifuge tube. Centrifuge for 2–3 min in a clinical centrifuge.
5. Transfer 5 mL of clarified extract to a glass vial and store in freezer until analysis.
6. Inject 5 μL of sample or working standard extract into the gas chromatographic column for ethyl methyl carbonate (EMC) analysis. Approximate retention times should be 6–7 min for EMC and 8 min for DEC.
7. Determine peak height ratios for EMC/DMC. Using ratios from injected working standards, prepare a standard curve of EMC/DEC ratios vs EMC concentrations (in mg/L). Compare the peak height ratios for wine sample extracts to the standard curve to determine their EMC concentrations.
8. Calculate the original dimethyldicarbonate addition level in the wine samples from the following equation:

$$\text{DMDC (mg/L)} = \frac{\text{EMC conc (mg/L)} \times 100}{(0.39) \times \text{\%ETOH (vol/vol)}}$$

IV. Supplemental Notes

1. The factor of 0.39 comes from linear regression data of the average amount of EMC produced vs. the % (vol/vol) ethanol in various wine samples (both red and white wines) when 100 mg/L of DMDC was added to those wines. The factor for white wines alone is 0.41; the factor for red wines alone is 0.37.

ETHANOL: DETERMINATION BY EBULLIOMETRY

Ebulliometry involves a measurement of the boiling point decrease caused by the presence of alcohol in a wine sample. Mechanically, the boiling point of the sample is measured relative to the boiling point of pure water, and the difference is related to percent ethanol. Sugar (>2% R.S.) is the major interference in this determination.

I. Equipment

Ebulliometer, mercury thermometer (°C), appropriate alcohol scale
Deionized water and cold tap water source
Microburner or alcohol lamp

II. Reagents

Sodium Hydroxide (1%) Cleaning Solution: Dissolve 10 g of sodium hydroxide in 990 mL tap water. Identify as "Ebulliometer Cleaning Solution."

III. Procedure (see Fig. 20–8)

(a) Determination of the boiling point of water:

1. Add approximately 30 mL of deionized water to boiling chamber "A." There is no need to add cold tap water to condenser "D" at this time.
2. Insert thermometer "C." Position instrument over flame. Before use, inspect thermometer, making certain that mercury column is not separated. If you elect to use a microburner rather than the alcohol lamp included with the kit, the ebulliometer should be positioned high enough above the flame to prevent excessive "bumping." The latter is a sign of overheating and impedes proper boiling point determination.
3. When thermometer reaches a stable point, allow 15–30 sec for minor fluctuations to occur. At this time, take boiling point reading and set inner scale opposite 0.0% alcohol on the "Degre Alcoholique Du Vin" outer scale.
4. Cool and drain instrument.

(b) Determination of the boiling point of wine:

5. Rinse boiling chamber with a few milliliters of wine to be analyzed and drain (This prevents dilution of sample).
6. Place approximately 50 mL of wine in boiling chamber.
7. Fill condenser with cold tap water.
8. Insert thermometer as shown in Fig. 20–8, and place instrument over heat source. Once again, check mercury column to ensure continuity of liquid.
9. When thermometer reaches a stable level, allow 15–30 sec for changes and take reading.
10. Locate the boiling point of wine on the inner "Degres du Thermometre" scale and record the corresponding alcohol content (% vol/vol) on the outer scale. *Alternatively:* Refer to Fig I-1 in Appendix I; the alcohol content can be read directly from the chart by knowing the difference in the boiling points for water and your wine sample.
11. Cool and thoroughly rinse the instrument.

IV. Supplemental Notes

1. Because the boiling point will vary with atmospheric pressure, it *must* be determined for the deionized water reference at least twice daily and more frequently during periods of unstable weather.
2. Reducing sugar levels exceeding 2% interfere with the procedure, yielding erroneously low boiling points (hence higher apparent alcohols). The effect of sugar may be overcome by carrying out an initial separation (distillation) and determining the boiling point of the collected distillate. Alternatively, one may simply dilute sweet

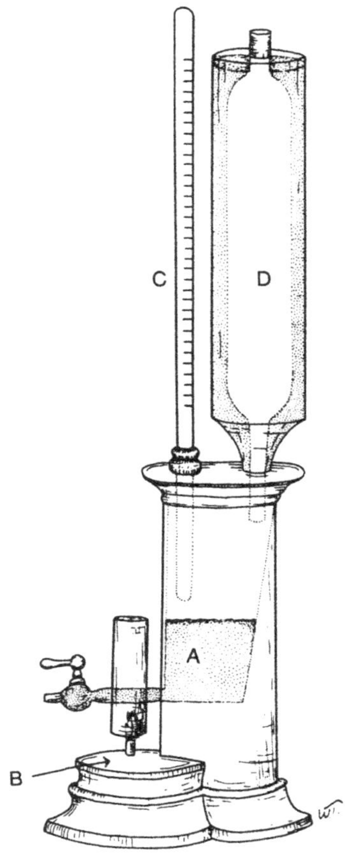

Fig. 20–8. Schematic representation of typical ebulliometer.

wines to a level of less than 2% reducing sugar before analysis, or use the correction factor (reducing sugar × 0.05) and subtract it from the apparent alcohol content.

3. To prevent premature loss of alcohol, cooling water in the condenser must be cold (ice water preferably). If the condenser water is allowed to warm, alcohol can be volatilized and lost, resulting in erroneously high boiling points and lower apparent alcohols.
4. In newly fermented wines, excessive foaming may be prevented by addition of antifoaming agents.
5. During extended use, debris will coat walls of boiling chamber, reducing efficiency. It is recommended that the boiling chamber be cleaned periodically with a boiling solution of 1% NaOH.

ETHANOL: GAS CHROMATOGRAPHIC ANALYSIS

Ethanol in a diluted wine sample can be separated from other wine components by gas chromatography. To improve quantitation, 2-propanol (used as an internal standard) solution is used to quantitatively dilute the sample. The peak area ratio for the two chromatographic peaks is compared with the area ratio obtained from injection of a standard ethanol-internal standard mixture.

I. Equipment

Gas chromatograph equipped with flame ionization detector(FID)
Electronic integrator (data processor) or strip-chart recorder
Hypodermic syringe (10 μL)

II. Reagents

Internal standard solution of 2-propanol (*iso*-propanol) (0.2% vol/vol): Dilute 2.0 mL reagent grade 2-propanol to 1,000 mL with deionized water. Mix well. Each standard or sample analyzed will require approximately 100 mL of this solution.

Ethanol standard solution (10%): Dilute 10.0 mL absolute ethanol to 100 mL in a volumetric flask at 20°C with deionized water. Standards can be similarly prepared to cover other anticipated sample concentration ranges (dessert wines, coolers, etc.).

III. Procedure

1. Install one of the following columns in the GC oven.
   (a) 6 ft × 2 mm i.d. glass, packed with 80–100 mesh Poropak QS.
   (b) 6 ft × 2 mm i.d. glass, packed with 0.2% Carbowax 1500 on 80–100 mesh Carbopack C.
   (c) 6 ft × 1/4 inch o.d. copper or stainless steel, packed with 3% Carbowax 600 on 40–60 mesh Chromosorb T.
2. Set the operating conditions as listed for the gas chromatograph (see Table 20–22).
3. Adjust the air and H_2 gas flows to those specified in instrument operation directions (approximately 300 and 30 mL/min, respectively). These may have to be altered somewhat to optimize conditions for the particular carrier gas flow used.

Table 20–22. Operational conditions for gas chromatograph.

Conditions	Col. a	Col. b	Col. c
Carrier gas	N_2	N_2	N_2
Flow rate (mL/min)	30	15	55
Oven temp, °C	200	105	88
Injector temp, °C	225	150	150
Detector temp, °C	225	150	150

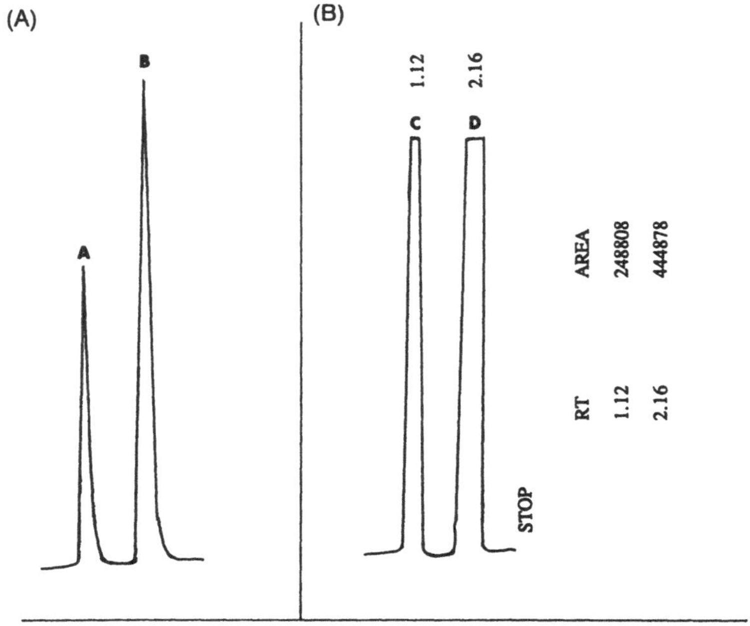

Fig. 20–9. (A) Typical strip-chart recorder and (B) integrator printout of chromatogram of wine sample. The average peak area ratio from three injections is 0.554, whereas the average peak area ratio from three injections of a 11.6% ethanol standard is 0.452. The calculated % ethanol for this sample is 14.2%.

4. Dilute the alcohol standard solution 1 + 99 with the 2-propanol internal standard solution. Inject at least three individual 1.0-μL aliquots into the gas chromatograph and record the resulting chromatograms. Determine the peak area (integrator) or peak height ratios (recorder) for the alcohol peak to the 2-propanol peak. Calculate the average of the three response ratios (*RR′*).
5. Dilute the sample 1 + 99 with the 2-propanol internal standard solution. Again inject at least three 1.0-μL aliquots into the gas chromatograph and record the resulting chromatograms. Determine the response ratios for the alcohol peak to internal standard peak. Calculate the average of the several response ratios (*RR*). See Fig. 20–9 for a typical chromatogram.
6. Calculate the percent alcohol in the wine sample:

$$\%\ \text{Alcohol} = \frac{RR \times \%\text{Alcohol in std}}{RR'}$$

IV. Supplemental Notes
1. Oven temperatures may be altered to a limited extent to optimize chromatographic conditions for the specific GC column being used.
2. A calibration curve may be constructed by preparing three or more alcohol standard solutions covering the range of expected sample alcohol concentrations. Follow the above procedure, and record the resulting chromatograms. The curve is constructed by plotting the average response ratios for each standard chromatographed against the concentration of alcohol in that standard. The unknown values are read directly off the calibration curve.

ETHANOL: DETERMINATION BY HYDROMETRIC ANALYSIS

Distillation of the alcohol from a wine sample followed by measurement of the specific gravity of the ethanol-water distillate can provide an accurate determination of the original alcohol content. Volatile acidity and sulfur dioxide can interfere with this analysis. Control of the temperature of the initial wine sample and final distillate is critical to accurate measurements.

I. Equipment
Distillation apparatus as seen below
200-mL Kohlrausch receiving flask
Alcohol hydrometer
Hydrometer cylinder
Thermometer
Deionized water

II. Reagents
Antifoaming agent (Tween 80 or equivalent)
Sodium Hydroxide (2 *N*): There is no need to standardize this reagent. Dissolve 82 g of NaOH in approximately 800 mL of deionized water and bring to 1 L final volume at room temperature.

III. Procedure
1. Using several milliliters of wine sample, rinse the 200-mL Kohlrausch receiving flask. Fill flask to within 1 cm of the volume mark with wine sample.
2. Adjust temperature to 20°C and bring to volume with wine.
3. Transfer contents to the Kjeldahl distillation flask, ensuring complete transfer with several rinses of deionized water. To prevent charring, add 40–50 mL of deionized water to distillation flask. In the case of young or sweet wines, a drop of antifoaming agent may be added to prevent excessive foam formation during the distillation.

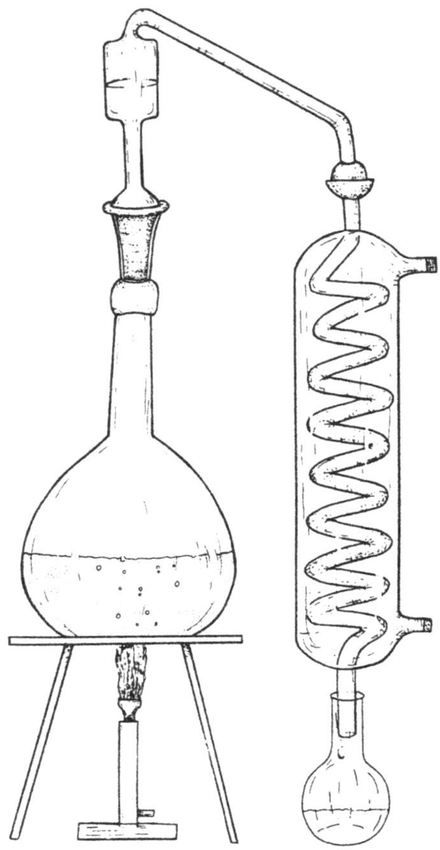

Fig. 20–10. Simple distillation apparatus for separation of alcohol.

4. Position and connect flask as shown in Fig. 20–10, ensuring a tight seal between expansion bulb, flask, and condenser.
5. Turn on condenser water. Adjust heat input to yield a moderate boiling.
6. Collect approximately 195 mL of distillate. If a subsequent extract analysis is desired, save residue in the Kjeldahl flask. (Refer to procedure for details of extract determination.)
7. Adjust temperature of distillate to 20°C, and bring to volume with deionized water. Mix thoroughly.
8. Transfer distillate to a hydrometer cylinder and immerse hydrometer of the appropriate range into the contents.
9. After hydrometer reaches equilibrium, take reading. Record tem-

perature of distillate and make corrections as necessary using Tables I6 and I7 provided in Appendix I.

IV. Supplemental Notes

1. Volatile acidities exceeding 0.10% and sulfur dioxide level greater than 200 mg/L may interfere with this method. Therefore, wines with excessive amounts of either or both should be neutralized with 2 *N* NaOH before distillation. However, such neutralized wines *cannot* be used for subsequent extract analysis.
2. Most analysts prefer to calibrate new alcohol hydrometers against accuratly prepared standards.

ETHANOL: HPLC PROCEDURE FOR GRAPE INSPECTION (SEE ACETIC ACID ANALYSIS)

ETHANOL: DETERMINATION BY TITRAMETRIC (DICHROMATE) ANALYSIS

Acidified dichromate can oxidize the alcohol collected by distillation from a wine sample. Excess dichromate is then titrated with iron (II) solution. The ratio of volumes of titrant consumed by the wine sample (V_A) and the blank (V_B) is used to calculate the ethanol content. Careful control of the sample temperature, and volume of sample analyzed, as well as the concentration of dichromate solution are critical for accurate measurements.

I. Equipment

Micro-Kjeldahl distillation apparatus (see Fig. 20–11)
50-mL burette
Constant temperature water bath (60°–65°C)
50-mL Erlenmeyer flasks (preferably graduated)
500-mL Erlenmeyer flasks
Volumetric pipettes (1 and 25 mL)
Wash bottle
High-intensity light source

II. Reagents

Ferrous Ammonium Sulfate: In a 2-L volumetric flask, dissolve 270 g of $FeSO_4(NH_4)_2SO_4 \cdot 6H_2O$ in approximately 1,500 mL of deionized water. Carefully add 50 mL of concentrated H_2SO_4 and bring to volume with deionized water.

Potassium Dichromate Solution (0.1148M): In a 2-L volumetric flask, dissolve 67.536 g of $K_2Cr_2O_7$ in approximately 1,000 mL of deionized water. Carefully add 650 mL of concentrated H_2SO_4 and bring to volume with deionized water.

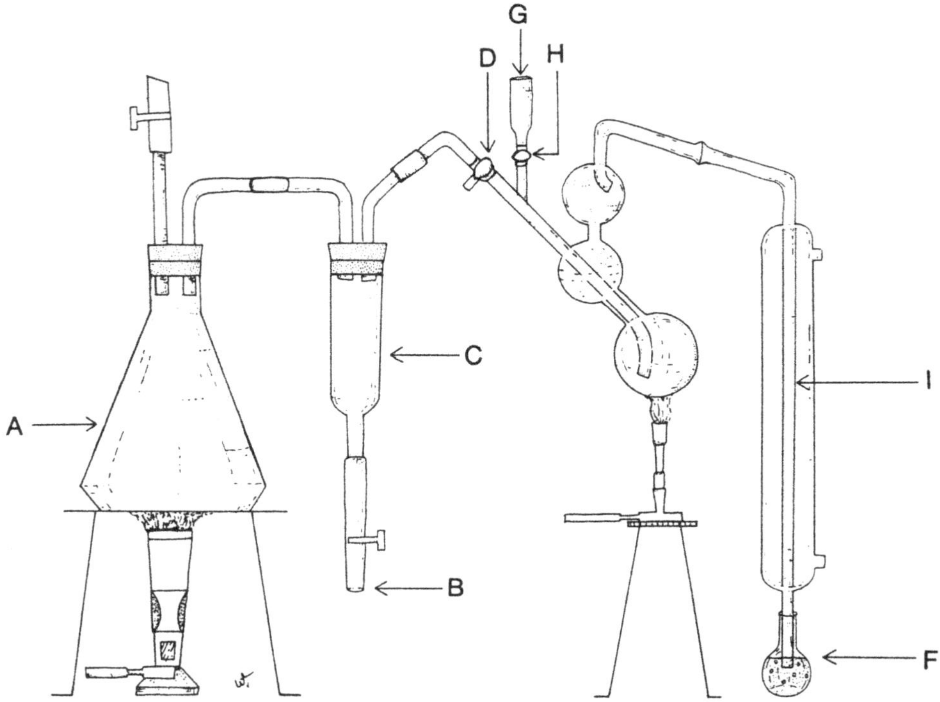

Fig. 20–11. Schematic representation of micro-Kjeldahl distillation apparatus. Lettered components are described in the text.

1,10-phenanthroline-ferrous sulfate indicator: In a 500-mL volumetric flask, dissolve 3.48 g ferrous sulfate ($FeSO_4 \cdot 7H_2O$) in approximately 250 mL of deionized water. Add 7.43 g of *o*-phenanthroline and dilute to volume with deionized water. Transfer to convenient-sized dropper bottles.

III. Procedure

(a) Preparing the Still:

1. At the beginning of distillation, water in steam generator "A" should be boiling and condensate passing through condenser exit "I."
2. Turn stopcock "D" such that steam from trap can escape through side exit.
3. Receiving flask "F" is placed such that the exit tube of condenser "I" is immersed in dichromate solution.
4. Sample stopcock "H" is in the closed position, and funnel "G" contains a small amount of deionized water.

(b) Sample Distillation:

5. Volumetrically transfer 1.0 mL of wine sample into funnel "G."
6. Open stopcock "H" to admit wine to still.
7. Rinse funnel 2–3 times with deionized water. Close stopcock "H."
8. Volumetrically transfer 25 mL of dichromate solution to a 50-mL Erlenmeyer receiving flask, and position under condenser. The tip of condenser should be immersed in the solution.
9. With stopcock "H" closed, turn stopcock "D" so that steam will flow from steam trap to receiver.
10. Position and ignite microburner. *Caution!* Use only a *very low flame* to prevent breakage. (*Note:* The above steps should be carried out in a sequential and continuous manner).
11. Collect distillate in the receiving flask to a final volume of approximately 45 mL.
12. When distillation is complete, lower the flask and rinse the outside of condenser tip with deionized water.
13. Remove microburner from its position under distilling bulb.

(c) Cleaning Operation:

14. Admit a little deionized water into the steam generator. This will cause dealcoholized residue in still bulb to siphon into steam trap "C."
15. Place 10–15 mL of deionized water into still through funnel "G" and repeat siphoning operation. Upon completion, leave funnel full of deionized water.
16. Drain steam trap at "B."

(d) Post-Distillation Sample Treatment:

17. Stopper flask and place in water bath at 60° – 65°C for 25 min.
18. Quantitatively transfer the contents of flask to 500-mL Erlenmeyer flask (using deionized water rinses), and titrate with ferrous ammonium sulfate. When the color of solution turns to an emerald green, add five drops of indicator. Continue the titration until color changes (sharply) from blue-green to brown. Some lab personnel prefer to use reflected light to assist in end point detection.
19. Record volume of FAS used in sample titration and relate it to the volume of FAS used to titrate the "blank" using the equation below. The blank consists of 1 mL of distilled water in 25 mL of dichromate incubated at 60° – 65°C for 25 min.

$$\text{Ethanol (\% vol/vol)} = 25 - 25\,[V_A/B_B]$$

where: V_A = volume of FAS used with wine sample
V_B = volume of FAS used with water blank

IV. Supplemental Notes

1. Alcohol in the distillate is quantitatively oxidized to acetic acid by an excess of standardized dichromate:

$$3C_2H_5OH + 2Cr_2O_7^{2-} + 16H^+ \rightarrow 3CH_3COOH + 4Cr^{3+} + 11H_2O$$

   Unreacted excess dichromate is then determined by titration with standardized FAS.

2. In order to monitor the quality/strength of the FAS, it is recommended that one blank titration be run in the morning and another in the afternoon.
3. When the end point is approached, titration should be on a drop-by-drop basis. If over-titration occurs, add 0.5 mL of dichromate solution and again titrate to proper end point. The additional volume of dichromate used should be added to the initial 25 mL.

ETHANOL: VISIBLE SPECTROMETRIC—DICHROMATE OXIDATION

Some laboratories use a modification of the micro-dichromate procedure presented in the previous section. However, instead of titration of the excess dichromate (after heating to 60°C) with FAS, the sample is measured spectrophotometrically at 600 nm. This result is compared with the absorbance of standard solutions of known alcohol levels. The following procedure is that of Caputi et al. (1968).

I. Equipment

See previous procedure for setup of distillation unit

Spectrophotometer

II. Reagents

Potassium dichromate solution: Prepare reagents as per acid dichromate procedure presented above. Preparation of standard alcohol series is presented in body of procedure.

Ethanol (200 proof)

III. Procedure

1. Prepare a series of alcohol standard solutions by diluting the indicated volumes of 200 proof (100%) ethanol to 250 mL final volume with deionized water (see Table 20–23).
2. Transfer 1.0 mL aliquots of the above standard solutions to individual 50 mL volumetric flasks containing 25 mL acid dichromate. Add approximately 20 mL of deionized water.

Table 20–23. Preparation of ethanol calibration standards from 200 proof ethanol.

Volume 200 proof ethanol (mL)	Final concentration (% vol/vol) in 250 mL volume
10	4
20	8
25	10
50	20

3. Stopper the flasks, and incubate in a 60°C water bath for 20 min.
4. Cool to standard temperature, and bring to volume with deionized water.
5. Mix the solutions thoroughly, and measure the absorbance relative to a blank of dichromate and deionized water at 600 nm.
6. Plot the concentration of each standard vs. its respective absorbance at 600 nm.
7. Distill the wine samples, and proceed as in steps 5–11 in the previous dichromate procedure, comparing the final absorbance to standard curve.

ETHYL ACETATE: GAS CHROMATOGRAPHIC ANALYSIS (SEE FUSEL OILS)

ETHYL CARBAMATE: GAS CHROMATOGRAPHIC ANALYSIS

Ethyl carbamate residues in wines and distillates can be determined using conventional extraction and capillary gas chromatographic techniques. The following procedure is adapted from the OIV (1990) procedure. It uses sodium sulfate to dewater the wine sample and enhance the extraction of ethyl carbamate into diethyl ether. By raising the pH of the wine sample, one avoids extracting volatile acids in the wine.

I. Equipment
   Gas chromatograph equipped with a NP or Mass Spectrometer (MSD) detector
   Capillary column: 30 meters × 0.32 mm i.d., coated with Carbowax 20M (polyethylene glycol), film thickness 0.25 μm
   Rotary vacuum evaporator
   Hamilton syringe, 10 μL
   Volumetric flasks, 20-mL and 100-mL
II. Reagents

Stock ethyl carbamate solution (1,000 mg/L): Dissolve 100 mg ethyl carbamate in 1:1 ethanol:water in a 100-mL volumetric flask. Mix to dissolve and bring to volume with ethanol:water.

Working standard of ethyl carbamate (10 mg/L): Transfer 1.0 mL stock ethyl carbamate to a 100-mL volumetric flask. Bring to volume with 1:1 ethanol:water.

Internal standard, methyl or propyl carbamate (20.0 mg/L): Dissolve 200 mg methyl or propyl carbamate in 1:1 ethanol:water in a 100-mL volumetric flask. Mix to dissolve and bring to volume with ethanol: water. Dilute 1.0 mL of this solution in a second 100-mL volumetric flask with 1:1 ethanol:water. Mix well and label as "20 mg/L internal standard." *Note:* methyl carbamate is used with the NP detector; propyl carbamate is preferred with the MSD.

Ethanol:water (1:1): Mix equal volumes of absolute ethanol and deionized water.

Disodium hydrogen phosphate ($Na_2HPO_4 \cdot 2H_2O$)

Sodium sulfate

Diethyl ether

III. Procedures

i) Sample preparation

1. Pipette a 10.0 mL wine sample into a 100-mL flask. Pipette another 10.0 mL wine sample and 10 μL ethyl carbamate standard into a second 100-mL flask. Add 100 μL of the internal standard solution to each flask and mix. To each flask: add 2 g disodium hydrogen phosphate and mix, add 40 mL diethyl ether and 40 g sodium sulfate.
2. Place magnetic stirring bar into each flask and stir for 5 minutes.
3. Decant the ether phase and place into separate 200-mL rotary evaporator flasks.
4. Wash the salt two times with 20 mL portions of diethyl ether. Combine the ether rinses with the material in each evaporator flask. Evaporate the ether phase cold (20°C) to about 2 mL.
5. Add 4 g sodium sulfate and 10 mL diethyl ether to the concentrated extracts. Mix for about 3 minutes.
6. Transfer each ether phase into a 25-mL evaporator flask. Wash each salt residue with 6 mL diethyl ether and add to the appropriate flask.
7. Concentrate each of the extracts to 1 mL.

ii) Chromatography

8. Following manufacturer's guidelines set up the following instrumental conditions: Splitless injector @ 210°C and a split time of 15 sec; NP detector @ 210°C with purified air and hydrogen flows set

to optimize sensitivity and stability; MSD set according to manufacturer's recommendations; carrier (helium) pressure set to maintain column flow rate @ 4 mL/min, a septum purge of ~0.5 mL/min, and recommended vent flow. The oven temperature is programmed from 60 to 200°C at 5°C/min, with a final hold time of 15 min.

9. Inject 2 μL of the sample extract into the gas chromatograph. Follow with a 2 μL injection of the "sample plus standard" extract.
10. Determine peak area ratios for ethyl carbamate/Internal standard from the "sample" and "sample plus standard" chromatograms, and designate as follows:

$$W = \frac{A_{(\text{Ethyl Carbamate})}}{A_{(\text{Internal Standard})}} \text{ for wine sample}$$

$$S = \frac{A_{(\text{Ethyl Carbamate + Standard})}}{A_{(\text{Internal Standard})}} \text{ for wine + standard sample}$$

11. Calculate the ethyl carbamate concentration as follows:

$$C_{(\text{Ethyl Carbamate})} = \frac{W}{S-W} 100$$

IV. Supplemental Notes

1. With brandies and spirits use 50 μL of a 100 mg/L internal standard to 5 mL of brandy. It may be necessary to increase the amount of standard ethyl carbamate for proper quantification.
2. With the MSD and propyl carbamate internal standard monitor the ions of m/e 60 and 89.

4-ETHYL PHENOL: GAS CHROMATOGRAPHY OF METABOLITE FROM *BRETTANOMYCES/DEKKERA* (SEE *BRETTANOMYCES*)

EXTRACT: DETERMINATION BY SPECIFIC GRAVITY

The *Official Methods of Analysis of the Association of Official Analytical Chemists* (1990) determines extract by taking the specific gravity of the original sample and subtracting the specific gravity of the aqueous ethanol solution distilled from the sample. After adding the specific gravity of water back in, the specific gravity is converted to appropriate units by multiplying by the corresponding °Brix.

I. Equipment
   Pycnometer or accurately graduated specific gravity hydrometer
   Distillation apparatus (see Fig. 20–10)

II. Reagents
   Deionized water

III. Procedure
   1. Determine the specific gravity (20°/20°) of initial wine sample using an accurate hydrometer or by pycnometry. An alternative way is to measure the °B and convert to specific gravity using Table I-2 in appendix I.
   2a. Volumetrically transfer a 200 mL wine sample to distillation flask. Add approximately 30 mL deionized water.
   2b. Collect approximately 195 mL of distillate. Adjust to volume with deionized water at standard temperature.
   2c. Determine specific gravity of distillate.
   3. Alternatively, one can determine the alcohol and convert the measured %ethanol (vol/vol) to specific gravity using Table I-12 in Appendix I.
   4. Calculate extract value according to the following equation:

$$D = S - S' + 1$$

   where:
   S = specific gravity of original sample
   S' = specific gravity of aqueous ethanol distillate
   D = specific gravity of dealcoholized sample

   5. Using Table I-2 in Appendix I, locate °B value corresponding to "D."
   Multiply this derived result by the value of "D":

$$\text{Extract value of wine (g/L)} = \text{Specific gravity (20/20)} \times \text{Corresponding °B (g/100 g)}$$

   6. To express the extract value in g/L, multiply the final result by 10.

IV. Supplemental Notes
   1. The sample distilled in step 2 may be treated with NaOH to eliminate the carryover of volatile acidity, sulfur dioxide, or carbon dioxide. This will not interfere with the measurement of specific gravity of the distillate.

2. Extract determinations have been used to detect falsification of wines. Gilbert (1976) proposed the following relationship for calculating a residual (corrected for major acid components) sugar-free extract:

$$E = R - 0.9T - 0.8A - 0.05\ ET$$

where:
E = sugar free extract (g/L)
R = apparent extract − reducing sugar (g/L)
T = titratable acidity (g/L)
A = ash (g/L)
ET = ethanol (g/L)

EXTRACT: DETERMINATION USING BRIX HYDROMETER

Extract is determined by distilling the alcohol out of the wine sample. The nonvolatile residue is brought back to original volume and temperature and its specific gravity (measured as °Brix) determined. A unit conversion produces an answer for extract in g/L.

I. Equipment
   Brix or Balling hydrometer (range: 0–8 °Brix)
   200-mL Kohlrausch receiving flask
   Hydrometer cylinder

II. Procedure
   1. Upon completion of alcohol distillation using 200 mL of initial sample, carefully pour the residue into a clean, dry 200-mL Kohlrausch flask. This step should be followed with several deionized water rinses to ensure complete transfer.
   2. Cool to 20°C and bring to volume with deionized water.
   3. Mix carefully to ensure sample uniformity.
   4. Transfer contents to a clean, dry hydrometer cylinder.
   5. Insert hydrometer into solution. When constant level is achieved, record results as g/100 g of solution (the definition of Brix). To convert results to g/100 mL (as per definition of extract) read the specific gravity corresponding to °Brix using Table I-2 in Appendix I and multiply this figure by the recorded °Brix. To express as g/L, multiply final result by 10.

IV. Supplemental Notes
   1. Extract determinations cannot be accurately run on wines that have been neutralized with NaOH before distillation, unless one uses a

double distillation technique. The use of concentrated NaOH (rather than 1–2 *N*) will minimize volume changes and mass increases.

2. It should be emphasized that the Marsh nomograph (Fig. I–3 in Appendix I) is restricted to dessert wines only.
3. Data derived from hydrometric procedures are reported to ±0.1 °Brix; so any report of extract values, using this method, should not exceed this limitation.

EXTRACT: NOMOGRAPHIC METHODS (SEE CHAPTER 6 AND FIGURES I-2 AND I-3 IN APPENDIX I)

FIBROUS MATERIALS: (SEE DIAGNOSTICS)

FLAVONOIDS: SEE PHENOLS: SPECTRAL ESTIMATION) (SEE PHENOLS & FOLIN-CIOCALTEU METHOD)

FLUORIDE: MEASUREMENT USING ION SELECTIVE ELECTRODE

The common technique currently for fluoride analysis involves the use of potentiometry and a solid membrane ion selective electrode. The fluoride electrode responds preferentially to fluoride ion. It requires a high quality expanded scale pH meter (mV scale capable of reading to the nearest 0.1 mV) or Select-Ion meter, and a reference electrode designed for use with selective ion electrodes.

I. Equipment

Expanded scale pH/mV meter or specific ion meter
Fluoride selective ion electrode
Single-junction reference electrode (calomel or Ag/AgCl)
Plastic beakers 100-mL
Pipettes, 25, 10, 5, and 1 mL
Magnetic stirrer and stir-bar

II. Reagents

Fluoride stock solution (1000 mg/L): Dissolve 2.210 g oven-dried (3–4 hours at 105°C) sodium fluoride in deionized water. Make up to 1 liter with deionized water. Store in a clean, dry plastic bottle.

Fluoride working standard (100 mg/L): Transfer 10.0 mL of fluoride stock to a 100-mL volumetric flask. Dilute to volume with deionized water. Transfer to plastic bottle for storage.

Total Ionic Strength Adjustor Buffer (TISAB): This is commercially available.

Sodium acetate (15% wt/vol): Dissolve 150 g sodium acetate with deionized water. Bring to 1 liter volume. Mix and store in a plastic bottle.

III. Procedure

1. Prepare fluoride calibration standards of 1 and 10 mg/L by transferring 1.0 and 10.0 mL of fluoride working standard to respective 100-mL volumetric flasks. Bring to volume with deionized water. Mix and store in plastic bottles. *Note:* the 1 and 10 mg/L calibration standards should be prepared fresh at the time of analysis.
2. Place electrodes in a solution containing 25 mL of the 1.0 mg/L calibration standard, 25 mL of 15% sodium acetate solution, and 5 mL of TISAB. Stir and allow meter reading to stabilize as per manufacturer's recommentation. Record the final mV reading.
3. Rinse electrodes, dry, and place in a solution containing 25 mL of the 10.0 mg/L calibration standard, 25 mL of 15% sodium acetate solution, and 5 mL of TISAB. Stir and allow meter reading to stabilize. Record the final mV reading.
4. Prepare a calibration curve by graphing millivolt readings vs. log concentration. This also may be accomplished by using semi-log paper (graph concentration values on the log axis). *Note:* The slope of the calibration curve should be between −50 and −60 mV.
5. Prepare wine samples the same as standards. Dilute 25 mL wine (or juice) with 25 mL of 15% sodium acetate and 5 mL TISAB in a 100 mL plastic beaker, and stir. Place clean, dry electrodes into the sample, allow the meter to stabilize, and record the MV reading. Refer to the calibration curve and read the fluoride concentration off the graph.

IV. Supplemental Notes

1. The electrodes should be recalibrated every few hours of use.

Most electrode operating manuals will also contain instructions for use with various ion meters, as well as for known addition techniques. The standard or known addition technique is generally used in order for the standards to match the matrix of the wine samples being analyzed. The electrode potential for the wine sample, with appropriate dilution buffer added, is first measured. A known amount of standard fluoride solution is added to this sample, and a second potential reading is taken. The difference in the two potential readings and the known concentration of the fluoride standard are used to calculate the fluoride concentration in the original wine sample.

FRUCTOSE: CHEMICAL ANALYSIS (SEE REDUCING SUGARS) HPLC ANALYSIS (SEE CARBOHYDRATES), SPECTROMETRIC ANALYSIS (ENZYMES) (SEE REDUCING SUGARS)

FUMARIC ACID: HPLC ANALYSIS (SEE ORGANIC ACIDS)

FUSEL OILS: GAS CHROMATOGRAPHIC ANALYSIS

The fusel oil content of distillates can be determined by gas chromatography. Qualitative information is obtained by comparing elution times for the various components with elution times of components in a mixture of standards. Quantitative information can be obtained by comparison of the standard and unknown chromatographic peak areas.

I. Equipment
   Gas chromatograph equipped with flame ionization detector (FID)
   Electronic integrator or strip-chart recorder
   Hypodermic syringe (10 µL)

II. Reagents
   Fusel oil standard stock solution: In a 100-mL volumetric flask, transfer 2 mL each of ethyl acetate, acetaldehyde, and methanol. To this mixture add 1 mL each of *n*-propyl alcohol, *n*-butyl alcohol, isobutyl alcohol, isoamyl alcohol, and 2-methyl-1-butanol ("active amyl" alcohol). Bring to volume using a solution of water-ethanol (1 + 1).
   Ethanol-water (1+1): Dilute 500 mL absolute ethanol to 1 L with deionized water.

III. Procedure
   1. Install one of the following columns in the GC oven.
      (a) 6 ft × 2 mm i.d. glass, packed with 6.6% Carbowax 20M on 80–100 mesh Carbopack B.
      (b) 6 ft × 2 mm i.d. glass, packed with 5% Carbowax 20M on 60–80 mesh Carbopack B.
   2. Set the conditions listed in Table 20–24 for operating the gas chromatograph.
   3. Adjust the air and H_2 gas flows to those specified in instrument operating directions (approximately 300 and 30 mL/min, respectively). These may have to be altered somewhat to optimize conditions for the particular instrument and column used.
   4. Prepare a working fusel oil standard by pipetting 5 mL of the standard stock solution into a 100-mL volumetric flask and diluting to the mark with 1 + 1 ethanol-water solution. See Fig. 20–12 for a typical chromatogram.

Table 20–24. Operational conditions for gas chromatograph (fusel oil analysis).

Conditions	Col. a	Col. b
Carrier gas	N_2	N_2
Flow rate, mL/min	40	40
Oven temp, °C	80	80
Program rate, °C/min	4	4
Final temp, °C	200	210
Injector temp, °C	250	250
Detector temp, °C	250	250

5. Inject 0.5 µL of the working standard into the gas chromatograph. Record the chromatogram and note the retention times of the various components. The components should elute in the order presented in the chromatogram (Fig. 20–12).
6. Inject 0.5 µL of distillate into the gas chromatograph. Record the chromatogram and note the retention times of the various components. Identify fusel oils present by comparison of retention times with those of the standards. An estimate of the amounts of individual fusel oils present may be made by comparison of the peak areas of the distillate sample to those of the standard solution.

IV. Supplemental Notes

1. Carbopack columns are capable of resolving complex mixtures of compounds in a variety of alcoholic beverages such as beer and wine. They can also be used to detect acetic, propionic, and other low molecular weight carboxylic acids that are usually adsorbed on other columns.
2. Carbopack columns work best with small or dilute samples. If too large or concentrated samples are used, tailing and poor separation efficiency will result. This is to be expected with the low percent stationary phase used on the packing.
3. Columns should be purchased with the ends packed with phosphoric acid treated glass wool. Untreated or silanized glass wool will adsorb acetic acid and inhibit its elution from the column.

GALACTURONIC ACID: HPLC ANALYSIS (SEE ORGANIC ACIDS)

GLUCANS: (SEE DIAGNOSTICS)

GLUCONIC ACID: HPLC ANALYSIS (SEE ORGANIC ACIDS)

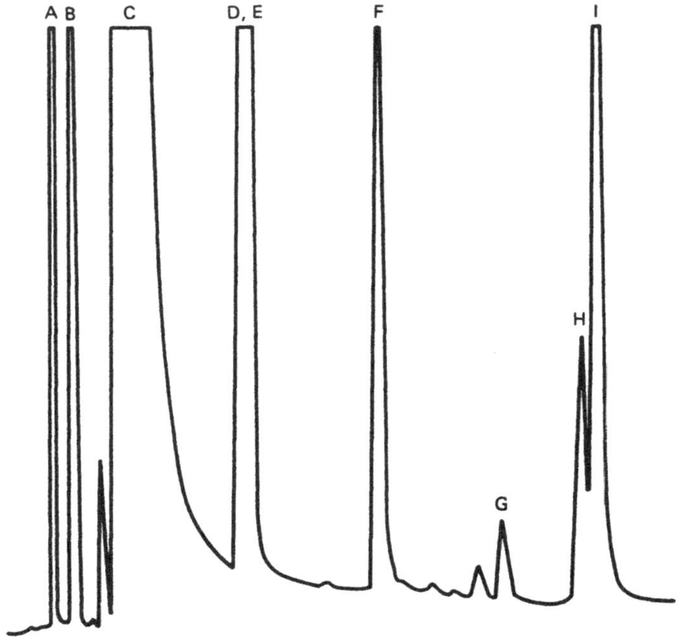

Fig. 20–12. Typical chromatogram of alcohols in 180 proof ethanol. Numbered peaks are: (A) acetaldehyde, (B) methanol, (C) ethanol, (D) ethyl acetate, (E) *n*-propanol, (F) isobutanol, (G) *n*-butanol, (H) active amyl alcohol, (2-methyl-1-butanol), (I) isoamyl alcohol, (3-mnethyl-1-butanol). The column used was 5% Carbowax 20M on Carbopack B 60/80 mesh.

GLUCOSE: CHEMICAL ANALYSIS (SEE REDUCING SUGARS), HPLC ANALYSIS (SEE CARBOHYDRATES), SPECTROMETRIC ANALYSIS (ENZYMES) (SEE REDUCING SUGARS)

GLYCEROL: HPLC ANALYSIS (SEE ACETIC ACID—HPLC PROCEDURE)

GLYCEROL: ENZYMATIC SPECTROMETRIC (VIS) ANALYSIS

Enzymatic procedures for the determination of glycerol have been developed, as reported in the OIV (1990). This procedure relies on the coupled enzymatic reactions:

$$\text{glycerol} + \text{ATP} \xleftrightarrow{\text{GK}} \text{glycerol-3-phosphate} + \text{ADP}$$

$$\text{ADP} + \text{PEP} \xleftrightarrow{\text{PK}} \text{ATP} + \text{pyruvate}$$

$$\text{pyruvate} + \text{NADH} + \text{H}^+ \xleftrightarrow{\text{LDH}} \text{lactate} + \text{NAD}^+$$

GK: glycerokinase
ATP: adenosine-5′-triphosphate
ADP: adenosine-5′-diphosphate
PEP: phosphoenol pyruvate
PK: pyruvate kinase
NADH: reduced nicotinamide adenine dinucleotide
LDH: lactate dehydrogenase

The formation of NAD^+ is stoichiometrically related to the quantity of glycerol, and thus the concentration of the latter may be determined by measurement of residual amounts of NADH present. Kits of reagents (e.g., see Boehringer-Mannheim catalogue) for conducting this procedure are now available to the analyst.

Refer to the product sheet that accompanies the glycerol kit of reagents for up-to-date information and procedural details specific for the analysis of glycerol. In general the glycerol procedure will require the following:

I. Equipment
A spectrometer capable of reading absorbance at 340, 334, or 365 nm plus suitable 1 cm cuvettes
Adjustable micropipettes (20-μL to 2.0 mL)

II. Reagents
As mentioned, the reagents used may be obtained from Boehringer-Mannheim or other supplier. They are sold in the form of a test kit of sufficient material for a number of glycerol determinations.

III. General procedure
1. Dilute a wine sample with deionized water so that the glycerol concentration is between 30 and 500 mg/L. A dilution of 2 mL wine to 100 mL is usually sufficient.
2. The diluted sample is then mixed with reagents (except glycerokinase, GK), and the absorbance values are measured.
3. The glycerokinase, GK, is then added, and the absorbance again is read after the reaction is complete.
4. A blank is run along with the samples, and the absorbance values are corrected for the blank reading.

5. The difference in the corrected absorbance values taken before and after the addition of glycerokinase, GK, is then used to calculate the concentration of glycerol using the equation given in the procedure.

HUBACH ANALYSIS: SEE CYANIDE

HYBRID WINE: FLUORESCENCE PROCEDURE FOR PRESENCE OF VARIOUS HYBRID CULTIVARS

In some countries, hybrid grapes are not authorized for wine production. Recht (1993) reported a screening test for the presence of the diglucoside of malvidol using fluorescence under UV light (365 nm). This procedure is capable of detecting the presence of as little as 1% hybrid wine in blends.

I. Equipment
   Ultraviolet light source (UV 365 nm)
   Test tubes

II. Reagents
   HCl 1 *N*
   190 proof ethanol in 5% ammonia
   Sodium nitrate solution, 1% (wt/vol)

III. Procedure
   1. Add 1 mL red wine to test tube. Add 1 drop HCl and 1 mL sodium nitrate
   2. Shake and set aside for several minutes
   3. Add 10 mL ethanol-ammonia solution
   4. In the dark, shine a UV light onto the tube. A strong green fluorescence is indicative of the presence of hybrids

IV. Supplemental Notes
   1. For comparison, *Vinifera* wine with addition of 15 mg/L malvidol diglucoside (commercially available) can be tested.
   2. For rosé wines it may be necessary to increase volume used to 2–5 mL.

HYDROGEN ION CONCENTRATION: pH METER MEASUREMENT

Hydrogen ion concentration is generally expressed as pH. It is measured using a pH meter and glass electrode setup. Standard buffer solutions are used to calibrate the pH meter, and sample pH values are then read by placing the electrode in an undiluted portion of juice, must, etc.

I. Equipment

pH meter (analog or digital unit) equipped with pH and reference electrodes or a combination electrode

50-mL beaker

II. Reagents

Buffers (pH 7.00 and 4.00): Prepared buffers with indicators for detection of dilution or breakdown are commercially available through chemical supply houses.

Buffer (pH 3.55): Add approximately 5 g potassium acid tartrate to 500 mL deionized water. Mix on magnetic stirring table for 5 min. Allow undissolved crystals to settle and decant, filtering as necessary. At 25°C, the pH of this solution is 3.55.

KCl internal filling solution: Purchase from chemical supply house.

Cleaning solution (75% methanol): Dilute 75 mL absolute methanol to 100 mL with deionized water. Keep container tightly capped.

III. Standardization of Meter

1. Consult operator's manual for standardization using two buffer solutions.
2. Rinse the beaker with sample. Place enough fresh sample in beaker to cover electrode junctions.
3. Place electrode(s) in the sample.
4. Allow meter reading to stabilize and record value.

IV. Supplemental Notes

1. Do not remove electrodes from buffer or sample solutions while instrument is responding. During changing of samples, the meter should be in the "stand-by" mode.
2. When not in use and between measurements, store electrodes according to manufacturer's recommendations.
3. The reference electrode should be checked routinely to ensure an ample supply of KCl filling solution. The level should not be allowed to drop much below the filling port. This port should be uncovered during use.
4. Electrodes should be periodically cleaned to ensure proper operation. This may be achieved by immersing them in the cleaning solution followed by several rinses in deionized water. Protein contamination is most easily removed by soaking electrodes in a solution of pepsin and 0.1 *N* HCl.
5. Ceramic junctions may be cleaned by soaking the electrode in hot (not boiling!) deionized water.
6. If single buffer standardization is used for routine work, a saturated

potassium bitrate solution or a buffer close to the pH of material being analyzed is preferred. Two buffer standardizations normally are used where a higher degree of accuracy is required.

7. Using expanded scale meters, results can be reported to three significant figures (i.e., 3.45) as compared with conventional meters with two-place accuracy (pH 3.4).

HYDROGEN SULFIDE: SENSORY SCREEN (INCLUDES MERCAPTANS)

The following procedure is a modification of the method suggested by Brenner et al. (1954). The technique involves separation of the two groups of volatiles based on sensory examination before and after treatment with copper- and cadmium-containing compounds. In the case of copper [as copper(II)sulfate] addition, reaction with both sulfides and mercaptans yields nonvolatile and relatively odorless products. Cadmium salts, by comparison, are chemically reactive toward hydrogen sulfide, but have little effect on mercaptans.

The following technique provides a quick, relatively simple and inexpensive diagnostic examination of wines suspected of having hydrogen sulfide and/or mercaptan problems.

NOTE!!! Wines evaluated using this technique *should not be tasted!* This procedure is designed for evaluation by smell only.

I. Reagents

Copper(II)sulfate (1% wt/vol): In approximately 90 mL of deionized water, dissolve 1 g $CuSO_4{\cdot}5H_2O$. Bring the solution to a final volume of 100 mL using deionized water.

Cadmium sulfate (1% wt/vol): In approximately 90 mL deionized water, dissolve 1 g $CdSO_4{\cdot}8H_2O$. Bring the solution to a final volume of 100 mL using deionized water.

Ascorbic acid (10% wt/vol): In approximately 90 mL deionized water, dissolve 10 g ascorbic acid. Bring the solution to a final volume of 100 mL using deionized water.

II. Procedure

1. Fill three *clean* glasses with 50 mL of wine to be evaluated.
2. Mark glass 1 as "Control" and set aside.
3. Mark glass 2 as "Copper." To the 50 mL of wine, add 1 mL of the copper sulfate solution and set aside.
4. Mark glass 3 as "Cadmium." Add 1 mL of 1% cadmium sulfate to the 50 mL of wine.
5. Mix each glass thoroughly.

Table 20–25. Interpretation of results from sensory evaluation of cadmium and copper-treated wines.

Glass 1 (Control)	Glass 2 (Copper)	Glass 3 (Cadmium)	Interpretation
Presence of offensive Odor	Odor is gone	Odor is gone	H_2S present
	Odor is gone	No change	Mercaptans
	Odor is gone	Odor is less but not gone	Both H_2S and mercaptans
	No change	No change	*a*

[a]The objectionable odor does not stem from either mercaptan or H_2S sources, *or* methyl mercaptan has been oxidized to dimethyl disulfide.

6. Smell the Control sample, noting the presence and relative intensity of offensive odors. In like manner, evaluate glasses 2 and 3. Attention should be directed toward decrease or absence of a reduced odor.
7. Interpret results of the test as follows with reference to Table 20–25.
8. If odor evaluation results indicate that neither H_2S nor mercaptans are responsible for the offensive odor (last selection in Table 20–21), further evaluation for the presence of dimethyl disulfide should be carried out (see steps 9–12 below).
9. Transfer 50 mL of untreated wine sample to a fresh, clean glass.
10. Add 0.5 mL of the 10% ascorbic acid solution. Mix, and let stand for several minutes.
11. Add 1 mL of copper solution. Mix, and evaluate for changes in the "nose."
12. If the offensive odor is less apparent than in Glass 2, the presence of dimethyl disulfide is suspected. Dimethyl disulfide does not react with copper. It must be reduced back to reactive species, methyl mercaptan (see Chapter 9).

III. Supplemental Notes
   1. The Control sample should not change appreciably over the short course of this test. However, if the objectionable character is diminished as a result of oxidative changes, this would suggest a possible cellaring technique for correction.

HYDROXYCINNAMATES: SPECTROMETRIC (UV) DETERMINATION IN ABSORBANCE UNITS OR AS CAFFEIC ACID EQUIVALENTS (SEE PHENOLS: SPECTRAL ESTIMATION)

IODINE: VISUAL QUALITATIVE TEST FOR RESIDUAL OXIDANT

At low pH (3.5–4.0) and in the presence of mercuric chloride, I_2 is stoichiometrically converted to hypoiodous acid (HOI). The latter reacts with leuco-crystal violet (N,N-dimethylaniline) yielding crystal violet. Upon full color development, the solution is measured spectrophotometrically at 592 nm. Inclusion of ammonium dihydrogen phosphate ($NH_4H_2PO_4$) in the buffer system converts any chlorine present into unreactive chloramine. Absorbance measurements should be made within 5 minutes of full color development.

Reported interferences include iodide and chloride at concentrations >50 and 200 mg/L, respectively. Oxidized manganese is also reported to interfere, producing erroneously high results.

I. Equipment
  Spectrophotometer capable of measurement at 592 nm
  100 mL glass-stoppered volumetric flasks
  Assorted pipettes

II. Reagents
  1. Standardized stock iodine solution: Weigh 20 g I_2 into approximately 300 mL deionized water. Allow to stand overnight. Decant and dilute 85 mL to 1 L with deionized water. Standardize with sodium thiosulfate.
  2. Citric acid buffer solution:
     (a) Citric acid: Weigh 192.2 g citric acid and volumetrically dilute to 1 L with deionized water.
     (b) NH_4OH (2 *N*): Dilute 131 mL stock NH_4OH to approximately 700 mL deionized water. Bring to 1 L vol with deionized water. Carefully and with mixing, add 350 mL 2 *N* NH_4OH to 670 mL citric acid solution. Add 80 g $NH_4H_2PO_4$. Dissolve by mixing.
  3. Leuco-crystal violet indicator (N,N-dimethylaniline):
     (a) In a brown glass container, carefully dilute 3.2 mL conc. H_2SO_4 into 200 mL deionized water. Add 1.5 g indicator with mixing.
     (b) In a separate container dissolve 2.5 g mercuric chloride ($HgCl_2$) in 800 mL deionized water. Transfer component B to component A with mixing. Using conc. H_2SO_4, adjust pH to <1.5. Store in brown glass bottle in dark area.
  4. Standardized sodium thiosulfate.

III. Procedure
  1. Iodine standards: Prepare I_2 standards by dilution from stock such that final concentration range is 0–6 mg/L:
  2. Transfer 50 mL of each standard to a 100-mL glass-stoppered volumetric flask.

3. To each, add: 1.0 mL citric acid buffer. Allow 30 sec for reaction. Add 1.0 mL indicator solution (with mixing). Dilute to volume with deionized water.
4. Measure absorbance of standards at 592 nm against a deionized water blank. Plot absorbance vs. concentration of I_2.
5. Using 50 mL rinse water to be tested, proceed per steps 2–4.

IV. Supplemental Notes
1. If samples contain >6 mg/L I_2, decrease the volume of sample used.

IRON: ATOMIC ABSORPTION SPECTROMETRIC ANALYSIS

As with the other metals, atomic absorption is finding increased application in analysis of iron in juice and wine.

I. Equipment
Atomic absorption spectrometer with iron hollow cathode lamp
Volumetric flasks
Volumetric pipettes

II. Reagents
Iron stock solution (1,000 mg/L): In a 1-L volumetric flask, dissolve 1.000 g of iron metal (available as wire or powder of defined purity) in several milliliters of concentrated HCl. When dissolved, bring to volume with deionized water. Alternatively, many laboratories use iron-containing salts (i.e., ferrous ammonium sulfate) or commercially available standards.
Ethanol (200° proof)
Glucose

III. Procedure
1. Prepare a working iron standard, 100 mg/L, by diluting 10 mL of iron stock to 100 mL in a volumetric flask.
2. Prepare 2.0–10.0 mg/L iron calibration standards in 100-mL volumetric flasks following the dilution scheme presented in Table 20–26.
3. Instrument Calibration. Consult operator's manual for instrument setup and calibration.
4. Run a blank consisting of 12% (vol/vol) ethanol in deionized water to set zero absorbance. Beginning with the standard solution of lowest concentration, aspirate through the sample inlet. Record absorbance (or read direct in concentration). Between samples, aspirate a sufficient volume of solvent to flush the system.
5. Aspirate filtered wine samples, recording absorbance (or concentration).

Table 20–26. Preparation of iron calibration standards from 100 mg/L working standard.

Volume working standard (mL)	Final iron concentration (mg/L) in 100 mL volume[a]
2.0	2.0
4.0	4.0
6.0	6.0
8.0	8.0
10.0	10.0

[a]Dilution is made with 12% (vol/vol) ethanol.

6. Plot absorbance of standards vs. their respective concentrations. Compare absorbance of unknown(s) to the standard curve, expressing the concentration in mg/L.

IRON: QUALITATIVE TEST FOR FERRIC CASSE (SEE DIAGNOSTICS)

IRON: SPECTROPHOTOMETRIC (VIS) ANALYSIS USING THIOCYANATE

This analysis depends on the quantitative reaction of thiocyanate anion (SCN^-) with Fe^{3+} and subsequent spectrophotometric measurement of the color formed. The colorimetric reaction depends on concentration of iron present, the concentration of thiocyanate, and the acidity of the solution. The method permits determination of both Fe^{2+} and Fe^{3+} via hydrogen-peroxide-induced oxidation of the iron II to iron III.

I. Equipment
   Pyrex test tubes (of more than 25 mL capacity)
   Volumetric pipettes (1.0, 2.0, 5.0, 10.0 mL)
   Volumetric flasks (100 mL)
   Whatman #41 filter paper
   Spectrophotometer (visible spectrum)
   Matched cuvettes
   Glass funnels

II. Reagents
   Iron standard solution (1,000 mg/L): Use same procedure as listed above in AA iron.
   Potassium Thiocyanate (8%): In a 100-mL volumetric flask, dissolve 8 g of KSCN in 95 mL of deionized water. Bring to volume when dissolved.

Table 20–27. Preparation of iron calibration standards from working standard (100 mg/L).

Volume (mL) of 100 mg/L standard solution	Final Fe^{3+} concentration (mg/L) in 100 mL volume
1.0	1.0
2.0	2.0
5.0	5.0
10.0	10.0

Hydrochloric Acid (5%): Dilute 12 mL of concentrated HCl to 100 mL with deionized water.

Amyl Acetate-Methanol Solution (2:1): Immediately before use, mix 2 volumes of amyl acetate to 1 volume of methanol. *Note:* The fumes are very volatile. Store this solution in a ventilated area (fume hood) until use.

Hydrogen Peroxide (3%): Dilute 10 mL of stock 30% H_2O_2 to 100 mL with deionized water immediately before use.

III. Procedure

1. Prepare a 100 mg/L iron working standard by diluting 10 mL of the stock solution (1,000 mg/L) to 100 mL in a volumetric flask.
2. Prepare 1.0, 2.0, 5.0, and 10.0 mg/L iron calibration standards in 100-mL volumetric flasks according to Table 20–27.
3. Volumetrically transfer 2.0 mL of each standard solution into separate Pyrex test tubes. The blank consists of 2.0 mL of deionized water.
4. To each tube, add with thorough mixing:
   (a) 1 mL of 5% HCl
   (b) 2 drops of H_2O_2
   (c) 1 mL of 8% KSCN (*CAUTION!!! Do not pipette by mouth!*)
5. Allow each to stand 1 min.
6. Add 15 mL amyl acetate-methanol solvent to each, and mix with mild agitation.
7. Allow phases to separate, and withdraw the upper organic phase with a 10-mL transfer pipette.
8. Filter the organic phase through Whatman No. 41 filter paper directly into a cuvette.
9. Determine absorbance or transmittance at 520 nm using the reagent blank to zero the spectrophotometer. Plot absorbance vs. concentration (mg/L).
10. Using 2.0 mL of wine, proceed from step 4 above, comparing the absorbance with the standard curve.

IV. Supplemental Notes

1. Post-filtration cloudiness may be cleared by addition of anhydrous sodium sulfate.
2. The amyl acetate-methanol solvent is employed to extract the colored complex $Fe(SCN)_6^{3-}$ from the aqueous phase, thus preventing interferences from wine pigments. The inclusion of methanol in the solvent mixture reduces the tendency for amyl acetate and wine to form an emulsion. In some instances, red wine pigments may partition into the solvent layer, thereby producing erroneously high results. One may compensate for the increased absorbance due to pigmentation by running a wine sample without the inclusion of thiocyanate. Subsequent subtraction of absorbance of this blank sample from that with iron development yields the true absorbance for comparison with the standard curve.
3. It should be noted that if the blank shows significant color development, reagents, water, and glassware may be contaminated. Exceptionally clean glassware is essential to success. It is highly recommended that laboratory personnel maintain separate glassware apart from routine laboratory glassware.

ISO-AMYL ALCOHOL: (SEE FUSEL OILS GAS CHROMATOGRAPHIC ANALYSIS)

ISO-BUTYL ALCOHOL: (SEE FUSEL OILS GAS CHROMATOGRAPHIC ANALYSIS)

LACCASE: SPECTROMETRIC (VIS) PROCEDURE FOR SYRINGALDAZINE (SEE *BOTRYTIS*)

LACTIC ACID: HPLC PROCEDURE (SEE ORGANIC ACIDS)

LACTIC ACID: PAPER CHROMATOGRAPHIC ANALYSIS (SEE MLF)

LACTIC ACID: ENZYMATIC SPECTROMETRIC (VIS) DETERMINATION

Enzymatic analysis procedures for the determination of L-lactic acid have been developed. The procedures rely on the coupled enzymatic reactions:

$$\text{(1)}\quad \text{L-Lactate} + \text{NAD}^+ \xleftrightarrow{\text{L-LDH}} \text{pyruvate} + \text{NADH} + \text{H}^+$$

$$\text{(2)}\quad \text{pyruvate} + \text{L-glutamate} \xleftrightarrow{\text{GPT}} \text{L-alanine} + \text{2-oxoglutarate}$$

L-LDH: L-lactate dehydrogenase
GPT: glutamate-pyruvate transaminase

The equilibrium position of reaction (1) is strongly in the direction of the reactants. However, when coupled with the reaction of pyruvate and L-glutamate (2) the equilibrium can be displaced toward the products. As seen in (1), formation of NADH is stoichiometrically related to the amount of L-lactic acid, and thus the concentration of the latter may be determined by measurement of NADH at the appropriate wavelength. Kits of reagents for conducting this procedure are available to the analyst.

Refer to the product sheet that accompanies the lactic acid kit of reagents for up-to-date safety information and a procedure specifically designed for the kit obtained. The lactic acid procedure will require the following:

I. Equipment

A spectrometer capable of reading absorbance at 340, 334, or 365 nm plus suitable, 1 cm, cuvettes.

Micropipettes (20-, 100-, and 200-μL), and adjustable volumetric pipettes capable of delivering 0.50–1.00 mL.

II. Reagents

The reagents used may be obtained from Boehringer-Mannheim or other suppliers. They are sold in the form of a test kit of sufficient material for approximately 25 L-lactic acid determinations.

III. General Procedure

1. Using a reflux apparatus, a wine sample is refluxed with 2 *N* NaOH for 15 minutes with (mixing) to hydrolyze any bound lactic acid.
2. The sample is then neutralized, diluted, and aliquots mixed with reagents (except the L-lactate dehydrogenase), and the absorbance values are measured.
3. The L-lactate dehydrogenase is added, and the absorbance is read again after the reaction is complete.
4. A blank is run along with the samples, and the absorbance values are corrected for the blank reading.
5. The difference in the corrected absorbance values taken before and after the addition of L-lactate dehydrogenase is then used to calculate the concentration of L-lactic acid.

IV. Supplemental Notes

1. The above procedure measures spectrophotometrically the change

in concentration of NADH at 340 nm (the absorption peak for NADH). Where photometers (equipped with a mercury vapor lamp) are used, measurements are made at 334 or 365 nm. The appropriate adsorption coefficients are given in product handouts.

2. In the authors' experience, enzyme kit procedures are frequently difficult to use and obtain quality results. When using these kits, readers are advised to run several known standard acid solutions to confirm their abilities to use the kit and to validate the constants provided in the kit procedure. In general, kit procedures will provide information on how one detects interferences and/or how to prepare suitable standards.

LACTIC ACID BACTERIA: SEE BACTERIA-YEAST

LEAD: ATOMIC ABSORPTION SPECTROMETRIC ANALYSIS

Atomic absorption is finding increased application in the analysis of lead in wine. To meet required sensitivity levels the technique uses flameless atomization.

I. Equipment

Atomic absorption spectrophotometer with electrothermal vaporization unit (graphite furnace) and background correction system
lead hollow cathode lamp
Volumetric flasks
Volumetric micropipettes

II. Reagents

Lead stock solution (1000 mg/L): In a 1-L volumetric flask, dissolve 1.600 g of lead(II)nitrate, $Pb(NO_3)_2$, in several milliliters of 1% (vol/vol) nitric acid. When dissolved, bring to volume with the acid solution. Alternatively, many laboratories use commercially available standards.
Phosphoric acid stock: 85%
Phosphoric acid solution: Dilute 6 mL stock phosphoric acid to 100 mL with deionized water
Nitric acid stock ($density_{20}$ = 1.38g/mL)
Nitric acid solution: Dilute 10 mL stock nitric acid to one liter with deionized water
Acid dilution mixture—1:1 Phosphoric acid solution and 1% nitric acid solution

III. Procedure

1. Prepare a working lead standard, 10 mg/L, by diluting 1.0 mL of

lead stock to 100 mL in a volumetric flask. Dilute using the acid dilution mixture (1:1 mixed acids)

2. Prepare 10, 20, and 30 μg/L lead calibration standards in 100-mL volumetric flasks following the dilution scheme presented below:

Table 20–28. Preparation of lead calibration standards from 100 mg/L working standard.

Volume working standard (mL)	Final lead concentration (μg/L) in 100 mL volume*
0.10	10.0
0.20	20.0
0.30	30.0

*Dilution is made with acid dilution mixture.

3. Instrument Calibration. Consult operator's manual for instrument set-up and calibration. Program the electrothermal vaporization unit as follows:

Table 20–29. Electrothermal vaporization (furnace) program for measurements of lead in wine

Step	Temperature	Time	N_2 L/min
1	75	2.0	3.0
2	95	2.0	3.0
3	140	15	3.0
4	300	8.0	3.0
5	450	7.0	3.0
6	480	10	3.0
7	900	20	3.0
8	900	1.0	0.0
9	2,250	0.7	0.0
10	2,250	1.0	0.0
11	2,250	2.0	3.0

4. Select the 283.3 nm wavelength lead emission line. Using doubly-distilled water (or equivalent) set zero absorbance.
5. Beginning with the standard solution of lowest concentration, inject 5 μL into the vaporization unit. Initiate the furnace program and record absorbance value. Repeat calibration process three times with each standard solution. Calculate the mean value of the absorbance for each standard.
6. Repeat the same triplicate determination process for each filtered wine sample recording absorbance values and calculating the mean absorbance values.

7. Plot absorbance of standards vs. their respective lead concentrations. Compare mean absorbance of unknown(s) to the standard curve, expressing the concentration in μg/L. Include any sample dilution factors in final expression of results.

MALIC ACID: ENZYMATIC SPECTROMETRIC (VIS) ANALYSIS

Enzymatic analysis procedures for the determination of L-malic acid have been developed. The procedures rely on the coupled enzymatic reactions:

$$\text{(1)}\quad \text{L-malate} + \text{NAD}^+ \xleftrightarrow{\text{LMDH}} \text{oxaloacetate} + \text{NADH} + \text{H}^+$$

$$\text{(2)}\quad \text{oxaloacetate} + \text{L-glutamate} \xleftrightarrow{\text{GOT}} \text{L-aspartate} + \alpha\text{-}\textit{keto}\text{glutarate}$$

LMDH: L-malate dehydrogenase
GOT: glutamate-oxaloacetate transaminase

In Reaction 1, the equilibrium is strongly in the direction of the reactants. In the procedure of Mayer and Busch (1963), the reverse reaction is prevented by inclusion of alkaline hydrazine solution. Alternatively, the coupled reaction of oxaloacetate and L-glutamate (reaction 2) can be used to drive the equilibrium toward the products. As seen from Reaction 1, formation of reduced NADH is stoichiometrically related to oxidation of L-malate, and thus the concentration of the latter may be determined by measurement of the NADH reduced. Kits of reagents for conducting this procedure are now available to the analyst.

Refer to the product sheet that accompanies the malic acid kit of reagents, for up-to-date information and a procedure specifically designed for the kit obtained. The malic acid procedure will require the following:

I. Equipment
   A spectrometer capable of reading absorbance at 340, 334, or 365 nm plus suitable cuvettes.
   Micropipettes (10-, 100-, and 200-μL), and adjustable volumetric pipettes capable of delivering 1.00–1.50 mL.

II. Reagents
   The reagents used may be obtained from Boehringer-Mannheim or other suppliers. They are sold in the form of a test kit of sufficient material for approximately 20 L-malic acid determinations.

III. General procedure
   1. Using a reflux apparatus, a wine sample is refluxed with 2 *N* NaOH for 30 min with (mixing) to hydrolyze any bound malic acid.

2. The diluted sample is then mixed with reagents (except L-malic dehydrogenase), and the absorbance values are measured.
3. The malate dehydrogenase is added, and the absorbance again is read after the reaction is complete.
4. A blank is run along with the samples, and the absorbance values are corrected for the blank reading.
5. The difference in the corrected absorbance values taken before and after the addition of L-malic dehydrogenase is used to calculate the concentration of L-malic acid.

MALIC ACID: HPLC ANALYSIS (SEE ORGANIC ACIDS)

MALIC ACID: PAPER CHROMATOGRAPHIC ANALYSIS (SEE MLF)

MALOLACTIC FERMENTATION: MONITORING BY PAPER CHROMATOGRAPHY (SEMIQUANTITATIVE MEASUREMENT)

Wine acids partition themselves in a chromatographic system according to their relative affinities for the mobile solvent and stationary phase. Usually the system is operated in an ascending manner; however, descending chromatography is possible, although the procedure involves purchase of additional components.

With the paper chromatogram presented below, the solvent used has the added advantage of not requiring subsequent development; it already contains an indicator for acid spots. Thus, after drying, results can be directly evaluated. Fresh solvent should be prepared on a weekly basis, as Kunkee (1968) noted that with old solvent there may be excessive trailing of spots, making interpretation of the chromatogram difficult. When not in use, solvent should be stored in a separatory funnel away from exposure to direct sunlight. Storage in a separatory funnel has the advantage that any remaining aqueous phase can be discarded before use.

The solvent mixture used contains the pH indicator bromocresol green, which undergoes a color change from yellow to blue in the pH range 3.8–5.4. The presence of an acid is indicated as a yellow spot on a blue background. To identify the acid, analysts may calculate relative front values (Rf), defined as the ratio of distances traveled by the spot to the distance traveled by the solvent front. Rf values vary with both solvent and solid phases. For the solvent system used in this separation, the appropriate

Table 20–30. Rf values for wine acids.

Acid	RF range
Tartaric	0.28–0.30
Citric	0.42–0.45
Malic	0.51–0.56
Lactic	0.69–0.78
Succinic	0.69–0.78

Source: Kunkee (1968).

ranges of values are identified in Table 20–30. Hence, the following order of acids would be expected (from solvent front to baseline) on a typical chromatogram: succinic and lactic, malic, citric (if present), and tartaric.

The analyst should run standard acid(s) as reference(s), to make spot identification easier. Standard acids are prepared at concentrations of 0.3%.

I. Equipment
   Whatman No. 1 chromatography paper
   Chromatography developing tank
   Separatory funnel
   Micropipettes (20 µL)

II. Reagents
   Wine acid standards (0.3%)
   Chromatography solvent: In an appropriate sized separatory funnel, add the following reagents:
      100 mL *n*-butanol
      100 mL deionized water
      10.7 mL stock formic acid
      15 mL indicator solution prepared by dissolving 1 g of water-soluble bromocresol green in 100 mL of deionized water.
   Shake solvent mixture thoroughly by repeated inversion of separatory funnel. Allow for phase separation, and discard lower phase.
   NOTE!! Solvent should be prepared on a weekly basis.

III. Procedure
   1. Taking care to handle chromatography paper only by the edges, cut a piece of appropriate size to fit into developing tank.
   2. Using a pencil, draw a line parallel to, and approximately 2.5 cm from the bottom edge of the paper.
   3. Using micropipettes or hematocrit tubes, spot standard acids and wine samples at equal intervals along baseline. Spots should be of as small a diameter as possible (less than 1 cm). It is recommended to respot at least twice in order to achieve this goal. Each spot should be at least 2.5–3.0 cm apart.

4. Transfer solvent to developing tank, allowing at least 30 min for vapor saturation to occur. A minimum depth of 0.75 cm of solvent is required for adequate development.
5. Immerse baseline side of paper into tank, taking care that solvent moves uniformly up the paper.
6. When the solvent has ascended to near the upper edge of paper, chromatogram may be removed and allowed to dry.
7. When dry, results may be interpreted by noting the positions of yellow spots (acids) on blue background (see Fig. 20–13) Identifi-

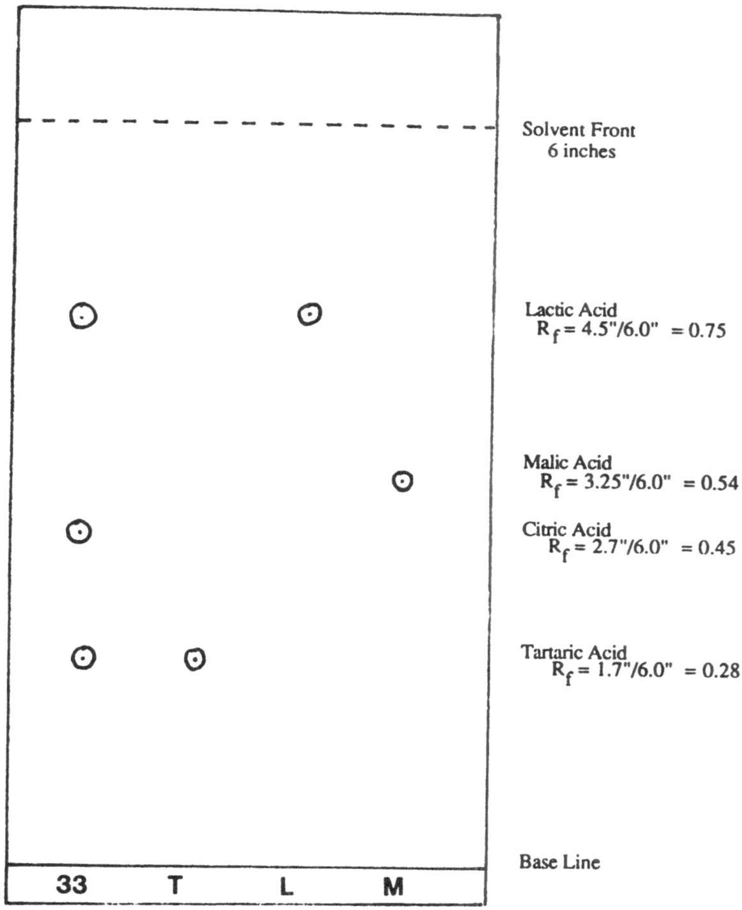

Fig. 20–13. Distribution of wine acids using conventional paper chromatography. Chromatogram shows separation of acids in white wine #33 compared with standards (T) tartaric (L) lactic and (M) malic.

cation of various wine acids may be made by comparison to standard acids or by calculation of Rf values. Sensitivity of the method for individual acids is about 100 mg/L.

IV. Supplemental Notes

1. Sensitivity levels for this separation are 100 mg/L. Thus, absence of a malic spot should not be taken to mean that MLF is complete. Follow-up analyses by either enzymatic methods or HPLC are required.
2. The time required for migration of solvent front is not critical. To facilitate maximum separation of wine acids, the solvent front should be allowed to move near the top edge.
3. In solvent preparation, formic acid is added to suppress ionization of acids, which would otherwise prevent their separation.
4. Homofermentative LAB produce lactic acid as a normal product of metabolism when growing on sugars. Therefore, the presence of a lactic acid spot on the chromatogram is not necessarily confirmation of an ongoing MLF.

MANNITOL SALT FORMATION: PROCEDURE FOR DIFFERENTIATION OF HETERO- FROM HOMOFERMENTATIVE LACTIC ACID BACTERIA (SEE BACTERIA: ISOLATION AND IDENTIFICATION)

MERCAPTANS: SENSORY SCAN (SEE PROCEDURE FOR HYDROGEN SULFIDE)

METAL INSTABILITIES: (SEE DIAGNOSTICS)

METHANOL: GAS CHROMATOGRAPHIC ANALYSIS (SEE FUSEL OILS)

MICROORGANISMS: SEE BACTERIA-YEAST AND CELL COUNTING

MOLD: SEE *BOTRYTIS*

MONOTERPENES: SEE TERPENES

NITRATE ION: DETERMINATION BY ION SELECTIVE ELECTRODE

Nitrate ion (NO_3^-) concentrations in water, waste streams, and so on, can be measured with the nitrate ion selective electrode. The electrode is constructed with an interior silver/silver chloride reference element separated from the outside by an ion selective membrane. The membrane consists of an organic liquid ion exchanger immobilized in a polyvinyl chloride matrix material. When the electrode is placed in a sample, nitrate is sensed preferentially at the ion exchange membrane. A double junction reference electrode is used in conjunction with the nitrate electrode. To measure nitrate levels, standards and samples should be maintained at the same ionic strength (total ion concentration). An ionic strength adjustment buffer is prepared and used for this purpose.

I. Equipment

Expanded-scale pH/mV meter or specific ion meter
Nitrate selective ion electrode
Double junction reference electrode
Magnetic stirrer and stir-bar

II. Reagents

Sodium Nitrate (10,000 mg/L NO_3^-): In a 100-mL volumetric flask, dissolve 1.38 g predried reagent grade sodium nitrate. Bring to volume with deionized water.

Ammonium Sulfate Ionic Strength Adjustment Buffer (2.0 M): Dissolve 26.41 g of ammonium sulfate $(NH_4)_2SO_4$ in approximately 75 mL deionized water and bring to 100 mL final volume.

III. Procedure

1. Prepare a working standard of 1,000 mg/L nitrate from the nitrate stock. Pipette 100 mL of stock solution into a 1-L volumetric flask, and dilute to volume.
2. Prepare calibration standards of 1.0, 10, and 100 mg/L from the nitrate working standard solution (see Table 20–31). Bring the solutions to final volume with deionized water.

Table 20–31. Preparation of nitrate calibration standards from 1,000 mg/L working standard.

Volume of working standard (mL)	Final NO_3^- concentration (mg/L) in 1-L volume
1.0	1.0
10	10
100	100

3. Transfer 50 mL of the 1.0-mg/L standard to a 150-mL beaker. Add 50 mL ionic strength adjustment buffer, place the electrodes (nitrate ISE and double-junction reference) in the standard. Begin gentle stirring; allow the mV reading to stabilize, and record.
4. Repeat the process with 10.0, 100.0, and 1,000.0 mg/L standards. Between readings, place the meter in the "stand-by" position, rinse the electrodes with a small amount of the next standard, and blot dry.
5. Construct a calibration curve of mV readings vs. log concentration. This can best be done using four-cycle semilog graph paper with the concentration (mg/L) on the log axis.
6. Treat the sample as in Step 3. Rinse the electrodes with a small amount of sample, and blot them dry before placing in the beaker. Record the reading, and determine nitrate concentration by reference to the calibration curve.
7. If the reading is off scale, dilute the sample before measuring nitrate concentration. Be sure to multiply the answer obtained from the calibration curve by the appropriate dilution factor.

NITROGEN: ESTIMATION OF SOLUBLE PROTEIN BY COOMASSIE BRILLIANT BLUE G250

A dye binding assay for the estimation of soluble proteins has been used for juice and wines. The method is based upon the Bradford (1976) procedure using Coomassie brilliant blue G250. When the dye binds to protein the absorbance maximum shifts from 465 to 595 nm with the corresponding shift from orange to blue measured spectrophotometrically. The following procedure has been modified by Murphey et al. (1989).

I. Equipment

Spectrometer capable of measurements at 595 nm

0.45-μm filters and housing

1 cm cuvettes

II. Reagents

Bovine serum albumin (BSA)

Coomassie brilliant blue G250 dye (Bio-Rad Laboratories, Richmond, California)

III. Procedure

1. Filter juice or wine through 0.45-μm filter.
2. Dilute sample (0.40–0.80 mL) with deionized water to 4.8 mL so the absorbance reading is <0.6

3. Add 1.20 mL of Coomassie brilliant blue.
4. Mix by inversion and incubate at room temperature (22°C) for 55 min.
5. Read adsorbance at 595 nm using a 1-cm path cuvette, and compare with a blank prepared with 4.80 mL deionized water in 1.20 mL dye.
6. A standard curve should be prepared using bovine serum albumin (BSA) to determine soluble protein concentration.

IV. Supplemental Note

Murphey et al. (1989) determined that phenols bound to proteins did not affect the reaction of the dye with proteins but that the incubation period (55 min) was critical.

NITROGEN: ESTIMATE OF FERMENTABLE NITROGEN BY FORMOL TITRATION

This is a rapid method for estimation of the fermentable nitrogen in juice, wine, or vinegar.

I. Equipment

pH meter
200 mL beaker
200 mL volumetric cylinder
Funnel
Filter paper, Whatman No. 1

II. Reagents

Sodium hydroxide (NaOH) 1 *N* and 0.1 *N*
Barium chloride (6.5 g/L)
Hydrochloric acid (HCl)
Formaldehyde 40%

III. Procedure

1. Place 100 mL of sample in 200 mL beaker, adjust to pH 8.0 with 1 *N* sodium hydroxide and add 10 mL of barium chloride.
2. Wait 15 minutes then transfer solution into a 200 mL volumetric flask. Bring to volume with deionized water.
3. Stir and filter solution through Whatman No. 1 filter paper.
4. Place 100 mL of the solution into a 200-mL beaker and readjust pH to 8.0.
5. Add 25 mL of formaldehyde solution and titrate with 0.1 *N* NaOH to a pH of 8.0.
6. The concentration of fermentable nitrogen is given as follows:
   Fermentable Nitrogen (mg N/L) = (mL of 0.1 NaOH titrated) × 28

IV. Supplemental Note

Barium chloride is included to precipitate sulfur dioxide. If juice is unsulfited, this addition may be omitted.

NONFLAVONOIDS: SPECTROMETRIC (VIS) ANALYSIS (SEE PHENOLS—FOLIN CIOCALTEU PROCEDURE)

NONSOLUBLE SOLIDS: DETERMINATION OF SUSPENDED SOLIDS IN WINE

Nonsoluble solids (NSS) play important roles in processing considerations, as well as in the sensory properties of wine. Aside from their importance in refrigeration needs and lees volume, particulates contribute to the nutritional status of must as well as to the formation of sensory impact compounds such as volatile acidity, ethyl acetate and fusel oils. High (>2.5%) levels of NSS can influence palate structure by imparting additional phenols into the wine. Solids may also contribute to the physical/chemical stability of a wine. In white wine production, most wineries try to reduce levels of NSS to <1.5% before fermentation. Depending upon the variety (and winemaking style), NSS levels before fermentation may be reduced to very low levels (<0.5%). This may be accomplished by use of various fining agents and/or centrifugation.

Measurements of NSS typically are done by centrifuging a defined volume or weight of juice and then determining the volume or weight of sediment (pellet) after clarification. Depending upon the facility, results may be expressed as %vol./vol. or %wt./wt.

I. Equipment
   1. Bench-top clinical (or other) centrifuge
   2. Graduated (0.1 mL increments) centrifuge tubes
   3. Balance capable of reading to 0.1 g.
   4. Pipettes

II. Procedure
   1. Transfer a representative juice/wine sample into a centrifuge tube. Note the volume and weigh.
   2. Prepare a duplicate (or different) sample as a counterbalance and weigh. To prevent breakage of tubes or damage to the centrifuge during operation, their weights should not vary by more than 0.1 g.
   3. Centrifuge for 5–10 minutes depending upon capabilities of unit. At end of run, allow centrifuge head to come to a complete stop before opening.

4. Remove sample(s), carefully decant clarified supernatant, and note the volume of sediment (pellet). During routine operations, one may choose to simply examine sample without pouring off supernatant.

   Where "tailing" of sediment up the side of tube is noted, sample should be recentrifuged for a longer period or at a higher speed. With most samples, some tailing is to be expected. In these cases, estimate the pellet volume at the midpoint of the shoulder.
5. Calculate the percent nonsoluble solids as:

$$\text{NSS (\% vol/vol)} = \frac{T_0 - T_1}{T_0}$$

T_0 = Volume of juice sample before centrifugation
T_1 = Volume of pellet after centrifugation

III. Supplemental Notes
1. One may express NSS as % wt./wt. In this case it is necessary to determine the weight of juice before and weight of pellet after centrifugation. This procedure necessarily involves more steps than those outlined above.
2. For wineries without a laboratory centrifuge, simply collect the juice sample in a 100-mL graduated cylinder and refrigerate. When clarification has occurred, the volume of sediment can be visually estimated. However, since the graduations of most cylinders are ± 1 mL this only provides a crude estimate of solids present.

ORGANIC ACIDS: HPLC ANALYSIS OF ACETIC, CITRIC, FUMARIC, GLUCONIC, LACTIC, MALIC, SUCCINIC, AND TARTARIC ACIDS

Organic acids can be separated on an HPLC column and quantitated. Sample preparation involves a preliminary ion exchange separation of sugars that would otherwise co-elute with the acids. The amounts of individual organic acids present in a sample are added to give a measure of its "total acid content."

I. Equipment

High performance liquid chromatograph with variable-wavelength (UV) detector
Bio-Rad Organic Acids column (or equivalent)
Guard column
Disposable polypropylene sample preparation columns

II. Reagents

Concentrated ammonium hydroxide

H_2SO_4 mobile phase (0.01 *N*): Dilute 0.28 mL concentrated sulfuric acid to 1 L with deionized water.

H_2SO_4 (1 + 4): Place 80 mL deionized water in beaker and carefully add 20 mL sulfuric acid. Mix well.

III. Procedure

1. Install the column and an appropriate guard column in the HPLC mobile phase flow stream.
2. Begin pumping H_2SO_4 mobile phase through the column at a flow rate of 0.6 mL/min.
3. Set the column oven temperature to 25°C, and allow the system to stabilize.
4. Turn on the variable wavelength UV detector and allow it to warm up. Set the wavelength to 210 nm.
5. Prepare and chromatograph the sample as follows:
   a. Volumetrically transfer the following to a capped vial
      (1) 2.00 mL of wine or grape must
      (2) 0.2 mL of concentrated ammonium hydroxide
   b. Mix the contents of the vial, and keep it closed until you are ready to place the contents on the anion exchange resin.
   c. Prepare an ion exchange column by slurrying 3 g Bio-Rex 5, 100–200 mesh anion exchange (Cl^- form) resin in 6 mL deionized water, and pour into a disposable polypropylene sample preparation column. Drain off water and wash column with an additional 3 mL deionized water. Do not allow the column to dry out before use.
   d. Add sample to the column and allow the solution to drain through. Wash the sugars off the column with deionized water (9 mL) and collect them in a 10 mL volumetric flask for subsequent analysis, if desired. (This solution should be brought to volume with deionized water and mixed well before injection onto the column).
   e. Carefully add 2 mL of (1 + 4) sulfuric acid and enough deionized water to collect 10 mL of eluent containing the acids. (Again mix this sample well).
   f. Prepare the sample by running about 3 mL through a Sep-pac that has been conditioned by running 3 mL of methanol followed by 5 mL of deionized water.
   g. Collect sample, filter through a 0.45-μm syringe filter, and inject.
6. Qualitative analysis may be accomplished by preparing standard solutions of the various acids expected in the sample. Compare

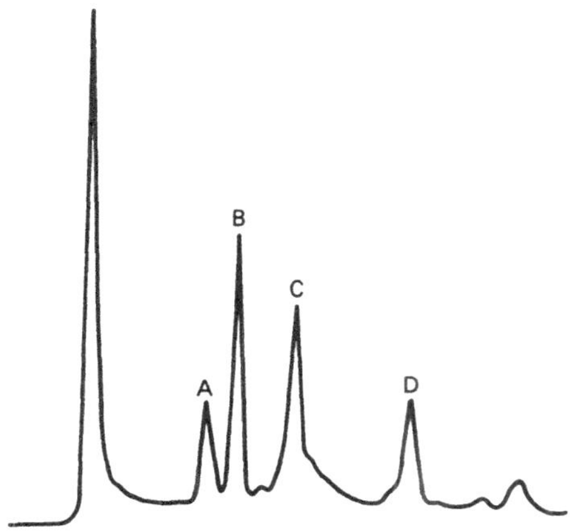

Fig. 20–14. Chromatogram of Fume Blanc wine. Labeled peaks are (A) Citric acid, (B) Tartaric acid, (C) Malic acid, and (D) Succinic acid.

observed retention times of peaks in the sample chromatogram with those in the standards chromatogram. (See Fig. 20–14 for a chromatogram of a typical juice sample.)

7. Chromatograph standard solutions of various acids at known concentrations. Peak height (recorder) or peak area (integrator) ratios are compared to those from wine samples, and the unknown concentrations can be read off the calibration curve or calculated as follows:

$$\text{Acid conc. (g/L)} = \frac{R}{R'} \times \text{acid conc. in std, g/L}$$

where:
R = response of unknown acid
R' = response of standard acid

IV. Supplemental Notes

1. To prepare the resin column for the next sample let approximately 3 mL deionized water flow through to rinse out the acid. Columns can be used for about four samples.
2. A similar method to that presented above has been developed by Gump and Kupina (1979) for the analysis of gluconic acid in juice or wine. Gluconic acid can be used as a marker compound for the occurrence of *Botrytis cineria* infections.

OXIDATIVE CASSE: VISUAL DETERMINATION OF OXIDATIVE POTENTIAL IN WHITE WINE

Müller-Späth (1992) developed a quick test for the estimation of browning potential in white wines. The principle of the test is that oxidizable phenols such as flavonoids can develop a yellow to reddish-brown color in an oxidative medium.

I. Equipment
   5 mL pipette
   50-mL graduated cylinders or pipette
   250-mL Erlenmeyer flasks (preferably wide-mouth)
   10-mL test tubes
   Water bath

II. Reagents
   3% hydrogen peroxide: Dilute 1 mL of 30% stock H_2O_2 to 10 mL using deionized water (or purchase 3% solution already prepared).

III. Procedure
   1. Add 50 mL of wine to control flask A.
   2. Add 50 mL of wine and 0.2 mL of fresh 3% hydrogen peroxide to flask B and mix.
   3. Place both flasks in a water bath at 60°–65°C for 1 hour.
   4. Compare the color of flask A and B directly or by decanting each into test tubes.
   5. If no color difference is noted between control sample A and treated sample B few oxidizable phenols are present and the wine is oxidatively stable. If a color difference is noted, oxidizable phenols are present and in increasing concentration as the color ranges from yellow to reddish-brown.

IV. Supplemental Notes
   1. The color change and the concentration of oxidizable flavonoids are directly correlated. If desired, a color scale can be established to help quantify results.
   2. Ascorbic acid interferes with the analysis.

OXYGEN: DETERMINATION USING A DISSOLVED OXYGEN METER

Oxygen in juice, must, and wine can readily be determined with an electrochemical measuring system. A typical oxygen measuring system is produced by Yellow Springs Instrument Company (YSI), consisting of an oxygen probe (sensing device) connected to a signal processor/readout unit (oxygen meter).

The YSI probe is a functioning electrochemical cell containing two electrodes with sufficient potential applied between them to reduce any oxygen present. This polarographic cell is isolated from the outside environment by a thin membrane that is permeable to gases. When the probe is placed in a liquid sample, oxygen in the sample passes through the permeable membrane at a rate proportional to the amount of oxygen in the sample. This oxygen is rapidly reduced at the cathode electrode in the cell, producing a current that is measured and displayed by the signal processing unit in terms of oxygen concentration.

I. Equipment
   YSI Dissolved Oxygen Meter
   YSI 5700 Series Probe

II. Reagents
   Half saturated KCl: Prepare approximately 100 mL saturated KCl solution. Pour off supernatent liquid and dilute to 200 mL with deionized water. Add four drops Kodak Photo Flo to this solution.

III. Procedures
   1. Following manufacturer's instructions, fill the probe body with KCl electrolyte and cover probe end with Teflon membrane.
   2. Follow instructions provided for calibrating the dissolved oxygen meter.
   3. Place probe in sample, allow to equilibrate, and read dissolved oxygen level from meter.
   4. When not in use store sensing probe according to manufacturer's instructions.

IV. Supplemental Notes
   1. Membranes will last indefinitely, if properly installed and treated with care during use. Poor membrane application or membrane damage results in erratic readings. Erratic behavior can be caused by loose, wrinkled, or fouled membranes, or by bubbles in the probe from electrolyte loss. The gold cathode in the probe may become plated with silver (from the anode), if the sensor is operated for extended periods of time with a loose or wrinkled membrane.
   2. Several gases are known interferents (SO_2, CO, H_2S, halogens, and neon). Suspect oxygen readings may be due to the presence of one or more of these gases.
   3. The gold cathode should always be bright and untarnished. Some gases contaminate the sensor, as evidenced by discoloration of the gold. Clean the gold cathode by vigorous wiping with a soft cloth, lab wipe, or hard paper. Do not use abrasives or chemical cleaners!

PECTINS/GUMS: VISUAL RAPID DIAGNOSIS FOR PECTINS, GUMS, AND OTHER POLYSACCHARIDES (SEE DIAGNOSTICS)

PH: SEE HYDROGEN ION CONCENTRATION

PHENOLS: JUICE, WINE, AND PER BERRY BY SPECTRAL ESTIMATION (ALSO SEE PIGMENTS: HPLC DETERMINATION)

Phenol components of juices and wines are known to absorb radiation in the UV and visible spectrum. Absorption readings can therefore be used to estimate the concentration of total phenols, total anthocyanins, colored anthocyanins, percentage of anthocyanins in the colored form, total hydroxycinnamates, and caffeic acid equivalents.

I. Equipment
   UV-VIS spectrophotometer covering wavelength range from 280 to 520 nm
   Quartz cuvettes 1, 2, 5, and 10 mm pathlength
   Volumetric pipettes
   Micropipettes covering range of 20–200 μL
   Membrane filters (0.45 μm)
   Centrifuge and tubes
   Scale and laboratory blender (for evaluation on a per berry basis)

II. Reagents
   Sodium metabisulfite solution [20% (wt/vol)]: Dissolve 2 g sodium metabisulfite in 10 mL deionized water.
   Aqueous acetaldehyde solution [10% (wt/vol)]: Dilute 1.26 mL freshly distilled acetaldehyde to 10 mL with cold deionized water. Store mixture in a refrigerator.
   Hydrochloric acid (1 M): See Table 20–2.
   Polyvinylpyrrolidone or Polyclar.
   Acidified, 50% ethanol (for per berry evaluations—see procedure).

III. Procedure
   1. Centrifuge juice and hazy wine samples at 4000 × g for 10 min. Bring samples to 25°C in a water bath.
   2. Prepare a blank from a saturated solution of potassium hydrogen tartrate with 11% ethanol (wine) or 20% glucose (juice). Treat blank same as samples.
   3. For red juice or wine samples:
      a. Place sample in 1, 2, or 5 mm pathlength quartz cuvette. Measure absorbances at 280, 420, and 520 nm against prepared blank.
      b. Add 20 μL sodium metabisulfite solution to sample, mix by inversion for 1 min, and remeasure absorbance at 520 nm.

c. Add 20 μL acetaldehyde solution to a fresh sample of juice or wine. Let sit for 45 min at 25°C, then measure absorbance at 520 nm.

d. Dilute 100 μL sample (200 μL, if light colored) to 10.0 mL with 1 M HCl. Let sit for 3 to 4 hours at 25°C, then measure absorbance at 520 nm in a 10-mm pathlength cuvette. Correct absorbance value for dilution by multiplying reading by 10.0 mL/0.10 mL or 10.0 mL/0.20 mL.

e. Correct all absorbance readings to a 10-mm pathlength cuvette (multiply absorbance value by appropriate value $\frac{10}{1}$, $\frac{10}{2}$, $\frac{10}{5}$ depending on path length of cuvette used).

4. Calculate red wine color parameters as follows:

a. Intensity or density of wine color = $A_{(420\text{ nm})} + A_{(520\text{ nm})}$

b. Wine hue or tint = $A_{(420)}/A_{(520)}$

c. Total phenolics (absorbance units) = $A_{(280)} - 4$

d. Measure of total anthocyanins (mg/L)

$$= 20[A^{HCl}{}_{(520\text{ nm})} - (5/3)A^{SO_2}{}_{(520\text{ nm})}]$$

e. Measure of colored (ionized) anthocyanins (mg/L)

$$= 20[A_{(520\text{ nm})} - A^{SO_2}{}_{(520\text{ nm})}]$$

f. Percent of total anthocyanins in colored (ionized) form

$$= \frac{A_{(520\text{nm})} - A^{SO_2}{}_{(520\text{ nm})}}{A^{HCl}{}_{(520\text{ nm})} - (5/3)A^{SO_2}{}_{(520\text{ nm})}} \times 10^2$$

g. Measure of extent that total anthocyanins have been decolorized by binding with SO_2

$$= A^{CH_3CHO}_{(520\text{nm})} - A_{(520\text{ nm})}$$

h. Percent total anthocyanins present in colored form corrected for effect of SO_2 upon wine color

$$= \frac{A^{CH_3CHO}{}_{(520\text{nm})} - A^{SO_2}{}_{(520\text{nm})}}{A^{HCl}{}_{(520\text{nm})} - (5/3)A^{SO_2}{}_{(520\text{nm})}} \times 10^2$$

i. Chemical age factors (degree to which polymeric pigment forms have replaced monomeric pigment forms)

$$= A^{SO_2}{}_{(520\text{ nm})}/A^{HCl}{}_{(520\text{ nm})}$$

5. For white juice or wine samples:
   a. Measure absorbance of undiluted samples at 280 and 320 nm in 1- or 2-mm cuvettes against prepared blank.
6. Correct all absorbance readings to 10-mm pathlengths
   a. Total phenolics (absorbance units)
   b. Total hydroxycinnamates (absorbance units)

$$= A_{(320\ nm)} - 1.4$$

   c. Total hydroxycinnamates as mg/L caffeic acid equivalents (mg/L CAE)

$$= \frac{A_{(320\ nm)} - 1.4}{0.90} \times 10\ mg/L$$

   d. Total flavonoids (absorbance units)

$$= [A_{(280\ nm)} - 4] - (2/3)[A_{(320\ nm)} - 1.4]$$

7. For Berries:
   a. Collect fresh or frozen berries and remove seeds.
   b. Weigh sample and homogenize using a laboratory blender.
   c. Centrifuge at 2000 × g for 20 min.
   d. Decant juice and save.
   e. Extract pellet with 1 M HCl and then with acidified, 50% ethanol (10 mL/g). The acidified ethanol is prepared by adjusting 50% ethanol to pH 2.8 with 1 M HCl.
   f. The anthocyanin content of the juice, HCl, and ethanol fractions can be determined by adsorbance at 520 nm at pH <1 as described Section 4. Total phenols can also be estimated as described above. The juice and pellet values can be combined to give the total content of each component on a per berry basis.

IV. Supplemental Notes
1. Total phenolics measurements include correction for the 280-nm absorbance of nonphenolic compounds present in juice and wine.
2. At low pH, simple anthocyanins are in their colored flavylium form causing an increase in absorbance. When treated with SO_2, monomeric anthocyanins are decolorized. Because polymeric pigments also contribute to a limited extent to increased color at low pH, a correction factor of 5/3 is used to get a more accurate measure of monomeric anthocyanin content in the juice or wine.

3. At wine pH the difference between absorbance values with and without SO_2 provides a measure of the amount of colored (flavylium ion) anthocyanins. Because polymeric pigments are more resistant to bleaching upon addition of SO_2, no correction for these components is made.
4. Acetaldehyde binds more strongly with SO_2 than do anthocyanins. Treatment of wine with acetaldehyde should therefore release those anthocyanins decolorized by any SO_2 treatments during fermentation, and give a better measure of anthocyanins present.
5. White wines with abnormally high absorbance values at 280 nm may have sorbic acid present as a preservative. Sorbic acid can be completely removed by extraction with iso-octane (see Sorbic acid: Extraction and UV Analysis) to eliminate this interference.
6. Hydroxycinnamates are the major nonflavonoid components of white wines and juices. These absorb radiation at 320 nm, as do a number of nonphenolic compounds. A spectral correction for UV absorbing nonphenolics of 1.4 is used.
7. Approximate measurements of the flavonoid content of a white wine or juice requires a correction factor for the 280 nm absorbance of the hydroxycinnamates. Their absorbance at 280 nm is approximately two-thirds of that measured at 320 nm.

PHENOLS: SPECTROMETRIC MEASUREMENT USING FOLIN-CIOCALTEU REAGENT

Juices and wines contain a large variety of phenolic compounds, and there is concern about the ability to differentiate between total phenols, and specific phenols. Although significant progress has been made in developing high performance liquid chromatographic techniques to accomplish the latter, the most common analyses performed are for total phenols. Like many other "general" analytical methods, these lack specificity. UV and visible absorbance measurements suffer from the fact that different phenolic substances can have significantly different molar absorptivities. In addition, nonphenolic substances can be responsible for the major fraction of the absorbance of white wines (Somers and Ziemelis 1972).

The reduction of phenolic substances by the Folin-Ciocalteu reagent involves reaction with a mixture of phosphotungstic acid ($H_3PW_{12}O_{40}$) and phosphomolybdic acid ($H_3PMo_{12}O_{40}$) and subsequent spectrophotometric comparison at 765 nm. The procedure uses gallic acid as a standard reference compound, and results are correspondingly expressed as gallic acid equivalents (GAE). Folin-Ciocalteu reagent also reacts with monohydroxy phenols and other readily oxidized substances (including ascorbic acid, sulfur dioxide, and aromatic amines). In sweet wines and musts a more serious interference occurs with reducing sugars.

I. Equipment
   Spectrophotometer (visible range)
   Cuvettes, 10 mm
   Glass reflux apparatus
   Volumetric flasks (100 mL)
   Pipettes (1, 2, 3, 5, 10 mL)

II. Reagents
   Phenol Standard (5,000 mg/L gallic acid): Prepare phenol stock solution by dissolving 500 mg of gallic acid in deionized water and bringing to 100 mL final volume.
   Folin-Ciocalteu Reagent (also commercially available): Transfer 700 mL deionized water to a 2-L round-bottom boiling flask. Add the following:
      (1) 100 g sodium tungstate
      (2) 25 g sodium molybdate
   When dissolved, add:
      (3) 50 mL stock phosphoric acid
      (4) 100 mL concentrated hydrochloric acid.
   Add several glass boiling beads, connect the reflux condenser with cooling water, and reflux for 10 hours. After 10 hours of refluxing, cool the flask and rinse reflux column into boiling flask with several mL of deionized water.
   Add 150 g of lithium sulfate (monohydrate) and several drops of bromine. Under a fume exhaust hood, boil contents for 15 min. Cool the reagent and bring to 1 L final volume with deionized water. Filter and store in amber bottle.
   Note: Final color of reagent should be a definite yellow. There should be no indication of earlier green coloration. To reoxidize older solutions of Folin-Ciocalteu reagent that may have blue-green-orange tinges, add several drops of bromine and reboil under hood.
   Sodium Carbonate Solution: Into approximately 700 mL of deionized water, add 200 g of anhydrous sodium carbonate. Bring to volume (1 L) and boil until completely dissolved. When dissolved, cool, add 2–3 g additional sodium carbonate, and hold for 24 hours. Filter before use.

III. Procedures
   1. Prepare calibration standards of 0, 50, 100, 150, 250, and 500 mg/L from phenol stock solution as shown in Table 20–32.
   2. Into a series of 100-mL volumetric flasks, pipette 1 mL of each

Table 20–32. Preparation of calibration standards from phenol stock (5,000 mg/L GAE).

Volume phenol standard (mL)	Final concentration (mg/L GAE) in 100-mL volumetric flask
1	50
2	100
3	150
5	250
10	500

standard, *or* 1 mL of each white wine sample, or 1 mL of each 1 + 9 diluted red wine sample.

3. To each flask add 60 mL deionized water and mix the solution; add 5 mL of Folin-Ciocalteu reagent and mix for 30 sec; add 15 mL sodium carbonate solution (add after the 30 sec mixing but before 8 min). Mix the solution again, and bring to the mark with deionized water.
4. Allow standard and sample solutions to sit for 2 hours at 20°C, and then measure the absorbance of each at 765 nm against the blank.
5. Construct a calibration curve from absorbance values of the standards. Read concentrations of white wine samples directly from the curve. Read concentrations of red wine samples from the curve, and multiply by the dilution factor (100/10) to obtain the phenol concentration in the original sample.

IV. Supplemental Notes

1. This method is nonspecific and measures the number of $-OH$ (potentially oxidizable phenolic groups) present in the sample. Different tannins will exhibit different responses. The values obtained are in GAE, as that is the calibration curve standard used.
2. Folin-Ciocalteu reagent uses lithium sulfate to reduce precipitation problems in the reagent. Lithium salts are more soluble than those of other common cations.
3. Sodium carbonate is used to make the reaction mixture alkaline. The reduction of Mo(VI) and W(VI) requires the presence of the phenolate anion. The reduced heteropoly molybdenum and tungsten molecules are blue; the unreduced molecules are yellow.
4. Reducing sugars (R.S.) are capable of reducing the heteropoly molecules in alkaline media. For R.S. values from 1.0 to 2.5 g/100 mL, divide the total phenol values by 1.03 to correct for this interference. For R.S. values from 2.5 to 10.0 g/100 mL, divide the total phenol values by 1.06. No correction is necessary for dry wines.

V. Nonflavonoid Phenolic Fraction

One can treat a filtered sample of wine with formaldehyde, and quantitate the remaining phenolics by Folin-Ciocalteu procedure. Subtracting this value from the original total phenolics value provides a measure of the flavonoid phenolic fraction. The procedure involves treating 10 mL of filtered wine (0.45 μm) with 10 mL of HCl (1 + 3) and 5 mL of formaldehyde solution (8,000 mg/L), allowing it to stand for 24 hours, refiltering, and measuring remaining nonflavonoid phenols (Folin-Ciocalteu). When this value is subtracted from that obtained on a second aliquot of the original wine, the difference is a measure of the flavonoid phenols (also reported in mg/L gallic acid).

PHENOLS: PERMANGANATE INDEX (POLYPHENOL INDEX)

The first widely applied oxidation method for phenol determination in wine was that of Lowenthal-Neubauer and involved titration with potassium permanganate ($KMNO_4$) to an indigocarmine end point. Traditionally the analysis was run before and after activated carbon treatment; the difference value was compared with a standard and expressed as tannin plus pigments. Permanganate can react with sugars and tartaric acid and does not react with monophenols or *meta*-polyphenols (Singleton 1988).

Some winemakers perform a quick screening test to aid in determining when to dejuice red fermenters, etc. The permanganate index test that follows provides a crude evaluation of the phenolic content of wines. Low-phenolic wines have index values from 35 to 50. Wines with a high phenol content may reach index values of 100 or above and are frequently "hard" and bitter.

I. Equipment

500-mL Erlenmeyer flask
50-mL volumetric pipette
2-mL volumetric pipette
50-mL burette

II. Reagents

Potassium permanganate (0.01 *N*): In a 1-L volumetric flask, dissolve 316 mg $KMnO_4$ in approximately 500 mL deionized water. Bring the solution to volume with deionized water.

Indigo carmine solution: Transfer 150 mg indigo carmine to approximately 500 mL deionized water in a 1-L volumetric flask. Add 50 mL dilute sulfuric acid (1 + 2) solution. Bring to volume with deionized water.

Sulfuric acid (1+2): To two volumes of water, add one volume of stock sulfuric acid.

III. Procedure

1. Pipette 50 mL of indigo carmine solution into a 500-mL Erlenmeyer flask.
2. Titrate with potassium permanganate to an endpoint color change of blue to straw (or orange). Record the volume (A) of permanganate used.
3. Pipette 50 mL of indigo carmine solution into another 500-mL Erlenmeyer flask; add 2 mL wine to the flask.
4. Titrate with potassium permanganate to end point. Record the volume (B) of titrant used.
5. Calculate the permanganate index as follows:

$$\text{Permanganate Index} = 5(B - A)$$

6. A permanganate index value of 25 is approximately equivalent to 320 mg/L GAE. A permanganate index value of 95 is equivalent to 2,390 mg/L GAE.

PHENOLS: VISUAL ESTIMATION USING FE(II) AMMONIUM SULFATE (pH TEST)

Schanderl (1962) proposed a simple screening test for the estimation of the phenol content of white wines and sparkling wine *cuveés*. The test involves the addition of ferrous ammonium sulfate to a pH 7 adjusted sample; an estimation of the phenol content is based on visual assessment of the color complexes formed.

I. Equipment

15–20-mL glass test tubes
pH meter
Dropper or glass pipette

II. Reagents

1–2% (wt/vol) ferrous ammonium sulfate
1 *N* NaOH

III. Procedure

1. Adjust 10 mL of clarified juice or white wine to pH 7.
2. Add 2 drops of 1–2% ferrous ammonium sulfate solution.
3. Mix and observe color. Pale yellow indicates a low phenol concentration. Black or dark violet indicates gallic acid present; red or red brown color indicates the presence of catechins. Red-brown precipitates suggest the presence of ellagic acid. Standards can be prepared for comparison as well as to help quantify the phenol content.

PHENOLS: VISUAL ESTIMATION USING FOLIN-CIOCALTEU REAGENT (SEE DIAGNOSTICS: PROTEIN/PHENOLS: VISUAL ESTIMATION USING AMIDO BLACK 10-B PROTEIN STAIN)

PHOSPHORUS: ATOMIC ABSORPTION SPECTROMETRIC ANALYSIS

This method involves an indirect determination of phosphorus (Chow and Gump 1987). Flame absorption spectrometry is not sensitive enough to directly measure the phosphorus content in wine, but phosphorus will form a compound with molybdenum (molybdophosphoric acid), and the molybdenum (equivalent to the phosphorus) can be determined with good sensitivity. Acidified ammonium molybdate reagent forms molybdophosphoric acid with phosphorus in wine. This acid is soluble in ether and is thereby removed from other potentially interfering compounds in wine. In the presence of an alkaline buffer, molybdophosphoric acid loses a proton, and the molybdophosphate ion transfers out of the ether layer and into the aqueous layer.

I. Equipment

Atomic absorption spectrophotometer with molybdenum hollow cathode lamp.

Volumetric flasks (100 mL)

Volumetric pipettes (1, 5, 10 mL)

II. Reagents

Phosphorous standard solution (1,000 mg/L): In a 1-L volumetric flask, dissolve 4.4 g of dried (2 hours at 110°C) reagent grade potassium dihydrogen phosphate in deionized water. Bring to volume with deionized water. Store in amber-colored glassware to prevent deterioration. Prepare fresh weekly.

Ammonium Molybdate Reagent (10% wt/vol): In a 250 mL volumetric flask, dissolve 25 g of reagent grade ammonium molybdate in deionized water. Bring to volume and store in amber-colored glassware. Prepare fresh weekly.

Basic Buffer: In 500 mL of deionized water, dissolve 53.3 g of ammonium chloride. Add 70 mL ammonium hydroxide and bring to 1 L with deionized water.

III. Procedure

(a) Instrument setup

1. Set up the atomic absorption instrument with an air-acetylene flame (strongly reducing), a molybdenum hollow cathode lamp (resonance line at 313.3 nm), and a 0.2-nm slit width.

Table 20–33. Preparation of phosphorus calibration standards from 1,000 mg/L phosphorous stock.

Volume phosphorous stock (mL)	Final phosphorous concentration (mg/L) in 100-mL volume
5.0	50
10.0	100
25.0	250
50.0	500
No dilution	1,000

(b) Preparation of wine samples

1. Pipette 10 mL wine into a 125-mL separatory funnel.
2. Add 1 mL 1 + 1 HCl, and adjust the volume to about 50 mL.
3. Add about 4 mL molybdate solution, and allow the mixture to stand for 10 min.
4. Add an additional 5 mL HCl, and allow the mixture to stand for another 5 min.
5. Add 45 mL diethyl ether, and shake the funnel vigorously for 4–5 min.
6. Allow the phases to separate, and discard the aqueous (lower phase). Rinse the tip of the separatory funnel with deionized water.
7. Wash the ether phase with 10 mL, 1 + 9 HCl. Discard the aqueous acid layer (contains excess molybdenum reagent), again rinsing the tip of the separatory funnel with deionized water.
8. Add 30 mL buffer to the separatory funnel, shake for 30 sec, and collect the aqueous (lower) layer in a 50-mL volumetric flask. Repeat this process with an additional 15-mL buffer.
9. The combined aqueous layers are diluted to the 50-mL mark with deionized water and mixed.

(c) Preparation of phosphorous standards

10. Prepare phosphorus calibration standards from the 1,000 mg/L stock solution in 100-mL volumetric flasks as indicated in Table 20–33.
11. Extract the phosphorus standards, following the same procedure as presented for wine samples.
12. Aspirate phosphorus standards into the AA instrument. Construct a calibration curve from the absorbance values of the standards.
13. Aspirate wine samples into AA instrument. Read the concentration of phosphorus in the extracted wine samples directly from the calibration curve.

IV. Supplemental Note

1. This method is also successful using a nitrous oxide-acetylene flame.

Because this is a hotter flame than air-acetylene, the method is more sensitive, and the standards and wine samples are diluted 1 + 9 before the extraction step.

PIGMENTS: HPLC DETERMINATION OF PHENOLIC PIGMENTS (ALSO SEE PHENOLS: SPECTRAL ESTIMATION)

Successful HPLC separation of anthocyanidins can be accomplished using a reverse phase column and a gradient elution program. Detection of the various glucosides and diglucosides is accomplished with a UV-VIS detector operated at 520 nm.

I. Equipment

HPLC capable of running programmed mobile phase gradients and having a UV-VIS detector monitoring absorbance at 520 nm
Analytical reversed phase (C_{18}) column
Guard column (C_{18})

II. Reagents

Mobile phase I: Acetic acid-water (15:85): Dilute 150 mL glacial acetic acid to 1 L with deionized water. Sparge with an inert gas (helium) before use.

Mobile phase II: Water-acetic acid-methanol (65:15:20): Dilute 150 mL glacial acetic acid plus 200 mL methanol to 1 L with deionized water. Sparge with helium before use.

Anthocyanidin standards (1 mg/mL): Prepare a stock solution (1 mg/mL) by dissolving 100 mg of the appropriate standard anthocyanidin in 100 mL deionized water. Dilute 1 mL of stock to 1 L to prepare working standards (1 mg/L).

III. Procedure

1. Turn on HPLC pumps and detector. Set mobile phase flow rate at 0.2 mL/min using a mixture of 95% mobile phase I plus 5% mobile phase II.
2. Filter a sample of juice through a 0.45-μm filter.
3. Inject 20 μL of the filtered juice sample onto the column. Chromatograph the sample with a mobile phase gradient that increases the percentage of mobile phase II from 5 to 100% over a 20-min period. This gradient should be nonlinear in that the increase in % mobile phase II is less during the first 10 min of the run and greater during the last 10 min.
4. Note retention times or volumes as the various pigment peaks elute from the column and detector. For identification purposes, compare these times/volumes to those of standards run under the same conditions.

POLYSACCHARIDES: SEE DIAGNOSTICS-PECTINS AND GLUCANS

POTASSIUM: ATOMIC ABSORPTION SPECTROMETRIC ANALYSIS

Analyses of potassium in wine are frequently carried out using flame spectroscopic (flame emission and flame absorption) techniques.

I. Equipment
Atomic absorption spectrometer
Potassium (or Na/K) hollow cathode lamp

II. Reagents
Potassium Standard Stock Solution (10^3 mg/L): In a 1-L volumetric flask, dissolve 1.907 g of dried reagent grade KCl in approximately 800 mL of deionized water and dilute to volume. *Alternatively,* premade stock potassium solutions are commercially available and may be conveniently utilized.

III. Procedure

1. Prepare a 100-mg/L working standard solution from the potassium stock by pipetting 10 mL of stock into a 100-mL volumetric flask and diluting contents to the mark with deionized water.
2. Prepare a series of calibration standards of 1, 2, 5, and 10 mg/L potassium by pipetting 1, 2, 5, and 10 mL of the working standard into four 100-mL volumetric flasks. Dilute to volume with deionized water.
3. Install the potassium hollow cathode lamp. Turn on and set the lamp power supply to the milliamp (ma) value listed in the operating instructions. Set the wavelength and slit width to the appropriate values.
4. Adjust acetylene and air pressures to values stated in instrument operating instructions. Likewise adjust fuel and support gas (oxidant) settings. Align burner head in optical path and ignite flame. Adjust flame to lean conditions (small blue flame cone), and while aspirating deionized water, adjust the lamp emission signal to 100%T. While aspirating one of the standard potassium solutions, adjust the burner height to maximize the measured absorbance reading.
5. Prepare a standard curve by aspirating each standard in turn from lowest to highest concentration and recording their respective indicated absorbance values. Plot concentration vs. absorbance readings. Alternatively, if the instrument has a "concentration" mode,

one may calibrate the instrument to read directly in concentration units.

6. Dilute 1 mL of wine sample to 100 mL with deionized water. Aspirate the sample into burner and read absorbance (or concentration). If value obtained from wine is not within the range of standard values, prepare a second sample at a more appropriate dilution level.
7. Read the potassium concentration from the calibration curve. Multiply by 100 (or other dilution value, if used) to obtain potassium concentration in original wine sample.

IV. Supplemental Note

Samples with high concentrations of sodium, such as may arise from cation exchange treatment, may produce high values for potassium concentration, an effect due to the small but significant amount of ionization of sodium atoms in the air-acetylene flame. One way to eliminate or swamp out this interference is to make up all standards and samples such that they contain from 2,000 to 5,000 mg/L of added sodium. In this case,the added sodium acts as an ionization buffer.

POTASSIUM: ANALYSIS BY FLAME EMISSION SPECTROMETRY

Flame emission techniques gradually are being replaced by atomic absorption and plasma emission techniques. However, they are quite adequate for measurements of potassium levels in juices and wines.

I. Equipment

Flame photometer (or atomic absorption spectrometer)

II. Reagents

Stock potassium solution (1,000 mg/L): See atomic absorption procedure for preparation.

III. Procedure

1. From stock potassium solution (1,000 mg/L), volumetrically transfer 10 mL into a 100-mL volumetric flask and bring to volume with deionized water. The concentration of K^+ in this working solution is 100 mg/L.
2. Prepare a series of calibration standards of 1, 2, 5, and 10 mg/L potassium by pipetting 1, 2, 5, and 10 mL of the working standard into four 100-mL volumetric flasks. Dilute to volume with deionized water.
3. A blank consisting of deionized water is used to zero the instrument to 100%T.
4. Consult operator's manual for instrument setup and calibration.

5. Introduce each standard solution to burner. Allow 15–30 sec for reading to stabilize and record %T.
6. Prepare a standard curve by plotting %T against the appropriate concentration.
7. Dilute wine sample 100× with deionized water and introduce into flame.
8. Using the standard curve, determine concentration. Remember, the final concentration of sample must reflect its dilution.

IV. Supplemental Notes

1. In quantitative flame photometry, one must be careful to minimize fluctuations during analysis. Assuming a properly operating instrument, two techniques are commonly employed to reduce the possibility of analytical interference:

   a. Frequently, quantitative flame emission methods call for use of internal standards of known concentration to compensate for uncontrollable variables. These internal standards are added to standard solutions as well as unknowns. Ideally, such internal standards should have characteristics similar to those of the element analyzed. Most important, the excitation line(s) should be in in the same region as the element of interest so that flame temperature fluctuations affect both in a similar manner. In K^+ and Na^+ analyses, lithium frequently is used as the internal standard.

   b. To stabilize aspiration rates of standard solutions and unknowns, procedures may call for inclusion of detergents (sold under several proprietary names) at levels of less than 0.1%.

2. On a daily basis, many workers prefer not to prepare a standard curve. Instead, the concentration of an unknown may be related to that of a single standard via the following relationship:

$$\text{Concentration} = \frac{\%T\,(\text{unknown})}{\%T\,(\text{standard})} \times \text{conc. (standard) mg/L}$$

   This procedure is usually limited to instruments that employ filters rather than more sophisticated methods of wavelength selection.

3. Because the emission lines of Ca^{2+} are close to those of Na^+, significant interference may be expected in flame photometric analyses of the former unless a monochromator is used for wavelength selection. By comparison, the emission lines of Na^+ and K^+ are sufficiently far apart such that interferences are not commonly encountered even with the use of a relatively crude filter.

POTASSIUM: ANALYSIS BY ION SELECTIVE ELECTRODE

An alternative technique for potassium analysis involves the use of potentiometry and an ion selective electrode. The potassium electrode is constructed so that it responds preferentially to potassium ion. It requires a high-quality expanded scale pH meter (mV scale capable of reading to the nearest 0.1 mV) or select-ion meter, and a reference electrode designed for use with selective ion electrodes.

I. Equipment
   Expanded scale pH/mV meter or specific ion meter
   Potassium selective ion electrode
   Single-junction reference electrode
   Magnetic stirrer and stir-bar

II. Reagents
   Stock potassium solution (1,000 mg/L): See Potassium by Atomic Absorption.
   Ionic Strength Adjustor (ISA): In a 100-mL volumetric flask, dissolve 35.1 g of reagent quality NaCl in deionized water and bring to volume.
   Reference Electrode Filling Solution: Transfer 2 mL ISA to a 100 mL flask and dilute to 100 mL with deionized water. Add silver nitrate (1 M) dropwise until a cloud persists.

III. Procedure
   1. Prepare working standards of 100 and 10 mg/L from potassium stock solution. Prepare the 100-mg/L standard by pipetting 10 mL of the stock solution and 2 mL of the ISA solution into a 100 mL volumetric flask. Dilute to the mark with deionized water. Prepare the 10-mg/L standard in a similar fashion using the 100-mg/L standard.
   2. Place electrodes in the 10-mg/L standard solution. Stir and allow meter reading to stabilize. Record the final mV reading.
   3. Rinse electrodes, dry, and place in the 100-mg/L standard solution. Stir and allow meter reading to stabilize. Record the final mV reading.
   4. Prepare a calibration curve by graphing millivolt readings vs. log concentration. This also may be accomplished by using semi-log paper (graph concentration values on the log axis).
   5. Dilute 10 mL wine (or juice) to 100 mL in a volumetric flask. Transfer the diluted sample into a 150-mL beaker, add 2 mL ISA, and stir. Place clean, dry electrodes into the sample, allow the meter to stabilize, and record the mV reading. Refer to the calibration curve and read the potassium concentration off the graph.
   6. Multiply the value obtained in Step 5 by a factor of 10 (dilution factor) to obtain the potassium concentration of the original sample.

IV. Supplemental Notes

1. The electrodes should be recalibrated every few hours of use.
2. Most electrode operating manuals will also contain instructions for use with various ion meters, as well as for known addition techniques. The standard or known addition technique is generally used in order to match the standards to the matrix of the wine samples being analyzed. The electrode potential for the wine sample, with appropriate dilution buffer added, is first measured. A known amount of standard potassium solution is added to this sample, and a second potential reading is taken. The difference in the two potential readings and the known concentration of the potassium standard are used to calculate the potassium concentration in the original wine sample.

PROLINE: SPECTROMETRIC DETERMINATION USING NINHYDRIN/FORMIC ACID

Of the various amino acids present in grape musts and wines, proline is present in the greatest concentration. Although its average level in grapes is reported to be 742 mg/L (Amerine and Ough 1980), its concentration is known to vary with variety as well as growing season. Cabernet Sauvignon and Chardonnay musts typically have higher proline levels than other varieties (Ough and Stashak 1974). Higher levels of the amino acid also are found in fruit produced in cooler growing seasons, compared with fruit harvested in warmer years (Winkler et al. 1974).

Under the conditions normally followed in wine production, this amino acid probably is not used by yeast, for two reasons. First, the two enzymes involved in the uptake and utilization (proline permease and proline oxidase) are repressed by the presence of ammonium ion in must. The permease is irreversibly inactivated by ammonium ion. Second, oxidase enzymes require oxygen for activity. Ingeldew et al. (1987) reported that wine yeasts may use proline as a nitrogen supplementation, and incorporation of air into yeast starters may prove valuable in reducing the frequency of stuck fermentations.

Proline may be determined via its reaction with ninhydrin in the presence of formic acid, yielding a colored product that absorbs visible radiation at 517 nm (Amerine and Ough 1980).

I. Equipment

Spectrometer capable of reading at 517 nm

130 × 15 mm screw-cap test tubes

100-mL volumetric flasks

Boiling water bath
Various pipettes from 0.25 to 10 mL

II. Reagents

Stock proline standard (575 mg/L): In a 100-mL volumetric flask, dissolve 57.5 mg proline in deionized water and bring to volume.

Methyl Cellosolve (2-Methoxyethanol)-Ninhydrin Solution: Dissolve 3 g of ninhydrin in 100 mL methyl cellosolve.

Formic acid (stock).

III. Procedures

1. Pipette 0, 1, 2, 3, 5, 7, and 10 mL of the proline standard into 100-mL volumetric flasks. Dilute to the mark with deionized water. This will produce proline working standards of 0, 5.75, 11.50, 17.25, 28.75, 40.25, and 57.50 mg/L.
2. Pipette 1 mL of wine or grape juice into a 50-mL volumetric flask, and dilute to the mark with deionized water.
3. Transfer 0.5 mL of each proline working standard and wine or juice sample into 130 × 15 mm screw-cap test tubes.
4. Add 0.25 mL formic acid and 1 mL of the ninhydrin-methyl cellosolve solution to each tube, mix the contents well, cap the test tubes, and place the tubes in the boiling water bath for exactly 15 min.
5. After 15 min, remove the test tubes, and place them in a 20°C water bath to cool. Add 5 mL of the isopropanol-water mixture to each tube, mix the solutions, and transfer the contents to appropriate cuvettes.
6. Read the absorbance of each standard and sample at 517 nm against the water blank carried through the procedure. Plot absorbance vs. concentrations of the samples directly from the graph.
7. Calculate the proline concentrations in the original wine or juice sample by multiplying the value taken from the calibration plot times the dilution factor (50 mL/1 mL if sample was diluted according to step 2).

IV. Supplemental Notes

1. If the sample of wine or grape juice has an absorbance value greater than that of the most concentrated standard, dilute the sample 1 + 1 with the isopropanol-water mixture and repeat the determination.
2. Loss of color for the solutions is reported to be about 2% per hour. For greatest accuracy, one should run the calibration curve with each batch of samples.

N-PROPYL ALCOHOL: GAS CHROMATOGRAPHIC ANALYSIS (SEE FUSEL OILS)

PROTEIN: SEE NITROGEN-COOMASSIE BLUE METHOD

PROTEIN/PHENOLS: VISUAL ESTIMATION USING AMIDO BLACK 10-B AND EOSIN Y PROTEIN STAINS (SEE DIAGNOSTICS)

PROTEIN STABILITY: VISUAL EXAMINATION—SATURATED AMMONIUM SULFATE TEST

Koch and Sajak (1961) recommend a testing procedure involving the addition of 5-mL of saturated ammonium sulfate to 95 mL of wine. Upon addition, the sample is heated to 55°C (131°F) for 7 hours and subsequently cooled in an ice bath for 15 mins. Precipitation indicates the presence of heat-labile proteins. This test is not a good predictor of bentonite requirements (Troost and Fetter 1960).

PROTEIN STABILITY; VISUAL EXAMINATION—BENTOTEST WITH AND WITHOUT NEPHELOMETRIC EVALUATION

The Bentotest uses phosphomolybdic acid prepared in hydrochloric acid to denature and precipitate juice or wine proteins by forming cross linkages with the molybdenum ion (Jakob 1962). The test has the advantages of being quick and not requiring heat. Haze development is proportional to protein content and may be used to determine bentonite addition levels. Rankine and Pocock (1971) demonstrated that the Bentotest is more sensitive than heat tests using 70°C for 15 min. It is also more sensitive than either the TCA test or storage at elevated temperature. The test is rather severe; therefore, one should decrease the suggested fining level in wine by 5–15% and to as much as 25% in the case of juice fining.

I. Equipment
   100-mL volumetric cylinders with stoppers
   Buchner or glass funnel
   Whatman No. 1 filter paper of equivalent
   0.45-µm membrane and filter housing
   High-intensity light source or turbidimeter (nephelometer)
   20-mL test tubes
II. Reagents
   Bentotest reagent (phosphomolybdic acid): Dissolve 5 g phosphomolybdic acid, 5 g sodium sulfate, and 0.25 g glucose in 15 g stock sulfuric

acid and enough deionized water to bring the volume up to 1 liter. *Note:* this reagent can be purchased from Fritz-Merkel Co. in Germany.

III. Procedure

A Protein fining trial may be established by placing 100 mL of sample into a series of 100-mL glass cylinders. To all but the control cylinder pipette the required volume of a 5% bentonite solution to establish a fining series (see Table 16.2), stopper cylinders, and mix thoroughly. Bentonite fining levels for wine usually range from 6 to 36 g/hL (1/2–3 lb/1,000 gal); and higher for juice. Allow at least 15 min for the fining trials to settle and decant off clear liquid. Filter approximately 20 mL of each trial.

1. Filter enough sample through 0.45-μm membrane to attain approximately 15 mL.
2. Pipette 10 mL of each fining level into separate test tubes.
3. Add 1 mL of Bentotest reagent into each tube and mix.
4. Examine each tube under high-intensity light and note the degree of light scattering (e.g., haze compared with control). The tube(s) without haze are considered protein stable.
5. The degree of haze formation may be quantified using a nephelometer.

IV. Supplemental Notes

1. A green/blue color may be formed as a result of the reaction of SO_2 and tannins with the Bentotest reagent. *Test results are based on haze formation, not color.*
2. This test may be used to estimate the concentration of bentonite required in the fermenter to attain a protein stable or near stable wine. See Chapters 8 and 16 for consideration of fermenting with bentonite.

PROTEIN STABILITY: VISUAL EXAMINATION—ETHANOL PRECIPITATION

This test involves slowly mixing 100% ethanol with an equal volume of wine, and noting the formation of haze visually or with the aid of a turbidimeter.

PROTEIN STABILITY: EXPOSURE TO HEAT AND VISUAL EXAMINATION

A clarified sample of wine is subjected to an elevated temperature (49°C, 162°F) for 24 hours. At the end of this time period, the sample is visually inspected for haze formation relative to a control (unheated) sample of the same wine. There are several variables on this procedure in terms of both temperature and time. Additionally, some wineries subject the wine to low-temperature storage as well.

I. Equipment

Two 4-oz clear glass screw-cap or stoppered bottles
Incubator or water bath at 49°C
0.45-μm membrane and filter housing
High-intensity light source

II. Procedures

1. Membrane-filter a sufficient quantity of wine to fill two 4-oz sample bottles.
2. Fill the same bottles, the first as "room temperature" and the second as "one day at 49°C."
3. Examine each sample under high-intensity light, and record your impressions of initial clarity.
4. Place the "one day at 49°C" sample in the incubator or water bath, noting the temperature.
5. At 24 hours, examine each sample carefully.
6. Clouding and/or precipitate formation in the heated sample vs. a clear control sample is indicative of protein instability.

III. Supplemental Note

Before evaluation of results, the wine should be brought to room temperature or chilled. Interpretation depends on the winemaker's experience. Although the absence of haze is judged to indicate stability, a faint haze may be considered acceptable based on prior experience with the test and wine type. As noted in Chapter 16, temperature and time regimens may be adjusted to suit individual needs.

PROTEIN STABILITY: TRICHLOROACETIC ACID TEST (TCA)—VISUAL OR NEPHELOMETRIC EVALUATION

This procedure, adapted from Berg and Akiyoshi (1961), involves passage of light through a turbid medium resulting in scattering and apparent energy loss in the incident beam. This scattering effect may be measured at any angle relative to the plane of incident light.

The degree of scattering depends primarily on particulate number, size, and shape. These parameters themselves depend on several variables, including temperature, pH, concentration of reagents, and mixing procedures.

I. Equipment
   Boiling water bath
   Pyrex test tubes (20-mL capacity)
   High-intensity light source
   Pipettes (1 mL)
   Turbidimeter (nephelometer) or visual comparison

II. Reagents
   Trichloracetic acid (55%): Dissolve 55 g of trichloroacetic acid in deionized water and bring the final volume to 100 mL.

III. Procedure
   1. Fill two test tubes with 10 mL each of filtered wine sample.
   2. Examine both sample tubes for clarity under a high-intensity lamp.
   3. To one sample add 1 mL of 55% trichloroacetic acid and transfer the sample tube to a boiling water bath for 2 min.
   4. At the end of the reaction period, remove the sample tube and visually compare its clarity with that of the control (untreated) sample. Haze in the heated sample is indicative of protein instability.
   5. If haze is not present, after removing sample from boiling water bath, wait 15 min for reaction to be complete and re-inspect.
   6. Consult the operator's manual for turbidimeter (nephelometer) setup and operation.
   7. Determine "nephlos units" of sample(s) or visually compare the clarity of the control and treated wine using a high-intensity light source.

IV. Supplemental Notes
   1. Rankine and Pocock (1971) suggest using a nephlos value <19 as the *upper* limit for protein stability in table wines. It has been the experience of the authors that this level is too high. High phenol white wines that undergo oxygen exposure during bottling may show protein-tannin co-precipitation. We suggest an upper nephlos limit in the low teens.
   2. Some wineries prefer to compare the absorbance of heated vs. control samples, using a spectrophotometer at a wavelength around 430 nm. Others simply evaluate relative clarity differences visually between control and treated wines using a high-intensity light source.

PROTEIN STABILITY: VISUAL EXAMINATION AFTER TANNIC ACID PRECIPITATION

A quick screening test can be performed by taking advantage of the ability of tannins to bind protein. A drop of 10% tannic acid solution is added to

10 mL of wine and heated slightly over a Bunsen burner flame. A haze or precipitate in the treated sample compared with the control indicates protein instability. A variation recommended by Krug (1967) involved the addition of 5 mL of 1% tannin solution. This test can also be used to determine the extent of overfining with protein fining agents.

REDUCING SUGAR: VISUAL RAPID ESTIMATION (CLINITEST)

Rapid routine reducing sugar measurements may be run with a variety of reducing sugar kits originally developed for use by diabetics. By reference to a color code, the reducing sugar content of a wine is determined within a range of 0–1%. Pentoses will prevent the reducing sugar level of wine from reaching zero; however, these pentoses are not fermentable by yeast.

In the case of proprietary products such as Clinitest, sensitivity levels are reported as 0.05% (The Ames Company 1978). The mechanism of reaction is generally the same as in copper reduction methods except that the heat required is provided internally by the neutralization reaction of NaOH and citric acid. The major limiting factor in the use of reducing sugar kits is that the sugar level must be less than 1.0%. Thus, these kits are primarily helpful in determining completion of fermentation.

REDUCING SUGARS: SPECTROMETRIC ANALYSIS USING ENZYMES

McCloskey (1978) reports an enzymatic analysis for the two major reducing sugars (glucose and fructose) in wine. This method does not detect the pentoses included in traditional analyses of reducing sugar. The results at low levels (< 1.0 g/L) are superior to those obtained by the Lane-Eynon procedure. McCloskey's procedure uses premeasured enzyme reagents sold under the name, Glucose Stat-Pak (Calbiochem). In that fructose is not an active substrate, a third enzyme, phosphoglucose isomerase, is necessary. The reaction sequence is presented below:

$$\text{Glucose} + \text{ATP} \overset{\text{HK}}{\rightleftharpoons} \text{G-6-Phosphate} + \text{ADP}$$

$$\text{Fructose} + \text{ATP} \rightleftharpoons \text{F-6-Phosphate} + \text{ADP}$$

$$\text{F-6-Phosphate} \overset{\text{PGI}}{\rightleftharpoons} \text{G-6-Phosphate}$$

$$\text{G-6-Phosphate} + \text{NADP} \xrightleftharpoons{\text{G6P-DH}} \text{Gluconate-6-Phosphate} + \text{NADPH} + \text{H}^{+}$$

where:

HK = hexose kinase
PGI = phosphoglucose isomerase
G6P-DH = glucose-6-phosphate dehydrogenase

The reader is referred to the product sheet that accompanies the carbohydrate kit of reagents. In this way, one will have the most recent update and a procedure specifically designed for the kit obtained.

The carbohydrate procedure will require the following:

I. Equipment
Spectrometer capable of reading absorbance at 340, 365, or 334 nm.
Single or multiple-range micropipettes (25 and 100 μL)
A water bath (22°–30°C)

II. Reagents
The reagents used in this analysis are most conveniently purchased in kit form. Such kits may be acquired as Calbiochem Glucose Stat-Pak and Boehringer-Mannheim Glucose/Fructose.

III. Procedure
The general thrust of the procedure is as follows:
1. Samples are generally diluted 1 + 9 to keep sugar concentration in the 0.1–1 g/L range.
2. An aliquot is mixed with reagents (except hexokinase/glucose-6-phosphate dehydrogenase and phosphoglucose isomerase) and the absorbance is read.
3. The hexokinase/glucose-6-phosphate dehydrogenase is added to initiate the reaction and the absorbance is again read upon completion.
4. Phosphoglucose isomerase is added and the absorbance again read after the reaction ceases.
5. Glucose and fructose levels are calculated by taking differences in the various absorbance levels.

REDUCING SUGARS: TITRAMETRIC-MODIFIED LANE-EYNON PROCEDURE

The Lane-Eynon procedure calls for determining the quantity of standard glucose solution (titrant) required to react with a known volume of alkaline copper sulfate under specified heating conditions. One milliliter of wine is added to a second volume of alkaline copper sulfate and the quantity of standard sugar solution required to complete the reduction determined. The difference in volumes needed to titrate the wine sample and blank are then related directly to the reducing sugar content of the sample.

I. Equipment
   500-mL wide-mouth Erlenmeyer flask
   50-mL side delivery burette
   Volumetric pipettes (1.0 and 10 mL)
   100-mL graduated cylinder
   Electric burner (not a hot plate!)
   High-intensity light source
   Glass boiling beads

II. Reagents
   Fehling's A Solution: In a 2-L volumetric flask, dissolve 138.6 g of cupric sulfate ($CuSO_4 \cdot 5H_2O$) in approximately 1,800 mL of deionized water. Bring to volume at defined temperature. Fehling's A solution should be stored in the refrigerator when not in use.
   Fehling's B Solution: In a 2-L flask, dissolve 692 g of sodium potassium tartrate and 200 g of sodium hydroxide in deionized water. Bring to volume at room temperature. One need not use an analytical balance for preparation of this reagent.
   Dextrose Solution (0.5%): In a 2-L flask, dissolve 10.00 g of anhydrous dextrose in deionized water. Bring to volume at defined temperature. Store in refrigerator. Alternately, add 2 g sodium benzoate and 1 g citric acid before bringing to volume.
   Methylene Blue Indicator: Dissolve 1 g of methylene blue in 95 mL of deionized water. Bring to volume. Store in dropper bottle.

III. Procedure
   (a) Blank Determination
   1. Add 70 mL deionized water and several glass boiling beads to a 500-mL wide-mouth Erlenmeyer flask.
   2. Volumetrically transfer 10 mL of Fehling's A solution to flask.
   3. Add 10 mL of Fehling's B solution and mix well. Note: This addition need not be volumetric and can be achieved with the use of any pipette.
   4. Place flask on preheated electric burner and immediately titrate with approximately 18 mL of 0.5% dextrose solution.
   5. When the solution comes to a rapid boil, add 5 drops of methylene blue indicator and titrate drop by drop until the end point is reached. The end point is detected as the first disappearance of blue indicator and appearance of red in the boiling solution.
   Note: This titration should not take more than 3 min to complete.

Theoretically, the blank titration should use 21.8 mL of 0.5% dextrose. However, this is not always the case.

6. Blank titrations should be repeated until end points vary by no more than 0.2 mL.
7. Record volume of dextrose required for blank "B."

(b) Sample Determination:

8. Proceed as in blank determination, adding 1 mL of wine (step 3).
9. Titrate rapidly until approximately 2 mL before the estimated end point. At full boil, add 5 drops of methylene blue and continue titration on a drop by drop basis until end point is reached.
10. Record the volume of dextrose required for sample titration "W." Calculate reducing sugar using the following relationship:

$$\text{R.S. (g/L)} = \frac{(B - W)\ (0.005\ \text{g/mL})\ (1{,}000\text{mL/L})}{\text{sample volume (mL)}}$$

where B = volume of 0.5% dextrose solution required to titrate blank
W = volume of 0.5% dextrose solution required to titrate sample
When using 1 mL sample volumes, the above equation can be simplified:

$$\text{R.S. (g/L)} = 5(B - W)$$

IV. Supplemental Notes

1. Some laboratories elect to decolorize highly pigmented wines before analysis. However, this is generally not necessary. Wines with a heavy suspended-solids content should be clarified before analysis.
2. During titration, the tip of the burette should be positioned within the neck opening of the flask to reduce oxygen contact. Continuous evolution of steam during titration also helps in reducing access of air. Interruptions in steam flow from the titration flask may introduce errors.
3. Several practice titrations may be necessary for consistent end point detection. Because different people may interpret end points differently, it is essential that the same person run the blank and samples.
4. End point detection is the disappearance of methylene blue indicator (formation of leuco-form). This is best observed in the bubbles and around the edges of the flask by use of a high-intensity light source.
5. Wines with a residual sugar content of more than 5% should be diluted accordingly before analysis.

6. In dry wine types, accuracy is reported as ±0.05%. Routinely, one should strive for a reproducibility of ±0.2 mL between replications.
7. When not in use, Fehling's A solution and dextrose should be stored in the refrigerator. These solutions should be at standard temperature when used.
8. Refer to the Rebelein procedure (Supplemental Notes) for precautionary comments relating to the use of Fehling's type solutions.

REDUCING SUGARS: TITRAMETRIC-REBELEIN (GOLD COAST) METHOD

In the Rebelein procedure, the excess copper remaining after reaction with sugar is subsequently reduced with excess iodide ion to produce an equivalent amount of iodine. The iodine is then titrated with sodium thiosulfate. To avoid having to standardize reagents, a deionized water blank determination (no reducing sugar) is run. The result is then calculated by comparing the sample titration to that of the blank.

I. Equipment
   Volumetric pipettes (10 mL)
   Erlenmeyer flasks (200 mL)
   Burner
   Burette (50 mL)
   Glass boiling beads
II. Reagents
   Alkali Rochelle Salt Solution:
   Component A: In approximately 40 mL of deionized water, dissolve 250 g of sodium potassium tartrate.
   Component B: In approximately 400 mL of deionized water, dissolve 80 g sodium hydroxide. Carefully, combine Component A and Component B in a 1-L volumetric flask and bring to volume with deionized water.
   Copper Sulfate Solution: In a 1-L volumetric flask, dissolve 41.92 g of copper sulfate ($CuSO_4 \cdot 5H_2O$) in approximately 600 mL of deionized water. Add 10 mL of 1 N H_2SO_4, mix, and bring to volume with deionized water.
   Potassium Iodide Solution: In a 1-L volumetric flask, mix 100 mL of 1 N sodium hydroxide and 300 g of potassium iodide. Add approximately 300 mL of deionized water. Upon complete dissolution, bring to volume with deionized water.
   Sodium Thiosulfate: In a 1-L volumetric flask, add 50 mL of 1 N NaOH and 13.777 g sodium thiosulfate ($Na_2S_2O_3 \cdot 5H_2O$). Add approximately

200 mL of deionized water and dissolve. Bring to volume with deionized water.

Starch Solution:

Component A: In 500 mL of boiling water, dissolve 10 g of soluble starch.

Component B: In 500 mL of deionized water, dissolve 20 g of potassium iodide. Add 10 mL of 1 *N* NaOH. Carefully combine Components A and B.

Sulfuric Acid (16% vol/vol): Carefully, and with mixing, add 175 mL of 95% sulfuric acid to 825 mL deionized water.

Sulfuric Acid (1 *N*): See Table 20–3.

III. Procedure

1. If reducing sugar is to be determined on a sweet wine, the sugar level must first be diluted, to less than 2.8% (28 g/L).
2. Volumetrically, transfer 10 mL of the copper sulfate solution into a 200-mL Erlenmeyer flask.
3. Add 5 mL alkali Rochelle salt solution, 2–3 boiling beads, and 2.0 mL wine sample.
4. Bring to a rapid boil for 1.5 min. Cool quickly.
5. When cool, add:
   a. 10 mL potassium iodide solution
   b. 10 mL sulfuric acid solution (16%)
   c. 10 mL starch solution.
6. Mix, and titrate with sodium thiosulfate to a cream-white end point.
7. Calculate reducing sugar using the following relationship:

$$\text{R.S. (g/L)} = 28 - 28\,(V_A/V_B)$$

where: V_A = Volume titrant used with sugar sample
V_B = Volume titrant used for blank sample

Determination of a blank titration value is accomplished by using the protocol presented in steps 2–6 above using 2.0 mL deionized water instead of wine sample.

IV. Supplemental Notes

1. A blank titration value should be determined each time a new batch of samples is run.
2. Red wines should be decolorized by pretreatment with decolorizing carbon or PVPP ("Polyclar AT") before analysis.
3. Once the potassium iodide, sulfuric acid, and starch are added, the thiosulfate titration should be run as quickly as possible.
4. Several variables must be carefully controlled, if success is to be achieved and the calculation (equation above) is to be valid.

a. Reagents used for sugar analyses, as ordinarily prepared, consist of two solutions, one containing $CuSO_4$ and the other alkaline Rochelle salt (sodium potassium tartrate). Classic procedure requires the two be mixed immediately before use. Mixed reagents cannot be kept satisfactorily due to the ease with which the mixture decomposes. The component solutions themselves will eventually decompose over prolonged periods of storage and it is recommended that the copper sulfate solution be stored in the refrigerator until time for use.

b. Temperature and Duration of Heating. The reduction reaction is accelerated with increases in temperature up to 90°C. At temperatures greater than 100°C, however, significant autoreduction of the copper in the reagents may occur, resulting in appreciable error.

c. The concentration of copper sulfate also affects reduction. Ideally, the ratio of alkali (OH^-) to copper should be 5 to 1. Beyond this, reduction decreases and the possibility of autoreduction increases.

REDUCING SUGAR (INVERT)-TITRAMETRIC ANALYSIS AFTER HYDROLYSIS

The addition of sugar to fermenting and finished wines is permitted in certain instances. In California, these additions are restricted to formula wines such as champagnes and special naturals. In such cases, the sugars used are generally in the form of concentrates, dry dextrose, sucrose, or syrups of the latter two. Sucrose creates difficulties from an analytical point of view in that it is not a reducing sugar. Where such analyses are required, a preliminary sample treatment is necessary. This is usually accomplished by acid hydrolysis of the nonreducing disaccharide to its component reducing monosaccharides glucose and fructose, which are reducing sugars. In an acid medium such as wine, this hydrolysis proceeds normally with time. In like manner, yeast-elaborated enzymes (invertase) bring about the hydrolysis in fermenting must. Thus, in time the acid nature of wine or the fermentative action of yeast will bring about the required hydrolysis, making laboratory hydrolysis unnecessary. Once hydrolysis has occurred, one may proceed with the reducing sugar analysis by one of the procedures above, keeping in mind that there will be a correction factor for dilution of the sample.

I. Equipment
   Water bath (60°C± 2.0)
   50-mL volumetric flask

Ice water bath
Timer
25-mL volumetric pipette

II. Reagents
Ammonium Hydroxide Solution (1 + 1.5): Dilute 400 mL of stock NH_4OH to 1 L with deionized water.
HCl Solution (1 + 1): Carefully mix 100 mL of stock HCl with 100 mL of deionized water.

III. Procedure
1. Volumetrically transfer 25 mL sample into a 50-mL volumetric flask.
2. Add approximately 10 mL of 1 + 1 HCl.
3. Place flask in 60°C water bath for 15 min.
4. After 15 min, cool flask in ice bath to 20°C.
5. Carefully add 10 mL of 1 + 1.5 NH_4OH with mixing.
6. Make to volume with deionized water at 20°C.
7. Mix thoroughly. Determine reducing sugar using one of the above procedures.
8. Using the appropriate equation, calculate reducing sugar (g/L). Keep in mind the 1:2 dilution factor required in the hydrolysis step.

RESIDUAL OXIDANTS: SEE CHLORINE (RESIDUAL), SEE IODINE

SODIUM: ATOMIC ABSORPTION SPECTROMETRIC ANALYSIS

Sodium (Na^+) in non-ion-exchanged wines is present at relatively low concentrations: 10–172 mg/L. Its level in wine is attributed, in part, to the proximity of grapes to saline water or soils. Although generally not of importance in wines, Na^+ becomes of concern in certain ion exchange applications.

I. Equipment
Atomic absorption spectrometer with sodium (or Na/K) hollow cathode lamp

II. Reagents
Standard Sodium Solution (as NaCl): Dissolve 2.542 g predried NaCl in deionized water. Bring to 1 L final volume at defined temperature. This solution is equivalent to a Na^+ concentration of 1,000 mg/L. Alternatively, prestandardized sodium standards are commercially available.

III. Procedure
1. Prepare a 100-mg/L working sodium standard by diluting 10 mL of the sodium stock solution to 100 mL in a volumetric flask.

Table 20–34. Preparation of calibration standards from 100 mg/L sodium working standard.

Volume of stock standard (mL)	Final concentration (mg/L) in 100 mL volume
1.0	1.0
2.0	2.0
5.0	5.0
10.0	10.0

2. Prepare calibration standards of 1, 2, 5, and 10 mg/L from the working standard as seen in Table 20–34. All dilutions are made in deionized water.
3. Install the sodium lamp. Set appropriate wavelength, slit width, and power supply current as listed in instrument operating instructions.
4. Ignite burner and adjust flame, optimize analytical conditions, and prepare calibration curve by aspiration of the standards prepared above.
5. Dilute 1 mL wine to 50 mL with deionized water. Aspirate in flame and read absorbance (or concentration) value. If necessary, prepare second sample at a more appropriate dilution.
6. Read the sodium concentration from calibration curve and multiply by 50 (or other dilution factor, if used) to obtain sodium concentration in original wine sample.

IV. Supplemental Notes
1. Samples with high potassium concentrations may produce high values for the sodium concentration. This effect is caused by the small but significant amount of ionization of potassium atoms in the air-acetylene flame. One way to eliminate or swamp out this interference is to make up all standards and samples to contain about 2,000–5,000 mg/L of added potassium (which acts as an ionization buffer).
2. Analyses conducted at the low mg/L concentration level require careful attention to cleanliness and careful pipetting and diluting of standards and samples.

SODIUM: FLAME EMISSION SPECTROMETRIC ANALYSIS

Sodium in wine lends itself to analysis by traditional flame photometric (atomic emission) methods. Emission analyses can be made with an atomic absorption spectrometer (without the sodium hollow cathode lamp) or with a flame photometer. As in other emission techniques, the sodium signal is read on the percent transmittance scale (more sodium, produces more emission, and thus more apparent transmittance on the read-out scale).

I. Equipment

Flame photometer (or atomic absorption spectrometer)

II. Reagents

Stock sodium chloride solution (1,000 mg/L as Na^+): See AA procedure above.

III. Procedure

1. Using a dilution series identical to that prepared for Na^+ analysis by atomic absorption (see above), prepare a standard curve of sodium concentration vs. %T.
2. Dilute wine sample 50× with deionized water and aspirate in flame.
3. Using the standard curve, determine the concentration of Na^+ in sample. Remember to multiply this value by the appropriate dilution factor.

IV. Supplemental Notes

Sodium analyses by flame emission are quite similar in type to those for potassium. Refer to the potassium procedure for relevant comments.

SOLUBLE SOLIDS: DETERMINATION OF °BRIX, °BALLING, °BAUME, OR ÖECHSLE BY HYDROMETRY

Hydrometry measures the specific gravity of a test solution at defined temperature. Brix hydrometers are calibrated against known wt/wt % sucrose solutions so that they can be read directly in °B rather than in typical density units. Glassware that is clean and dry (or rinsed with the sample) is required to avoid dilution and/or contamination with water or previous samples. The glass cylinder containing the sample should be about twice the diameter of the hydrometer in order to avoid frictional effects between the hydrometer and walls of the cylinder. Entrained air or carbon dioxide should be removed from the sample before hydrometer measurements are made.

I. Equipment

Brix, Balling, or equivalent hydrometer of varying scales
Hydrometer cylinder of approximately 250 mL capacity
Thermometer
Laboratory tissues

II. Procedure

1. Collect a homogeneous representative sample, removing gross par-

ticulates. Carefully transfer it to the hydrometer cylinder, taking care to create as few air bubbles as possible.

2. Record the temperature of the sample.
3. Select a hydrometer of appropriate scale range and immerse it into the solution with a gently spinning motion.
4. When a constant flotation level is reached, take a reading at the bottom of the meniscus, as shown in Fig. 20–15.
5. If necessary, make temperature compensations using Table I-3 in Appendix I. If such a table is not available, an approximate correction factor of ±0.06 for each degree above or below 20°C may be used. Alternatively the sample may be brought to 20°C before the measurement.

IV. Supplemental Notes

1. In unfermented juice (or fermenting wine), the buoying effects of entrapped air or carbon dioxide may act to produce erroneously high readings. Air and/or carbon dioxide can be reduced significantly by vacuum filtration of the sample. Readings should be taken as soon as the hydrometer reaches a constant level. The soluble solids content of a juice should not be taken to include excessive particulate matter. With highly turbid samples, accuracy can be improved by settling and decanting the sample before measurement.
2. Accuracy can also be improved by using whenever possible, the hydrometer with the shortest scale range.
3. Because of the ever-present problem of faulty data manipulation, the authors recommend, whenever possible, adjusting the sample to the temperature at which the hydrometer is calibrated. When a water bath is used for this purpose, temperature stratification may be encountered. Therefore, samples should be thoroughly mixed before analysis.
4. Even though Brix hydrometers are calibrated to 0.1 °B, interpolation between markings is not recommended, and readings are made only to the nearest 0.1 °B. Practically speaking, then, there is no value in reporting results beyond the first decimal place, even though temperature correction tables may indicate that more accurate results are possible.
5. A good practice is for laboratory personnel to calibrate new hydrometers against known sugar solutions to ensure accuracy.
6. Should °Brix hydrometers not be available, a specific gravity hydrometer calibrated "for liquids heavier than water" may be used. For conversion to °Brix, see Table I-1 in Appendix I.

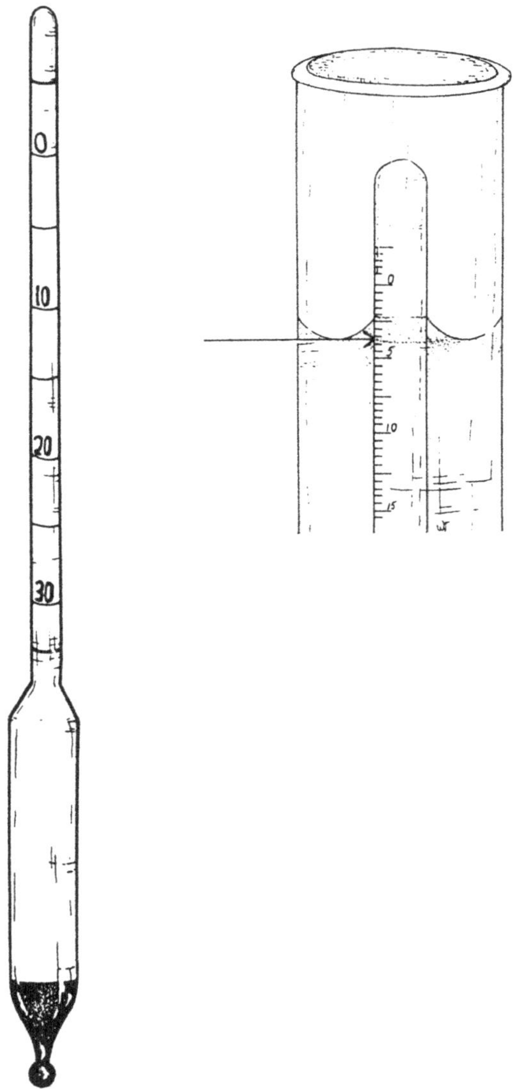

Fig. 20–15. Glass hydrometer designed to float upright in liquid. Scale can be calibrated to read specific gravity or concentration of some specific solute at the defined temperature of calibration. Hydrometers are read at the bottom of the miniscus.

SOLUBLE SOLIDS: DETERMINATION BY REFRACTOMETRY

As the concentration of soluble solids increases, so too does the refractive index of the sample. The principle here is that the more molecules there are in a sample, the more the light passing through the sample is bent (refracted). Most refractometers have a °Brix scale (calibrated against sucrose solutions) along with the refractive index scale; so one may read °Brix directly in juice samples. Because all soluble solids (especially ethanol) in a juice sample affect the measured refractive index, fermenting samples give °Brix measurements that are erroneously high.

I. Equipment
   Hand-held refractometer, Abbe refractometer, or equivalent digital refractometer (range: 1.300–1.700)
   Circulating water source if required
   Light source
   Wash bottle with deionized water
   Cotton swabs
   Laboratory tissue and lens paper

II. Procedure
   1. On instruments using water-cooled prisms, connect the laboratory water line to the intake nipple on the instrument and return water to the drain via the exit port.
   2. Circulate water to constant temperature through the illuminating prism.
   3. Follow the manufacturer's instructions for setup and operation of the instrument used.
   4. Field-type refractometers must have their reading temperatures corrected to 20°C (68°F). In the absence of temperature-corrected refractometers, refer to Table I-4 in Appendix I.
   5. All units must be standardized with deionized water such that the refractive index reads 1.330 with a corresponding sugar concentration of 0 °B.
   6. With the blotter and lens paper provided, dry the prisms.
   7. Using a cotton swab, apply several drops of juice or appropriately diluted concentrate to the lens, and repeat the operations. *Caution:* The user should be aware that particulate matter in the sample may scratch the prisms.
   8. Should the refractometer not be calibrated in °Brix, refractive index readings can be converted to equivalent % sucrose values (see Table I-5 in Appendix I).

III. Supplemental Notes
   1. The refractive index is critically dependent on the temperature of

the solution measured. Unless corrected, small differences from the reference temperature should be expected to result in significant error. For convenience, laboratory refractometers normally have a thermometer incorporated into the body of the instrument. Field units without temperature compensation capacity provide only a rough approximation of the soluble solids content.

2. Samples for analysis must be representative, homogeneous, and free of particulate matter that could damage prisms or impair accuracy.
3. In addition to temperature, alcohol is a major interference in accurate measurements of the refractive index. Therefore, a refractometer should not be used in soluble solids determinations of fermenting must or wine.

SORBIC ACID: COLORIMETRIC ANALYSIS AFTER DISTILLATION

As developed by Nury and Bolin (1962) and later modified by Caputi et al. (1974), this procedure follows the initial separation of sorbic acid by steam distillation with its oxidation to malonaldehyde. Reaction of the latter with thiobarbituric acid yields a highly colored condensation product that is measured at 530 nm (see Fig. 20–16).

I. Equipment
   Cash volatile acid assembly or equivalent
   Visible spectrometer
   Boiling water bath
   500-mL volumetric flasks
   100-mL volumetric flasks
   Volumetric pipettes (2, 5, 10, 15 mL)
   Several 15-mL Pyrex test tubes
II. Reagents
   Sorbic acid stock (100 mg/L): Accurately weigh 134 mg of potassium sorbate. Transfer to a 1-L volumetric flask, dissolve, and bring to volume with deionized water. This solution is equivalent to 100 mg/L sorbic acid.
   Sulfuric Acid (0.3 *N*): Dilute 15 mL of 2 *N* H_2SO_4 with deionized water to a final volume of 100 mL.
   Potassium Dichromate Solution: Dissolve 147 mg of $K_2Cr_2O_7$ in deionized water, bringing the final volume to 100 mL.
   Thiobarbituric Acid (0.5%): In a 50-mL volumetric flask, dissolve 250 mg of thiobarbituric acid in 5 mL of 0.5 *N* NaOH. Warm contents under a stream of hot water to speed dissolution. When complete, add 20 mL of deionized water adjusted with 3 mL of 1 *N* HCl. Dilute to

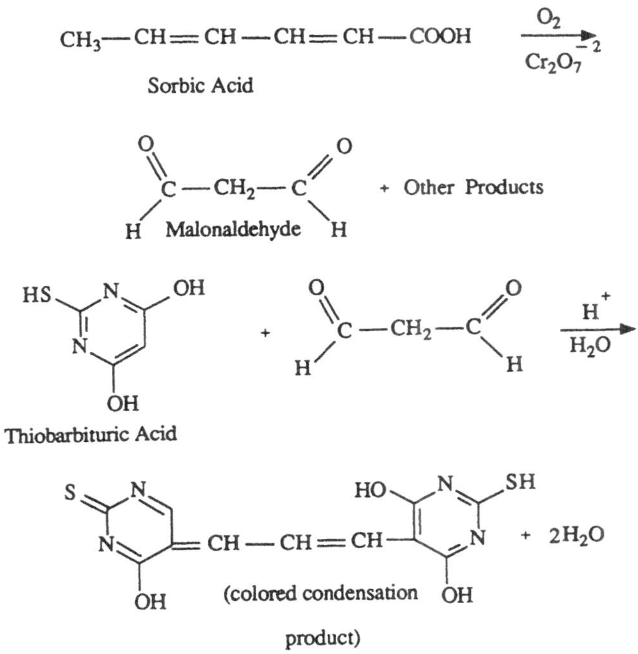

Fig. 20–16. Oxidation of sorbic acid and reacton with thiobarbituric acid to form a colored condensation product.

volume with deionized water. *NOTE*!! This solution must be prepared daily.

III. Procedure

a. Standard Curve Preparation

1. Prepare a series of sorbic acid calibration standards from the 100 mg/L stock solution according to Table 20–35.
2. Volumetrically transfer 2 mL of each standard to separate test tubes. Blank preparation consists of 2 mL of deionized water.

b. Preparation of Wine Sample

3. Volumetrically transfer 2 mL of wine to receiving funnel on cash still. Rinse several times with deionized water to ensure complete transfer.
4. Collect approximately 190 mL of distillate into a 200-mL volumetric flask. Dilute to volume with deionized water.
5. Volumetrically pipette 2 mL of distillate into a test tube.

c. Analysis of Samples (to be done simultaneously with standards)

Table 20–35. Preparation of standard sorbic acid solutions from 100-mg/L stock.

Volume of sorbic acid standard (mL)	Final concentration (mg/L) in 100 mL Volume
1	1
2	2
3	3

6. To each test tube, add:
   1 mL of 0.3 N H_2SO_4
   1 mL of potassium dichromate solution
7. Heat in boiling water bath for exactly 5 min.
8. Remove tubes and cool in ice water bath.
9. Pipette 2 mL of 0.5% thiobarbituric acid reagent into each tube. Return to boiling water bath for 10 min.
10. Cool to room temperature. Determine absorbance of each standard at 532 nm and plot against the appropriate concentration (in mg sorbic acid/L).
11. Using the standard curve, determine the concentration of unknown. Remember, this value must be multiplied by 200 mL/2 mL (the reciprocal of the original 2 mL to 200 mL dilution factor) to yield a result in mg/L.

SORBIC ACID: DISTILLATION AND DIRECT ULTRAVIOLET ANALYSIS

The distillate produced in the above procedure can also be analyzed directly at 260 nm without forming a colored derivative. Owing to its state of conjugation, sorbic acid absorbs light in the UV area of the spectrum. Using this to their advantage, Melnick and Luckman (1954) developed this rapid procedure. It is necessary to acidify the distillate to prevent the sorbic acid from dissociating into its nonabsorbing ionic form.

I. Equipment
   UV spectrophotometer
II. Reagents
   Sorbic acid stock (100 mg/L): Accurately weigh 134 mg of potassium sorbate. Transfer to a 1-L volumetric flask, dissolve, and bring to volume with deionized water. This solution is equivalent to 100 mg/L sorbic acid.
   HCl (0.1 N): Table 20–2 for preparation.

III. Procedure

(a) Standard Curve Preparation

1. Prepare sorbic acid calibration standards from 100-mg/L stock as in the colorimetric procedure above. Before bringing the dilutions to volume, add 0.5 mL of 0.1 *N* HCl.

(b) Preparation of Wine Samples

2. Volumetrically transfer 2 mL of wine to receiving funnel of cash volatile acid still.
3. Collect approximately 190 mL of distillate in a 200-mL volumetric flask containing 0.5 mL of 0.1 *N* HCl. Dilute to volume with deionized water.

(c) Colorimetric Analysis

4. Determine absorbance of each standard and sample at 260 nm (using a blank of 0.5 mL HCl in the 199.5 mL of deionized water). Plot absorbances of standards against the respective concentrations (in mg/L sorbic acid).
5. Compare absorbances of samples to the standard curve. Multiply answer by 200 mL/2 mL (the reciprocal of the dilution factor), express in units of mg/L.

SORBIC ACID: EXTRACTION AND DIRECT ULTRAVIOLET ANALYSIS

Due to potential problems of component loss during the preliminary distillation step, Ziemelis and Somers (1978) developed a direct extraction procedure. In this procedure, 0.25-mL aliquots of wine sample are extracted in iso-octane (2,2,4-trimethyl pentane) and absorbance measured at 255 nm relative to an iso-octane blank.

I. Equipment

Micropipette

UV spectrophotometer

Quartz cuvettes

Screw-cap test tubes (25 mL) and glass boiling beads (2 mm)

II. Reagents

Sorbic acid stock (500 mg/L): Accurately weigh 670 mg potassium sorbate and dissolve in 12% (vol/vol) ethanol reagent to a final volume of 1 L in a volumetric flask. This solution is equivalent to 500 mg/L sorbic acid.

Iso-octane: Used without further preparation.

Phosphoric Acid: Used without further preparation.

Ethanol (12% vol/vol): Dilute 120 mL absolute ethanol to a final volume of 1 L with deionized water. In cases where denatured or less

Table 20–36. Preparation of sorbic acid calibration standards from 500 mg/L stock solution.

Volume of sorbic acid stock (mL)	Final concentration (mg/L) in 100 mL volume[a]
5	25
10	50
15	75
20	100
30	150

[a]Bring to mark with 12% (vol/vol) ETOH reagent.

than absolute ethanol is to be used, preparation of ethanol solvent should account for differences by increasing the volume by proportionate amounts.

III. Procedure
1. Prepare a series of calibration standards from the 500 mg/L sorbic acid standard accordingly as per Table 20–36.
2. Using a micropipette, transfer 0.25 mL of each standard solution to a screw-cap test tube.
3. Similarly transfer 0.25-mL aliquots of wine samples into screw-cap test tubes.
4. Add 0.1 mL H_3PO_4, 10.0 mL iso-octane, and 2–3 glass beads to each standard and sample.
5. Replace caps and shake vigorously for 2 min.
6. Allow 3–5 min for phase separation.
7. Determine absorbance of each standard and sample against an iso-octane blank at 255 nm.
8. Prepare a standard curve by plotting absorbance of standards vs. concentration (mg/L).
9. Compare absorbance values of samples to the standard curve.
   NOTE!! No dilution factor is involved in expression of final results.

IV. Supplemental Notes
1. Iso-octane is the solvent of preference because of its limited extraction properties for wine phenolics that also absorb in the UV. It has the further advantage of reduced volatility.
2. H_3PO_4 is used to enhance separation between aqueous and organic phases.
3. Ziemelis and Somers (1978) reported sorbic acid recovery ranging from 98.7 to 101.9% of that added to wine samples. For red wines, these workers report background error (that attributed to interferences) of 5 mg/L and for white wines, 3 mg/L.
4. Prepared standards are more concentrated than those used in Cash

still methods, because the 2 mL sample to 200 mL distillate dilution step (Cash still distillation) is not part of *this* procedure.

SORBIC ACID: HPLC DETERMINATION (SEE HPLC BENZOIC ACID)

SPECTRAL EVALUATION: SPECTROMETRIC (UV-VIS) ANALYSIS OF PIGMENTS (SEE PHENOLS-SPECTRAL EVALUATION)

STAIN: VISUAL MORPHOLOGICAL DIFFERENTIATION USING NIGROSIN (SEE BACTERIA: ISOLATION AND IDENTIFICATION)

STAIN: VISUAL BACTERIAL DIFFERENTIATION USING GRAM STAIN (SEE BACTERIA: ISOLATION AND IDENTIFICATION)

STARCH: SEE DIAGNOSTIC TESTS

SUCCINIC ACID: HPLC DETERMINATION (SEE ORGANIC ACIDS)

SUCCINIC ACID: PAPER CHROMATOGRAPHIC METHOD (SEE MLF)

SUGAR (INVERT): SEE REDUCING SUGAR (INVERT)

SULFUR DIOXIDE: ENZYMATIC SPECTROMETRIC DETERMINATION

Enzymatic analysis of sulfur dioxide (sulfites) relies on the coupled enzymatic reactions:

$$\text{(1)}\quad SO_3^{2-} + O_2 + H_2O \overset{SO_2\text{-OD}}{\leftrightarrow} SO_4^{2-} + H_2O_2$$

$$\text{(2)}\quad H_2O_2 + \text{NADH} + H^+ \overset{\text{NADH-POD}}{\leftrightarrow} 2H_2O + \text{NAD}^+$$

SO_2-OD: sulfite oxidase
NADH-POD: NADH-peroxidase

The hydrogen peroxide formed in reaction (1) is reduced by the enzyme, NADH-peroxidase, and NADH in reaction (2). As seen in (2), oxidation of NADH is stoichiometrically related to the amount of sulfite or aldehyde-bound sulfite. The remaining concentration of NADH can be measured at the appropriate wavelength. Kits of reagents for conducting this procedure are available to the analyst.

Refer to the product sheet that accompanies the sulfite kit of reagents, for up-to-date safety information and a procedure specifically designed for the kit obtained. The sulfur dioxide procedure will require the following:

I. Equipment

A spectrometer capable of reading absorbance at 340, 334, or 365 nm plus suitable, 1 cm, cuvettes.

Micropipette (10-, 50-, and 200-μL), and adjustable volumetric pipettes capable of delivering from 1.00 to 2.00 mL.

II. Reagents

The reagents used may be obtained from Boehringer-Mannheim or other suppliers. They are sold in the form of a test kit of sufficient material for approximately 30 sulfite determinations.

III. General Procedure

1. White wine may be analyzed directly. Dilute and treat red wine samples with NaOH to hydrolyze any bound sulfur dioxide.
2. The sample is mixed with reagents (except the sulfite oxidase), and the absorbance values are measured.
3. The sulfite oxidase is added, and the absorbance value is read again after the reaction is complete.
4. A blank is run along with the samples, and the absorbance values are corrected for the blank reading.
5. The difference in the corrected absorbance values taken before and after the addition of sulfite oxidase is used to calculate the concentration of total sulfur dioxide.

IV. Supplemental Notes

1. The above procedure, as presented, measures spectrophotometrically the change in concentration of NADH at 340 nm (the absorption peak for NADH). Where photometers (equipped with a mercury vapor lamp) are used, measurements are made at 334 or 365

nm. The appropriate adsorption coefficients are given in the product handouts.

SULFUR DIOXIDE: RIPPER TITRAMETRIC METHOD USING IODINE

In this method, standard iodine is used to titrate free sulfur dioxide. The end point of the titration is traditionally monitored using starch indicator solution. A number of commercial electrochemical detector systems are also available and in common use today. Free sulfur dioxide is determined directly. Total sulfur dioxide can be determined by first treating the sample with sodium hydroxide to release bound sulfur dioxide. Commercial kits for conducting approximate analyses of sulfur dioxide by the Ripper technique are readily available.

I. Equipment
   250-mL Erlenmeyer flask (preferably wide-mouth)
   10-mL burette
   25-mL volumetric pipette
   Rubber stopper of appropriate size
   High-intensity light source

II. Reagents
   Stock Iodine Solution: Dissolve 12.9 g of iodine and 25 g of potassium iodide into approximately 100 mL of deionized water. When dissolved, transfer to a 1-L volumetric flask and bring to volume with deionized water at defined temperature.
   Working Iodine Solutions: To prepare iodine solution of approximate normality desired dilute with deionized water to a final volume of 1 L according to Table 20–37.
   Standardization of Iodine Solutions: Because these dilutions yield only approximate concentrations, it is necessary to standardize working solutions against primary standard sodium thiosulfate. Sodium thiosulfate may be purchased in liquid, prestandardized form. In the case of those wishing to prepare their own primary standard, the following procedure is employed:

Table 20–37. Dilution procedures for preparation of working solutions from stock iodine.

Approximate normality desired	Dilution
0.01	100 mL stock + 900 mL distilled water
0.02	200 mL stock + 800 mL distilled water

(a) Accurately weigh approximately 300 mg of solid reagent grade sodium thiosulfate and transfer to a 100-mL volumetric flask.
(b) Dissolve in approximately 50 mL deionized water. Add a pinch of sodium bicarbonate and dilute to volume with deionized water.
(c) The normality (molarity) of this solution may be calculated using the following formula:

$$N = \frac{\text{Wt } Na_2S_2O_3(\text{mg}) \times \% \text{ Purity}}{158.1 \text{ mg/mmol} \times 100}$$

Sodium Hydroxide (1 *N*): Dissolve 1 equivalent weight (41 g) of sodium hydroxide into 1 L of deionized water.
Sulfuric Acid (1 + 3): Carefully dilute 1 vol of concentrated acid into 3 vol of deionized water.
Starch Indicator (1%): Mix 10 g of soluble starch and 1 L of deionized water. Heat to incipient boiling. Cool, store in refrigerator. Note: Discard any starch indicator that appears turbid.
Sodium bicarbonate

III. Procedure: Free Sulfur Dioxide

1. Volumetrically transfer 25 mL of wine or must to a clean 250-mL Erlenmeyer flask.
2. Add approximately 5 mL of starch indicator and a pinch of bicarbonate.
3. Add 5 mL 1 + 3 H_2SO_4.
4. Rapidly titrate with standard iodine solution to a blue end point that is stable approximately 20 sec.
5. Calculate the free SO_2 concentration (in mg/L):

$$SO_2 \text{ (mg/L)} = \frac{(\text{mL iodine})\ (N\text{ iodine})\ (32)\ (1{,}000)}{\text{mL wine sample}}$$

IV. Procedure: Total Sulfur Dioxide

1. Volumetrically transfer 25 mL of wine sample to a clean 250-mL Erlenmeyer flask.
2. Add 25 mL of 1 *N* NaOH, swirl, stopper. Allow 10 min for hydrolysis reaction(s) to occur.
3. Add approximately 5 mL starch indicator and a "pinch" of bicarbonate.
4. Add 10 mL 1 + 3 H_2SO_4.
5. Rapidly titrate with standard iodine solution to a blue end point that remains stable for approximately 20 sec. The use of a high-intensity light source is useful for end point detection in red wines.
6. Calculate the total SO_2 concentration (mg/L) using the equation above.

IV. Supplemental Notes

1. The analysis for free and total SO_2 is dependent on the redox reaction:

$$H_2SO_3 + I_2 \rightarrow H_2SO_4 + 2HI$$

Completion of this reaction is signaled by the presence of excess iodine in the titration flask. This excess iodine can complex with starch (blue-black end point) or be sensed with a platinum electrode. A number of commercial electrode detection systems are available (Beckman, Fisher, etc.) to take advantage of the electrochemical properties of iodine.

2. Because iodine solutions oxidize quickly, the concentration should be checked on a regular basis. When not in use, iodine solutions should be stored in amber glass bottles.
3. Standardize iodine solutions with standard sodium thiosulfate.
   a. Volumetrically transfer 10 mL of thiosulfate solution to a 250-mL Erlenmeyer flask.
   b. Add starch and titrate with iodine to a blue end point.
   c. Calculate the normality of iodine solution:

$$N\,\text{Iodine} = \frac{(\text{mL thiosulfate})\ (N\,\text{thiosulfate})}{\text{mL iodine solution}}$$

4. In analyses of total SO_2 it is first necessary to hydrolyze bisulfite addition compounds. This hydrolysis is pH dependent. Tomada (1927) reported less than 5% completion in the pH range 6–8, whereas in the range of pH 8–10, 50% of the addition product is hydrolyzed. Beyond pH 12, almost total dissociation occurs. The analysis of total SO_2 is a timed reaction; therefore, allowing additional time to elapse may introduce error.
5. In the determination of free SO_2, the sample is acidified to reduce oxidation of wine polyphenols by iodine and drive the reaction equilibrium to H_2SO_4.
6. Burroughs (1975) pointed out that acidification and resultant drop in pH liberates SO_2 bound to anthocyanins. Furthermore, because SO_2 is volatile at low pH, upon addition of 1 + 3 H_2SO_4 (the last step before titration), it is imperative that titration be carried out as rapidly as possible to reduce loss from volatilization. As titration continues (removing SO_2 from solution), further dissociation of anthocyanin-bisulfite occurs. As a result, data for free SO_2 in red wines may be expected to be erroneously high.

7. To aid in end point detection in red musts and wine, a high-intensity lamp should be positioned such that light is transmitted through the sample.
8. In deeply colored red wines, a quantitative dilution of sample with deionized water is often made. This facilitates easier detection of the starch end point. However, results must be multiplied by the appropriate dilution factor. Furthermore, dilution may affect equilibrium between free and total SO_2 in solution.
9. The starch indicator solution is subject to rapid microbial decomposition. Therefore, it should be stored in the refrigerator when not in use. If the integrity of the indicator is in question, place a few milliliters into a test tube and add a drop or two of iodine solution. The contents should turn an immediate and intense purple. If an amber color is noted, discard the starch.
10. The anticipated error using the procedure outlined is ±7 mg/L (for total SO_2), which is generally acceptable for routine winery analyses. Results of this analysis should be rounded to the nearest whole number.
11. In SO_2 determinations on products where ascorbic acid has been added or is naturally present in significant amounts, iodine titration leads to erroneously high results. This effect is largely the result of competitive oxidation of ascorbic acid and SO_2 by the iodine titrant. Thus in determinations of SO_2 in such products, it is necessary to either correct for the presence of the acid or use a procedure not involving redox titrations. The aeration oxidation method has been used successfully in these cases.

SULFUR DIOXIDE: RIPPER TITRAMETRIC METHOD USING IODATE

Schneyder and Vlcek (1977) proposed the use of iodate solutions rather than iodine as a titrant in sulfur dioxide determinations. The advantage of such a substitution is stability of the iodate solution compared with iodine solutions. In this procedure an excess of iodide is added to the juice or wine sample and it is then titrated with iodate as above.

The iodate reacts with the iodide and sulfuric acid in the sample and produces iodine *in situ*:

$$IO_3^- + 5I^- + 6H^+ \rightarrow 3I_2 + 3H_2O$$

The iodine produced reacts with the SO_2 as before. The titration proceeds as in the Ripper procedure above, and the results are calculated using the equation given above.

I. Equipment
Same as in the Ripper procedure above.

II. Reagents
Potassium Iodate/Iodine Standard Solution (0.02 *N*): In a 1 L volumetric flask, dissolve 713.3 mg of dried reagent grade potassium iodate in approximately 800 mL of deionized water. Add 50 mL of 1 + 3 sulfuric acid with mixing and dilute to volume with deionized water.
Potassium Iodide-Starch Solution: In a 1-L volumetric flask, dissolve 10 g of potassium iodide and 2.5 g soluble starch in approximately 800 mL of deionized water. Bring to volume with deionized water.
H_2SO_4 (1 + 2): Add 1 vol stock H_2SO_4 to 2 vol of deionized water.

III. Procedure
1. Follow instructions of the Ripper procedure for free and total SO_2, substituting the above reagents for comparable reagents specified in the above procedure.
2. Calculate results using equation given above.

SULFUR DIOXIDE: RAPID ESTIMATION USING TITRITE KIT METHOD

Commercial ampules of reagents for rapid SO_2 measurements are available. These utilize Ripper method chemistry and are therefore subject to the same limitations.

SULFUR DIOXIDE: AERATION OXIDATION DISTILLATION AND TITRATION PROCEDURE

Sulfur dioxide in wine or juice can be distilled (with nitrogen as a sweeping gas or with air aspiration) from an acidified sample solution into a hydrogen peroxide trap where the volatilized SO_2 is oxidized to H_2SO_4:

$$H_2O_2 + SO_2 \rightarrow SO_3^{2-} + H_2O \rightarrow H_2SO_4$$

The volume of 0.01 *N* NaOH required to titrate the acid formed to an end point is measured, and used to calculate SO_2 levels. Two types of apparatus are presented below. These are typical of a number of commercial glassware systems used in this procedure. An efficient condenser and cold coolant water are effective in preventing distillation of volatile acid components in total SO_2 determinations. In routine production operations where only approximate values for free SO_2 levels are adequate, glassware setups similar to those described below, but without a condenser, have been used.

I. Equipment
   Recirculating ice bath
   Microburner
   Volumetric pipettes (10 and 20 mL)
   Glass distillation unit (see Figure 20–17):

II. Reagents
   Sodium Hydroxide (0.01 *N*): See Table 20–1.
   Hydrogen Peroxide (0.3%): Immediately before use, volumetrically transfer 5.0 mL of stock H_2O_2 (30% wt/wt) to approximately 400 mL of deionized water in a 500-mL volumetric flask. Bring to volume using deionized water at defined temperature.
   Phosphoric Acid (1 + 3): Using *o*-phosphoric acid (85% stock) carefully prepare a 1 + 3 solution with deionized water.
   Indicator Solution: Dissolve 50 mg of methylene blue and 100 mg of methyl red in approximately 90 mL of 50% ethanol-water solvent. When dissolved, dilute to a final volume of 100 mL using the solvent.

III. Glassware Assembly
   Refer to Fig. 20–17 for assembly. Connect the 3-neck 100-mL round bottom flask to the 300 mm condenser. Place a Pasteur pipette through the thermometer adapter and connect it to the round bottom flask. Adjust the height of the Pasteur pipette so that it comes within 1 mm of the bottom of the flask. Place the stopper in the third neck of the flask.

   Connect the adapter to the top of the condenser. Connect a Pasteur pipette using a short piece of Tygon tubing to the central glass tube of the vacuum adapter. Adjust the length of the pipette so that it will reach to within 1 mm of the bottom of the pear-shaped receiver flask. Connect the pear shaped flask to the vacuum adapter.

   Run a vacuum hose between a water aspirator and the side arm of the vacuum adapter. Connect a recirculating ice bath to the condenser.

IV. Procedure for Free Sulfur Dioxide
   1. Place an ice bath below the round bottom flask.
   2. Rinse all glassware with deionized water, and turn on pump supplying ice water to the condenser. To the pear shaped flask add 10 mL 0.3% hydrogen peroxide solution and 6 drops indicator. Titrate from the initial violet color to a turquoise end point with the 0.01 *N* NaOH (usually one drop with a fine tip burette). Be sure and note color, since it will be the same color as the final end point of the

determination. Connect the pear-shaped flask to the vacuum adapter.

3. Distillation of Sample. Remove sample bottle from refrigerator. Add exactly 20.0 mL sample to the round bottom flask immersed in an ice water bath. Add 10 mL of 1 + 3 phosphoric acid to the same flask. Replace glass stopper in round bottom flask. Begin aspirating vigorously, and continue aspirating sample for exactly 10 min taking care that none of the peroxide solution is lost from the pear-shaped flask.
4. Titration. At the end of the 10 min period turn off aspiration. Remove pear-shaped flask. Using deionized water, rinse the inside of the vacuum adapter, and the outside of the Pasteur pipette connected to it, into the pear-shaped flask.
5. Using a 10-mL burette, titrate the contents of this flask with 0.01 *N* NaOH to the end point noted above. Read the titration volume to the nearest 0.01 mL. Report the sulfur dioxide concentration to the nearest ± 1 mg/L. *Note:* It is preferable to set up and begin running the total SO_2 determination (next section) before titrating the "free" SO_2 sample. This first titration can be done during the aspiration time of the subsequent run.
6. Calculate SO_2 as follows:

$$SO_2\ (\text{mg/L}) = \frac{\text{mL NaOH} \times N\,\text{NaOH} \times 32 \times 1{,}000}{20\ \text{mL (sample size)}}$$

V. Procedure for Bound Sulfur Dioxide

1. Initial Setup. Follow the procedure listed above for free sulfur dioxide except place a microburner below the round bottom flask.
2. Distillation. The same sample in the round bottom flask as used above for the free analysis is now used for bound SO_2. The contents of the round bottom flask (sample and phosphoric acid) are heated to a boil with the microburner. A few drops of an inert antifoam may be necessary to control excessive foaming during distillation. When boiling commences, aspirate vigorously for exactly 15 min taking care that none of the peroxide solution is lost from the pear-shaped flask.
3. Titration. At the end of the 15-min period turn off aspiration and burner. Remove pear-shaped flask. Using deionized water rinse the inside of the vacuum adapter, and the outside of the Pasteur pipette connected to it, into the pear-shaped flask. Using the 10-mL burette, titrate the contents of this flask with 0.01 *N* NaOH to the end point

noted above. Read the titration volume to the nearest 0.01 mL. Report calculated sulfur dioxide values to the nearest ± 1 mg/L.

4. Calculate bound sulfur dioxide (mg/L) using the above equation.

V. Calculation for Total Sulfur Dioxide

Take the sum of the volumes used in the free and bound titrations. Calculate the mg/L SO_2 as above.

VI. Supplemental Notes

1. Potential sources of error in the procedure include:
   a. Carryover of CO_2 and volatile acids resulting from high initial VA, inefficient cooling condenser, or cooling water not being cold enough.
   b. Acidification of sample may liberate some SO_2 bound to wine anthocyanins. Thus this should be the last step before beginning operation.
   c. In that increased temperatures accelerate dissociation of bound SO_2, analysis for the free form should be carried out at less than 20°C (hence the need for an ice bath).
2. Nitrogen gas may be used to push the sample through the apparatus; the results will be comparable to those obtained using aspiration.
3. Following completion of the free SO_2 determination, the analyst may prefer to use a new wine sample in the reaction flask. The analytical value obtained in this instance is the total sulfur dioxide level in the sample.
4. Glassware kits designed specifically for this analysis are commercially available.

SULFUR DIOXIDE: TITRAMETRIC MODIFIED MONIER-WILLIAMS PROCEDURE

This procedure has been the official AOAC method for total SO_2 for a number of years. The chemistry involved is essentially the same as that used in the aeration-oxidation procedure; SO_2 is distilled out of the sample, oxidized with hydrogen peroxide, and titrated as sulfuric acid against standard sodium hydroxide. When other titratable compounds are distilled over with the sulfur dioxide, a gravimetric precipitation of the sulfate with barium can be employed to correct for this interference.

I. Equipment

Refer to a current volume of the AOAC *Official Methods* for appropriate SO_2 distillation glassware.

II. Reagents

Refer to a current volume of the AOAC *Official Methods* for appropriate reagents and their preparation.

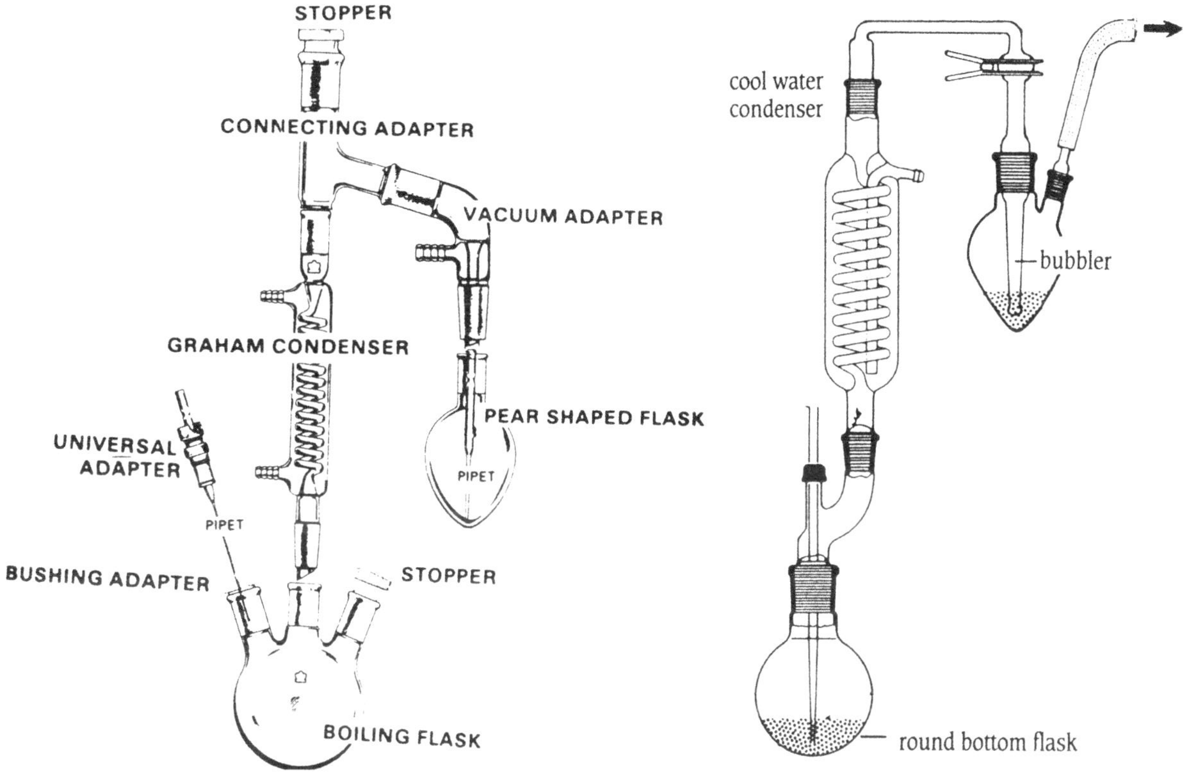

Fig. 20–17. Recommended glassware for aeration oxidation procedure.

III. Procedure (Total Sulfur Dioxide)

Refer to latest addition of *Official Methods* (AOAC) for specifics of the procedure.

1. Transfer 100–300 mL wine sample to distillation flask. Dilute to approximately 400 mL with deionized water.
2. Add 90 mL HCl (1 + 2) to flask.
3. Start nitrogen gas flow and heat flask to boiling. Reflux for 1.75 hours with steady flow of bubbles.
4. Titrate contents of collection vessel with 0.01 *N* NaOH using methyl red as indicator. Calculate mg/L SO_2 (ppm) using Equations given in Aeration-Oxidation procedure.
5. If desired a gravimetric determination can be carried out by acidifying sample and precipitating the sulfate ion with $BaCl_2$ solution. The precipitate is washed, dried, and weighed to determine the amount of $BaSO_4$ collected.

TANNIN (TOTAL): DETERMINATION OF TOTAL TANNINS AND THE GELATIN INDEX OF RED WINES

This method for total tannins is based on the transformation of proanthocyanidins to anthocyanidins by heating in an acid medium [Ribereau-Gayon and Glories (1986)].

I. Equipment

100°C water bath
Condenser
Tight-fitting glass stoppered vial
Centrifuge
Spectrophotometer (visible range)
1-cm pathlength cuvettes

II. Reagents

Gallic acid stock solution (5,000 mg/L): In a 100-mL volumetric flask dissolve 500 mg gallic acid in approximately 50 mL of 12% (vol/vol) ethanol. Bring to volume with the same ethanol solution
HCl, 12 *N*
Ethanol, 95%
Gelatin

III. Procedure

1. Prepare gallic acid standards from gallic acid stock of 5,000 mg/L, as presented in Table 20–38.
2. In each of two tubes add 4 mL of red wine diluted 1 + 49, 2 mL of deionized water, and 6 mL 12 *N* HCl.

Table 20–38.

Volume gallic acid stock solution (5,000 mg/L)	Final concentration (mg/L GAE) in 100-mL volumetric flask
2	100
5	250
10	500
15	750
20	1,000

3. Place one of the tubes in a flask fitted with a "cold" refluxing condenser and place in a 100°C water bath for 30 min.
4. Let sample cool and add 1 mL 95% ethanol to both tubes.
5. Measure the absorbance of both tubes at 550 nm using a 1 cm cuvette, against a water blank.
   6. The concentration in g/L is determined by the difference between the absorbance reading of the unknown and that of a standard curve according to the formula:

$$\text{g/L total tannins} = 19.33\ (A_{STD}\text{-}A_{UNK})$$

Gelatin Index

7. The tannin concentration determined above can be used to calculate the gelatin index, which expresses the affinity of tannins for proteins and their astringency. The procedure is as follows:
   1. Add 5 mL of gelatin (70 g/L) to 50 mL of wine (equivalent to a 7-g/L addition).
   2. Mix sample, place into an airtight vessel sparged with nitrogen.
   3. Store for 3 days. The sample is centrifuged, and the supernatant diluted 1 to 50. Tannin is determined as described above. The gelatin index is determined according to the following:

$$\text{G.I.} = \frac{C - A}{C \times 100}$$

where:
C = g/L tannin in control
A = tannin in supernatant after gelatin reaction

TARTARIC ACID: HPLC ANALYSIS (SEE ORGANIC ACIDS)

TARTARIC ACID: PAPER CHROMATOGRAPHIC METHOD (SEE MLF)

TARTARIC ACID: METAVANADATE SPECTROMETRIC (VIS) ANALYSIS (CARBON DECOLORIZATION)

Several methods of tartaric acid determination are currently being used in winery laboratories. The metavanadate method involves a colorimetric reaction between sodium (or ammonium) metavanadate and tartaric acid in acetic acid solution.

This methodology frequently requires initial treatment of wine samples and standards with activated charcoal for decolorizing, prior to reaction of the extract with metavanadate. Several studies have shown that variations in this first step lead to significant differences in the final measurement of tartrate (Clark et al. 1985). Carbon removes different amounts of tartaric acid from the standards than from the wine samples. Thus, the metavanadate procedure (with carbon pretreatment) generally will give high results.

Hill and Caputi (1970) reported that the presence of sugar and lactic acid interferes with the analysis. These workers used a preliminary separation of the tartaric acid on acetate-charged resin before analysis (see next procedure).

Clark's remedy for the problem with carbon was to use a high-performance liquid chromatographic technique (see HPLC Organic Acids procedure). By avoiding the use of carbon both Caputi's and Clark's methods for tartrate analysis should produce accurate results.

I. Equipment
   125-mL Erlenmeyer flasks
   100-mL volumetric flasks
   Electric hot plate
   Volumetric pipettes (2, 4, 25 mL)
   Graduated cylinder (100 mL)
   Spectrophotometer (visible)
   Matched cuvettes
   Whatman #1 filter paper
   Glass funnels
   Wash bottles

II. Reagents
   Tartaric acid standard solution (10 g/L): In a 100-mL volumetric flask dissolve 1.00 g reagent grade tartaric acid in deionized water. After dissolution is complete, bring to volume with deionized water.
   Metavanadate Solution (5% wt/vol): In a 100-mL volumetric flask, dissolve 5 g of $NaVO_3$ in hot (less than 70°C) deionized water. Bring to volume with deionized water. Sodium metavanadate does not completely dissolve, so filter solution, using Whatman #2 paper, before use.

Table 20–39. Preparation of standard solutions of tartaric acid from 10 g/L stock.

Standard solution (mL)	Volume H_2O (mL)	Final concentration H_2T (g/L)
0	20	0.00
2	18	1.00
4	16	2.00
6	14	3.00
8	12	4.00

HCl (1 *N*): See Table 20–2.
Decolorizing carbon
Boiling chips

III. Procedure

(a) Preparation of Standard Curve:

1. Using 125-mL Erlenmeyer flasks, prepare a series of standards according to Table 20–39.
2. With these 20-mL standards, proceed with steps 4–11 below. Graph concentration (g H_2T/L) vs. absorbance at 520 nm.

(b) Wine Samples

3. To a clean 125-mL Erlenmeyer flask, volumetrically transfer 20 mL of wine.
4. Add 2 mL 1 *N* HCL, 30 mL deionized water, 0.50 g (accurately weighed) decolorizing carbon, and several boiling chips.
5. Boil the solutions on hot plate 2–3 min, cool to room temperature.
6. Filter the solutions into 100-mL volumetric flasks using Whatman #1 filter paper.
7. Examine clarity. If necessary, refilter. Take care to thoroughly rinse Erlenmeyer flask and filter paper with deionized water. Multiple rinses using small amounts of water are recommended.
8. Bring to 100-mL volume with deionized water.
9. Transfer 25 mL decolorized solution to another 100-mL volumetric flask. Add 2 mL concentrated acetic acid, 10 mL filtered metavanadate solution, and bring to volume.
10. Allow 30 min for color development. Record absorbance on spectrometer at 520 nm.
11. With reference to a standard curve, determine tartaric acid content of the sample.

IV. Supplemental Notes

1. As noted, the above procedure suffers from interferences resulting

from use of carbon in the decolorization step as well as the presence of sugar and lactic acid.
2. Keep developed solution away from direct sunlight.
3. For wines that are being stabilized at low temperatures, it is essential that the following routine be observed in regard to collection and analysis of H_2T: While the wine is under refrigeration, the sample is collected at either middle or top "draws" (avoid collection at bottom valve!) into a cool sample bottle. It is then rushed to the laboratory and filtered while cold before analysis. Allowing the sample to warm before filtration may result in resolubilization of some bitartrate crystals.

TARTARIC ACID: METAVANADATE SPECTROMETRIC ANALYSIS (VIS) (ION EXCHANGE SAMPLE PREPARATION)

In the following procedure of Hill and Caputi (1970), the sample is placed on an acetate-charged resin and eluted with sulfate ion. The tartaric acid-containing fraction, now isolated from potential interferences, is then analyzed in a manner similar to the preceding method.

I. Equipment
   Spectrophotometer (visible) and matched cuvettes
   Chromatography column (500 × 12 mm) with sintered glass plate and needle control valve.
   Strong anion exchange resin (Duolite A101D)
   50-mL volumetric flasks
   Volumetric pipettes
II. Reagents
   Tartaric acid standard solution (25 g/L)
   Acetic Acid (30% vol/vol): Into a 1-L flask containing approximately 300 mL deionized water, add 300 mL glacial acetic acid. Bring to volume with deionized water.
   Acetic Acid (0.5% vol/vol): As above, except dilute 5 mL of glacial acetic acid to 1 L final volume using deionized water.
   Sodium Sulfate (0.5 M): Dissolve 71 g of Na_2SO_4 in approximately 500 mL of deionized water. Bring to a final volume of 1 L when dissolution is complete.
   Sodium Metavanadate (3% wt/vol): As per above procedure except use 3 g $NaV0_3$ per 100 mL deionized water.
III. Procedure
   (a) Separation:
   1. Pack 20-mL anion exchange resin into column.
   2. Using 20-bed volumes of 30% acetic acid, charge resin.

3. Follow with a wash of 30 mL acetic acid (0.5%) and 50 mL deionized water. Steps 2 and 3 (and all following steps) should be carried out at a flow rate of 2.5–5.0 mL/min.
4. Add 10 mL of wine and 20 mL of 0.5% acetic acid sequentially to the column. Follow with 50 mL deionized water.
5. Add 0.5 M sodium sulfate to elute acids. Eluates are collected into graduate cylinders. The tartaric acid fraction elutes from the 46th through 75th mL.

(b) Analysis:

6. Transfer 25 mL of tartaric acid fraction to a 50-mL volumetric flask.
7. Add 2 mL of metavanadate and 0.5 mL glacial acetic acid. Bring the solution to volume with deionized water.
8. Immediately prepare a reagent blank, according to steps 6–7, using 25 mL of 0.5 M sodium sulfate.
9. Thoroughly mix each sample and blank and hold at room temperature for 80 min.
10. Measure the absorbance of sample(s) at 480 nm vs. blank and compare to standard curve.

(c) Standard Curve Preparation:

11. Using 50-mL volumetric flasks, prepare the following series from stock tartaric acid solution (25 g/L). The primary and each secondary standard are brought to volume using 0.5 M sodium sulfate (see Table 20–40).
12. Using 25-mL aliquots of these standards, proceed per steps 6–10 above.
13. Plot concentration (g/L tartaric acid) vs. absorbance at 480 nm.

IV. Supplemental Notes

1. The above analysis has the added benefit that, under the conditions of separation, lactic acid present in the sample is eluted in the first 30 mL.
2. Once developed, color is reported to be stable for 1 hour (Hill and Caputi 1970).

Table 20–40. Preparation of standard solutions of tartaric acid from stock (25 g/L H_2T).

Stock solution (mL)	Final concentration H_2T (g/L) in 50-mL volume
2	1.00
4	2.00
6	3.00
8	4.00
0	0.00

3. Hill and Caputi did not specify resin particle size in their paper. However, for best results, the smallest available particle size should be used.
4. Because resins and packing methods differ, it is advisable to test the ion exchange column by running a standard tartaric acid solution. Ten milliliter fractions preceding and following the tartaric acid fraction should be collected.

TARTRATE DEPOSITS: MICROSCOPIC ANALYSIS (SEE DIAGNOSTICS-IDENTIFICATION OF HAZES AND PRECIPITATES)

TARTRATE DEPOSITS: SEE POTASSIUM BY AA, SEE CALCIUM BY AA

TERPENES: SPECTROMETRIC (VIS) ANALYSIS USING CASH OR MARKHAM STILLS (FRUIT MATURITY)

A technique developed by Dimitriadis and Williams (1984) to estimate the quantity of free volatile terpenes (FVT) and potential volatile monoterpenes (PVT) in grape juice can provide an index of the flavor and flavor potential of the fruit. Steam distillation of pH-neutral juice yields free aroma compounds, whereas steam distillation of low-pH juice yields monoterpenes derived from the polyol and glucosidically bound forms. Linalool is used as the reference monoterpene for constructing a calibration curve, and results are obtained in linalool equivalents. Essentially quantitative recovery of free volatile terpenes is obtained with this method. However, recoveries of potential volatile monoterpenes are reported to be in the 55 to 88% range.

I. Equipment

Cash, Markham, or equivalent still for steam distillation of sample
Waring blender (or equivalent)
Water bath (60° ± 1°C)
Visible spectrometer (608 nm)
Pyrex test tubes (2 × 150 nm) with silicon septum lined screw caps

II. Reagents

Linalool stock solution (1 mg/mL): Accurately weigh approximately 50 mg of reagent grade linalool into a 50-mL volumetric flask. Add 10 mL absolute ethanol and swirl to dissolve. Bring to mark with deionized water.

Linalool standard solution (0.1 mg/mL): As needed, dilute linalool stock 1 + 9 with deionized water.

Vanillin:Sulfuric acid reagent (2% wt/wt): Weigh 2 g reagent grade vanillin and dissolve in concentrated sulfuric acid in a 100-mL volumetric flask. Store in brown glass bottle at 0°–4°C. *Note:* this reagent is corrosive. Addition to aqueous solutions generates considerable heat. Safety glasses and gloves are recommnded when using this reagent!

Sodium hydroxide (20% wt/vol): Add 20 g reagent grade sodium hydroxide pellets to a beaker. Dissolve in approximately 80 mL deionized water and cool. When cool bring volume to approximately 100 mL. Store in plastic bottle.

Phosphoric acid (20% vol/vol): Add 20 mL reagent grade phosphoric acid to approximately 80 mL deionized water to make 100 mL solution. Store in glass bottle.

III. Procedure (Modified Method)

(a) Distillation of sample

1. Homogenize approximately 500 g destemmed berries in a Waring blender for 10–20 sec. Filter the homogenate through cheesecloth.
2. Immediately before distillation, adjust the pH of the homogenate to 6.6–6.8 with the 20% (wt/vol) sodium hydroxide solution.
3. Transfer 100 mL of neutral juice to the inner chamber of a Cash or Markham still for steam distillation of monoterpenes (see Fig. 20–2).
4. Distill and collect the first 25 mL of condensate.
5. Immediately add 10 mL of 50% (vol/vol) phosphoric acid reagent to the still inner chamber to drop the sample pH (2.0–2.2).
6. Continue steam distillation, and collect the next 40 mL of condensate.
7. Mix each collected condensate sample to ensure homogeneity.

(b) Colorimetric development

1. Into a series of test tubes pipette 0.2, 0.5, 1.0, 1.5, and 2.0 mL of linalool standard. Bring the volume of each standard to 10 mL with deionized water. The amount of linalool present in each tube is listed in Table 20–41.
2. Into additional test tubes, pipette 10-mL aliquots of each FVT and PVT distillate. Prepare a deionized water blank at the same time.
3. To each precooled test tube add 5 mL of the vanillin: sulfuric acid (2% wt/vol) reagent while agitating the tube in an ice bath.
4. Place test tubes for standards and FVT and PVT samples in a 60° ± 1°C water bath for 20 minutes.

Table 20–41. Preparation of linalool calibration standards from 0.1 mg/mL standard solution.

Volume linalool standard (mL)	Final amount of linalool (μg) in calibration standard
0.2	20
0.5	50
1.0	100
1.5	150
2.0	200

5. Cool the tubes at 26°C for 5 min and read the absorbance on a spectrophotometer at 608 nm. *Note:* Absorbance readings should be taken within 20 min following the cooling of the standards.
6. Prepare a calibration curve of μg linalool vs. absorbance for the standards.
7. Read the amount of linalool equivalents contained in each distillate aliquot directly from the calibration curve.
8. Calculate amounts of linalool equivalents in original samples using the following equation:

$$\text{FVT or PVT } (\mu\text{g/ml}) = \mu\text{g linalool} \times \frac{V\,(\text{distillate})}{V\,(\text{aliquot})} \times \frac{1}{100 \text{ mL}}$$

where: μg linalool = amount read from calibration curve
V (distillate) = 26 or 40 mL for FVT and PVT, respectively
V (aliquot) = 10 mL aliquot taken from distillate
100 mL = original juice sample on which test is run

IV. Supplemental Notes

1. Sulfur dioxide, at normal use levels, produces low values for PVT. To avoid this interference, add an equivalent amount of hydrogen peroxide to oxidize SO_2 before distillation.
2. Ethanol present in wine samples produces an enhanced color reaction. If wine samples are to be run by this procedure, standards should be prepared so that their ethanol content will match that of the distillates.
3. The potential volatile monoterpenes analyzed here include only volatile products derived from polyols and glycosides. Thus, values obtained are subject to the degree of dehydration, hydrolysis, etc., under the specific experimental conditions used.

TITRATABLE ACIDITY: TITRAMETRIC PROCEDURE USING NAOH

Titratable acidity (TA) is a measure of the organic acid content of the juice, must, or wine sample being analyzed. The titration determines the amount of organic acids that can be titrated with a dilute alkali solution to a pH of 8.2.

I. Equipment
   pH meter
   Magnetic stirring table and stir-bar
   50-mL burette
   5- and 10-mL volumetric pipettes

II. Reagents
   0.067 *N* or other standardized NaOH of less than 0.1 *N*: (see Table 20–1)
   Phenolphthalein Indicator (1% wt/vol): Dissolve 1 g of indicator in approximately 70 mL of 95% ethyl alcohol. Add sufficient dilute NaOH <0.1 *N* to achieve a very light pink solution. Dilute to 100 mL with deionized water.
   Standard buffers of pH 7.00 and 4.00 (or 3.55): See procedure on pH determination

III. Procedure: (Production Modification)
   1. Standardize pH meter per operator's manual.
   2. Add approximately 100 mL of boiled and cooled deionized water to a 250-mL beaker. Place the stir-bar in the beaker and position it on a stir plate.
   3. *Carefully* immerse electrode(s) into solution so that they are away from the stir-bar.
   4. Add 2–3 mL of wine sample (*NOTE*! This is not a quantitative transfer.) Rapidly titrate with standard base to pH 8.2.
   5. Transfer 5.0 mL of wine into solution using a volumetric pipette. Titrate to pH 8.2. Record the volume of base used for this titration.
   6. Calculate results using following equation:

$$\text{Titratable Acidity (g/L tartaric acid)} = \frac{(\text{mL base})\ (N\,\text{base})\ (0.075)\ (1{,}000)}{\text{mL sample}}$$

TITRATABLE ACID: AOAC (TITRAMETRIC) PROCEDURE

The currently recognized procedure of the AOAC is presented below. Equipment and reagents used are the same as presented above.

I. Procedure

1. Add 1 mL phenolphthalein indicator to 200 mL of hot, boiled deionized water. Neutralize to pink with standard NaOH.
2. Volumetrically transfer 5.0 mL of degassed wine or must sample (see Production Modification procedure or Supplemental Note 2).
3. Titrate to pH 8.2 with standardized NaOH according to prior instructions.
4. Calculate results according to above equation.

IV. Supplemental Notes

1. In practice, up to four samples may be run before preparing fresh solution at Step 2.
2. CO_2 in must or sparkling wine, as well as some table wines, serves as a major source of error. Where CO_2 is known to be present, hold approximately 10 mL of sample in a 60°C water bath for several minutes. After the gas has been removed, cool the solution and volumetrically transfer 5 mL to the titration beaker. One may hold the sample under vacuum for several minutes before analysis.
3. Carbon dioxide present in the water diluent may also present problems in end point detection. In this case, the end point will appear to "fade" (as a result of slow hydration of the CO_2). Therefore the use of boiled (and cooled) deionized water for sample preparation is recommended.
4. Sodium hydroxide is not a primary standard; carbonate ion present in solution acts as a buffer and, as such, alters the actual concentration. Therefore, solutions of NaOH must be standardized against some primary standard. The effects of carbonate interferences can be reduced/ eliminated by preparing a saturated solution of NaOH and allowing the carbonate to precipitate over a period of several days. The carbonate-free supernatant can then be diluted to the approximate concentration desired and standardized. Standardization normally is carried out by titration of the primary standard acid, potassium acid phthalate (KHP). Only one proton per molecule of KHP is acidic, and the titration of KHP is not complete until approximately pH 8.2. Therefore, phenolphthalein is the indicator of choice.
5. With reference to the normality of NaOH used in the titration, one should avoid using concentrations much greater than 0.1 *N*, since small volume differences in burette readings reflect large relative errors. Where production and speed are of the essence, NaOH may be prepared at 0.0667 *N*. At this concentration, the quantity of base used yields total acidity in the appropriate units. Thus a total volume of 5.0 mL of base (at 0.0667 *N*) corresponds directly to a TA of 5.00 g/L.

6. For routine work, one may expect a probable error of ± 0.01%. Results of TA analyses are usually expressed to two decimal places (e.g., 7.50 g/L), rounding to the hundredths place.

TRICHLOROANISOL: SENSORY EVALUATION OF CORK DEFECTS (SEE CORK)

UREA: SPECTROMETRIC (VIS) DETERMINATION IN WINE AND JUICE

Douglas and Bremner (1970) and Nagel and Weller (1989) reported a colorimetric procedure to determine urea content in juices and wine at concentration as low as 1 mg/L. This method involves measurement of the absorbance at 527 nm of the red color formed when urea is heated with diacetyl monoxime and thiosemicarbazide under acidic conditions. Nitrogen compounds other than carbamido compounds do not interfere. Sugars that interfere with color development are removed by ion exchange.

I. Equipment

Visible spectrometer capable of measurement at 527 nm
Cuvettes
Dowex 50X8, hydrogen form (100–200 mesh, 1.0 cm i.d., 2.5 mL bed volume)
Water baths (20°C; 100°C) and ice bath
Venting hood
Pipette (1–5 mL)
Screw cap culture tubes (20 mL)

II. Reagents

Diacetyl Monoxime (DAM) solution, 2.5% (w/v) in deionized water
Thiosemicarbazide (TSC) solution, 0.25% (w/v) in deionized water
Acid reagent: Mix 300 mL of 85% phosphoric acid with 10 mL of concentrated sulfuric acid, and dilute the mixture to 500 mL with deionized water.
Color reagent. Add 25 mL of DAM solution and 10 mL of TSC solution to 500 mL of acid reagent, prepare immediately before use.
Standard urea solution, 40 mg/L stock solution

III. Procedure

1. Regenerate Dowex 50X8 with 1 M NaOH followed by washing and treatment with 1 M HCl to convert to the H^+ form. Rinse column with deionized water.

2. Run 10 mL of sample through Dowex 50X8 column, and elute with 10 mL of 1 M sodium acetate solution after washing with a small volume of water to remove sugars, which interfere with color development.
3. Add 8 mL of color reagent to 2 mL of standard or sample solution in 20-mL screw cap culture tube immersed in an ice bath, mix well, and seal.
4. Transfer the tube to a 20°C water bath for 1 min then to a boiling water bath under a hood for exactly 27 min.
5. Cool the sample rapidly by placing the tube in an ice bath.
6. Equilibrate the sample to 20°C by placing it in a water bath before absorbance is read at 527 nm.

UREA/AMMONIA: ENZYMATIC SPECTROMETRIC (VIS) ANALYSIS

Gutmann and Bergmeyer (1974) reported an enzymatic assay for trace ammonia and urea in wine. The sensitivity of the method is reliable to 1 mg/L. The procedure relies on the coupled enzymatic reactions:

$$\text{Urea} + H_2O \xrightarrow{\text{UA}} 2\ NH_3 + CO_2$$

$$\text{2-Oxoglutarate} + \text{NADH} + NH_4^+ \xrightarrow{\text{GIDH}} \text{L-Glutamate} + \text{NAD}^+\ H_2O$$

where:

UA = urease
NADH = reduced nicotinamide-adenine dinucleotide
GIDH = glutamate dehydrogenase

The amount of NADH oxidized in the above reaction is stoichiometric with half the amount of urea present. NADH is determined by means of its absorbance at 340 nm. A kit of reagents for conducting this procedure is now available to the analyst. For up-to-date information and a procedure specifically designed for the kit obtained, refer to the product sheet that accompanies the urea/ammonia UV-method kit. The procedure was modified for maximum assay capacity and efficiency.

Samples should be filtered through a 0.45-μm filter. Colored samples should be treated with PVPP (0.75 g per 10 mL of wine) or activated carbon (2% wt/vol) then centrifuged or settled and filtered. In general, the urea procedure will require the following:

I. Equipment
A spectrophotometer capable of reading absorbance at 340 or 365 nm and cuvettes
Adjustable micropipettes (20-, 200–1,000, and 1,000–5000 μL)

II. Reagents
Urea/ammonia UV-method kit (Boehringer-Mannheim), sufficient for 35 urea determinations

III. Procedure
1. Add 1.8 mL of deionized water and 1.0 mL of NADH to cuvette containing 0.2 mL of sample (or water blank).
2. Mix, wait 5 min at 20°–25°C then read absorbance at 340 nm (A1).
3. Add 20 μL of glutamate dehydrogenase (kit solution #4).
4. Mix, wait 20 min at 20°–25°C, then read absorbance at 340 nm (A2).
5. Add 20 μL of urease (kit solution #3).
6. Mix, wait 20 min at 20°–25°C, then read absorbance at 340 nm (A3).
7. A blank is run along with the samples, and absorbance values are corrected for the blank reading.
8. The difference in the corrected absorbance values taken before and after the addition of urease is then used to calculate the concentration of urea.
9. The difference in the corrected absorbance values taken before and after the addition of glutamate dehydrogenase is then used to calculate the concentration of ammonia.

$$\text{Urea (mg/L)} = [(A2_{\text{sample}} - A2_{\text{blank}}) \times 3.02/3.04 - (A3_{\text{sample}} - A3_{\text{blank}})] \times 72.9$$

$$\text{Ammonia (mg/L)} = [(A1_{\text{sample}} - A1_{\text{blank}}) \times 3/3.02 - (A2_{\text{sample}} - A2_{\text{blank}})] \times 40.8$$

VOLATILE ACIDITY: (SEE ACETIC ACID DISTILLATION-TITRATION, GAS CHROMATOGRAPHIC, AND ENZYME PROCEDURES)

YEAST IDENTIFICATION: PHYSIOLOGICAL AND VISUAL EVALUATION OF *BRETTANOMYCES* (SEE BACTERIA: ISOLATION AND IDENTIFICATION)

YEAST IDENTIFICATION: PHYSIOLOGICAL AND VISUAL EVALUATION OF *ZYGOSACCHAROMYCES* (SEE BACTERIA: ISOLATION AND IDENTIFICATION)

YEAST VIABILITY: PHYSIOLOGICAL AND VISUAL EVALUATION USING PONCEAU S (SEE BACTERIA: ISOLATION AND IDENTIFICATION)

YEAST VIABILITY: PHYSIOLOGICAL AND VISUAL EVALUATION USING METHYLENE BLUE (SEE BACTERIA: ISOLATION AND IDENTIFICATION)

Appendix I. Tables of Constants, Conversion Factors

Table I–1. Weight of sucrose solutions at 20°C (68°F).

Sucrose by weight	Specific gravity at 20°/20°C	Total pounds per gallon	Pounds solids per gallon	Pounds water per gallon
1.0	1.00387	8.35379	0.08354	8.27026
2.0	1.00777	8.38626	0.16773	8.21854
3.0	1.01170	8.41898	0.25257	8.16641
4.0	1.01565	8.45186	0.33807	8.11379
5.0	1.01964	8.48508	0.42425	8.06083
6.0	1.02366	8.51855	0.51111	8.00744
7.0	1.02771	8.55218	0.59865	7.95353
8.0	1.03179	8.58615	0.68689	7.89926
9.0	1.03589	8.62029	0.77583	7.84446
10.0	1.04002	8.65468	0.86547	7.78921
11.0	1.04418	8.68923	0.95582	7.73342
12.0	1.04836	8.72404	1.04688	7.67715
13.0	1.05257	8.75909	1.13868	7.62041
14.0	1.05681	8.79440	1.23122	7.56318
15.0	1.06110	8.83003	1.32451	7.50553
16.0	1.06541	8.86592	1.41855	7.44738
17.0	1.06975	8.90206	1.51335	7.38871
18.0	1.07412	8.93845	1.60892	7.32953
19.0	1.07853	8.97509	1.70527	7.26983
20.0	1.08297	9.01207	1.80241	7.20966
21.0	1.08743	9.04921	1.90033	7.14888
22.0	1.09193	9.08660	1.99905	7.08755
23.0	1.09645	9.12424	2.09858	7.02567
24.0	1.10101	9.16222	2.19893	6.96329

(continued)

Table I–1. (*continued*)

Sucrose by weight	Specific gravity at 20°/20°C	Total pounds per gallon	Pounds solids per gallon	Pounds water per gallon
25.0	1.10562	9.20053	2.30013	6.90040
26.0	1.11025	9.23909	2.40216	6.83693
27.0	1.11492	9.27790	2.50503	6.77287
28.0	1.11962	9.31705	2.60877	6.70827
29.0	1.12435	9.35644	2.71337	6.64307
30.0	1.12913	9.39617	2.81885	6.57732
31.0	1.13392	9.43607	2.92518	6.51089
32.0	1.13876	9.47630	3.03241	6.44388
33.0	1.14363	9.51686	3.14056	6.37630
34.0	1.14854	9.55767	3.24961	6.30807
35.0	1.15348	9.59882	3.35959	6.23923
36.0	1.15845	9.64022	3.47048	6.16974
37.0	1.16348	9.68204	3.58235	6.09968
38.0	1.16852	9.72402	3.69513	6.02889
39.0	1.17362	9.76642	3.80890	5.95752
40.0	1.17875	9.80907	3.92363	5.88544
41.0	1.18390	9.85197	4.03931	5.81266
42.0	1.18910	9.89520	4.15599	5.73922
43.0	1.19433	9.93877	4.27367	5.66510
44.0	1.19961	9.98267	4.39238	5.59030
45.0	1.20491	10.02683	4.51207	5.51476
46.0	1.21026	10.07131	4.63280	5.43851
47.0	1.21564	10.11613	4.75458	5.36155
48.0	1.22106	10.16121	4.87738	5.28383
49.0	1.22653	10.20669	5.00128	5.20541
50.0	1.23202	10.25243	5.12622	5.12622
51.0	1.23756	10.29850	5.25224	5.04627
52.0	1.24313	10.34483	5.37931	4.96552
53.0	1.24874	10.39157	5.50753	4.88404
54.0	1.25439	10.43856	5.63682	4.80174
55.0	1.26008	10.48588	5.76723	4.71865
56.0	1.26575	10.53312	5.89855	4.63457
57.0	1.27147	10.58070	6.03100	4.54970
58.0	1.27729	10.62910	6.16488	4.46422
59.0	1.28321	10.67835	6.30023	4.37812
60.0	1.28908	10.72726	6.43635	4.29090
61.0	1.29500	10.77650	6.57367	4.20284
62.0	1.30096	10.82608	6.71217	4.11391
63.0	1.30695	10.87591	6.85182	4.02409
64.0	1.31298	10.92615	6.99274	3.93341
65.0	1.31905	10.97665	7.13482	3.84183
66.0	1.32516	11.02748	7.27814	3.74934
67.0	1.33131	11.07864	7.42269	3.65595
68.0	1.33749	11.13005	7.56844	3.56162
69.0	1.34371	11.18189	7.71550	3.46638
70.0	1.34997	11.23397	7.86378	3.37019

SOURCE: From Hoynak and Bollenback 1966.

Table I–2. Interconversion between Specific Gravity, °Brix (Balling), Baumé, Percent sugar (wt/vol) and Öechsle (°0).

°Brix	Baumé	Specific gravity at 20/20	Specific gravity at 20/4	Specific gravity at 15/4	% Sugar (wt/vol)
0.0	0.0	1.0000	0.9982	1.0009	0.0
0.4	0.2	1.0016	0.9998	1.0025	0.4
0.8	0.4	1.0031	1.0013	1.0040	0.8
1.2	0.7	1.0047	1.0029	1.0056	1.2
1.6	0.9	1.0062	1.0045	1.0071	1.6
2.0	1.1	1.0078	1.0060	1.0087	2.0
2.4	1.3	1.0094	1.0076	1.0103	2.4
2.8	1.6	1.0109	1.0091	1.0118	2.8
3.2	1.8	1.0125	1.0107	1.0134	3.2
3.6	2.0	1.0141	1.0123	1.0150	3.6
4.0	2.2	1.0157	1.0139	1.0166	4.1
4.4	2.4	1.0173	1.0155	1.0182	4.5
4.8	2.7	1.0189	1.0171	1.0198	4.9
5.2	2.9	1.0204	1.0186	1.0213	5.3
5.6	3.1	1.0021	1.0203	1.0262	6.6
6.0	3.3	1.0237	1.0219	1.0246	6.1
6.4	3.6	1.0253	1.0235	1.0262	6.6
6.8	3.8	1.0269	1.0251	1.0278	7.0
7.2	4.0	1.0285	1.0267	1.0294	7.4
7.6	4.2	1.0301	1.0283	1.0310	7.8
8.0	4.4	1.0318	1.0300	1.0327	8.2
8.4	4.7	1.0334	1.0316	1.0343	8.7
8.8	4.9	1.0350	1.0332	1.0359	9.1
9.2	5.1	1.0367	1.0349	1.0376	9.5
9.6	5.3	1.0383	1.0365	1.0392	10.0
10.0	5.6	1.0400	1.0382	1.0409	10.4
10.4	5.8	1.0416	1.0398	1.0425	10.8
10.8	6.0	1.0433	1.0415	1.0442	11.2
11.2	6.2	1.0450	1.0432	1.0459	11.7
11.6	6.4	1.0466	1.0448	1.0475	12.1
12.0	6.7	1.0483	1.0464	1.0492	12.6
12.4	6.9	1.0500	1.0481	1.0509	13.0
12.8	7.1	1.0517	1.0498	1.0526	13.4
13.2	7.3	1.0534	1.0515	1.0543	13.9
13.6	7.6	1.0551	1.0532	1.0561	14.3
14.0	7.8	1.0568	1.0549	1.0578	14.8
14.4	8.0	1.0585	1.0566	1.0595	15.2
14.8	8.2	1.0602	1.0683	1.0612	15.7
15.2	8.4	1.0619	1.0600	1.0629	16.1
15.6	8.7	1.0636	1.0617	1.0646	16.6
16.0	8.9	1.0653	1.0634	1.0663	17.0
16.4	9.1	1.0671	1.0652	1.0681	17.5
16.8	9.3	1.0688	1.0669	1.0698	17.9

(*continued*)

Table I–2. (*continued*)

°Brix	Baumé	Specific gravity at 20/20	Specific gravity at 20/4	Specific gravity at 15/4	% Sugar (wt/vol)
17.2	9.6	1.0706	1.0687	1.0716	18.4
17.6	9.8	1.0723	1.0704	1.0733	18.8
18.0	10.0	1.0740	1.0721	1.0750	19.3
18.4	10.2	1.0758	1.0739	1.0768	19.8
18.8	10.4	1.0776	1.0757	1.0786	20.2
19.2	10.7	1.0793	1.0774	1.0803	20.7
19.6	10.9	1.0811	1.0792	1.0821	21.2
20.0	11.1	1.0829	1.0810	1.0839	21.6
20.4	11.3	1.0846	1.0827	1.0856	22.1
20.8	11.6	1.0864	1.0845	1.0874	22.6
21.2	11.8	1.0882	1.0863	1.0892	23.0
21.6	12.0	1.0900	1.0881	1.0910	23.5
22.0	12.2	1.0918	1.0899	1.0928	24.0
22.4	12.4	1.0927	1.0908	1.0937	24.4
22.8	12.7	1.0945	1.0926	1.0955	24.9
23.2	12.9	1.0964	1.0945	1.0974	25.4
23.6	13.1	1.0982	1.0963	1.0992	25.9
24.0	13.3	1.1009	1.0990	1.1019	26.4
24.4	13.6	1.1028	1.1009	1.1038	26.9
24.8	13.8	1.1046	1.1026	1.1056	27.3
25.2	14.0	1.1064	1.1044	1.1074	27.8
25.6	14.2	1.1083	1.1063	1.1093	28.3
26.0	14.4	1.1101	1.1081	1.1111	28.8

The German scale of degree Ōechsle represents the difference in weight between 1 L of must and 1 L of water. This can be read as the first three digits following the decimal point in the specific gravity column. Thus: SpGr = 1.0829 (20 °B) is equivalent to 83 °Ō.

For potential alcohol values multiply °B by 0.55 − 0.63 (see text).

Table I–3. Correction factors for hydrometer Brix measurements taken at temperatures other than 68°F (20°C).

		°Brix								
	Temp (°C)	0	5	10	15	20	25	30	35	40
To be subtracted from the indicated degree	0	0.30	0.49	0.65	0.77	0.89	0.99	1.08	1.16	1.24
	5	0.36	0.47	0.56	0.65	0.73	0.80	0.86	0.91	0.97
	10	0.32	0.38	0.43	0.48	0.52	0.57	0.60	0.64	0.67
	11	0.31	0.35	0.40	0.44	0.48	0.51	0.55	0.58	0.60
	12	0.29	0.32	0.36	0.40	0.43	0.46	0.50	0.52	0.54
	13	0.26	0.29	0.32	0.35	0.38	0.41	0.44	0.46	0.48
	14	0.24	0.26	0.29	0.31	0.34	0.36	0.38	0.40	0.41
	15	0.20	0.22	0.24	0.26	0.28	0.30	0.32	0.33	0.34
	16	0.17	0.18	0.20	0.22	0.23	0.25	0.26	0.27	0.28
	17	0.13	0.14	0.15	0.16	0.18	0.19	0.20	0.20	0.21
	18	0.09	0.10	0.10	0.11	0.12	0.13	0.13	0.14	0.14
	19	0.05	0.05	0.05	0.06	0.06	0.06	0.07	0.07	0.07
	20°C									
To be added to the indicated degree	21	0.04	0.05	0.06	0.06	0.06	0.07	0.07	0.07	0.07
	22	0.10	0.10	0.11	0.12	0.12	0.13	0.14	0.14	0.15
	23	0.16	0.16	0.17	0.17	0.19	0.20	0.21	0.21	0.22
	24	0.21	0.22	0.23	0.24	0.26	0.27	0.28	0.29	0.30
	25	0.27	0.28	0.30	0.31	0.32	0.34	0.35	0.36	0.38
	26	0.33	0.34	0.36	0.37	0.40	0.40	0.42	0.44	0.46
	27	0.40	0.41	0.42	0.44	0.46	0.48	0.50	0.52	0.54
	28	0.46	0.47	0.49	0.51	0.54	0.56	0.58	0.60	0.61
	29	0.54	0.55	0.56	0.59	0.61	0.63	0.66	0.68	0.70
	30	0.61	0.62	0.63	0.66	0.68	0.71	0.73	0.76	0.78
	35	0.99	1.01	1.02	1.06	1.10	1.13	1.16	1.18	1.20

SOURCE: Adapted from Charlottenberg, 1900. Physikalische-technische reichsanstalt. Wiss. Abhandl. *Kaiserliche Normal-Eichungs-Kommission* 2:140.

Table I–4. Correction factors for refractive index measurements taken at temperatures other than 20°C (68°F).

International Temperature Correction Table (1936) for the Normal Model of Refractometer Above and Below 20° C.

Temp. °C	Per cent Sucrose														
	0	5	10	15	20	25	30	35	40	45	50	55	60	65	70
	Subtract from the per cent Sucrose														
10	0.50	0.54	0.58	0.61	0.64	0.66	0.68	0.70	0.72	0.73	0.74	0.75	0.76	0.78	0.79
11	0.46	0.49	0.53	0.55	0.58	0.60	0.62	0.64	0.65	0.66	0.67	0.68	0.69	0.70	0.71
12	0.42	0.45	0.48	0.50	0.52	0.54	0.56	0.57	0.58	0.59	0.60	0.61	0.61	0.63	0.63
13	0.37	0.40	0.42	0.44	0.46	0.48	0.49	0.50	0.51	0.52	0.53	0.54	0.54	0.55	0.55
14	0.33	0.35	0.37	0.39	0.40	0.41	0.42	0.43	0.44	0.45	0.45	0.46	0.46	0.47	0.48
15	0.27	0.29	0.31	0.33	0.34	0.34	0.35	0.36	0.37	0.37	0.38	0.39	0.39	0.40	0.40
16	0.22	0.24	0.25	0.26	0.27	0.28	0.28	0.29	0.30	0.30	0.30	0.31	0.31	0.32	0.32
17	0.17	0.18	0.19	0.20	0.21	0.21	0.21	0.22	0.22	0.23	0.23	0.23	0.23	0.24	0.24
18	0.12	0.13	0.13	0.14	0.14	0.14	0.14	0.15	0.15	0.15	0.15	0.16	0.16	0.16	0.16
19	0.06	0.06	0.06	0.07	0.07	0.07	0.07	0.08	0.08	0.08	0.08	0.08	0.08	0.08	0.08
	Add to the per cent Sucrose														
21	0.06	0.07	0.07	0.07	0.07	0.08	0.08	0.08	0.08	0.08	0.08	0.08	0.08	0.08	0.08
22	0.13	0.13	0.14	0.14	0.15	0.15	0.15	0.15	0.15	0.16	0.16	0.16	0.16	0.16	0.16
23	0.19	0.20	0.21	0.22	0.22	0.23	0.23	0.23	0.23	0.24	0.24	0.24	0.24	0.24	0.24
24	0.26	0.27	0.28	0.29	0.30	0.30	0.31	0.31	0.31	0.31	0.31	0.32	0.32	0.32	0.32
25	0.33	0.35	0.36	0.37	0.38	0.38	0.39	0.40	0.40	0.40	0.40	0.40	0.40	0.40	0.40
26	0.40	0.42	0.43	0.44	0.45	0.46	0.47	0.48	0.48	0.48	0.48	0.48	0.48	0.48	0.48
27	0.48	0.50	0.52	0.53	0.54	0.55	0.55	0.56	0.56	0.56	0.56	0.56	0.56	0.56	0.56
28	0.56	0.57	0.60	0.61	0.62	0.63	0.63	0.64	0.64	0.64	0.64	0.64	0.64	0.64	0.64
29	0.64	0.66	0.68	0.69	0.71	0.72	0.72	0.73	0.73	0.73	0.73	0.73	0.73	0.73	0.73
30	0.72	0.74	0.77	0.78	0.79	0.80	0.80	0.81	0.81	0.81	0.81	0.81	0.81	0.81	0.81

SOURCE: Bausch and Lomb Optical Company.

Table I–5. International scale of refractive indexes of sucrose solutions at 20°C (68°F).

International Scale (1936) of Refractive Indices of Sucrose Solutions at 20° C.

Index	Per cent	Index	Per cent	Index	Per cent	Index	Per cent
1.3330	0	1.3723	25	1.4200	50	1.4774	75
1.3344	1	1.3740	26	1.4221	51	1.4799	76
1.3359	2	1.3758	27	1.4242	52	1.4825	77
1.3373	3	1.3775	28	1.4264	53	1.4850	78
1.3388	4	1.3793	29	1.4285	54	1.4876	79
1.3403	5	1.3811	30	1.4307	55	1.4901	80
1.3418	6	1.3829	31	1.4329	56	1.4927	81
1.3433	7	1.3847	32	1.4351	57	1.4954	82
1.3448	8	1.3865	33	1.4373	58	1.4980	83
1.3463	9	1.3883	34	1.4396	59	1.5007	84
1.3478	10	1.3902	35	1.4418	60	1.5033	85
1.3494	11	1.3920	36	1.4441	61		
1.3509	12	1.3939	37	1.4464	62		
1.3525	13	1.3958	38	1.4486	63		
1.3541	14	1.3978	39	1.4509	64		
1.3557	15	1.3997	40	1.4532	65		
1.3573	16	1.4016	41	1.4555	66		
1.3589	17	1.4036	42	1.4579	67		
1.3605	18	1.4056	43	1.4603	68		
1.3622	19	1.4076	44	1.4627	69		
1.3638	20	1.4096	45	1.4651	70		
1.3655	21	1.4117	46	1.4676	71		
1.3672	22	1.4137	47	1.4700	72		
1.3689	23	1.4158	48	1.4725	73		
1.3706	24	1.4179	49	1.4749	74		

SOURCE: Bausch and Lomb Optical Company 1952.

Table I–6. Temperature corrections of alcohol hydrometers callibrated at 15.56°C (60°F) in volume % ethanol.

Observed alcohol content (vol. %)	To or from the observed																	
	Add			Subtract														
	at 57°F 13.9°C	at 58°F 14.4°C	at 59°F 15.0°C	at 61°F 16.1°C	at 62°F 16.7°C	at 63°F 17.2°C	at 64°F 17.8°C	at 65°F 18.3°C	at 66°F 18.9°C	at 67°F 19.4°C	at 68°F 20.0°C	at 69°F 20.6°C	at 70°F 21.1°C	at 72°F 22.2°C	at 74°F 23.3°C	at 76°F 24.4°C	at 78°F 25.6°C	at 80°F 26.7°C
1	0.14	0.10	0.05	0.05	0.10	0.16	0.22	0.28	0.34	0.41	0.48	0.55	0.62	0.77	0.93			
2	0.14	0.10	0.05	0.05	0.11	0.17	0.23	0.29	0.35	0.42	0.48	0.56	0.63	0.78	0.94	1.10	1.28	1.46
3	0.14	0.10	0.05	0.06	0.12	.081	0.24	0.30	0.36	0.43	0.50	0.57	0.64	0.80	0.96	1.13	1.31	1.50
4	0.14	0.10	0.05	0.06	0.12	0.19	0.25	0.32	0.38	0.45	0.52	0.59	0.67	0.83	1.00	1.17	1.35	1.54
5	0.15	0.10	0.05	0.07	0.13	0.20	0.26	0.33	0.40	0.47	0.54	0.62	0.70	0.86	1.03	1.21	1.40	1.60
6	0.17	0.11	0.06	0.07	0.14	0.20	0.27	0.34	0.42	0.50	0.57	0.66	0.74	0.90	1.09	1.27	1.46	1.66
7	0.18	0.12	0.06	0.07	0.14	0.21	0.29	0.36	0.44	0.52	0.60	0.68	0.77	0.94	1.13	1.32	1.52	1.73
8	0.19	0.13	0.06	0.08	0.16	0.23	0.31	0.39	0.47	0.55	0.64	0.73	0.81	0.99	1.18	1.38	1.59	1.80
9	0.21	0.14	0.07	0.08	0.16	0.24	0.32	0.41	0.50	0.58	0.67	0.76	0.86	1.04	1.25	1.46	1.67	1.89
10	0.23	0.16	0.08	0.08	0.17	0.25	0.34	0.43	0.52	0.61	0.71	0.80	0.90	1.10	1.32	1.54	1.76	1.99
11	0.25	0.16	0.08	0.09	0.18	0.27	0.37	0.46	0.56	0.65	0.75	0.85	0.96	1.16	1.39	1.61	1.84	2.09
12	0.27	0.18	0.09	0.10	0.20	0.29	0.39	0.49	0.59	0.70	0.80	0.91	1.02	1.23	1.46	1.70	1.94	2.20
13	0.29	0.19	0.10	0.10	0.21	0.31	0.42	0.52	0.63	0.74	0.85	0.97	1.08	1.31	1.55	1.80	2.05	2.31
14	0.32	0.21	0.11	0.11	0.22	0.32	0.44	0.55	0.66	0.78	0.91	1.02	1.14	1.39	1.65	1.91	2.17	2.44
15	0.35	0.23	0.12	0.12	0.24	0.35	0.48	0.60	0.71	0.84	0.97	1.10	1.23	1.50	1.76	2.03	2.30	2.58
16	0.37	0.24	0.12	0.13	0.26	0.38	0.52	0.65	0.77	0.90	1.03	1.17	1.31	1.60	1.88	2.16	2.44	2.72
17	0.40	0.26	0.13	0.14	0.27	0.41	0.54	0.68	0.82	0.96	1.10	1.25	1.40	1.70	1.99	2.28	2.58	2.87
18	0.44	0.29	0.14	0.14	0.29	0.44	0.58	0.73	0.88	1.03	1.18	1.33	1.49	1.80	2.10	2.41	2.72	3.02
19	0.47	0.32	0.16	0.15	0.30	0.46	0.62	0.78	0.94	1.10	1.26	1.42	1.58	1.90	2.22	2.54	2.86	3.17
20	0.51	0.34	0.17	0.16	0.32	0.49	0.66	0.82	0.98	1.15	1.33	1.48	1.65	2.00	2.32	2.65	2.98	3.33
21	0.53	0.35	0.18	.017	0.34	0.51	0.68	0.85	1.02	1.20	1.38	1.54	1.72	2.06	2.41	2.76	3.10	3.45
22	0.56	0.38	0.19	0.17	0.36	0.53	0.71	0.90	1.07	1.25	1.44	1.61	1.78	2.13	2.48	2.84	3.20	3.56
23	0.58	0.40	0.20	0.18	0.37	0.55	0.74	0.92	1.11	1.30	1.49	1.66	1.84	2.20	2.56	2.93	3.30	3.67
24	0.60	0.40	0.20	0.18	0.38	0.56	0.77	0.96	1.16	1.54	1.72	1.91	2.27	2.65	3.03	3.40	3.8	

SOURCE: U.S. Internal Revenue Service: "Regulations No. 7". U.S. Govt. Printing Office, Washington, D.C. 1945.

Table I–7. Temperature corrections of alcohol hydrometers calibrated at 20°C (68°F) in volume % ethanol.

Temperature (°C) — Add	Apparent degree of alcoholic strength at 20°C																	
	0	1	2	3	4	5	6	7	8	9	10	11	12	13	14	15	16	17
0	0.76	0.77	0.82	0.87	0.95	1.04	1.16	1.31	1.49	1.70	1.95	2.26	2.62	3.03	3.49	4.02	4.56	5.11
1	0.81	0.83	0.87	0.92	1.00	1.09	1.20	1.35	1.52	1.73	1.97	2.26	2.59	2.97	3.40	3.87	4.36	4.86
2	0.85	0.87	0.92	0.97	1.04	1.13	1.24	1.38	1.54	1.74	1.97	2.24	2.54	2.89	3.29	3.72	4.17	4.61
3	0.88	0.91	0.95	1.00	1.07	1.15	1.26	1.39	1.55	1.73	1.95	2.20	2.48	2.80	3.16	3.55	3.95	4.36
4	0.90	0.92	0.97	1.02	1.09	1.17	1.27	1.40	1.55	1.72	1.92	2.15	2.41	2.71	3.03	3.38	3.75	4.11
5	0.91	0.93	0.98	1.03	1.10	1.17	1.27	1.39	1.53	1.69	1.87	2.08	2.33	2.60	2.89	3.21	3.54	3.86
6	0.92	0.94	0.98	1.02	1.09	1.16	1.25	1.37	1.50	1.65	1.82	2.01	2.23	2.47	2.74	3.02	3.32	3.61
7	0.91	0.93	0.97	1.01	1.07	1.14	1.23	1.33	1.45	1.59	1.75	1.92	2.12	2.34	2.58	2.83	3.10	3.36
8	0.89	0.91	0.94	0.98	1.04	1.11	1.19	1.28	1.39	1.52	1.66	1.82	2.00	2.20	2.42	2.65	2.88	3.11
9	0.86	0.88	0.91	0.95	1.01	1.07	1.14	1.23	1.33	1.44	1.57	1.71	1.87	2.05	2.24	2.44	2.65	2.86
10	0.82	0.84	0.87	0.91	0.96	1.01	1.08	1.16	1.25	1.35	1.47	1.60	1.74	1.89	2.06	2.24	2.43	2.61
11	0.78	0.79	0.82	0.86	0.90	0.95	1.01	1.08	1.16	1.25	1.36	1.47	1.60	1.73	1.88	2.03	2.20	2.36
12	0.72	0.74	0.76	0.79	0.83	0.88	0.93	0.99	1.07	1.15	1.24	1.34	1.44	1.56	1.69	1.82	1.96	2.10
13	0.66	0.67	0.69	0.72	0.76	0.80	0.84	0.90	0.96	1.03	1.11	1.19	1.28	1.38	1.49	1.61	1.73	1.84
14	0.59	0.60	0.62	0.64	0.67	0.71	0.74	0.79	0.85	0.91	0.97	1.04	1.12	1.20	1.29	1.39	1.49	1.58
15	0.51	0.52	0.53	0.55	0.58	0.61	0.64	0.68	0.73	0.77	0.83	0.89	0.95	1.02	1.09	1.16	1.24	1.32
16	0.42	0.43	0.44	0.46	0.48	0.50	0.53	0.56	0.60	0.63	0.67	0.72	0.77	0.82	0.88	0.94	1.00	1.06
17	0.33	0.33	0.34	0.35	0.37	0.39	0.41	0.43	0.46	0.48	0.51	0.55	0.59	0.62	0.67	0.71	0.75	0.80
18	0.23	0.23	0.23	0.24	0.25	0.26	0.27	0.29	0.31	0.33	0.35	0.37	0.40	0.42	0.45	0.48	0.51	0.53
19	0.12	0.12	0.12	0.12	0.13	0.13	0.14	0.15	0.16	0.17	0.18	0.19	0.20	0.21	0.23	0.24	0.25	0.27

Table I–7. (*continued*)

Temperature (°C) Deduct	Apparent degree of alcoholic strength at 20°C																	
	0	1	2	3	4	5	6	7	8	9	10	11	12	13	14	15	16	17
21	0.13	0.13	0.13	0.14	0.14	0.15	0.16	0.17	0.18	0.18	0.19	0.19	0.20	0.22	0.23	0.25	0.26	0.28
22	0.26	0.27	0.28	0.29	0.30	0.31	0.32	0.34	0.36	0.36	0.37	0.39	0.41	0.44	0.47	0.49	0.52	0.55
23	0.40	0.41	0.42	0.44	0.45	0.47	0.49	0.51	0.54	0.54	0.57	0.60	0.63	0.66	0.70	0.74	0.78	0.82
24	0.55	0.56	0.58	0.60	0.62	0.64	0.67	0.70	0.73	0.73	0.77	0.81	0.85	0.89	0.94	0.99	1.04	1.10
25	0.69	0.71	0.73	0.76	0.79	0.82	0.85	0.89	0.93	0.93	0.97	1.02	1.07	1.13	1.19	1.25	1.31	1.37
26	0.85	0.87	0.90	0.93	0.96	1.00	1.04	1.08	1.13	1.13	1.18	1.24	1.30	1.36	1.43	1.50	1.57	1.65
27		1.03	1.07	1.11	1.15	1.19	1.23	1.28	1.34	1.34	1.40	1.46	1.53	1.60	1.68	1.76	1.84	1.93
28		1.21	1.25	1.29	1.33	1.38	1.43	1.49	1.55	1.55	1.62	1.69	1.77	1.85	1.93	2.02	2.11	2.21
29		1.39	1.43	1.47	1.52	1.58	1.63	1.70	1.76	1.76	1.84	1.92	2.01	2.10	2.19	2.29	2.39	2.50
30		1.57	1.61	1.66	1.72	1.78	1.84	1.91	1.98	1.98	2.07	2.15	2.25	2.35	2.45	2.56	2.67	2.78
31		1.75	1.80	1.86	1.92	1.98	2.05	2.13	2.21	2.21	2.30	2.39	2.49	2.60	2.71	2.83	2.94	3.07
32		1.94	2.00	2.06	2.13	2.20	2.27	2.35	2.44	2.44	2.53	2.63	2.74	2.86	2.97	3.09	3.22	3.36
33			2.20	2.27	2.34	2.42	2.50	2.58	2.67	2.67	2.77	2.88	2.99	3.12	3.24	3.37	3.51	3.65
34			2.41	2.48	2.56	2.64	2.72	2.81	2.91	2.91	3.02	3.13	3.25	3.38	3.51	3.65	3.79	3.94
35			2.62	2.70	2.78	2.86	2.95	3.05	3.16	3.16	3.27	3.39	3.51	3.64	3.78	3.93	4.08	4.23
36			2.83	2.91	3.00	3.09	3.19	3.29	3.41	3.41	3.53	3.65	3.78	3.91	4.05	4.21	4.37	4.52
37				3.13	3.23	3.33	3.43	3.54	3.65	3.65	3.78	3.91	4.04	4.18	4.33	4.49	4.65	4.82
38				3.36	3.47	3.57	3.68	3.79	3.91	3.91	4.03	4.17	4.31	4.46	4.61	4.77	4.94	5.12
39				3.59	3.70	3.81	3.93	4.05	4.17	4.17	4.30	4.44	4.58	4.74	4.90	5.06	5.23	5.41
40				3.82	3.94	4.06	4.18	4.31	4.44	4.44	4.57	4.71	4.86	5.02	5.19	5.36	5.53	5.71

Table I-8. Changes in percent bitartrate (HT) in total tartrates with changes in pH and alcohol (% vol/vol) at 20°C

alcohol (%)																					
10		11		12		13		14		16		17		18		19		20		21	
pH	%HT	pH	%HT	pH	%HT	pH	%HT	pH	%HT	pH	%HT	pH	%HT	pH	%HT	pH	%HT	pH	%HT	pH	%HT
2.81	37.7	2.83	38.0	2.84	38.3	2.85	38.6	2.87	38.9	2.89	39.5	2.91	39.8	2.92	40.0	2.93	40.3	2.95	40.6	2.96	40.9
2.91	42.9	2.93	43.2	2.94	43.5	2.95	43.8	2.97	44.1	2.99	44.7	3.01	45.0	3.02	45.3	3.03	45.6	3.05	45.9	3.06	46.2
3.01	48.1	3.03	48.4	3.04	48.7	3.05	49.0	3.07	49.3	3.09	49.9	3.11	50.3	3.12	50.5	3.13	50.8	3.15	51.1	3.16	51.4
3.11	53.0	3.13	53.3	3.14	53.6	3.15	53.9	3.17	54.2	3.19	54.8	3.21	55.1	3.22	55.4	3.23	55.7	3.25	56.0	3.26	56.3
3.21	57.4	3.23	57.7	3.24	58.0	3.25	58.3	4.37	58.6	3.29	59.1	3.31	59.4	3.32	59.7	3.33	60.0	3.35	60.3	3.36	60.6
3.31	61.1	3.33	61.4	3.34	61.7	3.35	61.9	3.37	62.2	3.39	62.8	3.41	63.1	3.42	63.4	3.43	63.6	3.45	63.9	3.46	64.2
3.41	63.9	3.43	64.2	3.44	64.4	3.45	64.7	3.47	65.0	3.49	65.5	3.51	65.8	3.52	66.1	3.53	66.4	3.55	66.6	3.56	66.9
3.51	65.6	3.53	65.9	3.54	66.2	3.55	66.4	3.57	66.7	3.59	67.2	3.61	67.5	3.62	67.8	3.63	68.0	3.65	68.3	3.66	68.6
3.61	66.2	3.63	66.5	3.64	66.7	3.65	67.0	3.67	67.3	3.69	67.8	3.71	68.1	3.72	68.3	3.73	68.6	3.75	68.9	3.76	69.1
3.71	65.6	3.73	65.9	3.74	66.1	3.75	66.4	3.77	66.7	3.79	67.2	3.81	67.5	3.82	67.8	3.83	68.0	3.85	68.3	3.86	68.6
3.81	63.9	3.83	64.1	3.84	64.4	3.85	64.7	3.87	65.0	3.89	65.5	3.91	65.8	3.92	66.1	3.93	66.4	3.95	66.6	3.96	66.9
3.91	61.1	3.93	61.4	3.94	61.7	3.95	61.9	3.97	62.2	3.99	62.8	4.01	63.1	4.02	63.4	4.03	63.6	4.05	63.9	4.06	64.2
4.01	57.4	4.03	57.7	4.04	58.0	4.05	58.3	4.07	58.6	4.09	59.1	4.11	59.4	4.12	59.7	4.13	60.0	4.15	60.3	4.16	60.6
4.11	53.0	4.13	53.3	4.14	53.6	4.15	53.9	4.17	54.2	4.19	54.8	4.21	55.1	4.22	55.4	4.23	55.7	4.25	66.0	4.26	56.3
4.21	48.1	4.23	48.4	4.24	48.7	4.25	49.0	4.27	49.3	4.29	49.9	4.31	50.2	4.32	50.5	4.33	50.8	4.35	61.1	4.36	51.4

SOURCE: Berg, H.W., and R.M. Keefer. Analytical determination of tartrate stability in wine. I. Potassium bitartrate. *American Journal of Enology* 9:180–183 (1958).

Table I-9. Changes in percent tartrate ($T^=$) in total tartrates with changes in pH and alcohol (% vol/vol) at 20°C.

Percent alcohol by volume																					
10		11		12		13		14		16		17		18		19		20		21	
pH	% $T^=$	pH	% $T^=$	pH	% $T^=$	pH	% $T^=$	pH	% $T^=$	pH	% $T^=$	pH	% $T^=$	pH	% $T^=$	pH	% $T^=$	pH	% $T^=$	pH	% $T^=$
2.81	1.5	2.83	1.5	2.84	1.5	2.85	1.5	2.87	1.5	2.89	1.5	2.91	1.5	2.92	1.5	2.93	1.5	2.95	1.5	2.96	1.5
2.91	2.2	2.93	2.2	2.94	2.2	2.95	2.2	2.97	2.1	2.99	2.1	3.01	2.1	3.02	2.1	3.03	2.1	3.05	2.1	3.06	2.1
3.01	3.1	3.03	3.1	3.04	3.0	3.05	3.0	3.07	3.0	3.09	3.0	3.11	3.0	3.12	2.9	3.13	2.9	3.15	2.9	3.16	2.9
3.11	4.3	3.13	4.3	3.14	4.2	3.15	4.2	3.1	4.2	3.19	4.1	3.21	4.1	3.22	4.1	3.23	4.0	3.25	4.0	3.26	4.0
3.21	5.8	3.23	5.8	3.24	5.8	3.25	5.7	3.27	5.7	3.29	5.6	3.31	5.6	3.32	5.5	3.33	5.5	3.35	5.4	3.36	5.4
3.31	7.8	3.33	7.8	3.34	7.7	3.35	7.7	3.37	7.6	3.39	7.5	3.41	7.4	3.42	7.4	3.43	7.3	3.45	7.2	3.46	7.2
3.41	10.3	3.43	10.2	3.44	10.1	3.45	10.1	3.47	10.0	3.49	9.8	3.51	9.7	3.52	9.7	3.53	9.6	3.55	9.5	3.56	9.4
3.51	13.3	3.53	13.2	3.54	13.1	3.55	13.0	3.57	12.9	3.59	12.7	3.61	12.6	3.62	12.5	3.63	12.4	3.65	12.3	3.66	12.2
3.61	16.9	3.63	16.8	3.64	16.6	3.65	16.5	3.67	16.4	3.69	16.1	3.71	16.0	3.72	15.8	3.73	15.7	3.75	15.6	3.76	15.4
3.71	21.1	3.73	20.9	3.74	20.8	3.75	20.6	3.77	20.4	3.79	20.1	3.81	19.9	3.82	19.8	3.83	19.6	3.85	19.4	3.86	19.3
3.81	25.9	3.83	25.7	3.84	25.5	3.85	25.3	3.87	25.1	3.89	24.7	3.91	24.5	3.92	24.3	3.93	24.1	3.95	23.9	3.96	23.7
3.91	31.1	3.93	30.9	3.94	30.7	3.95	30.4	3.97	30.2	3.99	29.8	4.01	29.5	4.02	29.3	4.03	29.1	4.05	28.8	4.06	28.6
4.01	36.8	4.03	36.5	4.04	36.3	4.05	36.0	4.07	35.8	4.09	35.3	4.11	35.0	4.12	34.8	4.13	34.5	4.15	34.3	4.16	34.0
4.11	42.8	4.13	42.5	4.14	42.2	4.15	41.9	4.17	41.7	4.19	41.1	4.21	40.8	4.22	40.6	4.23	40.3	4.25	40.0	4.26	39.8
4.21	48.9	4.23	48.6	4.24	48.3	4.25	48.0	4.27	47.7	4.29	47.1	4.31	46.9	4.32	46.6	4.33	46.3	4.35	46.0	4.36	45.7

SOURCE: Berg, H.W., and R.M. Keefer. Analytical determination of tartrate stability in wine. II. Calcium tartrate. *American Journal of Enology* 10:105–109 (1959).

Table I–10. Summary of approved enological materials and maximum allowances or limitations by OIV and BATF.

Enological material	OIV	BATF	Remarks
Acacia	30 g/hL	24 g/hL	
Activated carbon (vegetal or animal)	100 g/hL	300 g/hL	If over 300 g/hL, BATF requires notice
Albumen of egg	Yes	0.25 L/hL	
Albumen (blood)	Yes	No	
Alginates	Yes	Yes	
Bentonite and kaolin	Yes	Yes	Limitations as to use by OIV
Ammonium carbonate	Yes	0.25 g/hL	BATF does not permit natural fixed acids to be reduced below 5 g/L
Ammonium phosphate	Yes	96 g/hL	
Ascorbic acid	10 g/hL	Yes	OIV limits use to bottling
Calcium carbonate	Yes	Yes	BATF does not permit natural fixed acids to be reduced below 5 g/L
Calcium phytate	Yes	No	
Carbon dioxide	max. of 0.100 g/100 mL	max. of 0.392 g/100 mL	
Casein	Yes	Yes	
Citric acid	max. 1 g/L	max. 0.7 g/L	
Copper sulfate	No	max. 0.5 ppm	BATF max. residual level of copper not to exceed 0.5 ppm
Defoaming agents		100% active: 1.8 g/hL 30% active: 6 g/hL	If silicon dioxide, BATF limits amount remaining in wine to 10 ppm
Diatomaceous earth	Yes	Yes	
Enzymes	Yes, for most	Yes, for most	
Ferrocyanide of potassium	Yes	No	
Ferrous sulfate	No	to 22 ppm	For clarifying and stabilizing wine
Fumaric acid	Not recommended	to max. 2.4 g/hL	
Gelatin	Yes	Yes	
Isinglass	Yes	Yes	
Lactic acid	No	Yes	
Malic acid	No	Yes	
Malolactic bacteria	Yes	Yes	
Maltol	No	to 250 ppm	For stabilization
Nitrogen gas	Yes	Yes	
PVPP	to 80 g/hL	to 72 g/hL	

(*continued*)

Table I–10. *(continued)*

Enological material	OIV	BATF	Remarks
Potassium carbonate, bicarbonate, and potassium tartrate	Yes	Yes	BATF does not permit natural fixed acids to be reduced below 5 g/hL
Potassium metabisulfite	max. 250–350 ppm	max. 350 ppm	BATF requires warning labels for sulfites
Silica gel	Yes	max. 240 g/hL at 30%	
Sorbic acid	max. 200 mg/L	max. 300 mg/L	
Sulfur dioxide	max. 250–300 ppm	max. 350 ppm	BATF requires warning labels for sulfites
Tannin	Yes	white-max. 0.8 g/hL; red-max. 3 g/hL	
Tartaric acid	Yes	Yes	BATF limits acidification to 9 g/hL for normal wines and 11 g/hL for sweeter wines having extracts over 8 g/100 mL
Thiamine	60 mg/hL	60 mg/hL	
Yeast hulls	40 g/hL	36 g/hL	

SOURCE: Recht, 1992.

Table I-11. Temperature Conversions—Fahrenheit to Celsius.

°F	°C	°F	°C	°F	°C	°F	°C	°F	°C	°F	°C
−10	−23.33	31	−0.56	72	22.22	113	45.00	154	67.78	195	90.56
−9	−22.78	32	0.00	73	22.78	114	45.66	155	68.33	196	91.11
−8	−22.22	33	0.56	74	23.33	115	46.11	156	68.89	197	91.67
−7	−21.67	34	1.11	75	23.89	116	46.67	157	69.44	198	92.22
−6	−21.11	35	1.67	76	24.44	117	47.22	158	70.00	199	92.78
−5	−20.56	36	2.22	77	25.00	118	47.78	159	70.56	200	93.33
−4	−20.00	37	2.78	78	25.56	119	48.33	160	71.11	201	93.89
−3	−19.44	38	3.33	79	26.11	120	48.89	161	71.67	202	94.44
−2	−18.89	39	3.89	80	26.67	121	49.44	162	72.22	203	95.00
−1	−18.33	40	4.44	81	27.22	122	50.00	163	72.78	204	95.56
0	−17.78	41	5.00	82	27.78	123	50.56	164	73.33	205	96.11
1	−17.22	42	5.56	83	28.33	124	51.11	165	73.89	206	96.67
2	−16.67	43	6.11	84	28.89	125	51.67	166	74.44	207	97.22
3	−16.11	44	6.67	85	29.44	126	52.22	167	75.00	208	97.78
4	−15.56	45	7.22	86	30.00	127	52.78	168	75.56	209	98.33
5	−15.00	46	7.78	87	30.56	128	53.33	169	76.11	210	98.89
6	−14.44	47	8.33	88	31.11	129	53.89	170	76.67	211	99.44
7	−13.89	48	8.89	89	31.67	130	54.44	171	77.22	212	100.00
8	−13.33	49	9.44	90	32.22	131	55.00	172	77.78	213	100.56
9	−12.78	50	10.00	91	32.78	132	55.56	173	78.33	214	101.11
10	−12.22	51	10.56	92	33.33	133	56.11	174	78.89	215	101.67
11	−11.67	52	11.11	93	33.89	134	56.67	175	79.44	216	102.22
12	−11.11	53	11.67	94	34.44	135	57.22	176	80.00	217	102.78
13	−10.56	54	12.22	95	35.00	136	57.78	177	80.56	218	103.33
14	−10.00	55	12.78	96	35.56	137	58.33	179	81.11	219	103.89
15	−9.44	56	13.33	97	36.11	138	58.89	179	81.67	220	104.44
16	−8.89	57	13.89	98	36.67	139	59.44	180	82.22	221	105.00
17	−8.33	58	14.44	99	37.22	140	60.00	181	82.78	222	105.56
18	−7.78	59	15.0	100	37.78	141	60.56	182	83.33	223	106.11
19	−7.22	60	15.56	101	38.33	142	61.11	183	83.89	224	106.67
20	−6.67	61	16.11	102	38.89	143	61.67	184	84.44	225	107.22
21	−6.11	62	16.67	103	39.44	144	62.22	185	85.00	226	107.78
22	−5.56	63	17.22	104	40.00	145	62.78	186	85.56	227	108.33
23	−5.00	64	17.78	105	40.56	146	63.33	187	86.11	228	108.89
24	−4.44	65	18.33	106	41.11	147	63.89	188	86.67	229	109.44
25	−3.89	66	18.89	107	41.67	148	64.44	189	87.22	230	110.00
26	−3.33	67	19.44	108	42.22	149	65.00	190	87.78	231	110.56
27	−2.78	68	20.00	109	42.78	150	65.56	191	88.33	232	111.11
28	−2.22	69	20.56	110	43.33	151	66.11	192	88.89	233	111.67
29	−1.67	70	21.11	111	43.89	152	66.67	193	89.44	234	112.22
30	−1.11	71	21.67	112	44.44	153	67.22	194	90.00	235	112.78

Table I-12. Percent alcohol by volume vs. specific gravity at 20°C.[1,2]

Specific gravity	Alcohol (v/v%)	Specific gravity	alcohol (v/v%)
1.00000	0.00	0.98530	11.00
0.99851	1.00	0.98471	11.50
0.99704	2.00	0.98412	12.00
0.99560	3.00	0.98297	13.00
0.99419	4.00	0.98182	14.00
0.99281	5.00	0.98071	15.00
0.99149	6.00	0.97960	16.00
0.99020	7.00	0.97850	17.00
0.98894	8.00	0.97743	18.00
0.98771	9.00	0.97638	19.00
0.98711	9.50	0.97532	20.00
0.98650	10.00	0.97425	21.00
0.98590	10.50	0.97318	22.00

[1]Specific gravity determined with reference to water at 20°C. To convert the specific gravity with reference to water at 4°C, multiply the above values by 0.99908.

[2]Table based on data in U.S. Bureau of Standards Circular No. 19 (1924).

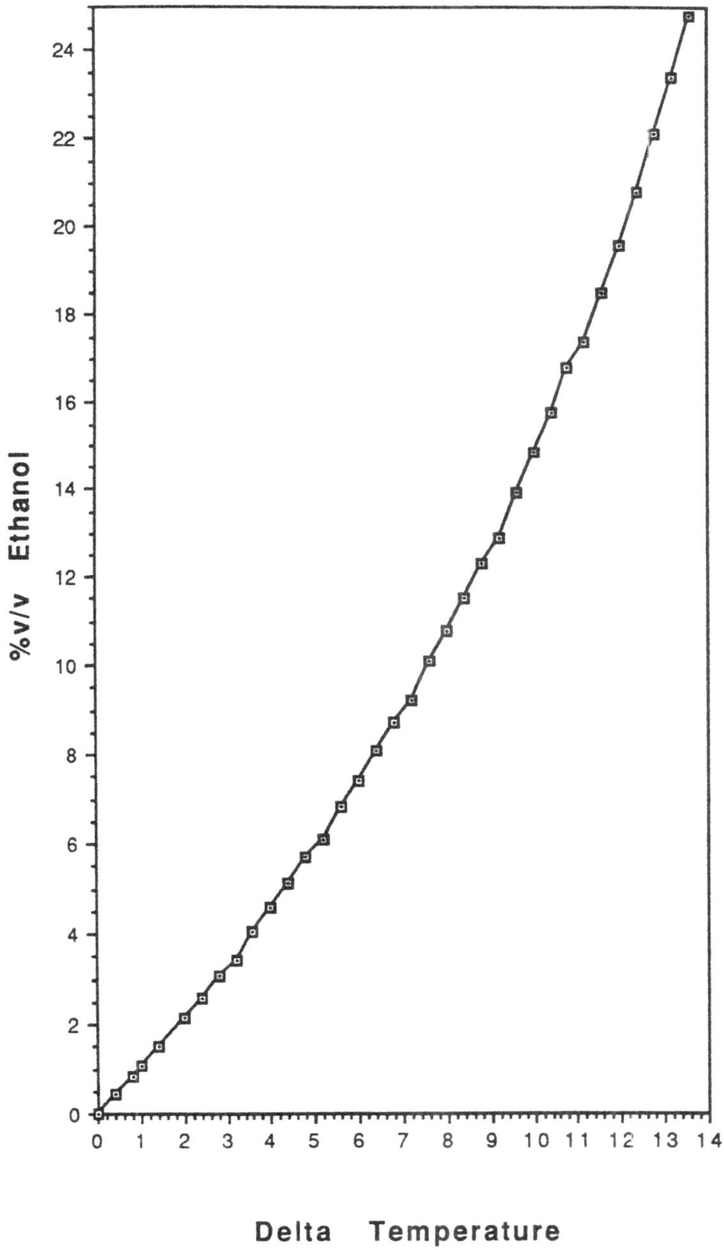

Fig. I–1. Ebulliometric temperature difference readings vs. percent ethanol (vol/vol).

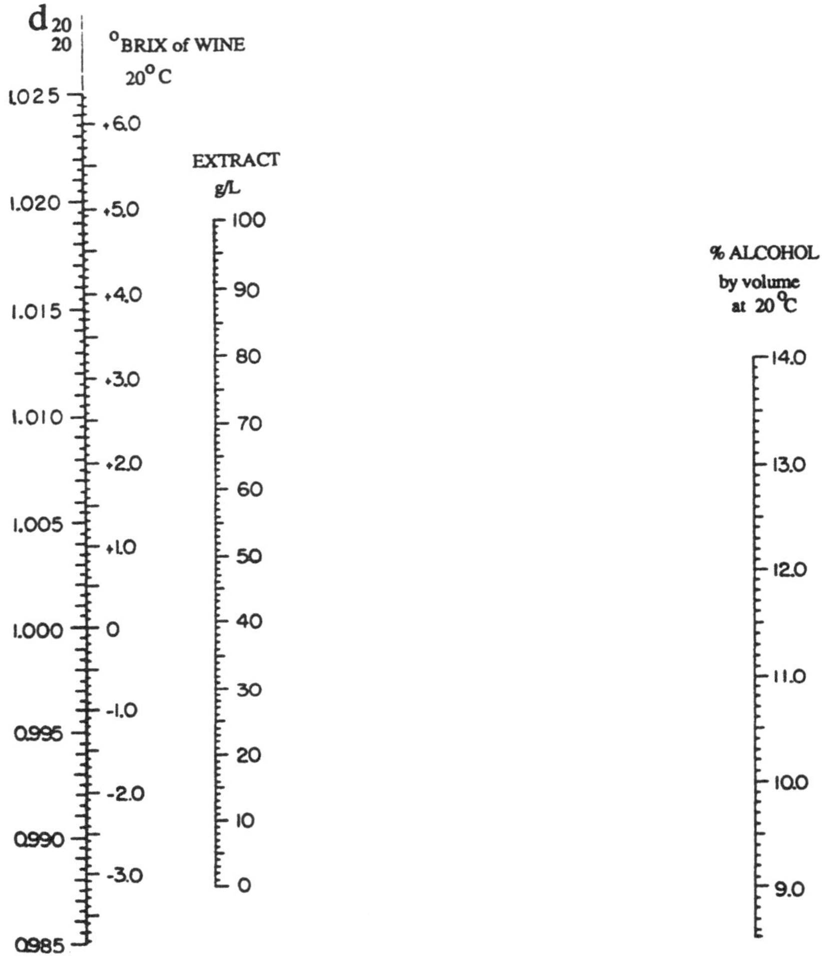

Fig. I–2. Nomograph for table wine after J.M. Vahl (1979). Measured values of specific gravity (20°/20°) and % ethanol are used to determine extract (g/L) values for table wines.

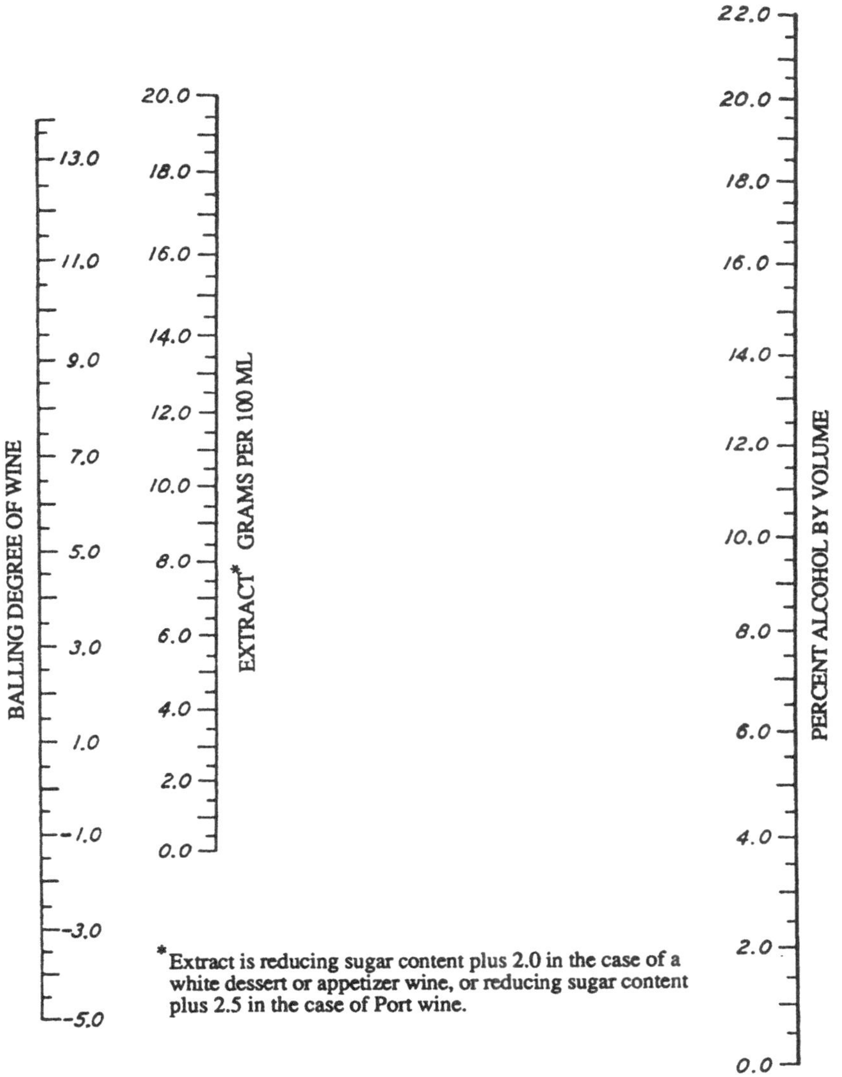

Fig. I-3. Marsh nomograph for determining extract values of sweet wines.

APPENDIX II. LABORATORY SAFETY

Modern laboratory practice requires the use of an extensive list of chemicals and reagents. Some of these materials are potentially dangerous, if mishandled. It is incumbent upon laboratory supervisory personnel to provide safety information to their employees, as well as instruction in the safe handling of such substances and the residues resulting from laboratory analyses. Chemical supply houses are required to send Material Safety Data Sheets (MSDS)with any chemical shipped to you. These must be kept in a binder and be available to laboratory personnel. Finally, laboratory management should be aware of federal, state, and local regulations regarding the safe disposal of all laboratory materials.

Table II–1 lists the chemicals used as reagents and in the various experimental preparations. The authors have attempted to compile important safety information regarding these chemicals. For some compounds, limits have not been set or the information was not available in the sources cited at the end of the table.

Name	OSHA PEL	ACGIH TLV	LD50 (mg/kg)	Detect odor
Acetaldehyde	200 ppm	100 ppm	1,930	0.21 ppm
Acetic acid	10 ppm	10 ppm	3,310	1.0 ppm
Acetone	1,000 ppm	750 ppm	9,750	140 ppm
Acetonitrile	40 ppm	40 ppm	2,730	40 ppm
Activated PVPP			12,000	
Amido black				
p-Aminobenzoic acid				
Ammonia (Ammonium hydroxide)	50 ppm	25 ppm	350	40 ppm
Ammonium chloride	10 mg/M^3	10 mg/M^3	1,650	
Ammonium molybdate	5 ppm	5 mg/M^3	333	
Ammonium sulfate			3,000	
Amyl acetate	100 ppm	100 ppm	6,500	0.08 ppm
Isoamyl alcohol	100 ppm		5,700	35 ppm
Analine	5 ppm	2 ppm	250	1 ppm
L-Arginine				
L-Arginine monohydrochloride				
Ascorbic acid			2,500	
Basic fuchin indicator				
Bentonite				
Benzoic acid			2,530	
Biotin				
Bromine	0.1 ppm	0.1 ppm	14	3.5 ppm
Bromocresol green			860	
Bromothymol blue				
n-Butanol	100 ppm skin	50 ppm skin	790	11 ppm
iso-Butanol	100 ppm	50 ppm	2,460	40 ppm
2-methyl-1-butanol			3,950	
Cadmium sulfate	200 μg/M^3 as Cd	50 μg/M^3 as Cd	7	
Calcium carbonate	15 mg/M^3	10 mg/M^3	1,000 (TLm 96)	
Calcium pantothenate			10,000	
Carbon disulfide	20 ppm	10 ppm	14 (LDLo)	1.2 ppm
Charcoal			40	
Citric acid			11,700	
Copper sulfate	1 ppm		300	
Crystal violet			420	
Cycloheximide			2	
Dextrose			25,800	
Diacetyl			1,580	
Diethyl carbonate				
Diethyl ether	400 ppm	400 ppm	1,215	0.83 ppm
Dipotassium phosphate				
Disodium EDTA			2,000	
Eosin Y				
Ethanol	1,000 ppm	1,000 ppm	7,060	10 ppm

Name	OSHA PEL	ACGIH TLV	LD50 (mg/kg)	Detect odor
Ethyl acetate	400 ppm	400 ppm	11,300	50 ppm
Ethyl methyl carbonate				
Ferrous ammonium sulfate	1 mg/M^3 (Fe)	1 mg/M^3 (Fe)	3,250	
Ferrous sulfate	1 mg/M^3	1,520		
Folic acid				
Formalin	3 ppm	1 ppm	800	
Formic acid		5 ppm	1,100	21 ppm
Fructose				
Gallic acid			5,000	
Gelatin				
Glucose				
Glycerol	15 mg/M^3	10 mg/M^3	12,600	
Hydrochloric acid	5 ppm	5 ppm ceiling	3,124 (LC50)	1–5 ppm
Hydrogen peroxide	1.4 mg/M^3	1.5 mg/M^3	6,220 (TDlo)	
Hydroxylamine hydrochloride			400	
Indigo carmine			93	
Inositol			400	
Iodine	0.1 ppm ceiling	0.1 ppm ceiling	14,000	
Iron	350 mg/M^3			
Iso-octane				
Linalool			2,790	
Lithium sulfate			1,190	
Litmus	0.2 mg/M^3	0.15 mg/M^3	13,000 (TDLo)	
Magnesium sulfate			5,000 (LDLo)	
Maltose			25,200	
Mercuric chloride		0.05 mg/M^3	10	
Methanol	200 ppm skin	200 ppm skin	5,628	2,000 ppm
Methylene blue				
Methyl cellosolve	25 ppm		2,460	60 ppm
Methyl red			12,000 (TDLo)	
Niacin			7,000	
Nigrosin				
Ninhydrin			250 (TDLo)	
Nitric acid	2 ppm	2 ppm	430 (LDLo)	
Paraffin		2 mg/M^3		
n-Pentyl alcohol			3,030	
Peptone				
Pyridoxine hydrochloride				
Petroleum ether	100 ppm	100 ppm	27 (LDLo)	
1,10-Phenthroline	100 ppm	100 ppm	132	
Phenolphthalein			500 (LDLo)	
Phloroglucinol			5,200	
Phosphoric acid	1 mg/M^3	1 ppm	1,530	
Ponceau S				

Name	OSHA PEL	ACGIH TLV	LD50 (mg/kg)	Detect odor
Potassium chloride			2,600	
Potassium ferrocyanide			1,600 (LDLo)	
Potassium hydrogen phthalate				
Potassium hydrogen tartrate				
Potassium hydroxide	2 mg/M^3	2 mg/M^3	365	
Potassium iodate			531 (LDLo)	
Potassium iodide			1,862 (LDLo)	
Potassium dichromate	0.1 mg/M^3	0.5 mg/M^3 as Cr	2,829 (LDLo)	
Potassium dihydrogen phosphate				
Potassium permanganate			1,090	
Potassium sorbate			420	
Potassium thiocyanate			854	
Proline				
n-Propanol	22 ppm		5,400	30 ppm
2-Propanol			5,840	
Raffinose				
Riboflavin			560	
Silver nitrate	0.01 mg/M^3 (Ag)	0.01 mg/M^3 (Ag)	50	
Sodium acetate			3,530	
Sodium benzoate			4,070	
Sodium bicarbonate			4,220	
Sodium bisulfite		5 mg/M^3	115	
Sodium borate	10 mg/M^3	5 mg/M^3	2,660	
Sodium carbonate			4,000 (LDLo)	
Sodium caseinate				
Sodium chloride			3,000	
Sodium diethyl dithiocarbamate			1,500	
Sodium hydroxide	2 mg/M^3	2 mg/M^3 ceiling	40	
Sodium metabisulfite		5 mg/M^3	115	
Sodium metavanadate			200 (LDLo)	
Sodium molybdate		5 mg/M^3 (Mo)	245 (LDLo)	
Sodium nitrate			200 (LDLo)	
Sodium potassium tartrate				
Sodium sulfite			115	
Sodium thiosulfate			>2,500	
Sodium tungstate			1,190	
Starch			6,600	
Sucrose	15 mg/M^3	10 mg/M^3	29,700	
Sulfosalicylic acid			2,450	

Name	OSHA PEL	ACGIH TLV	LD50 (mg/kg)	Detect odor
Sulfuric acid	1 mg/M^3	1 mg M^3	2,140	>1 mg/M^3
Syringaldazine				
Tannic acid			200	
Tartaric acid			485	
Thiamine			301	
Thiobarbituric acid			600	
Trichloroacetic acid		1 ppm	5,000	
2,4,6-Trinitro benzene sulfonic acid				
Tryptone				
Tween 80			1,790	
Vanillin			1,580	
Vaseline				
Zinc chloride	1 mg/M^3		350	

Note: Data are from the Occupational Health Guidelines for Chemical Hazards (NIOSH/OSHA); the Registry of Toxic Effects of Chemical Substances (RTECS) 1981–82 (NIOSH); SAFETY (The Sigma-Aldrich Library of Chemical Safety Data, Edition I, 1985); and Material Safety Data Sheets (MSDS) distributed by Aldrich Chemical Co., Fisher Scientific, J.T. Baker, Mallinckrodt, the U.S. Deptartment of Labor, Supelco, Bio-Rad Laboratory, Thiokol, EM Science, Sigma Chemical Co., Spectrum Chemical Manufacturing, Matheson Gas Products, Orion Research Inc., Great Western Chemical Co., Union Carbide, Liquid Carbonic Industries, Boehringer Mannheim Biochemicals, and US Biochemicals Corp. The odor concentrations are those most persons can detect under normal conditions. Permissible Exposure Limit values (PEL) are established by the Occupational Safety and Health Administration (OSHA) for allowed workplace concentrations that are expected not to cause harm to workers exposed during a working career. Threshold Limit Values (TLV) are established by the American Conference of Governmental Industrial Hygienists (ACGIH) and represent an exposure limit that will protect nearly everyone exposed to the chemical. When "Ceiling" appears in the ACGIH column, the ceiling value listed should not be exceeded. "M^3" means cubic meters, LD50 means lethal dose–50% and represents the lethal dose in mg/kg body weight of test animal for a population of test animals. "LDLo" means lowest dose that has caused death. "TDLo" means toxic dose, low, and is the lowest dose known to produce toxic effects.

BIBLIOGRAPHY

ABBOTT, N.A., COOMBE, G.G., and WILLIAMS, P.J. 1991. The contribution of hydrolyzed flavor precursors to quality differences in Shiraz juice and wines: an investigation by sensory descriptive analysis. *Am. J. Enol. and Vitic.* 42(3):167–69.

ACREE, T.E., SONOFF, E.P., and SPLITTSTOESSER, D.F. 1972. Effect of yeast strain and type of sulfur compounds on hydrogen sulfide production. *Am. J. Enol. Vitic.* 23:6–9.

ADKINSON, D., HOLLWARTH, M.E., BENOIT, J.N., PARKS, D.A., MCCORD, J.M., and GRANGER, D.N. 1986. Role of free radicals in ischemia-reperfusion injury to the liver. *Acta Physiol. Scand.* Suppl. 548:101–107.

AFANAS'EV, I.B., DOROZHKO, A.I., BRODSKII, A.V., KOSTYUK, V.A., and POTAPOVICH, A.I. 1990. Chelating and free radical scavenging mechanisms of inhibitory action of rutin and quercitin in lipid peroxidation. *Biochem. Pharmacol.* 39:1763–1769.

ALLEN, D.B. 1993. Balanced gas mixtures for winemaking. *Aust. Grapegrower and Winemaker* 352:69–71.

AMATI, A., ARFELLI, G., SIMONI, M., GANDINI, A., GERBI, V., TORTIA, C., and ZIRONI, R. 1992. Inibizione della fermentazione malolattica mediante il lisozima: Aspetti microbiologica e technologici. *Biologia Oggi* 1.2.

AMERICAN COLLOID COMPANY. 1983. Methylene blue determination of bentonite. Procedure No. 1016. American Colloid Co. Skokie, IL 60077.

AMERICAN COLLOID CO. "Microfine Bentonite" Skokie, IL 60077.

AMERICAN PUBLIC HEALTH ASSN., AM. WATER WORKS ASSN., WATER POLLUTION CONT. FED. 1989. *Standard Methods for the Examination of Water and Waste Water.* 17th ed. The Am. Public Health Assn., Washington, DC.

AMERINE, M.A. 1954. Composition of wines. I. Organic constituents. In *Advances in Food Research,* 5:354–466. New York: Academic Press.

AMERINE, M.A. 1958. Composition of wines. II. Inorganic constituents. In *Advances in Food Research,* 8:133–224. New York: Academic Press.

AMERINE, M.A. 1965. *Laboratory Procedures for Enologists.* Davis, CA: University of California, Department of Viticulture and Enology.

AMERINE, M.A. and CRUESS, W.V. 1960. *The Technology of Winemaking.* 2nd ed. Westport, CT: AVI Publishing Co.

AMERINE, M.A. and JOSLYN, M.A. 1970. *Table Wines, The Technology of Their Production.* 2nd ed. Berkeley, CA: The University of California Press.

AMERINE, M.A. and KUNKEE, R.E. 1968. The microbiology of winemaking. *Ann. Rev. Microbiol.* 22:323–358.

AMERINE, M.A. and OUGH, C.S. 1957. Studies on controlled fermentation. III. *Am. J. Enol. and Vitic.* 8:18–30.

AMERINE, M.A. and OUGH, C.S. 1974. *Wine and Must Analysis.* New York: John Wiley & Sons, Inc.

AMERINE, M.A. and OUGH, C.S. 1980. *Methods for Analysis of Musts and Wines.* New York: John Wiley & Sons, Inc.

AMERINE, M.A. and ROESSLER, E.B. 1958a. Methods for determining field maturity of grapes. *Am. J. Enol. and Vitic.* 9:37–40.

AMERINE, M.A. and ROESSLER, E.B. 1958b. Field testing of grape maturity. *Hilgardia* 28:93–114.

AMERINE, M.A. and ROESSLER, E.B. 1976. *Wines: Their Sensory Evaluation.* San Francisco: W.H. Freeman and Co.

AMERINE, M.A. and THOUKIS, G. 1958. The glucose-fructose ratio in California grapes. *Vitis* 1:224–229.

AMERINE, M.A. and WINKLER, A.J. 1940. Maturity studies with California grapes. I. Balling-acid ratio of wine grapes. *Proc. Am. Soc. Hort. Sci.* 38:379–397.

AMERINE, M.A. and WINKLER, A.J. 1942. Maturity studies with California grapes: II. The titratable acidity, pH and organic acid content. *Proc. Am. Soc. Hort. Sci.* 40:313–324.

AMERINE, M.A., BERG, H.W., and CRUESS, W.V. 1972. *The Technology of Winemaking.* 3rd ed. Westport. CT: AVI Publishing Co.

AMERINE, M.A., BERG, H.W., KUNKEE, R.E., OUGH, C.S., SINGLETON, V.L., and WEBB, A.D. 1981. *The Technology of Winemaking.* 4th ed. Westport, CT: AVI Publishing Co.

AMERINE, M.A., OUGH, C.S., and BAILEY, C.B. 1959a. Suggested color standards for wines. *Food Tech.* 13:170–175.

AMERINE, M.A., ROESSLER, E.B., and FILIPELLO, F. 1959b. Modern sensory methods of evaluating wines. *Hilgardia* 28:447–567.

AMERINE, M.A., ROESSLER, E.B., and OUGH, C.S. 1965. Acids and the acid taste. I. The effect of pH and titratable acidity. *Am. J. Enol. and Vitic.* 16:29–37.

AMES, B.M. 1983. Dietary carcinogens and anticarcinogens. *Science* 221:1256–1264.

AMES, B.M. 1989. Endogenous oxidative DNA damage, aging, and cancer. *Free Rad. Res. Commun.* 7:121–128.

AMES, B.N., SHIGENAGA, M.K. 1992. Oxidants are a major contributor to aging. *Ann. NY Acad. Sci.* 663:85–96.

AMES, B.N., SHIGENAGA, M.K., HAGEN, T.M. 1993. Oxidants, antioxidants, and the degenerative diseases of aging. *Proc. Natl. Acad. Sci. USA* 90:7915–7922.

AMES COMPANY. 1978. Dextrocheck Test for Reducing Sugar: Information Summary. Elkhart, IN: The Ames Company, A Division of Miles Laboratory, Inc.

AMON, J.M. and SIMPSON, R.F. 1986. Wine corks: a review of the incidence of cork related problems and the means of their avoidance. *Australian Grapegrower and Winemaker* 268:63–72.

AMON, J.M., VANDEPEER, J.M., SIMPSON, R.F. 1989. Compounds responsible for cork taint in wine. *Australian and New Zealand Wine Ind. J.* 4:62–69.

AN, D. and OUGH, C.S. 1993. Urea excretion and uptake by various wine yeasts as affected by various factors. *Am. J. Enol. Vitic.* 44:35–40.

ANDERSON, J.H. 1959. Carbon dioxide retention and academic theory. Part I. Discussion of theory. *Comm. of the Master Brewers Assn. of America.* 20(1,2):3–7.

ANDREWS, J.T., HEYMANN, H., and ELLERSIECK, M. 1990. Sensory and chemical analyses of Missouri Seyval blanc wines. *Am. J. Enol. Vitic.* 42(3):116–118.

ANELLI, G. 1977. The proteins of must. *Am. J. Enol. and Vitic.* 28:200–203.

ANON. 1990. Vinozym is changing art of winemaking. *Australian Grapegrower and Winemaker* 320:40–41.

ANON. 1993. Ethanol stimulates Apo A-1 secretion in human hepatocytes: a possible mechanism underlying the cardioprotective effect of ethanol. *Nutr. Rev.* 51:151–152.

ARCHER, T.E. and GAUER, W.O. 1979. Levels of cryolite on Thompson seedless grapes and raisins. *Am. J. Enol. Vitic.* 30:202–204.

ARENA, A. 1936. Alteraciones bacterianas de vinos Argentinos. *Rev. Facultad de Agric. y Vet. (Buenos Aires)* 8:155–315.

ARICHI, H., KIMURA, Y., OKUDA, H., BOBA, K., KOZAWA, M., and ARCHI, S. 1982. Effects of stilbene compounds on the roots of *Polygonum cuspedatum* on lipid metabolism. *Chem. Pharm. Bull.* 30:1766–1770.

ARIES, V. and KIRSOP, B.H. 1977. Sterol synthesis in relation to growth and fermentation by brewing yeasts inoculated at different concentrations. *J. Instit. Brew.* 83:220–223.

ARNOLD, R.A. and NOBLE, A.C. 1978. Bitterness and astringency of grape seed phenolics in model wine solutions. *Am. J. Enol. and Vitic.* 29:150–152.

ARNOLD, R.A., NOBLE, A.C., SINGLETON, V.L. 1980. *J. Agric. Food Chem.* 28:675–678.

ARUOMA, O.I. 1993. Free radicals and food. *Chem. Br.* 29:210–214.

AROUMA, O.I., MAURCIA, A., BUTLER, J., HALLIWELL, B. 1993. Evaluation of the antioxidant and prooxidant actions of gallic acid and its derivatives. *J. Agric. Food Chem.* 41:1880–1885.

ASAI, T. 1968. *Acetic Acid Bacteria: Classification and Biochemical Activities.* Tokyo: University of Tokyo Press.

ASHIDA, J., HIGASHI, N., and KIKUCHI, T. 1963. An electron microscopic study of copper precipitation by copper-resistant yeast cells. Protoplasma 57:27–32.

ASMUS, K.-D. 1990. Sulfur-centered free radicals. *Meth. Enzymol.* 186:168–180.

ASTM. 1968. *Manual on Sensory Testing Methods.* ASTM STP-434. Philadelphia, PA: The American Society for Testing and Materials.

ASTM. 1981. *Guidelines for the Selection and Training of Sensory Panel Members.* ASTM STP 758. Philadelphia, PA: The American Society for Testing and Materials.

ATLAS, R.M. and BARTHA, R. 1981. *Microbial Ecology: Fundamentals and Applications.* Reading, MA: Addison-Wesley Pub. Co.

AUERBACH, R.C. 1959. Sorbic acid as a preservative in wine. *Wine and Vines* 40(8): 26–28.

BABBS, C.F. 1990. Free radicals and the etiology of colon cancer. *Free Rad. Biol. Med.* 8:191–200.

BACHMANN, O. 1978. Verbreitung von Phenolcarbonsauren und Flavonoiden bei Vitaceen. *Vitis* 17:234–257.

BAKALINSKY, A.T. and BOULTON, R. 1985. The study of an immobilized acid protease for the treatment of wine proteins. *Am. J. Enol. Vitic.* 36:23–29.

BAKKER, J., BRIDLE, P., TIMBERLAKE, C.F., and ARNOLD,. 1986. The colours, pigments and phenol contents of young port wines: effects of cultivar, season and site. *Vitis* 25:40–52.

BAKKER, J., PRESTON, N.W., and TIMBERLAKE, C.F. 1986. The determination of anthocyanins in aging and wines: comparison of HPLC and spectral methods. *Am. J. Enol. Vitic.* 37:121–126.

BALAKIAN, S. and BERG, H.W. 1968. The role of polyphenols in the behavior of potassium bitartrate in red wines. *Am. J. Enol. and Vitic.* 19:91–100.

BALDWIN, G. 1993. Treatment and prevention of spoilage films on wines. *Australian Grapegrower and Winemaker.* 35:255–256.

BANDO, M. and OBAZAWA, H. 1988. Ascorbate free radical reductase and ascorbate redox cycle in the human lens. *Jpn. J. Ophthalmol.* 32:176–186.

BANDO, M. and OBAZAWA, H. 1990. Activities of ascorbate free radical reductase and H_2O_2-dependent NADH oxidation in senile cataractous human lenses. *Exp. Eye Res.* 50:779–784.

BATE-SMITH, E.C. 1954. Flavanoid compounds in foods. In *Advances in Food Research* 5:262–295. New York: Academic Press.

BATE-SMITH, E.C. and MORRIS, T.N. 1952. The nature of enzymatic browning. In *Food Science.* Cambridge MA: The University Press.

BATE-SMITH, E.C. and SWAN, T. 1953. Identification of leucoanthocyanins as "tannins" in foods. *Chem. and Ind.* 377.

BAUSCH & LOMB OPTICAL COMPANY. *Abbe-56 Refractometer Reference Manual.* New York: Bausch & Lomb Optical Company.

BAYLY, F.C. and BERG, H.W. 1967. Grape and wine proteins of white wine varietals. *Am. J. Enol. and Vitic.* 18:18–32.

BECKER, L.C. and AMBROSIO, G. 1987. Myocardial consequences of reperfusion. *Prog. Cardiovasc. Dis.* 30:23–44.

BECKMAN, J.S. and CROW, J.P. 1993. Pathological implications of nitric oxide, superoxide and peroxinitrite formation. *Biochem. Soc. Trans.* 21:330–334.

BECKMAN, J.S., BECKMAN, T.W., CHEN, J., MARSHALL, P.A., FREEMAN, B.A. 1990. Apparent hydroxyl radical production by peroxynitrite: implications for endothelial injury from nitric oxide and superoxide. *Proc. Natl. Acad. Sci. USA* 87: 1620–1624.

BECKMAN, J.S., ISCHIROPOULOS, H., ZHU, L., VAN DER WOERD, M., SMITH, C., CHEN, J., HARRISON, J., MARTIN, J.C., TSAI, M. 1992. Kinetics of superoxide dismutase- and iron-catalyzed nitration of phenolics by peroxynitrite. *Arch. Biochem. Biophys.* 298:438–445.

BECKWITH, J. 1936. Pure cultured yeast. *J. Dept. Agriculture S. Australia* 38:858–867.

BEECH, R.W., BURROUGHS, L.F., TIMBERLAKE, C.F., WHITING, G.C. 1979. Progres recents sur l'aspect chimique et l'actopm amto-microbienne de l'anhydride sulfureux. *Bull, O.I.V.* 586:1001–1022.

BEELMAN, R.B. 1982. Development and utilization of starter cultures to induce malolactic fermentations in red table wines. *Proceedings University of California, Davis, Grape Wine Centennial Symp.* A.D. Webb (ed.), pp 109–117. Dept. of Vitic, and Enology, University of Calif., Davis.

BEELMAN, R.B., KEEN, R.M., BANNER, M.J., and KING, S.W. 1982. Interactions between wine yeast and malolactic bacteria under wine conditions. *Dev. Ind. Microbiol.* 23:107–121.

BELL, T.A., ETCHELLIS, J.L., and BERG, A.F. 1958. *Influence of sorbic acid on growth of certain species of bacteria, yeast, and filamentous fungi.* Raleigh, NC: U.S. Food Laboratory, North Carolina Agr. Exp. Sta.

BELY, M., SABLAYROLLES, J.-M., and BARRE, P. 1990. Automatic detection of assimilable nitrogen deficiencies during alocholic fermentation in enological conditions. *J. Ferm. and Bioengineering* 70:246–252.

BELY, M., SABLAYROLLES, J.-M., and BARRE, P. 1991. Automatic detection and correction of assimilable nitrogen deficiency during alcoholic fermentation under enological conditions. In: *Proc. of the International Symposium on Nitrogen in Grapes and Wine* (Seattle, WA) Am. Soc. for Enol. and Vitic. J. Rantz (ed.). pp 211–214.

BENDA, I. 1984. Zum einflush von hefen auf den glukonsaure- gehalt des weines. *Wein Wissenschaft* 39:263–267.

Bendich, A., Machlin, L.J., Scandurra, O. 1986. The antioxidant role of vitamin C. *Adv. Free Rad. Biol. Med.* 2:419–444.

Bentley, R. 1990. The Shikimate pathway—a metabolic tree with many branches. *Crit. Rev. Biochem. Molec. Biol.* 25:307–384.

Bentley, R. and Meganathan, R. 1981. Geosmin and methylisoborneol biosynthesis in Streptomycetes. Evidence for an isoprenoid pathway and its absence in non-differentiating isolates. *FEBS Lett.* 125:220–222.

Berg, H.W. 1953. Special wine filtration procedures recommended for use following Cufex fining of wines. Wine Institute Tech. Advisory Committee.

Berg, H.W. 1959a. Investigations of defects in grapes delivered to California Wineries: 1958. *Am. J. Enol.* 10:61–69.

Berg, H.W. 1959b. The effects of several fungal pectic enzyme preparations on grape musts and wines. *Am. J. Enol. Vitic.* 10:130–134.

Berg, H.W. 1960. Stabilization studies on Spanish sherry and on factors influencing KHT precipitation. *Am. J. Enol. and Vitic.* 11:123–128.

Berg, H.W.1981. Personal communication.

Berg, H.W. and Akiyoshi, M. 1956a. Some factors involved in browning of white wines. *Am. J. Enol.* 7:1–7.

Berg, H.W. and M. Akiyoshi. 1956b. The effect of contact time of juice with pomace on the color and tannin content of red wines. *Am. J. Enol. Vitic.* 7:84–90.

Berg, H.W. and Akiyoshi, M. 1957. The effect of various must treatments on color and tannin content of red grape juices. *Food Res.* 22:373–383.

Berg, H.W. and Akiyoshi, M. 1961. Determination of protein stability in wine. *Am. J. Enol. and Vitic.* 12:107–110.

Berg, H.W. and Akiyoshi, M. 1962. Color behavior during fermentation and aging of wine. *Am. J. Enol. and Vitic.* 13:126–132.

Berg, H.W. and Akiyoshi, M. 1971. The utility of potassium bitartrate concentration product values in wine processing. *Am. J. Enol. and Vitic.* 22(3):127–134.

Berg, H.W. and Akiyoshi, M.A. 1975. On the nature of reactions responsible for color behavior in red wines: a hypothesis. *Am. J. Enol. and Vitic.* 26(3):134–143.

Berg, H.W. and Keefer, R.M. 1958. Analytical determination of tartrate stability in wine. I. Potassium bitartrate. *Am. J. Enol. and Vitic.* 9:180–193.

Berg, H.W. and Keefer, R.M. 1959. Analytical determination of tartrate stability in wine. II. Calcium tartrate. *Am. J. Enol. and Vitic.* 10:105–109.

Berg, H.W. and Marsh, G.L. 1954. Sampling deliveries of grapes on a representative basis. *Food Tech.* 9:104–108.

Berg, H.W., Akiyoski, M., and Amerine, M.A. 1979. Potassium and sodium content of California wines. *Am. J. Enol. and Vitic.* 30(1):55–57.

Berg, H.W., DeSoto, R.T., and Akiyoshi, M. 1968. The effect of refrigeration, bentonite clarification and ion exchange on potassium bitartrate behavior in wines. *Am. J. Enol. and Vitic.* 19:208–212.

BERG, H.W., FILIPELLO, F., HINREINER, E., and WEBB, A.D. 1955. Evaluation of thresholds and minimum difference concentrations for various constituents of wines: I. Water solutions of pure substances. *Food Tech.* 9:23–26.

BERGERET, J. 1963. Action de la gelatine et de la bentonite sur la couleur et l'astringence de quelques vins. *Ann. Technol. Agri. Paris* 12:15–25.

BERTRAND, A., PISSARD, R., SARRE, C. and SAPLS, J.C. 1976. Etude de l'influence de la des vins. *Connaissance Vigne Vin* 10:427–446.

BERTRAND, G.L., CARROLL, W.R., and FOLTYN, E.M. 1978. Tartrate stability of wines. I. Potassium complexes with pigments. *Am. J. Enol. and Vitic.* 29:25–29.

BEUTLER, E. 1989. Nutritional and metabolic aspects of glutathione. *Annu. Rev. Nutr.* 9:287–302.

BEZWODA, W.R., TORRANCE, J.D., BOTHWELL, T.H., MACPHAIL, A.P., GRAHAM, B., and MILLS, W. 1985. Iron absorption from red and white wines. *Scand. J. Haematol.* 34:121–127.

BIDAN, P. and ANDRE, L. 1958. Sur la composition en acides amines de quelques vins. *Ann. Technol. Agr.* 7:403–432.

BIELSKI, B.H.J., RICHTER, H.W., CHAN, P.C. 1975. Some properties of ascorbate free radical. *Ann. NY Acad. Sci.* 258:231–237.

BIOLETTI, F.T. and CRUESS, W.V. 1912. Enological investigations. *California Agr. Expt. Station Bull.* 230:1–118.

BISSON, L.F. 1993a. Personal communication. University of California, Davis.

BISSON, L.F. 1993b. *Alternatives to bentonite.* Davis, CA: Recent Adv. Vet. & Enol. Conf. (RAVE).

BLACKBURN, D. 1984. Malolactic bacteria: research and application. *Pract. Winery Vineyard* 5(3):44–48.

BLADE, H.W. and BOULTON, R. 1988. Adsorption of protein by bentonite in a model solution. *Am. J. Enol. Vitic.* 39:193–199.

BLEDSOE, A.M., KLIEWER, W.M., and MAROIS, J.J. 1988. Effects of timing and severity of leaf removal on yields and fruit composition of Sauvignon Blanc grape vines. *Am. J. Enol. Vitic.* 39:49–54.

BLONDIN, J., BARAGI, V., SCHWARTS, E.R., SADOWSKI, J.A., and TAYLOR, A. 1987. Dietary vitamin C delays UV-induced eye lens protein damage. *Ann. NY Acad. Sci.* 498:460–463.

BLOUIN, G., GUIMBERTEAU, G., and AUDOUIT, P. 1982. Prevention des precipitations tartriques dans le vin par le procede contact. *Connaissance Vigne Vin* 16(1): 63–67.

BOEHRINGER-MANNHEIM GMBH. 1980. *Acetic Acid. UV-Method for the Determination of Acetic Acid in Foodstuffs.* Indianapolis, IN.

BOHRINGER, P. 1960. Technologie der hefen. In *Die Hefen,* Nurenberg: Verlag, pp 157–270.

BOIDRON, J.N., CHATONNET, P., and PONS, M. 1988. Influence du bois sur certain substances odorants des vines. *Connaiss Vigne Vin.* 22:275–294.

BOILLOT, J.M. 1986. *Botrytis cinerea* in Chardonnay. Sym. Proc. Focus on Chardonnay. Sonoma-Cutrer Vineyards, Inc., pp 157–168.

BONASTRE, J. 1959. Contribution a l'etude des matieres minerales dans les produits vegetaux. Application au vin. *Ann. Technol. Agr.* 8:377–446.

BORGES, M. 1985. New trends in cork treatment and technology. *Beverage Rev.* 5(5):15–21.

BORS, W., HELLER, W., MICHEL, SARAN, M. 1990. Flavonoids as antioxidants: determination of radical scavenging efficiencies. *Meth. Enzymol.* 186:343–355.

BOULTON, R. 1980a. The relationship between total acidity, titratable acidity and pH in wines. *Am. J. Enol. and Vitic.* 31:76–80.

BOULTON, R. 1980b. The general relationship between potassium, sodium, and pH in grape juices and wines. *Am. J. Enol. and Vitic.* 31:182–186.

BOULTON, R. 1980c. A hypothesis for the presence, activity and role of potassium/hydrogen adenosine triphosphatases in grapevines. *Am. J. Enol. and Vitic.* 31: 283–287.

BOULTON, R. 1980d. The relationship between total acidity, titratable acidity and pH in grape tissue. *Vitis* 19:113–120.

BOULTON, R. 1980e. The nature of wine proteins. In: *Proc. of the Sixth Wine Industry Tech. Seminar.* Santa Rosa, CA: WITS, J.L. Jacobs, ed. pp 67–70.

BOULTON, R. 1981. Advances in Viticulture and Enology. University of California Extension Bulletin, 812E28.

BOULTON, R. 1982. Total acidity, titratable acidity and pH for winemakers and grape growers. In: *Proc. of the 1982 UCD Wine-Grape Day.*

BOULTON, R. 1983. The conductivity method or evaluating potassium bitartrate stability of wine. Part I. Enology Briefs 2(1):1–3. University of Calif. Extension. G. Cooke (ed.). Davis, CA.

BOULTON, R. 1985. Potassium balance in grapevines. *Pract. Winery and Vineyard* 5(5):40–41.

BOUSBOURAS, G.E. and KUNKEE, R.E. 1971. The effect of pH on malolactic fermentation in wine. *Am. J. Enol. and Vitic.* 22(3):121–126.

BOWLING, A.C., SCHULZ, J.B., BROWN, JR., R.H., BEAL, M.F. 1993. Superoxide dismutase activity, oxidative damage, and mitochondrial energy metabolism in familial and sporadic amyotrophic lateral sclerosis. *J. Neurochem.* 61:2322–2325.

BRADFORD, M.M. 1976. A rapid and sensitive method for quantification of microgram quantities of protein utilizing the principle of protein-dye binding. *Anal. Bio. Chem.* 72:248–249.

BRAVERMAN, J.B.S. 1953. The mechanism of interaction of sulfur dioxide with certain sugars. *J. Sci. Food Agr.* 4:540–547.

BRAVERMAN, J.B.S. 1963. *Introduction to the Biochemistry of Foods.* Amsterdam: Elsevier Publishing Co.

BRECHOT, P., CHAUVET, J., CROSON, M., and IRRMAN, R. 1966. Configuration optique de l'acide lactique apparu au cours de la fermentation malolactique pendant la vinification. *Comp. Rend. C.* 262:1605–1607.

BRECHOT, P., CHAUVET, J., DUPUY, P., CROSON, M., and RABATU, A. 1971. Acide oleanoique facteur de aroissance anaerobic de la levure du vin. *Cr. Acad. Sci.* 272:890–893.

BRENNER, M.W., OWADES, J.L., GUTCHO, M., and GOLYZNIAK, R. 1954. Determination of volatile sulfur compounds III. Determination of mercaptans. *Proc. Am. Soc. Brew. Chem.* 88–97.

BROOME, G., PAYNE, M., and SCIPIONE, J. 1985. Ultrafiltration: physical stability of wine. *Proceedings of the Australian Wine Research Institute,* Urrbrae, South Australia. Austr. Ind. Publishers.

BROWN, J.L. 1973. Mechanical dishwashing. In: *Current Concepts in Food Protection.* FDA, USDAHEW, Cincinnati, OH.

BRUNE, K. 1974. How aspirin might work: a pharmacokinetic approach. *Agents Actions* 4:230–232.

BRUNE, M., ROSSANDER, L., and HALLBERG, L. 1989. Iron absorption and phenolic compounds: importance of different phenolic structures. *Eur. J. Clin. Nutr.* 43: 547–558.

BRUSSONET, F. and MAUJEAN, A. 1993. Characterization of foaming proteins in Champagne base wines. *Am. J. Enol. Vitic.* 44:297–301.

BUCHANAN, R.E. and GIBBONS, N.E. 1984. *Bergey's Manual of Systematic Bacteriology, 9th ed.* J.G. Holt, ed. Baltimore, MD: The Williams and Wilkins Co.

BUECHENSTEIN, J.W. and OUGH, C.S. 1978. Sulfur dioxide determination by aeration-oxidation: a comparison with Ripper. *Am. J. Enol. and Vitic.* 29(3):161–164.

BULIT, J. and LAFON, R. 1970. Quelques aspects de la Biologie du *Botrytis cinerea* Per agent de la pourriture grise des raisins. *Connaissance Vigne Vin* 4:159–174.

BULLOCK, C. 1990. The biochemistry of alcohol metabolism. A brief review. *Biochem. Ed.* 18:62–66.

BULPITT, C.J. 1990. Vitamin C and blood pressure. *J. Hypertens.* 8:1071–1075.

BUNKER, V.W. 1992. Free radicals, antioxidants and aeging. *Med. Lab. Sci.* 49:299–312.

BURNETT, A.L., LOWENSTEIN, C.J., BREDT, D.S., CHANG, T.S.K., and SNYDER, S.H. 1992. Nitric oxide: a physiological mediator of penile erection. *Science* 257:401–403.

BURROUGHS, L.F. 1975. Determining free sulfur dioxide in red wines. *Am. J. Enol. and Vitic.* 26:25–29.

BURROUGHS, L.F. and WHITING, G.C. 1960. The sulfur dioxide combining power of cider. *Ann. Report Agr. Hort. Res. Station,* Long Ashton, Bristol, England, pp 144–147.

BUTLER, A.R. and WILLIAMS, D.L.H. 1993. The physiological role of nitric oxide. *Chem. Soc. Rev.* 22:233–242.

BUTLER, J. and HOEY, B.M. 1992. Reactions of glutathione and glutathione radicals with benzoquinones. *Free Rad. Biol. Med.* 12:337–345.

CABRERA, M.J., MORENO, J., ORTEGA, J.M., and MEDINA, M. 1988. Formation of ethanol, higher alcohols, esters and terpenes by five yeast strains in musts of Pedro Ximenez grapes at varying degrees of ripeness. *Am. J. Enol. and Vitic.* 39:283–287.

CADENAS, E. 1989. Biochemistry of oxygen toxicity. *Annu. Rev. Biochem.* 58:79–110.

CALIFORNIA DEPARTMENT OF FOOD AND AGRICULTURE. 1986. Wine grape inspection report.

CANAL-LLAUBERES, R.-M. 1993. Enzymes in wine making. In: *Wine Microbiology and Biotechnology*. Switzerland: Harwood Academic Publishers, pp 477–506.

CANO-MAROTA, C.R. and ARES PONS, J. 1961. Study of the malolactic fermentation in wines of Uruguay. *Rev. Soc. Quim. Mex.* 5:168.

CANTARELLI, C. 1963. Prevention des precipitations tartriques. *Ann. Technol. Agric.* 12:343–357.

CANTARELLI, C., BRENNA, O., GIOVANELLI, G., and ROSSI, M. 1989. Beverage stabilization through enzymatic removal of phenolics. *Food Biotechnology* 3:203–213.

CAPT, E. and HAMMEL, G. 1953. Le traitement des vins par l'acide carbonique. *Rev. romande Agr. Vitricul. Arboricult.* 9:41–43.

CAPUCCI, C. 1948. Observazone sulle prababite cause della unconsueta acidita volatile di alcuni vini delta regione Emiliano-Romagnola. *Riv. Viticolt. e Enol. (Conegliano)* 1:386–388.

CAPUTI, A. 1993. Ethyl carbamate/urease enzyme preparation. Reduction of urea in wine by the use of an acid urease. A compendium from the June 1993 Seminars.

CAPUTI, A., JR. and PETERSON, R.G. 1966. The browning problem of wines. *Am. J. Enol. and Vitic.* 16:9–13.

CAPUTI, A., JR., and UEDA, M. 1967. The determination of copper and iron in wine by atomic absorption spectrophotometry. *Am. J. Enol. and Vitic.* 18:66–70.

CAPUTI, A., JR., and WALKER, D.R. 1987. Titrimetric determination of carbon dioxide in wine: a collaborative study. *J.A.O.A.C.* 70(6):1060–1062.

CAPUTI, A., JR., UEDA, M., and BROWN, T. 1968. Spectrophotometric determination of ethanol in wine. *Am. J. Enol. and Vitic.* 19:160–165.

CAPUTI, A., JR., UEDA, M., and TROMBELLA, B. 1974. Determination of sorbic acid in wine. *J.A.O.A.C.* 57:951–953.

CARDWELL, T.J., CATTRALL, R.W., MRZLJAK, R.I. SWEENEY, T., ROBINS, L.M., and SCOLLARY, G.R. 1991. Determination of ionized and total calcium in white wine using a calcium ion-selective electrode. *Electroalysis* 3:575–576.

CARPENTER, C.E. and MAHONEY, A.W. 1992. Contributions of heme and nonheme iron to human nutrition. *Crit. Rev. Food Sci. Nutr.* 31:333–367.

CARTER, G.H., NAGEL, C.W., and CLORE, W.J. 1972. Grape sample preparation methods representative of must and wine analysis. *Am. J. Enol. Vitic.* 23(1):10–18.

CARTER, H.E. 1950. *Nitrogen Compounds of Yeasts. Proc. of the Symposium.* Champaign, IL: Garrard Press, pp 5–8.

CARTESIO, M.S. and CAMPOS, T.R. 1988. Malolactic fermentation in wine: improvement in paper chromatographic techniques. *Am. J. Enol. and Vitic.* 39(2):188–189.

CARTONI, G.P., COCCIOLI, F., and PONTELLI, L. 1991. Separation and identification of free phenolic acids in wines by high-performance liquid chromatography. *J. Chromatog.* 537:93–99.

CASINO, P.R., KILCOYNE, C.M., QUYYUMI, A.A., HOEG, J.M., and PANZA, J.A. 1993. The role of nitric oxide in endothelium-dependent vasodilation of hypercholesterolemic patients. *Circulation* 88:2541–2547.

CASTELLI, T. 1965. Ruolo della microbiologis nell'enolgia di oggi ed in quella di domani. *Atti. Accad. Ital. Vite Vino* 17:3–13.

CASTINO, M. 1965. L'azione riducente dell'acido ascorbico nella demetallizzazione dei vini con ferrocinauro potassico. *Ann. Accad. Ital. Viti. Vino* 17:143–151.

CASTINO, M., USSEGLIO-TOMASSET, L., and GANDINI, A. 1975. Factors which affect the spontaneous initiation of the malolactic fermentation in wines. The possibility of transmission by inoculation and its effect on organoleptic properties. In: *Lactic Acid Bacteria in Beverages and Foods.* J.G. Carr, C.V. Cutting, and G.C. Whiting (eds.), London: Academic Press, pp 139–148.

CASTONGUAY, A. 1993. Pulmonary carcinogenesis and its prevention by dietary polyphenolic compounds. *Ann. NY Acad. Sci.* 686:177–185.

CASTOR, J.G.B. and GUYMON, J.F. 1952. On the mechanism of formation of higher alcohols during alcoholic fermentation. *Science* 115:147–149.

CHATONNET, P. 1993a. The effects of heat on the composition of oak wood. *Int. Oak. Symp. Practical Winery and Vineyard.* San Francisco, CA. 1–30.

CHATONNET, P. 1993b. Barrel fermentation and aging of wines: research in Bordeaux in the 1990's. In: *Proc. Int. Oak Symp. Practical Winery and Vineyard.*

CHATONNET, P. and BOIDRON, J.N. 1988. Dosages de phenols volatils dans les vins par chromatographie en phase gazeuse. *Sci. des aliments* 8(4):479–488.

CHATONNET, P., DUBOURDIEU, D., BOIDRON, J.N., PONS, M. 1992. The origin of ethyl phenols in wines. *J. Sci. Food Agric.* 60:165–178.

CHATUNET, S., SUDRAUD, P., and JOUAN, T. 1986. La cryoextraction selective, premierres observations at perspectives. *Bulletin de l'office International der Vin* 667–668, 1021–1043.

CHEIDELIN, V.H. and KING, T.E. 1953. Nutrition of microorganisms. *Ann. Rev. Microbiol.* 7:113–142.

CHEYNIER, V., SOUQUET, J.M., and MOUTOUNET, M. 1989. Glutathione content and glutathione to hydroxycinnamic acid ratio in Vitis vinifera grapes and musts. *Am. J. Enol. and Vitic.* 40:320–324.

CHIBATA, I., TOSA, T., MORI, T., WATANABE, and SAKATA, N. 1986. Immobilized tannins—a novel adsorbent for protein and metal ion. *Enzyme Microbiol. Technol.* 8:130–136.

CHLEBEK, R.W. and LISTER, M.W. 1966. Ion pair effects in the reaction between potassium ferrocyanide and potassium persulfate. *Can. J. Chem.* 44:437–442.

CHOI, D.W. 1993. Nitric oxide: Foe or friend to the injured brain? *Proc. Natl. Acad. Sci. USA* 90:9741–9743.

CHOU, P.-T. and KHAN, A.U. 1983. L-ascorbic acid quenching of singlet delta molecular oxygen in aqueous media: generalized antioxidant property of vitamin C. *Biochem. Biophys. Res. Commun.* 115:932–937.

CHOW, H. and GUMP, B.H. 1987. Phosphorus in wine: comparison of atomic absorption spectrometry methods. *J.A.O.A.C.* 70(1):61–63.

CHURCH, D.F. and PRYOR, W.A. 1985. Free-radical chemistry of cigarette smoke and its toxicological implications. *Environ. Health Perspect.* 64:111–126.

CLARK, J.P., FUGELSANG, K.C., and GUMP, B.H. 1988. Factors affecting induced calcium tartrate precipitation from wine. *Am. J. Enol. and Vitic.* 39(2):155–161.

CLARK, J.P., GUMP, B.H., and FUGELSANG, K.C. 1985. Tartrate precipitation and stability. *Proc. of the Eleventh Wine Industry Tech. Symposium.* Santa Rosa, CA: WITS, J.L. Jacobs (ed.).

CLARK, W.G. and GEISSMAN, T.A. 1949. Potentiation of the effect of adrenaline by flavanoid (vitamin p-like) compounds. *J. Pharmacol. Exp. Ther.* 95:363.

COCHRANE, C.G. 1991. Mechanisms of oxidant injury to cells. *Molec. Aspects Med.* 12:137–147.

COFRAN, D.R. and MEYER, J. 1970. The effect of fumaric acid on malolactic fermentation. *Am. J. Enol. Vitic.* 21:189–192.

COHEN, S., TYRRELL, A.J., RUSSELL, M.A.H., JARVIS, M.J., SMITH, A.P. 1993. Smoking, alcohol consumption and susceptibility to the common cold. *Am. J. Publ. Health* 83:1277–1283.

COLLIER, J. and VALANCE, P. 1989. Second messenger role for NO widens to nervous and immune system. *Trends Pharmacol. Sci.* 10:426–431.

COMPORTI, M. 1989. Three models of free radical induced cell injury. *Chem. Biol. Interactions* 72:1–56.

CONE, C. 1987. Personal communication.

COOKE, G.M. and BERG, H. 1983. A re-examination of varietal table wine processing practices in California. I. Grape standards, grape and juice treatment and fermentation. *Am. J. Enol. and Vitic.* 34:249–256.

COOKE, G.M. and BERG, H.W. 1984. A re-examination of varietal table wine processing practices in California. II. Clarification, stabilization, aging and bottling. *Am. J. Enol. and Vitic.* 35(3):137–142.

COOKE, J.P. and TSAO, P.S. 1993. Cytoprotective effects of nitric oxide. *Circulation* (Edit.) 88:2451–2454.

COOPER, T.G. 1982. Nitrogen metabolism in *Saccharomyces cerevisiae*. In: *The Molecular Biology of Yeasts. Saccharomyces: metabolism and Gene Expression*. J. Strathern, E.W. Jones and J.R. Borach (eds.). Cold Springs Harbor Laboratory, Cold Springs, NY.

COOTES, R.L. 1983. Grape juice aroma and grape quality assessment used in vineyard classification. In: *Proc. of the Fifth Australian Wine Industry Tech Conf.* Adelaid. T.H. Lee and T.C. Somers, eds. Austr. Wine Research Instit. pp 275–292.

CORDONNIER, R. 1966. Etude des proteins et des substances azotees. Leur evolution au cours des traitements oenologique. Conditions de la stabilite proteique des vins. *Bull. O.I.V.* 39:1475–1489.

CORDONNIER, R. and DUGAL, A. 1968. Les activities proteolytiques du raisin. *Ann Technol Agricole* 17:189–206.

CORISON, C.A., OUGH, C.S., BERG, H., and NELSON, K.E. 1979. Must, acetic acid and ethyl acetate as mold and rot indicators in grapes. *Am. J. Enol. and Vitic.* 30(2): 130–134.

CORREA, I., POLO, M.C., AMIGO, L., and RAMOS, M. 1988. Separation des proteines des mouts de raison au moyen de techniques electrophoretiques. *Bull. O.I.V.* 39:1475–1489.

COWPER, E.P. 1987. Volatile Acidity in High Brix Fermentations. Masters Thesis, California State University, Fresno.

CRAWFORD, C. 1951. Calcium in dessert wines. In: *Proc. Am. Society Enol.* 76–69.

CREASY, L.L. and M. COFFEE. 1998. Phytoalexin production potential of grape berries. *J. Am. Soc. Hortic. Sci.* 113(2):230–234.

CRIPPEN, D.D. and MORRISON, J.C. 1986. The effects of sun exposure on compositional development of Cabernet Sauvignon berries. Am. J. Enol. Vitic. 37:235–242.

CROSBY, D.G. 1981. Environmental chemistry of pentachlorophenol. *Pure and Appl. Chem.* 53:1051–1080.

CROSS, C.E., HALLIWELL, B., BORISH, E.T., PRYOR, W.A., AMES, B.N., SAUL, R.L., MCCORD, J.M., AND Harman, D. 1987. Oxygen radicals and human disease. *Ann. Intern. Med.* 107:526–545.

CROTTY, B. 1994. Ulcerative colitis and xenobiotic metabolism. *Lancet* 343:35–38.

CROWELL, E.A. and GUYMON, J.F. 1975. Wine constituents arising from sorbic acid addition and identification of 2-ethoxyhexa-3,5-diene as a source of geranium-like odor. *Am. J. Enol. and Vitic.* 26:96–102.

CROWELL, E.A., OUGH, C.S., and BAKALINSKY, A. 1985. Determination of alpha amino nitrogen in musts and wines by TNBS method. *Am. J. Enol. and Vitic.* 36(2):175–177.

CRUESS, W.V. 1918. The fermentation organisms of California grapes. *Univ. California Publ. Agric. Sci.* 4(1):1–66.

CRUM, J.D. 1993. Chemistry of the constituents of Oak. *Int. Oak Symp. Practical Winery and Vineyard.* San Francisco, CA. pp. 1–10.

CULOTTA, E. and KOSHLAND, D.E. 1992. NO news is good news. *Science* 258:1862–1865.

CUMMINGS, M. 1994. Enhancement of wine aromas by enzymatic hydrolysis of glycosidic precursors from the grape. Third Annual WineTECH. San Jose, California.

CURVELO-GARCIA, A.S. 1987. Producto de solubilidade do tartarato de calcio em meios hidroalcoolicos ern funcao dos seus factores determinantes. *Ciencia Tec. Vitiv.* 6:19–28.

CUTLER, R.G. Antioxidants and aging. 1991. *Am. J. Clin. Nutr.* 53(Suppl.):373S–379S.

CUVELIER, M.-E., RICHARD, H., and BERSET, C. 1992. Comparison of the antioxidant activity of some acid-phenols: Structure-activity relationship. *Biosci. Biotech. Biochem.* 56:324–325.

DADIC, M. and BELLEAU, B. 1973. Polyphenolics and beer flavor. *Proc. Am. Soc. Brew. Chem.* pp 107–114.

DALY, N.M., LEE, T.H., and FLEET, G.H. (1984). Growth of fungi on wine corks and its contribution to corky taints in wine. *Food Tech. in Australia* 36:22–24.

DAUDT, C.E. and OUGH, C.S. 1973. A method for quantitative measurement of volatile acetate esters for wine. *Am. J. Enol. and Vitic.* 24(3):125–129.

DAVID, M.H., and KIRSOP, B.H. 1972. The varied response of brewing yeasts to oxygen and sterol treatment. *Proceedings of the Am. Soc. Brew. Chem.* 30:14–16.

DAVIS, C.R. and REEVES, M.J. 1988. Acid formation during fermentation and conservation of wine. *Proc. Sec. Int. Symp. Cool Climate Viticulture and Oenology.* R. Smart, R. Thornton, S. Rodriquez, and J. Young (eds.), pp 308–312.

DAVIS, C.R., FLEET, F.H., and LEE, T.H. 1981. The microflora of wine corks. *Australian Grapegrower and Winemaker* 208:42–44.

DAVIS, C.R., FLEET, G.H. and LEE, T.H. 1982. Inactivation of wine cork microflora by a commercial sulfur dioxide treatment. *Am. J. Enol. and Vitic.* 33:124–127.

DAVIS, C.R., WIBOW, D., ESCHENBRUCH, R., LEE, T.H., and FLEET, G.H. 1985. Practical implications of the malolactic fermentation: a review. *Am. J. Enol. and Vitic.* 36:290–301.

DAVIS, C.R., WIBOW, D., FLEET, G.H., and LEE, T.H. 1988. Properties of wine lactic bacteria. Their potential enological significance. *Am. J. Enol. and Vitic.* 39(2): 137–142.

DAVISON, C. 1971. Salicylate metabolism in man. *Ann. NY Acad. Sci* 179:249–268.

DE GAETANO, G., CELETTI, C., DJANA, E., and LATINI, R. 1985. Pharmacology of platelet inhibition in humans: implications of the salicylate-aspirin interaction. *Circulation* 72:1185–93.

DE LEY, J. 1958. Studies on the metabolism of *Acetobacter peroxydans.* Part I. General properties and taxonomic position of the species. *Antonie van Leeuwenhoek* 24: 281–297.

De Ley, J. 1961. Comparative carbohydrate metabolism and a proposal for a phylogenetic relationship of the acetic acid bacteria. *J. Gen. Microbiol.* 24:31–50.

De Ley, J. and Dochy, R. 1960. On the localization of oxidase systems in *Acetobacter* cells. *Biochim. Biophys. Acta* 40:277–289.

De Ley, J. and Schell, J. 1959. Oxiation of several substrates by *Acetobacter aceti. J. Bacteriol.* 77:445–451.

Delin, C.S. and Lee, T.H. 1992. Psychological concomitants of the moderate consumption of alcohol. *J. Wine Res.* 3:5–23.

DeMan, J.M. 1976. Enzymes. In: *Principles of Food Chemistry.* Westport, CT: AVI Publishing Co.

DeMora, S.J., Knowles, S.J., Eschenbruch, R., and Torrey, W.J. 1987. Dimethyl sulfide in some Australian red wines. *Vitis* 26:79–84.

De Rosa, T., Margheri, G., Moret, I., Scarponi, G., and Versini, G. 1983. Sorbic acid as a preservative in sparkling wine. Its efficacy and adverse flavor effect associated with ethyl sorbate formation. *Am. J. Enol. and Vitic.* 34(2):98–102.

DeSoto, R. and Warkentine, H. 1955. Influence of pH and total acidity on calcium tolerances of sherry wine. *Food Res.* 20:301–309.

DeSoto, R. and Yamada, H. 1963. Relationship of solubility products to long range tartrate stability. *Am. J. Enol. and Vitic.* 14:43–51.

Desrosier, N.W. 1963. *The Technology of Food Preservation.* Westport CT: AVI Publishing Co.

Desrosier, N.W. and Desrosier, J.N. 1977. *The Technology of Food Preservation.* Westport, Conn: AVI Publishing Co., Inc.

DeVilliers, J.P. 1961. The control of browning in white table wines. *Am. J. Enol. and Vitic.* 12:25–30.

Difco Laboratories, 1953. *Manual of Dehydrated Culture Media and Reagents for Microbiological and Clinical Laboratory Procedure.* 9th ed. Detroit, MI.

Dimitriadis, E. and Bruer, D.R.G. 1984. Using the Markham still for the assay of terpene flavorants in grapes. *Australian Grape Grower and Winemaker* Apr:61–64.

Dimitriadis, E. and Williams, P.J. 1984. The development and use of a rapid analytical technique for estimation of free and potentially volatile monoterpene flavorants of grapes. *Am. J. Enol. and Vitic.* 35(2):66–71.

Di Stefano, R. 1981. Terpene compounds of white muscat from Piemonte. *Vini Ital.* 23:29–43.

Dittrich, H.H. 1979. Anwendung von Trockenhefen bei der Weinbereitung. *Dtsch. Weinbau* 14:792–796.

Dittrich, H.H. 1987. *Mikrobiologie des Weines.* Stuttgart:Ulmer.

Dittrich, H.H. and Kerner, E. 1974. Diacetyl als Weinfehler, Ursache und Beseitigung des Milchsauretones. *Wein-Wissen.* 19:525–538.

Dittrich, H.H. and Sponholz, W.R. 1975. Die Aminosaureabnahme in *Botrytis*-infizierten Traubenbeeren und die Bildung hoherer Alkohole in diesen Mostein bei ihrer Vergarung. *Wein-Wissen.* 30:188–211.

DITTRICH, H.H., SPONHOLZ, W.R., GOBEL, H.G. 1975. Vergleichende Untersuchungen von Mosten und Weinen aus gesunden und aus *Botrytis*-infizierten Traubenbeeren II. Modellversuche zur Veranderung des Mostes durch *Botrytis*-Infektion und ihre Konsequenzen fur die Nebenproduktbildung bei der Garung. *Vitis* 13:336–347.

DITTRICH, H.H., SPONHOLZ, W.R., KAST, W. 1974. Vergleichende Untersuchungen von Mosten und Weinen aus gesunden und aus *Botrytis*-infizierten Traubenberren. I. Saurestoffwechsel, Zuckerstofwechselprodukte, Leucoanthocyangehalte. *Vitis* 13:36–49.

DITTRICH, H.H. and STAUDENMAYER, T. 1968. SO_2-Bildung, Bockserbildung und Bockserbeseitigung. *Deutsche Wein-Zeitung* 24:707–709.

DOELLE, H.W. 1975. *Bacterial Metabolism,* 2nd ed. New York: Academic Press.

DONECHE, B. 1990. Metabolisme de l'acide tartrique du raisin par Botrytis cinerea. Premiers resultats. *Science de Alements* 10:589–602.

DONECHE, B. 1993. Botrytized Weines. In: *Wine Microbiology and Biotechnology*. G.H. Fleet (ed.). Philadelphia, PA: Harwood Academic Publishers, pp 327–352.

DOURNEL, J.M. 1985. Recherches sur les combinaisons anthocyanes-flavonols. Influence de ces reactions sur la couleur dir vin rouge. Bordeaux: Universite' de Bordeaux II. These Doctorat oenologie.

DOZON, N.M. and NOBLE, A.C. 1989. Sensory study of the effect of fluorescent light on a sparkling wine and its base wine. *Am. J. Enol. and Vitic.* 40(4):265–267.

DRAWERT, F., HEIMANN, W., EMBERGER, R., and TRESSL, R. 1966. Uber die Biogenese von Aromastoffen bei Pflanzen und Fruchten. II. Enzymatische Bildung von Hexen-2-al-1 Hexanal und deren Vorstuen. *Liebigs Ann. Chem.* 694:200–202.

DRAWERT, F., TRESSL, R., HEIMANN, W., EMBERGER, R., and SPECK, M. 1973. Uber die Biogenese von Aromastoffen bei Pflanzen und Fructen. XV. Enzymatische-oxydative Bildung von C-6-Aldehyden und Alkoholen un deren Vorstufen bei Apfeln und Trauben. *Chem. Mikrobiol. Technol. Lebensm.* 2:10–14.

DRYSDALE, G.S. and FLEET, G.H. 1985. Acetic acid bacteria in some Australian wines. *Food Tech. Aust.* 37:17–20.

DRYSDALE, G.S. and FLEET, G.H. 1988. Acetic acid bacteria in winemaking: a review. *Am. J. Enol. and Vitic.* 39(2):143–154.

DRYSDALE, G.S. and FLEET, G.H. 1989. The growth and survival of acetic acid bacteria in wines at different concentrations of oxygen. *Am. J. Enol. and Vitic.* 40(2):99–105.

DUBERNET, M., BERTRAND, A., BIBEREAU-GAYON, P. 1974. Presence constante dans les vins d'erythritol, d'arabitol et de mannitol. *CR Acad. Sci. Paris Serie D* 279: 1561–1564.

DUBERNET, M., RIBEREAU-GAYON, P., LERNER, H.R., HAREL, E., and MAYER, A.M. 1977. Purification and properties of laccase from Botrytis cinerea. *Phytochemistry* 16:191–193.

DUBOIS, P. and DEKIMPE, J. 1982. Constituants volatils odorants des vins de Bourgogne éléves en fûts de chêne neufs. *Rev. Fr. Oenol.* 88:51–53.

DUBOURDIEU, D. 1978. Etude des polysaccharides secretes par *Botrytis cinerea* dans la baie de raisin. Incidence sur les difficultes de clarification des vins de vendages pourries. These de Doctorat, Univ. de Bordeaux II, France.

DUBOURDIEU, D. 1982. Recherches sur les polysaccharides secretes par *Botrytis cinerea* dans la baie de rasin. These de Doctrorat d'Etat, Univ. de Bordeaux II, France.

DUBOURDIEU, D. and CANAL-LLAUBERES, R.M. 1989. Influence of some colloids (polysaccharides and proteins on the clarification and stabilization of wines. In: *Proc. Seventh Aust. Wine Industry Tech. Conf.* P.J. Williams, D.M. Davidson, and T.H. Lee (eds.). pp 180–185.

DUBOURDIEU, D., GRASSIN, C., DERUCHE, C., and RIBEREAU-GAYON, P. 1984. Mise au point d'une mesure rapide de l'activite laccase, dans les mouts et dans les vins par methode a la syringaldazine. Application a l'appreciation de l'estat sanitaire es vendages. *Connaissance Vigne Vin* 18(4):237–252.

DUBOURDIEU, D., RIBEREAU-GAYON, P., and FOURNET, B. 1981. Structure of the extracellular B-D-glucan from *Botrytis cinerea. Carbohydrate Res.* 3:294–299.

DUERR, P. 1985. Wine quality evaluation. In: *Proceedings of the International Symposium on Cool Climate Viticulture and Enology.* D.A. Heatherbell, P.B. Lombard, F.W. Bodyfelt and S.F. Price (eds.). Oregon State Univ. pp 257–66.

DUNSFORD, P. 1979. The kinetics of potassium bitartrate crystallization from wine. Master's Thesis. University of California, Davis.

DUNSFORD, P. and BOULTON, R. 1981a. The kinetics of potassium bitartrate crystallization from table wines. I. Effect of particle size, particle surface area, and agitation. *Am. J. Enol. and Vitic.* 32:100–105.

DUNSFORD, P. and BOULTON, R. 1981b. The kinetics of potassium bitartrate crystallization from table wines. II. Effect of temperatures and cultivar. *Am. J. Enol. and Vitic.* 32:106–110.

DUPLESSIS, C.S. and DEWET, P. 1968a. Browning in white wines. I. Time and temperature effects upon tannin and leucoanthocyanidin uptake by musts from seeds and bushes. *S. Afr. J. Agric. Sci.* 11:459–468.

DUPLESSIS, C.S. and UYS, A.L. 1968b. Browning in white wines. II. The effect of cultivar, fermentation, husk, seed and stem contact upon browning. *S. Afr. J. Agric. Sci.* 11:637–640.

DUPUY, P. 1957a. Les facteurs du development de l'acescence dans le vin. *Ann. Technol. Agric.* 6:391–407.

DUPUY, P. 1957b. Les *Acetobacter* du vin identification de quelques souches. *Ann. Technol. Agric.* 6:217–233.

DUPUY, P. and MAUGENET, J. 1962. Oxidation de l'ethanol par *Acetobacter rancens. Ann. Technol. Agric.* 11:219–225.

DUPUY, P. and MAUGENET, J. 1963. Metabolism de l'acid lactique par *Acetobacter rancens. Ann. Technol. Agric.* 12:5–14.

DUPUY, P., NORTZ, M., and PUISAIS, J. 1955. Le vin et quelques causes de son enrichissement en fer. *Ann. Technol. Agric.* 4:101–112.

EDINGER, W.D. and SPLITTSTOESSER, D.F. 1986. Production by lactic acid bacteria of sorbic alcohol, the precursor of geranium odor compound. *Am. J. Enol. and Vitic.* 37:34–38.

EDWARDS, M.A. and AMERINE, M.A. 1977. Lead content of wines determined by atomic absorption spectrophotometry using flameless atomization. *Am. J. Enol. Vitic.* 29:239–240.

ENGLISH, J.T., BLEDSOE, A.M., MAROIS, J.J., and KLIEWER, W.M. 1990. Influence of grapevine canopy management on the evaporative potential in the fruit zone. *Am. J. Enol. and Vitic.* 41:137–141.

ENKELMANN, R. 1989. Spurenelement-Abgabe von Weinbehandlungsmittein. 2. Mitteilung: Aktivkohle. *Dtsch. Lebensm.-Rundsch.* 85:44–50.

ENKELMANN, R. 1990. Spurenelement-Abgabe von Weinbehandlungsmittein. 4. Mitteilung: Kieselgur. *Dtsch. Lebensm.-Rundsch.* 86:314–321.

ESAU, P. 1967. Pentoses in wine. I. Survey of possible sources. *Am. J. Enol. and Vitic.* 18(4):210–216.

ESAU, P. and AMERINE, M.A. 1964. Residual sugar in wine. *Am. J. Enol. and Vitic.* 15:187–189.

ESAU, P. and AMERINE, M.A. 1966. Quantitative estimation of residual sugar in wine. *Am. J. Enol. and Vitic.* 17:265–267.

ESCHENBRUCH, R. 1974. Sulfite and sulfide formation during winemaking—a review. *Am. J. Enol. and Vitic.* 25(3):157–161.

ESCHENBRUCH, R. 1983. Hydrogen Sulfide Formation—The Continuing Problem During Winemaking Fermentation Technology. *Australian Society of Viticulture and Oenology Proceedings.* T.H. Lee (ed.), pp 79–87. Glen Osmond, SA.

ESCHENBRUCH, R. and BONISH, P. 1976. Influence of pH on sulfite formation by yeasts. *Arch. Microbiol.* 107:229–231.

ESCHENBRUCH, R. and DITTRICH, H.H. 1986. Metabolism of acetic acid bacteria in relation to their importance in wine quality. *Zentralblatt fur Mikrobiol.* 141:279–289.

ESCHENBRUCH, R. and KLEYNHANS, P.H. 1974. The influence of copper-containing fungicides on the copper content of grape juice and on hydrogen sulphide formation. *Vitis* 12:320–324.

ESCHENBRUCH, R., BONISH, P., and FISHER, B.M. 1978. Production of hydrogen sulfide by pure culture wine yeasts. *Vitis* 17:67–74.

ESTERBAUER, H., PUHL, H., DIEBER-ROTHANEDER, M., WAEG, G., and RABL, H. 1991. Effects of antioxidants on oxidative modification of LDL. *Ann. Med.* 23: 573–581.

ETHIRAJ, S. and SURESH, E.R. 1988. *Pichia membranefaciens:* a benzoate resistant yeast from spoiled mango pulp. *J. Food Sci. Tech.* 25(2):63–66.

EWART, A.J.W. 1984. A study of cold stability in Australian white table wines. *Australian Grape Grower and Winemaker.* April: 104–107.

EWART, A.J.W. 1986. Polysaccharide instability in wine. In: *Physical Instability of Wine.* T.H. Lee (ed.). Australian Soc. of Vitic. and Enology, Glen Osmond, SA. pp 99–108.

EWART, A.J.W., HASELGROVE, N.J., SITTERS, J.H., and YOUNG, R. 1989. The effect of *Botrytis cinerea* on the color of *Vitis vinifera* CV Pinot Noir. In: *Proc. of the Seventh Australian Wine Industry Technical Conference.* Adelaide. P.J. Williams, D.M. Davidson, and T.H. Lee (eds.), Austr. Wine Res. Instit. pp 229–230.

FAIRLEY, J.L. and KILGOUR, G.L. 1966. *Essentials of Biological Chemistry,* 2nd ed. New York: Reinhold Publishing Co.

FAMUYIWA, O.O. and OUGH, C.S. 1991. Modification of acid urease activity by fluoride ions and malic acid in wine. *Am. J. Enol. and Vitic.* 42(1):79–80.

FEENEY, L. and BERMAN, E.R. 1976. Oxygen toxicity: membrane damage by free radicals. *Invest. Ophthalmol.* 15:789–792.

FELIX, R. and VILLETTAZ, J.C. 1983. Wine. In: *Industrial Enzymology.* New York: The Nature Press, pp 410–421.

FEMS MICROBIOLOGICAL LETTERS, FEB. 1988. Abstracted in: *The Australian Wine Res. Inst. Tech. Rev.* #52. T.H. Lee (ed).

FERENCZY, S. 1966. Etude des proteines et des substances azotees. Leur evolution au cours des traitements oenologiques. Conditions de la stabilite proteique des vins. *Bull. O.I.V.* 39:1311–1336.

FERRARI, G. and FEUILLAT, M. 1988. L'elevage sur lies des vins blancs de Bourgogne. l'ere Partie: etude des composes azotes, des acids gras et analyse sensorielle des vins. *Vitis* 27:183–193.

FESSLER, J.H. 1961. Erythorbic acid and ascorbic acid as antioxidants in bottled wines. *Am. J. Enol. and Vitic.* 12(1):20–24.

FEUILLAT, M. 1987. Expose sur la vinification en rouge en Bourgogne. *Le Vigneron Champenois* 6:340–352.

FITZPATRICK, D.F., HIRSCHFIELD, S.L., and COFFEY, R.G. 1993. Endothelium-dependent vasorelaxing activity of wine and other grape products. *Am. J. Physiol.* 265:H774–H778.

FLANZY, M. and DUBOIS, P. 1964. Etude du dosage des tocophenols totaux. Application a l'huile de pepins de raisin. *Ann. Technol. Agric.* 13:67–75.

FLANZY, M. and POUX, C. 1960. Application spectrophotometrique a l'etude de la casse oxydasique des vins. *Compt. Rend.* 251(18):1910–1911.

FLANZY, C., POUX, C., and FLANZY, M. 1964. Les levures alcooliques dans les vins. Proteolyse et proteogenese. *Ann. Technol. Agric.* 13:283–300.

FLEET, G.H. 1984. The physiology and metabolism of wine lactic acid bacteria. In: *Malolactic Fermentation.* The Australian Society of Viticulture and Oenology, Inc.

FLORENZANO, G. 1949. La microflora blastomiceta de mostre e dei vini di alcune zona Toscane. *Ann. Sper. Agrar.* 3:887–918.

FLORES, J.H., HEATHERBELL, D.A., HENDERSON, L.A., and MCDANIEL, M.R. 1991. Ultrafiltration of wine: effect of ultrafiltration on the aroma and flavor characteristics of White Riesling and Gewurtztraminer wines. *Am. J. Enol. and Vitic.* 42(2):91–93.

FLORES, J.H., HEATHERBELL, D.A., and MCDANIEL, M.R. 1990. Ultrafiltration of wine: effect of ultrafiltration on White Riesling and Gewurztraaminer wine composition and stability. *Am. J. Enol. and Vitic.* 41:207–214.

FORNACHON, J.C.M. 1943. Bacterial Spoilage of Fortified Wines. Adelaide: Australian Wine Board.

FORNACHON, J.C.M. 1957. The occurrence of malolactic fermentation in Australian wines. *Aust. J. Appl. Sci.* 8:120–129.

FORNACHON, J.C.M. 1963. Inhibition of certain lactic acid bacteria by free and bound sulfur dioxide. *J. Sci. Food and Agric.* 14:857–862.

FORNACHON, J.C.M. 1964. A *Leuconostoc* causing malolactic fermentation in Australian wines. *Am. J. Enol. and Vitic.* 15:184–186.

FORNACHON, J.C.M. 1965. Sulfur dioxide in winemaking. *Australian Wine Brew. Spirits Rev.* May:20–26.

FRANCIS, I.L., SEFTON, M.A., and WILLIAMS, P.J. 1992a. A study of sensory descriptive analysis of the effects of oak origin, seasoning and heating on the aromas of oak model wine extracts. *Am. J. Enol. and Vitic.* 43:23–30.

FRANCIS, I.L., SEFTON, M.A., and WILLIAMS, P.J. 1992b. Sensory descriptive analysis of hydrolyzed precursor fractions from Semillion, Chardonnay and Sauvignon Blanc grape juices. *J. Sci. Food Agric.* 59:511–520.

FRANCIS, I.L., SEFTON, M.A., and WILLIAMS, P.J. 1994. The sensory effects of pre- and post-fermentation thermal processing on Chardonnay and Semillion Wines. *Am. J. Enol. Vitic.* 45(2):243–251.

FRANKEL, E.N., WATERHOUSE, A.L., and KINSELLA, J.E. 1993. Inhibition of human LDL oxidation by resveratrol. *Lancet* (Lett.) 341:1103–1104.

FREEMAN, B.M. 1982. Regulation of potassium in grape vines (*Vitis vinifera*). Ph.D Thesis, University of California, Davis.

FREEMAN, B.M. 1983. Effects of irrigation and pruning of Shiraz grapevines on subsequent wine pigments. *Am. J. Enol. and Vitic.* 34:23–26.

FREEMAN, B.M. and KLIEWER, W.M. 1983. Effect of irrigation, crop level and potassium fertilization on Carnignane vines. II. Grape and wine quality. *Am. J. Enol. and Vitic.* 34:197–207.

FREI, B., ENGLAND, L., and AMES, B.N. 1989. Ascorbate is an outstanding antioxidant in human blood plasma. *Proc. Natl. Acad. Sci. USA* 86:6377–6381.

FREISLEBEN, H-J., and PACKER, L. 1993. Free-radical scavenging activities, interactions and recycling of antioxidants. *Biochem. Soc. Trans.* 21:325–330.

FRIDOVICH, I. 1986. Biological effects of superoxide radical. *Arch. Biochem. Biophys.* 247:1–11.

FRIEDMAN, G.D. and KLATSKY, A.L. 1993. Is alcohol good for your health? *N. Engl. J. Med.* (Lett.) 329:1882–1883.

FRITZ, J.S. and SCHENK, G.H., JR. 1974. *Quantitative Analytical Chemistry,* 3rd ed. Boston: Allyn and Bacon, Inc.

FRUTON, J.S. and SIMMONDS, S. 1961. *General Biochemistry.* New York: John Wiley & Sons, p 472.

FUGELSANG, K.C. 1987. Utilization of yeast starters in winemaking. *Practical Winery and Vineyard* July–Aug.:28–32.

FUGELSANG, K.C. and CALLAWAY, D. 1995. Cork quality assurance, a guide to sampling. How much is enough? *Practical Winery and Vineyard* Jan./Feb.:In press.

FUGELSANG, K.C. and ZOECKLEIN, B.W. 1993. MLF Survey. *Pract. Winery and Vineyard* May/June:12–19.

FUGELSANG, K.C., OSBORN, M.M., and MULLER, C.J. 1992. Involvements of *Brettanomyces* sp. in Winemaking. *Wine Ind. Tech. Symp.* Jan 1991. Rohnert Park, CA.

FUGELSANG, K.C., OSBORN, M.M., and MULLER, C.J. 1993. *Brettanomyces* and *Dekkera:* Implications in Winemaking. In: *Beer and Wine Production: Analysis, Characterization and Technological Advances.* B.H. Gump (ed.). Am. Chem. Society Symposium Series 536. ACS, Washington, DC. pp 110–131.

FUGELSANG, K.C., OSBORN, M.M., WAHLSTROM, V.E., and MULLER, C.J. 1994. Effect of fluoride on several strains of *Saccharomyces cerevisiae.* Presented at the Ann. Meeting of the Am. Soc. for Enol. Vitic. Anaheim, CA.

FUMI, M.D. and COLAGRANDE, O. 1988. Inactivation thermique de la microflore dans les bouchons de liege. *Rev. Oenologues* 50:28–30.

FUMI, M.D., TRIOLI, G., COLOMBI, M.G., and COLAGRANDE, O. 1988. Immobilization of *Saccharomyces cerevisiae* in calcium alginate gel and its application to bottle-fermented sparkling wine production. *Am. J. Enol. and Vitic.* 39(4):267–272.

FURCHGOTT, R.F. 1984. The role of endothelium in the responses of vascular smooth muscle to drugs. *Annu. Rev. Pharmacol. Toxicol.* 24:175–197.

FURCHGOTT, R.F. 1988. Studies on relaxation of rabbit aorta by sodium nitrite: the basis for the proposal that the acid-activatable inhibitory factor from retractor penis is inorganic nitrite and the endothelium relaxing factor is nitric oxide. In: *Vasodilation: Vascular Smooth Muscle, Peptides, Autonomic Nerves and Endothelium,* P.M. Vanhoutte (ed), New York: Raven Press, pp 401–414.

GABOURY, J., WOODMAN, R.C., GRANGER, D.N., REINHARDT, P., and KUBES, P. 1993. Nitric oxide prevents leukocyte adherence: role of superoxide. *Am. J. Physiol.* 265:H862–H867.

GAHAGAN, R. 1988. Personal communication.

GALLA, H.-J. 1993. Nitric Oxide, NO, an intercellular messenger. *Angew. Chem. Engl. Edtn.* 32:378–380.

GALLANDER, J. 1977. Deacidification of eastern table wines with *Schizosaccharomyces pombe. Am. J. Enol. and Vitic.* 28(2):65–68.

GANCEDO, C., GANCEDO, J.M., and SOLS, A. 1968. Glycerol metabolism in yeasts. Pathways of utilization and production. *Eur. J. Biochem.* 5:165–172.

GARHWAITE, J., CHARLES, S.L., and CHESS-WILLIAMS, R. 1988. Endothelium-derived relaxing factor release on activation of NMDA receptor suggests role as intracellular messenger in the brain. *Nature* 336:385–388.

GAZIANO, J.M., BURING, J.E., BRESLOW, J.L., GOLDHABER, S.Z., ROSNER, B., VAN-DENBURGH, M., WILLET, W., and HENNEKENS, C.H. 1993. Moderate alcohol intake, increased levels of high-density lipoprotein and its subfractions, and decreased risk of myocardial infarction. *N. Engl. J. Med.* 329:1829–1834.

GEHMAN, H. and OSMAN, E.M. 1954. The chemistry of the sugar-sulfite reaction and its relationship to food problems. In: *Adv. Food Res.* 5:53–91, New York: Academic Press.

GELVAN, D., SALTMAN, P., and POWELL, S.R. 1991. Cardiac reperfusion damage prevented by nitroxide free radical. *Proc. Natl. Acad. Sci. USA* 88:4680–4684.

GENERAL ANILIN AND FILM CORP. 1975. Polyclar AT Stabilizer in Winemaking. Bulletin 9653–003 New York.

GERBER, N.N. 1979. Volatile substances from Actinomycetes: their role in the odor pollution of water. *CRC Crit. Rev. Microbiol.* 7:191–214.

GESTRELIUM, S. 1982. Potential application of immobilized viable cells in the food industry: malolactic fermentation in wine. *Enzyme Eng.* 6:245–250.

GETTLER, A.O. and GOLDBAUM, L. 1947. Detection and estimation of microquantities of cyanide. Anal. Chem. 19:270–271.

GILBERT, E. 1976. Uberlegungen zur Berechnung und Beurteilung des Restexraktgehaltes bei Wein. *Wein Wirtschaft* 112(6):118–127.

GILLIS, M. 1978. Intra- and intergeneric similarities of the rRNA cistrons of *Acetobacter* and *Gluconobacter*. *Antonie van Leeuwenhoek* 44:117–118.

GILLIS, M. and DE LEY, J. 1980. Intra- and intergeneric similarities of the ribosomal RNA cistrons of *Acetobacter* and *Gluconobacter*. *Int. J. Syst. Bacteriol.* 30:7–27.

GINI, B. and VAUGHN, R.H. 1962. Characteristics of some bacteria associated with spoilage of California dessert wines. *Am. J. Enol. and Vitic.* 13:20–31.

GIUDICI, P. and GUERZONI, M. 1982. Sterol content as a character for selecting wine strains in enology. Vitis 21:5–14.

GLORIES, Y. 1987. Phenomenes oxydatifs lies a la conservation sous bois. *Connoissance Vigne Vin* Special issue:81–91.

GNAEGI, F. 1985. Fongicides viticoles et fermentation. *Rev. Franc. d'Enol. Paris* 25: 9–13.

GNAEGI, F. and SOZZI, T. 1983. Les bacteriophages de *Leuconostoc oenos* et leur importance oenologique. *Bull. O.I.V.* 56:352–357.

GNEKOW, B. and OUGH, C.S. 1976. Methanol in wines and must: source and amounts. *Am. J. Enol. and Vitic.* 27(1):1–6.

GOLDFARB, A. 1994. Wild or natural native yeasts have a role in modern winemaking. *Wine and Vines,* June:27–30.

GOLDSTEIN, S., MEYERSTEIN, D., and CZAPSKI, G. 1993. The Fenton reagents. *Free Rad. Biol. Med.* 15:435–445.

GONIAK, O.J. and NOBLE, A.C. 1987. Sensory study of selected volatile sulfur compounds in white wine. *Am. J. Enol. and Vitic.* 38:223–227.

GORDON AND STEWART. 1972. Effect of lipid status on cytoplasmic and mitochondrial protein synthesis in anaerobic cultures of *Saccharomyces cerevisiae. J. Gen. Microbiol.* 72:231–242.

GORINSTEIN, S., GOLDBLUM, A., KITOV, S., and DEUTSCH, J. 1984. Fermentation and post-fermentation changes in Israeli wines. *J. Food Sci.* 49(1):251–256.

GORINSTEIN, S., WEISZ, M., ZEMSER, M., TILIS, K., STILLER, A., FLAM, I., and GAT, Y. 1993. Spectroscopic analysis of polyphenols in white wines. *J. Ferm. Bioeng.* 75:115–120.

GORINSTEIN, S., GOLDBLUM, A., KITOV, S., DEUTSCH, J., LORINGER, C., COHEN, S., TABAKMAN, H., STILLER, A., and ZYKERMAN, A. 1984. The relationships between metals, polyphenols, nitrogenous substances and the treatment of red and white wines. *Am. J. Enol. and Vitic.* 35(1):9–15.

GOTTSCHALK, A. 1946. The mechanism of selective fermentation of d-fructose from invert sugar by sauternes yeast. *Biochem. J.* 40:621–626.

GOTTSCHALK, A. 1947. The effect of temperature on fermentation of d-mannose by yeast. *Biochem. J.* 41:276–280.

GRAF, E., MAHONEY, J.R., BRYANT, R.G., and EATON, J.W. 1984. Iron-catalyzed hydroxyl radical formation. *J. Biol. Chem.* 259:3620–3624.

GRANGER, D.N., HOLLWARTH, M.E., and PARKS, D.A. 1986. Ischemia-reperfusion injury: role of oxygen derived free radicals. *Acta Physiol. Scand.* Suppl. 548:47–63.

GRAZIANO, J.H., GRADY, R.W., and CERAMI, A. 1974. The identification of 2,3-dihydroxybenzoic acid as a potentially useful iron-chelating drug. *J. Pharmacol. Exp. Ther.* 190:570–575.

GREENFIELD, S. and CLAUS, G.W. 1972. Nonfunctional tricarboxylic acid cycle and the mechanism of glutamate biosynthesis in *Acetobacter suboxydans. J. Bacteriol.* 112:1295–1301.

GREENSHIELDS, R.N. 1978. Acetic acid: vinegar. In: *Primary Products of Metabolism, Economic Microbiology,* Vol. 2. A.H. Rose (ed.). London: New York, pp 121–186.

GROAT, M. and OUGH, C.S. 1978. Effect of insoluble solids added to clarified musts on fermentation rate, wine composition, and wine quality. *Am. J. Enol. and Vitic.* 29:112–119.

GROOTVELD, M., HALLIWELL, B. 1988. 2,3-dihydroxybenzoic acid is a product of human aspirin metabolism. *Biochem. Pharmacol.* 37:271–280.

GUERZONI, M.E., MATTIOLI, R., and GIUDICI, P. 1981. Acetic acid degradation in wines by film-forming yeasts. Criteria for selection. *Vigne et Vini* 8(11):43–47.

GUIMBERTEAU, G., DUBOUDREIU, D., SERRANO, M., and LEFEBRE, A. 1981. Clarification et mise en bouteilles. In *Actualities Oenological et Viticoles.* P. Ribereau-Gayon and P. Sudraud (eds.). Paris: Dunod.

GUINARD, J.-X. and CLIFF, M. 1987. Descriptive analysis of Pinot Noir wines from Carneros, Napa and Sonoma. *Am. J. Enol. and Vitic.* 38(3):211–214.

GULSON, B.L., MIZON, K.J., KORSCH, M.J., ESCHNAUER, H.R., and LEE, T.H. 1990. Are tin-lead capsules a source of lead in wine? *Australia and New Zealand Wine Ind. J.* Nov.:274–276.

GUMP, B.H. and KUPINA, S.A. 1979. Analysis of gluconic acid in botrytized wine. In: *Liquid Chromatographic Analysis of Foods and Beverages,* Vol. 2. New York: Academic Press, pp 331–353.

GUMP, B.H., SAGUANDEEKUL, S.M., MURRAY, G., and VILLAR, J.T. 1985. Determination of malic acid in wines by gas chromatography. *Am. J. Enol. and Vitic.* 36:248–251.

GUNATA, Y.Z., BAYONOVE, C.L., BAUMES, R.L., and CORDONNIER, R.E. 1985. The aroma of grapes. I. Extraction and determination of free and glycosidically bound fractions of some grape aroma components. *J. Chromatogr.* 331:83–90.

GUNATA, Y.Z., BIRON, C., SAPIS, J.C., and BAYONORA, C. 1989. Glycosidase activities in sound and rotten grapes in relation to hydrolysis of grape monoterpenyl glycosides. *Vitis* 28:191–197.

GUNATA, Y.Z., BITTEUR, S., BRILLOUET, J-M., BAYONOVE, C., and CORDONNIER, R. 1988. Sequential enzymic hydrolysis of potentially aromatic glycosides from grapes. *Carbohydr. Res.* 184:139–149.

GURR, M.I. 1992. Wine and coronary heart disease. *Lancet* (Lett.) 340:313.

GUTTERIDGE, J.C. 1992. Ageing and free radicals. *Med. Lab. Sci.* 49:313–318.

GUTTERIDGE, J.C. 1993. Free radicals in disease processes: a compilation of cause and consequence. *Free Rad. Res. Commun.* 19:141–158.

GUYMON, J.F. and CROWELL, E.A. 1972. G.C. separated brandy components derived from French and American oaks. *Am. J. Enol. and Vitic.* 23:114–120.

GUYMON, J.F., INGRAHAM, J.L., and CROWELL, E.A. 1961. Influence of aeration upon formation of higher alcohols by yeasts. *Am. J. Enol. and Vitic.* 12:60–66.

HAGEN, M. and LEMBLE, J. 1990. Respect du au liege face a l'evolution des technologies. *Rev. Fr. Oenologie* 30(127):47–50.

HAHN, G.D. and POSSMAN, P. 1977. Colloidal silicon dioxide as a fining agent for wine. *Am. J. Enol. and Vitic.* 28:108–112.

HAIGHT, K.G. and GUMP, B.H. 1994. The use of macerating enzymes in grape juice processing. *Am. J. Enol. Vitic.* 45:113–116.

HALE, C.R. 1977. Relationship between potassium and the malate and tartrate content of grape berries. *Vitis* 16:9–19.

HALL, G.W. *Procedures for Winery Sanitation.* Santa Rosa, CA: Geanall, Inc.

HALLIWELL, B. 1990. How to characterize a biological antioxidant. *Free Rad. Res. Commun.* 9:1–32.

HALLIWELL, B. 1993a. Free radicals and vascular disease: how much do we know? *Br. Med. J.* 307:885–886.

HALLIWELL, B. 1993b. The role of oxygen radicals in human disease, with particular reference to the vascular system. *Haemostasis* 23(Suppl. 1):118–126.

HALLIWELL, B. 1993c. Cigarette smoking and health: a radical view. *J. Roy. Soc. Health* 13:91–96.

HALLIWELL, B. 1993d. The chemistry of free radicals. *Toxicol. Ind. Health* 9:1–21.

HALLIWELL, B. 1993e. Antioxidants in wine. *Lancet* (Lett.) 341:1538.

HALLIWELL, B. and CROSS, C.E. 1991. Reactive oxygen species, antioxidants, and acquired immunodeficiency syndrome. Sense or speculation? *Arch. Int. Med.* 151:29–31.

HALLIWELL, B. and GUTTERIDGE, J.M.C. 1985a. Oxygen radicals and the nervous system. *Trends Neurosci.* 8:22–26.

HALLIWELL, B. and GUTTERIDGE, J.M.C. 1985b. The importance of free radicals and catalytic metal ions in human diseases. *Molec. Aspects Med.* 8:89–193.

HALLIWELL, B. and GUTTERIDGE, J.M. 1986. Oxygen free radicals and iron in relation to biology and medicine: some problems and concepts. *Arch. Biochem. Biophys.* 246:501–514.

HALLIWELL, B. and GUTTERIDGE, J.M.C. 1990. Role of free radicals and catalytic metal ions in human disease: an overview. *Meth. Enzymol.* 186:1–85.

HALLIWELL, B. and GUTTERIDGE, J.M.C. 1992. Biologically relevant metal ion-dependent hydroxyl radical generation. An update. *FEBS Lett.* 307:108–112.

HALLIWELL, B., GUTTERIDGE, J.M.C., and BLAKE, D. 1985. Metal ions and oxygen radical reactions in human inflammatory joint disease. *Phil Trans. R. Soc. London Ser. B* 311:659–671.

HANDBOOK OF CHEMISTRY AND PHYSICS. 1975. 56th ed. Cleveland, OH: The Chemical Rubber Company Press.

HANDSON, P.D. 1984. Lead and arsenic levels in wines produced from vineyards where lead arsenate sprays are used for caterpillar control. *J. Sci. Food Agric.* 35:215–218.

HARMAN, D. 1956. Aging: A theory based on free radical and radiation chemistry. *J. Gerontol.* 11:298–300.

HARRISON, J.J. and GRAHAM, J.C. 1970. Yeasts in distillary practice. In: *The Yeasts*. Vol. III. A.H. Rose and J.S. Harrison (eds.). London: Academic Press, pp 283–348.

HATHAWAY, D.E. and SEAKINS, J.W. 1957. Enzymic oxidation of catechin to a polymer structurally related to some phlobatannins. *Biochem. J.* 67:239–245.

HAUGE, J.G., KING, T.E., and CHELDELIN, V.E. 1955. Oxidation of dihydroxyacetone via the pentose cycle in *Acetobacter suboxydans*. *J. Biol. Chem.* 214:11–26.

HAWKER, J.S., RUFFNER, H.P., and WALKER, R.R. 1976. The sucrose content of some Australian grapes. *Am. J. Enol. and Vitic.* 27:125–129.

HAYNES, D.R., WRIGHT, P.F.A., GADD, S.J., WHITEHOUSE, M.W., and VERNON-ROBERTS, B. 1993. Is aspirin a prodrug for antioxidant and cytokine-modulating oxymetabolites? *Agents Actions* 39:49–58.

HEARD, G.M. and FLEET, G.H. 1988a. The effect of sulfur dioxide on yeast growth during natural and inoculated wine fermentation. *Australian and New Zealand Wine Ind. J.* 3:57–60.

HEARD, G.M. and FLEET, G.H. 1988b. The effects of temperature and pH on growth of yeast species during the fermentation of grape juice. *J. Appl. Bacteriol.* 65:23–28.

HEATHERBELL, D.A. and FLORES, J.H. 1988. Unstable proteins in grape juice and wine and their removal by ultrafiltration. In *Proceedings Second International Cool Climate Viticulture and Oenology Symposium,* Auckland, New Zealand, p. 243. New Zealand Society for Viticulture and Oenology, Auckland West, New Zealand.

HEATHERBELL, D., NGABA, P., FOMBIN, J., EATSON, B., BARCIA, Z., FLORES, J., and HSU, J. 1985. Recent developments in the application of ultrafiltration and protease enzymes to grape juice and wine processing. In: *Proceedings of the International Symposium on Cool Climate Viticulture and Enology.* D.A. Heatherbell, P.B. Lombard, F.W. Bodyfelt, and S.F. Price (eds.). Oregon State Univ., Corvallis, OR. pp 418–45.

HEERDE, E. and RADLER, F. 1978. Metabolism of anaerobic formation of succinic acid by *Saccharomyces cerevisiae. Arch. Microbiol.* 117:269–76.

HEINTZE, K. 1976. *Ind. Obst. Gemueseverwert* 61:555–556. (referenced from Amerine and Ough, 1974). *Wine and Must Analysis.* NY: John Wiley and Sons, Inc.

HENDRICKS, H.F.J., VEENSTRA, J., VELTHUIS-TE WIERIK, E.J.M., SCHAAFSMA, G., and KLUFT, C. 1994. Effect of moderate dose of alcohol with evening meals on fibrinolytic factors. *Br. Med. J.* 308:1003–1006.

HENDRICKSON, J.B., GRAM, D.J., and HAMMOND, G.S. 1970. *Organic Chemistry.* 3rd ed. New York: McGraw-Hill.

HENNING, K. 1958. Das chemische bild des mostes und weines. *Weinfach Kalender* 194–209.

HENICK-KLING, T. 1988. Yeast and bacteria control in winemaking. In: *Modern Methods of Plant Analysis: Wine Analysis.* H.F. Linskens and J.F. Jackson (eds). Heidelberg: Springer-Verlag. pp 276–316.

HENICK-KLING, T. 1991. Malolactic flavor and Chardonnay. *Proc. of Focus on Eastern U.S. Chardonnay Symposium.* Front Royal, VA. T.K. Wolf and B.W. Zoecklein (eds.). Virg. Polytech. Instit. pp 18–121.

HENICK-KLING, T., COX, D.J., and OLSEN, E.B. 1991. Production de l'energie durant la fermentation malolactique. *Rev. Francaise d'Oenologie* 132:63–66.

HENICK-KLING, T. and STOEWSAND, G.S. 1993. Lead in wine. *Am. J. Enol. Vitic.* 44(4):459–463.

HENNING, K. and BURKHARDT, R. 1960. Detection of phenolic compounds and hydroxyacids in grapes, wines, and similar beverages. *Am. J. Enol. and Vitic.* 11:64–79.

HENSCHKE, P.A. and JIRANEK, V. 1991. Hydrogen sulfide formation during fermentation: effect of nitrogen composition in model grape musts. In: *International*

Symposium on Nitrogen in Grapes and Wine. Am. Soc. for Enol. Vitic. June, Seattle, WA. Am. Soc. for Enol. and Vitic. J. Rantz, ed. pp. 172–84.

HENSCHKE, P.A. and JIRANEK, V. 1993. Yeasts—Metabolism of Nitrogen Compounds. In: *Wine Microbiology and Biotechnology.* Graham H. Fleet (ed.). Harwood Academic Publishers. pp. 77–164.

HERBERT, V. 1994. Iron worsens high-cholesterol-related coronary artery disease. *Am. J. Clin. Nutr.* 60:299–302.

HEYMANN, H. and NOBELE, A.C. 1987. Descriptive analysis of commercial Cabernet Sauvignon wines from California. *Am. J. Enol. and Vitic.* 38(1):41–44.

HIGGNES, S. 1991. Table wines found to contain lead. Ind. Circular. BATF No. 91–11. Washington, DC.

HILL, G.K. 1987. *Botrytis* control-current knowledge and future prospects. In: *Proceedings of the Sixth Australian Wine Ind. Tech. Conf.* T.H. Lee (ed.). The Australian Wine Research Institute, Adelaide, SA. pp 175–79.

HILL, G. and CAPUTI, A., JR. 1970. Colorimetric determination of tartaric acid in wine. *Am. J. Enol. and Vitic.* 21:153–161.

HIRANO, S., IMAMURA, T., UCHIAMA, K., and YU, T.L. 1962. Formation of acrolein by *Clostridium perfringens. Chem. Abstracts* 59:3096c.

HODGES, T.K. 1976. ATPases associated with membranes of plant cells. In: *Encyclopedia of Plant Physiol.* Vol. 2A. U. Luttge and M.G. Pitman (eds.). Berlin: Springer Verlag, pp 260–283.

HOGG, N., DARLEY-USMAR, V.M., GRAHAM, A., and MONCADA, S. 1993. Peroxynitrite and atherosclerosis. *Biochem. Soc. Trans.* 21:358–362.

HOLZER, H., BERNHARD, W., and SCHNEIDES, S. 1963. Zur Glycerinhldung in Backerhofe. *Biochem. Z.* 336(6):495–499.

HOOD, A.V. 1984. Possible factors affecting SO_2 inhibition of lactic acid bacteria in newly-fermented wines. *Malolactic Fermentation.* Aust. Soc. of Vitic. and Oenology Inc.

HOOTMAN, R.C. (ED.). 1992. *Manual of Descriptive Testing for Sensory Evaluation.* ASTM MNL 13. Philadelphia, PA: The American Society for Testing and Materials.

HORNING, D. 1975. Distribution of ascorbic acid, metabolites and analogues in man and animals. *Ann. NY Acad. Sci.* 258:103–118.

HOROWITZ, W. 1975. *Official Methods of Analysis of the Association of Official Analytical Chemists.* 12th ed. Washington, DC.

HOSSACH, J.A. and ROSE, A.H. 1976. Fragility of the plasma membrane in *Saccharomyces cerevisiae* enriched with different sterols. *J. Bacteriol.* 127:67–75.

HOWARD, K.L. 1970. The use of Mono-flex contact plates in the meat industry. *Proc. 13th Rutgers Meat Science Institute.*

HSIA, C.L., PLANCK, R.W., and NAGEL, C.W. 1975. Influence of must processing on iron and copper content of experimental wines. *Am. J. Enol. and Vitic.* 26:57–61.

HSU, J.C. and HEATHERBELL, D.A. 1987. Heat-unstable proteins in wine. I. Characterization and removal by bentonite fining and heat treatment. *Am. J. Enol. and Vitic.* 38(1):6–10.

HSU, J.C., HEATHERBELL, D.A., FLORES, J.H., and WATSON, B.T. 1987. Heat-unstable proteins in grape juice and wine. II. Characterization and removal by ultrafiltration. *Am. J. Enol. and Vitic.* 38:17–22.

HUBACH, C.E. 1948. Detection of cyanides and ferrocyanides in wines. *Anal. Chem.* 20:115–116.

HUNTER, K., and ROSE, A.H. 1971. Yeast lipids and membranes. In: *The Yeasts,* Vol. II. A.H. Rose and J.H. Harrison (eds.). London: Academic Press.

HURRELL, R.F. and FINOT, P.-A. 1984. Nutritional consequences of the reactions between proteins and oxidized polyphenolic acids. *Adv. Exp. Biol. Med.* 177:423–435.

IGNARRO, L.J. 1990. Biosynthesis and metabolism of endothelium-derived nitric oxide. *Annu. Rev. Pharmacol. Toxicol.* 30:535–560.

IGNARRO, L.J. 1991. Heme-dependent activation of guanylate cyclase by nitric oxide: a novel signal transduction mechanism. *Blood Vessels* 28:67–73.

ILAND, P. 1990. Personal communication.

INGLEDEW, W.M. and KUNKEE, R.E. 1985. Factors influencing sluggish fermentations of grape juice. *Am. J. Enol. and Vitic.* 36:65–76.

INGLEDEW, W.M., MAGNUS, C.A., and PATTERSON, J.R. 1987. Yeast foods and ethyl carbamate formation in wine. *Am. J. Enol. and Vitic.* 38(4):332–335.

INGRAHAM, J.L. and COOKE, G.M. 1960a. A survey of the incidence of malo-lactic fermentation in California table wines. *Am. J. Enol. and Vitic.* 11:160–163.

INGRAHAM, J.L. and WOOD, W.A. 1961. Pentose dehydrogenase in *Saccharomyces rouxii. J. Bact.* 89:1186.

INGRAHAM, J.L., VAUGHN, R.H., and COOKE, G.M. 1960b. Studies on the malo-lactic organisms isolated from California wines. *Am. J. Enol. and Vitic.* 11:1–4.

INGRAM, M., OTTOWAY, F.J.H., and COPPOCK, J.B.M. 1956. The preservation action of acid substances in food. *Chem. and Ind.* 1154–1163.

INSTITUTE FOOD TECHNOLOGISTS. 1981. Sensory evaluation guide for testing food and beverage products. *Food Technol.* 35(11):50–59.

INTERNATIONAL ORGANIZATION FOR STANDARDIZATION. 1993. Circular ISO-10718. Cork stoppers—enumeration of colony-forming units of yeasts, moulds and bacteria capable of growth in alcoholic medium. ISO. Geneva, Switzerland.

JACKSON, D.I. 1991. Environmental and hormonal effects on the development of early bunch-stem necrosis. *Am. J. Enol. and Vitic.* 42:290–294.

JACKSON, D.I. and LOMBARD, P.B. 1993. Environmental and management practices affecting grape composition and wine quality—a review. *Am. J. Enol. and Vitic.* 44:409–430.

JACKSON, R.L. 1993. Anti-oxidants for the treatment and the prevention of atherosclerosis. *Biochem. Soc. Trans.* 21:650–651.

JACKSON, R.L., KU, G., THOMAS, C.E. 1993. Antioxidants: a biological defense mechanism for the prevention of atherosclerosis. *Med. Res. Rev.* 13:161–182.

JAIN, A., MARTENSSON, J., STOLE, E., AULD, P.A.M., and MEISTER, A. 1991. Glutathione deficiency leads to mitochondrial damage in brain. *Proc. Natl. Acad. Sci. USA* 88:1913–1917.

JAKOB, L. 1962. Bentotest. *Das Weinblatt.* (No. 34/35, Sept).

JAMES, T.H. and WEISSBERGER, A. 1939. Oxidation processes: XIII. The inhibitory action of sulfite and other compounds in auto-oxidation of hydroquinone and its homologs. *J. Am. Chem. Soc.* 61:442–450.

JANDA, W. 1983. Fruit Juice. In: *Industrial Enzymology.* New York: The Nature Press, pp 315–320.

JANSER, E. 1994. Personal communication.

JAY, J.M. 1970. *Modern Food Microbiology.* New York: Van Nostrand Reinhold Co.

JEANDET, P., BESSIS, R., and GAUTHERON, B. 1991. The production of resveratrol (3,5,4-trihydroxystilbene) by grape berries in different developmental stages. *Am. J. Enol. Vitic.* 42:41–46.

JENKINS, R.R. 1993. Exercise, oxidative stress, and antioxidants: a review. *Int. J. Sport Nutr.* 3:356–375.

JENNINGS, W.G. 1965. Theory and practice of hard-surface cleaning. *Adv. Food Res.* 14:325–459.

JIRANEK, V., LANGRIDGE, P., and HENSCHKE, P.A. 1989. Nitrogen requirement of yeast during wine fermentation. In: *Proc. of the Seventh Australian Wine Industry Tech. Conference* (Adelaide). P.J. Williams, D.M. Davidson, and T.H. Lee (eds.). Austr. Ind. Publ. pp 166–171.

JIRANEK, V., LANGRIDGE, P., and HENSCHKE, P.A. 1991. Yeast nitrogen demand: Selection criteria for wine yeasts for fermenting low nitrogen musts. In: *Proceedings of the International Symp. on Nitrogen in Grapes and Wine.* Seattle, WA. J. Rantz, (ed). Am. Soc. Enol. Vitic. pp. 266–69.

JOHNSON, T. and NAGEL, C.W. 1976. Composition of central Washington grapes during maturation. *Am. J. Enol. and Vitic.* 27:15–20.

JONES, A.R., DOWLING, E.J., and SKRABA, W.J. 1953. Identification of some organic acids by paper chromatography. *Anal. Chem.* 3:35.

JONES, D.D. and GREENSHIELDS, R.N. 1971. *J. Inst. Brew.* 77:160.

JONES, R.S. and OUGH, C.S. 1985. Variation in the percent ethanol (v/v) per Brix conversions of wines from different climatic regions. *Am. J. Enol. and Vitic.* 36(4): 268–270.

JORDAN, A.D. and CROSER, B.J. 1983. Determination of grape maturity by aroma/flavor assessment. *Proceedings of the Fifth Austrialian Wine Industry Technical Conference.* (Adelaide) T.H. Lee and T.C. Somers, (eds.) Australian Wine Res. Instit. pp 261–274.

JORDAN, A.D. and NAPPER, D.H. 1987. Some aspects of the physical chemistry of bubble and foam phenomena in sparkling wine. In: *Proc. Sixth Australian Wine Ind. Tech. Conf* (Adelaide) T.H. Lee (ed.) Austr. Ind. Publs. pp. 237–246.

JOSLYN, M.A. 1950a. *Methods in Food Analysis.* New York: Academic Press, 528 pp.

JOSLYN, M.A. 1950b. Hard chrome plating and avoidance of metal pick-up. *Wine and Vines* 31(4):67–69.

JOSLYN, M.A. 1953. The theoretical aspects of clarification of wine by gelatin fining. *Proc. Am. Soc. Enol.* 4:39–68.

JOSLYN, M.A. and AMERINE, M.A. 1964. *Dessert, Appetizer and Related Flavored Wines; the Technology of Their Production.* Berkeley, Division of Agricultural Sciences, University of California. 482 pp.

JOSLYN, M.A. and BRAVERMAN, J.B.S. 1954. The chemistry and technology of the pretreatment and preservation of fruit and vegetable products with sulfur dioxide and sulfites. In: *Adv. Food Research* 5:97–160.

JOSLYN, M.A. and DUNN, R. 1938. Acid metabolism of wine yeasts. I: The relation of volatile acid formation to alcoholic fermentation. *J. Am. Chem. Soc.* 60:1137–1141.

JOSLYN, M.A. and GOLDSTEIN, J.L. 1964. Astringency of fruit and fruit products in relation to phenolic content. In: *Adv. Food Research* 13:179–217.

JOYEUX, A., LAFON-LAFOURCADE, S., and RIBEREAU-GAYON, P. 1984a. Evolution of acetic acid bacteria during fermentation and storage of wine. *Appl. Environ. Microbiol.* 48:153–156.

JOYEUX, A., LAFON-LAFOURCADE, S., and RIBEREAU-GAYON, P. 1984b. Metabolism of acetic acid bacteria in grape must: consequences on alcoholic and malolactic fermentation. *Sciences Alim.* 4:247–255.

JUHASZ, O., KOZMA, P., AND POLYAK, D. 1984. Nitrogen status of grapevines as reflected by the arginine content of the fruit. *Acta Agron. Acad. Sci. Hung.* 33:3–17.

JURD, L. 1964. Reactions involved in sulfite bleaching of anthocyanins. *J. Food Sci.* 29:16–19.

JURD, L. 1969. Review of polyphenol condensation reactions and their possible occurrence in the aging of wines. *Am. J. Enol. and Vitic.* 20:191–195.

JURD, L. and ANSEN, S. 1966. The formation of metal and "copigment" complexes of cyanidin-3-glucoside. *Phytochemistry* 5(6):1263–1271.

KANNER, J., FRANKEL, E., GRANIT, R., GERMAN, B., KINSELLA, J.E. 1994. Natural antioxidants in grapes and wines. *J. Agric. Food Chem.* 42:64–69.

KANNER, J., GERMAN, J.B., and KINSELLA, J.E. 1986. Initiation of lipid peroxidation in biological systems. *Crit. Rev. Food Sci. Nutr.* 25:317–364.

KANTZ, K. and SINGLETON, V.L. 1991. Isolation and determination of polymeric polyphenols in wines using Sephadex LH-20. *Am. J. Enol. and Vitic.* 42:309–316.

KASIMATIS, A. 1984. Viticultural practices for varietal winemaking. University of California-Davis Extension Short Course Series.

KASIMATIS, A. and VILAS, E.P. 1985. Sampling for degrees Brix in vineyard plots. *Am. J. Enol. Vitic.* 36:207–213.

KAUFMANN, A. 1992. Messing, eine mogliche Ursache fur erhohte Bleikonzentrationen in Wein. *Mitt. Geb. Lebensm. Hyg.* 83:204–210.

KAUR, H. and HALLIWELL, B. 1994. Detection of Hydroxyl Radicals by Aromatic Hydroxylation. *Meth. Enzymol.* 233:67–82.

KEAN, C.E. 1954. Chemical Composition of Copper Complexes Causing Cloudiness in Various Wines. PhD Thesis, University of California, Davis.

KEAN, C.E. and MARSH, J.L. 1956a. Investigation of copper complexes causing cloudiness in wines. I: Chemical Composition. *Food Res.* 21:441–447.

KEAN, C.E. and MARSH, J.L. 1956b. Investigation of copper complexes causing cloudiness in wines. II: Bentonite treatment of wines. *Food Tech.* 10:355–359.

KELLY-TREADWELL, P.H. 1988. Protease activity in yeast: its relationship to autolysis and champagne character. *Australian Grapegrower and Winemaker* 292:58–66.

KELM, M. and SCHRADER, J. 1990. Control of coronary vascular tone by nitric oxide. *Circulation Res.* 66:1561–1575.

KENNEDY, N.P. and TIPTON, K.F. 1990. Ethanol metabolism and alcoholic liver disease. *Essays Biochem.* 25:137–195.

KERSHENOBICH, D., HADDAD, L., LORENZANA-JIMENEZ, M., VARGAS, F., DE LA FUENTE, J.R., and ZAPATA, L. 1993. Alcohol metabolism in healthy subjects. *Gastroenterology* 105:308–309.

KIELHOFER, E. 1941. Troubles albuminoides du vin. *Bull. Office Inter. Vin.* 16:7–10.

KIELHOFER, E. 1960. Neue erkenntnisse uber die schweflige saure imwein und ihren ersatz durch ascorbinsaure. *Deut. Wein-Zkg.* 96:14–24.

KIELHOFER, E. 1963. Etat et action de l'acide sulfureux dans les vins; regles de son emploi. *Ann. Technol. Agric.* 12:77–89.

KIELHOFER, E. and WURDING, G. 1960a. Die an Aldehyd gebundene Schweflige saure im Wein. I. Acetaldehyd-bildung durch Enzymatishce und nicht Enzymatische alkohol-Oxydation. *Weinberg-Keller* 7:16–22.

KIELHOFER, E. and WURDIG, G. 1960b. Die an aldehyd gebundene Schweflige Ssaure im Wein. II. Acetaldehydbildung bei der garung. *Weinberg Keller* 7:50–61.

KIMURA, Y., H. OKUDA, and S. ARICHI. 1985. Effect of stilbene derivatives on leukocyte arachidonic acid metabolism. *Wakan Lyaka Gakkaishi* 2(3):516–517.

KING, S.W. and BEELMAN, R.B. 1986. Metabolic interactions between *Saccharomyces cerevisiae* and *Leuconostoc oenos* in a model grape juice/wine system. *Am. J. Enol. and Vitic.* 37(1):53–60.

KING, T.E. and CHELDELIN, V.H. 1952. Oxidations in *Acetobacter suboxydans. Biochim. Biophys. Acta* 14:108–116.

KING, T.E., KAWASAKI, E.H., AND CHELDELIN, V.H. 1956. Tricarboxylic acid cycle activity in *Acetobacter pasteurianum. J. Bacteriol.* 72:418–421.

KINGSTON, C.M. and VAN EPENHUIJSEN, C.W. 1989. Influence of leaf area on fruit development and quality of Italia glass-houre table grapes. *Am. J. Enol. and Vitic.* 40:130–134.

KINSELLA, J.E., KANNER, J., FRANKEL, E., and GERMAN, B. 1992. Wine and health: the possible role of phenolics, flavonoids and other antioxidants. In: *Proceedings Potential Health Effects of Components of Plant Foods and Beverages in the Diet,* L.F. Bisson (ed.). University of California at Davis, pp 107–121.

KITAMOTO, K., YOSHIZAWA, K., OSHUMI, Y., and ANRAKU, Y. 1988. Dynamic aspects of vacuolar and cytosolic amino acid pools of *Saccharomyces cerevisiae. J. Bacteriology* 170:2683–2686.

KLATSKY, A., and AMSTRONG, M. Alcoholic beverage choice and risk of coronary artery disease. Mortality: do red wine drinkers fare best? *Am. J. Cardiol.* 71:467–469.

KLIEWER, W.M. 1967a. Glucose-fructose ratio of *Vitis vinifera* grapes. *Am. J. Enol. and Vitic.* 17:33–41.

KLIEWER, W.M. 1967b. Concentration of tartrates, malates, glucose and fructose in fruits of the genus *Vitis. Am. J. Enol. and Vitic.* 18:87–96.

KLIEWER, W.M. 1971. Effect of day temperature and light intensity on coloration of *Vitis vinifera* grapes. *J. Am. Soc. Hort. Sci.* 95:693–697.

KLIEWER, W.M. 1980. Vineyard canopy management—a review. In: *Proceedings of the Univ. of California, Davis, Grape and Wine Symposium.* A.D. Webb (ed.). U.C. Press. pp 342–352.

KLIEWER, W.M. and BENZ, J. 1985. Personal communication.

KLIEWER, W.M., and LIDER, L.A. 1968. Influence of cluster exposure to sunlight on the composition of Thompson Seedless fruit. *Am. J. Enol. Vitic.* 19:175–184.

KLINGSHIRN, L.M., LIU, J.R., and GALLANDER, J.F. 1987. Higher alcohol formation in wines as it relates to particle size profiles of juice insoluble sollids. *Am. J. Enol. and Vitic.* 38(3):207–210.

KOCH, J. 1963. Proteines des vins blancs. Traitements des precipitations proteiques par chauffage et a l'aide de la bentonite. *Ann. Technol. Agric.* 12:297–313.

KOCH, J. and SAJAK, E. 1959. A review of some studies on grape proteins. *Am. J. Enol. Vitic.* 10:114–123.

KODAMA, S., SUZUKI, T., FUJINAWA, S., DEL LA TEJA, P., and YOTSUZUKA, F. 1991. Prevention of ethyl carbamat formation in wine by urea degradation using acid urease. In: *Proc. of the International Symposium on Nitrogen in Grapes and Wine.* Seattle, WA. J.M. Rantz (ed.). Am. Soc. for Enol. and Vitic., pp 270–273.

KONTOS, H.A. 1993. Nitric oxide and nitrosothiols in cerebrovascular and neuronal regulation. *Stroke* 24(Suppl.I):I-155–I-158.

KOPIN, I.J. 1993. The pharmacology of Parkinson's disease therapy: an update. *Annu. Rev. Pharmacol. Toxicol.* 32:467–495.

KOPPENOL, W.H. 1985. Energetics of interconversion reactions of oxyradicals. *Adv. Free Rad. Biol. Med.* 1:91–131.

KOPPENOL, W.H. 1990. Oxyradical reactions: from bond-dissociation energies to reduction potentials. *FEBS Lett.* 264:165–167.

KOPPENOL, W.H. and BARTLETT, D. 1993. *The nitration and hydroxylation of phenol and salicylic acid by peroxynitrite.* Presented at the 1st Annual Meeting of the Oxygen Society, Charleston, S.C., Nov. 12–17, Abstract 4:6.

KOPPENOL, W.H., MORENO, J.J., PRYOR, W.A., ICHIROPOULOS, H., and BECKMAN, J.S. 1992. Peroxynitrite, a cloaked oxidant formed by nitric oxide and superoxide. *Chem. Res. Toxicol.* 5:834–842.

KRAMLING, T.E. and SINGLETON V.L. 1969. An estimate of nonflavonoid phenols in wines. *Am. J. Enol. Vitic.* 20:86–92.

KRAUS, J.K., SCOPP, R., and CHEN, S.L. 1981. Effect of rehydration of dry wine yeast activity. *Am. J. Enol. and Vitic.* 32(2):132–134.

KRUG, K. 1967. Eiweiasnachtrubungen in Wein und ihre Verhutung. Deut. Wein-2Tg. 103:1029–1035.

KRUMPERMAN, P.H. and VAUGHN, R.H. 1966. Some Lactobacilli associated with decomposition of tartaric acid in wine. *Am. J. Enol. and Vitic.* 17(3):185–190.

KUBES, P. and GRANGER, D.N. 1992. Nitric oxide modulates microvascular permeability. *Am. J. Physiol.* 262:H611–H616.

KUMAR, K.V. and DAS, U.N. 1993. Are free radicals involved in the pathology of human essential hypertension? *Free Rad. Res. Commun.* 19:59–66.

KUNKEE, R.E. 1967. Control of malo-lactic fermentation induced by *Leuconostoc citrovorum. Am. J. Enol. and Vitic.* 18:71–77.

KUNKEE, R.E. 1968. Simplified chromatographic procedure for detection of malo-lactic fermentation. *Wine and Vines* 49(3):23–24.

KUNKEE, R.E. and AMERINE, M.A. 1970. Yeasts in winemaking. In: *The Yeasts III.* eds. A.H. Rose and J.S. Harrison (eds.) New York: Academic Press.

KUNKEE, R.E. and NERADT, F. 1974. A rapid method for detection of viable yeasts in bottled wines. *Wine and Vines* 55(12):36–39.

KUNKEE, R.E., OUGH, C.S., and AMERINE, M.A. 1964. Induction of malo-lactic fermentation by innoculation of must and wine with bacteria. *Am. J. Enol. and Vitic.* 15:178–183.

KUPINA, S.A. 1984. Simultaneous quantitation of glycerol, acetic acid and ethanol in grape juice by high performance liquid chromatography. *Am. J. Enol. and Vitic.* 35(2):59–62.

LABATUT, E., CARUANA, C., BRTRAND, A., and LAFON-LAFOURCADE, S. 1984. Action des acides octanoique et decanoique a l'engard de quelquos levures en croissance dan le mout de raisin. *Institute d'Oenologie, Univ. de Bourdeaux II.*

LACEY, J. 1973. The air spora of a Portuguese cork factory. *Ann Occupat Hyg.* 16:223–230.

LAFON-LAFOURCADE, S. 1985. Role des microorganismes dans la formation de substances combinant le SO_2. *Bull. OIV* 652–653:590–604.

LAFON-LAFOURCADE, S. and RIBEREAU-GAYON, P. 1976. Premiers observations sur l'utilisation des levures seches en vinification en blanc. *Connaissance Vigne Vin* 10:277–292.

LAFON-LAFOURCADE, S., CARRE, E., and RIBEREAU-GAYON. 1983. Occurrence of lactic acid bacteria at different stages of vinification and conservation of wine. *Appl. Environ. Microbiol.* 46:874–880.

LAFON-LAFOURCADE, S., GENEIX, C., and RIBEREAU-GAYON, P. 1984. Inhibition of alcoholic fermentation of grape must by fatty acids produced by yeasts and their elimination by yeast ghosts. *Appl. Environ. Microbiol.* 47:1246–1249.

LANGHANS, E. and SCHLOTTER, H.A. 1987. die Hefeschonung zur Reduzierung des Kupfer-Gehaltes von Weinen. *Wein-Wissen.* 42:202–210.

LARGE, P.J. 1986. Degradation of organic nitrogen compounds by yeasts. *Yeast* 2:1–34.

LARUE, R., LAFON-LAFOURCAD, S., and RIBEEREAU-GAYON, P. 1980. Relationship between the sterol content of yeast cells and their fermentation activity in grape musts. *Appl. Environ. Microbiol.* April:808–811.

LAUGHTON, M.J., HALLIWELL, B., EVANS, P.J., and HOULT, J.R.S. Antioxidant and pro-oxidant actions of the plant phenolics quercitin, gossypol and myricetin. Effects on lipid peroxidation, hydroxyl radical generation and bleomycin-dependent damage to DNA. *Biochem. Pharmacol.* 1989. 38:2859–2865.

LAW, B.A. and KOLSTAD, J. 1983. Proteolytic systems in lactic acid bacteria. *Antonie Van Leeuwenhoek* 49:225–245.

LEA, A.G.H. and ARNOLD, G.M. 1978. The phenolics of ciders: bitterness and astringency. *J. Sci. Agric.* 29:478–483.

LEAF, C.D., WISHNOK, J.S., and TANNENBAUM, S.R. 1989. L-arginine is a precursor for nitrate biosynthesis in humans. *Biochem. Biophys. Res. Commun.* 163:1032–1037.

LEE, A. 1978. Bacteriophages associated with Lactobacilli isolated from wine. In: *Proceedings of the Fifth Oenol. Symposium.* (Auckland, NZ) E. Lemperle and J. Frank (eds.). Breisach: International Assn. of Modern Winery Technol. and Mgmt. pp. 287–295.

LEE, A. 1990. In: *Bitterness in Food and Beverages.* R. Rousseff, (ed.). New York: Elsevier, pp. 123–143.

LEE, T. 1985. Protein instability: nature, characterization and removal by bentonite. In: *Physical Stability of Wine.* T.H. Lee (ed.). pp. 23–40, Seminar by Australian Society of Viticulture and Oenology, Inc.

LEE, T.H. 1985. Protein instability: nature, characterization and removal by bentonite. *Proceedings of the Australian Society of Viticulture and Oenology.*

LEE, T.H., HIN, M.W., OUGH, C.S., AND BERG, H.W. 1977. Effects of fermentation variables on the color and sensory quality of Pinot Noir wines. *Proc. 3rd Aust. Wine Ind. Tech. Conference,* Albany, p. 77.

LEE, T.H., SIMPSON, R.F., VANDEPEER, J.M., FLEET, G.H., DAVIS, C.R., DALY, N.M., and YAP, A.S.J. 1984. Microbiology of wine corks. In: *Proc. of the Fifth Australian*

Wine Ind. Tech. Conf. T.H. Lee and T.C. Somers (eds.). Adelaide: The Aust. Wine Res. Inst., pp. 435–450.

LEE, C.Y., SMITH, N.L., and NELSON, R.R. 1979. Relationship between pectin methyl esterase activity and formation of methanol in concord grape juice. *Food Chem.* 4(2):143–148.

LEE-RUFF, E. 1977. The organic chemistry of superoxide. *Chem. Soc. Rev.* 2:195–214.

LEFEBVRE, A., RIBOULET, J.M., BOIDRON, J.N., RIBEREAU-GAYON, P. 1983. Incidence des micro-organisms du liege sur les alterations Llfactives du vin. *Sciences Aliments* 3:265–278.

LEGLISE, M. 1980. Vinification de Pinot Noir. In: *Seminaire de Oenologie, les Vinivication a Caractere Special.* I.N.R.A. Colomar 13 MARS 1980.

LEHRINGER, A.L., REYNAFARJE, B., ALEXANDRE, A. 1979. The stoichiometry of H^+ movements coupled to electron transport and ATP synthesis in mitochondria. In: *Cation Flux Across Biomembranes.* Y. Mukohata and L. Packer (eds.). New York: Academic Press, pp. 243–254.

LEMPERLE, E. 1988. Fungicide residues in must and wines. In: *Proceedings of the Second Int. Sym. Cool Climate Viticulture and Oenology Conference.* R. Smart, R. Smart, R. Thornton, S. Rodriquez, and J. Young (eds.). Auckland, NZ. Soc. for Vitic. and Enol. pp. 211–18.

LEMPERLE, E. and KERNER, K. 1974. Workstoffrucks-stande und Garbeeinflu Bungen nach Anwendung systematischer Fungiside im Weinbau. *Die Wein Wissenschaft* 29:92–103.

LEMPERLE, E. and LAY, H. 1989. Zusammensetzung und Beurteilung der Weine. In: *Chemie des Weines.* G. Wurdig and R. Woller (eds.). Stuttgart: Ulmer.

LEPPANEN, O.A., DENSLOW, J., and RONKAINED, P.P. 1979. A gas chromatographic method for the accurate determination of low concentrations of volatile sulfur compounds in alcoholic beverages. *J. Instit. Brew.* 85:350–353.

LEVINE, S.M. 1992. role of reactive oxygen species in the pathogenesis of multiple sclerosis. *Med. Hypoth.* 39:271–274.

LEVY, G. 1979. Pharmacokinetics of salicylate in man. *Drug Metab. Rev.* 9:3–19.

LEWIS, V.M., ESSELEN, W.B., and FELLERS, C.R. 1949a. Nonenzymatic browning of foods. Production of carbon dioxide. *Ind. Eng. Chem.* 41:2587–2591.

LEWIS, V.M., ESSELEN, W.B., and FELLERS, C.R. 1949b. Nitrogen-free carboxylic acids in browning reactions. *Ind. Eng. Chem.* 41:2591–2594.

LEWITT, P.A. 1993. Neuroprotection by anti-oxidant strategies in Parkinson's disease. *Eur. Neurol.* 33(Suppl. 1):24–30.

LIND, C., HOCHSTEIN, P., and ERNSTER, L. 1982. DT-diaphorase as a quinone reductase: a cellular control device against semiquinone and superoxide radical formation. *Arch. Biochem. Biohys.* 216:178–185.

LINDBLOOM, B. 1993. A basic format for conducting oak trials in the winery. In: *Proceedings of the International Oak Symp. San Francisco State Univ.* pp 1–5. International Wine Academy and Pract. Winery & Vineyard.

LIPTON, S.A., CHOI, Y-B., PAN, Z-H., LEI, S.Z., CHEN, H-S., V., SUCHER, N.J., LOSCALZO, J., SINGEL, D.J., AND STAMLER, J.S. 1993. A redox-based mechanism for the neuroprotective and neurodestructive effects of nitric oxide and related nitroso-compounds. *Nature* 364:626–632.

LIU, G-T., ZHANG, T-M., WANG, B., AND WANG, Y-W. 1992. Protective action of seven natural phenolics compounds against peroxidative damage to biomembranes. *Biochem. Pharmacol.* 43:147–152.

LODDER, J. 1970. *The Yeasts: A Taxonomic Study*. 2nd ed. Amsterdam: North Holland Publishing Co.

LOHMANN, W. 1987. Ascorbic acid and cataract. *Ann. NY Acad. Sci.* 498:307–311.

LONG, Z. 1984a. Monitoring sugar per berry. *Pract. Winery and Vineyard* 5(2):52–54.

LONG, Z. 1984b. Viticultural practices or varietal winemaking. *University of California Davis* Short Course.

LONG, Z.R. 1987. Manipulation of grape flavour in the vineyard: California, North Coast region. In: *Proceedings of the Sixth Australian Wine Industry Conference,* Adelaide. T.H. Lee (ed.) Austr. Ind. Publs. pp. 82–88.

LONG, Z. and LINDBLOOM, B. 1986. A report on Zelma Long's work at Simi. Juice oxidation experiment. *Wine and Vines* (Nov.): 44–49.

LOSCALZO, J. 1992. The Relation Between Atherosclerosis and Thrombosis. *Circulation* 86(Suppl III):III-95–III-99.

LOWENSTEIN, C.J. and SNYDER, S.H. 1992. Nitric oxide, a novel biologic messenger. *Cell* 70:705–707.

LOWENSTEIN, C.J., DINERMAN, J.L., and SNYDER, S.H. 1994. Nitric oxide: a physiologic messenger. *Ann. Intern. Med.* 120:227–237.

LUCIA, S.P. 1954. *Wine as Food and Medicine* Blakiston, NY:

LUCIA, S.P. 1993. *A History of Wine as Therapy*. Philadelphia: Lippincott.

LUCRAMET, V. 1981. Quelques proprietes des bacteries lactiques. In: *Actualities Oenologiques et Viticoles.* J. Ribereau-Gayon and P. Sudraud (eds.). Universite de Bordeaux, Talence, France. pp. 239–243.

LUIS, S. 1984. A study of proteins during grape maturation, juice preparation and wine processing. Dissertation, University of New S. Wales, Australia. [Diss. Abs. Intl. B, 43(1):118].

LUTHI, H. 1957. Symbiotic problems relating to the bacterial deterioration of wines. *Am. J. Enol. and Vitic.* 8:176–181.

LUTHI, H. 1959. Microorganisms in noncitric juices. In: *Advances in Food Research* 9:221–273. New York: Academic Press.

LUTHI, H. and SCHLATTER, C.H. 1993. Biogene Amine in Lebensmitteln: Zur Wirkung von Histamin, Tyramin und B-phenylethylamin auf den Mesnchen. *Z. Lebensm. Unter. Forsch.* 177:439–443.

LUTHY, J., and SCHLATTER, C. 1983. Biogene amine in Lebenamitteln: zur Wirkung von Histamin, Tyramin und Beta-Phenylethylamin auf den Menschen. *Z. lebensm. Unter. Forsch.* 177:43–43.

MACLURE, M. 1994. Alcohol intake and risk of myocardial infarction. *New Eng. J. Med.* 330:1241–1242.

MAGA, J.A. 1985. Flavor contribution of wood in alcoholic beverages. Adda, J. (ed.). *Progress in Flavour Research; Proceedings of the 4th Weurman Flavour Research Symposium;* May 9–11, 1984; Dourdan, France. Amsterdam: Elsevier Science Publishers, pp. 409–416.

MAHLER, S., EDWARDS, P.A., and CHISHOLM, M.G. 1988. HPLC identification of phenols in Vidal Blanc wine using electrochemical detection. *J. Agric. Food Chem.* 36:946–951.

MALAMY, J. and KLESSIG, D.F. 1992. Salicylic acid and plant disease resistance. *Plant J.* 2:643–654.

MARAIS, J. 1983. Terpenes in the aroma of grapes and wines: a review. *S. Afr. J. Enol. Vitic.* 4:49–58.

MARAIS, P.G., and KRUGER, M.M. 1975. Fungus contamination of corks responsible for unpleasant odors in wine. *Phytophylactica* 7:115–117.

MARGALITH, P.Z. 1981. In: *Flavour Microbiology* C.G. Thomas (ed.). pp 73–224.

MARMOT, M.G., ELLIOT, P., SHIPLEY, M.J., DYER, A.R., USEHIMA, H., BEEVERS, D.G., STAMLER, R., KESTELOOR, H., ROSE, G., and STAMLER, J. 1994. Alcohol and blood pressure: the INTERSTALT study. *Br. Med. J.* 308:1263–1267.

MARSH, G.L. 1951. *Calculation of Proof Gallon Equivalents per Ton of Grapes.* Wine Institute Technical Advisory Committee.

MARSH, G.L. 1952. New compound ends metal clouding. A report on the Fessler compound. *Wine and Vines* 33:19–21.

MARSH, G.L. and GUYMON, J.F. 1959. Refrigeration in winemaking. *Am. Soc. Refrig. Eng. Data Book,* Vol. 1. Chapter 10.

MARSHALL, W. 1977. Thermal decomposition of ethylenbisdithiocarbamate fungicide to ethylenthiourea in aqueous media. *J. Agric. and Food Chem.* 25:357–361.

MARTELL, A.E. 1982. Chelates of ascorbic acid. Formation and catalytic properties. *Am. Chem. Soc. Ser.* 200:153–178.

MASKOS, Z., RUSH, J.D., and KOPPENOL, W.H. 1990. The hydroxylation of the salicylate anion by a Fenton reaction and T-radiolysis: a consideration of the respective mechanisms. *Free Rad. Biol. Med.* 8:153–162.

MASUDA, M., NISHIMURA, K., and FAGACEAE. 1971. Branched nonalactones from some *Quercus* species. *Pytochem.* 10:1401–1402.

MATHEIS, G., SHERMAN, M.P., BUCKBERG, G.D., HYBRON, D.M., YOUNG, H.H., and IGNARRO, L.J. 1992. Role of L-arginine-nitric oxide pathway in myocardial reoxygenation injury. *Am. J. Physiol.* 262:H616–H620.

MATTICK, L.R. 1984. A method for the extraction of grape berries used in total acid, potassium and individual acid analyses. Technical Note. *Am. J. Enol. and Vitic.* 34(1):49.

MAUJEAN, A., MILLERY, P., and LEMARSQUIER, H. 1985. Explications biocheimiques et metaboliques de la confusion entre gout de bouchon et gout de moisi. *Rev. Franc Oenol.* 99:55–61.

MAURICIO, J.C., MORENO, J.J., VALERO, E.M., ZEA, L., MEDINA, M., ORTEGA, J.M. 1993. Ester formation and specific activity of *in vitro* alcohol acetyltransferase and esterase by *Saccharomyces cerevisiae* during grape must fermentation. *J. Agric. Chem.* 41:2086–2091.

MAW, G.A. 1960. Utilization of sulfur compounds by brewer's yeast. *J. Instit. Brew.* 66:162–67.

MAW, G.A. 1961. Effects of cysteine and other thiols on the growth of brewer's yeast. *J. Instit. Brewing* 67:57–63.

MAXCY, R.B. 1969. Residual microorganisms in cleaned-in-place systems for handling milk. *J. Milk and Food Tech.* 32:140–143.

MAXWELL, S.R.J., JAKEMAN, P., THOMASON, H., LEGUEN, C., and THORPE, G.H.G. 1993. Changes in plasma antioxidant status during eccentric exercise and the effect of vitamin supplementation. *Free Rad. Res. Commun.* 19:191–202.

MAYER, B., JOHN, M., HEINZEL, B., WERNER, E.R., WACHTER, H., SCHULTZ, G., and BOHME, E. 1991. Brain nitrix oxide synthase is a biopterin- and flavin-containing multi-functional oxido-reductase. *FEBS Lett.* 288:187–191.

MAYER, J. and HERNANDEZ, R. 1970. Seed tannin extraction in Cabernet sauvignon. *Am. J. Enol. and Vitic.* 21(4):184–188.

MAYER, K. 1974. Nachteilige Auswirkungen auf die Weinqualitat be ungunstig verlaufenem biologischen Saureabbau. *Schweiz Zeitschrift Obst- u. Weinbau.* 110: 385–391.

MAYER, K. and BUSCH, I. 1963. Uber enine Enzymatische Apfelsaurebestimmung in Wein und Traubensaft. *Mitt. Gebiete. Lebensm. Hyg.* 54:60–65.

MCCARTHY, M.G., CIRAMI, R.M., and FURKALIER, D.C. 1987. The effect of crop load and vegetative growth control on wine quality. In: *Proceedings of the Sixth Australian Wine Industry Technical Conference,* (Adelaide) T.H. Lee (ed.). Austr. Ind. Publ. pp. 75–77.

MCCLEVERTY, J.A. 1979. Reactions of nitric oxide coordinated to transition metals. *Chem. Rev.* 79:53–76.

MCCLOSKEY, L.P. 1974. Gluconic acid in California wines. *Am. J. Enol. and Vitic.* 25(4):198–201.

MCCLOSKEY, L.P. 1976. An acetic acid assay for wine using enzymes. *Am. J. Enol. and Vitic.* 27(4):176–180.

MCCLOSKEY, L.P. 1978. An enzymatic assay for glucose and fructose. *Am. J. Enol. and Vitic.* 29(3):226–227.

MCCLOSKEY, L.P. 1980. An improved enzymic assay for acetic in juice and wine. *Am. J. Enol. and Vitic.* 31(2):170–173.

MCCORD, J.D., TROUSDALE, E., and RYU, D.D.Y. 1984. An improved sample preparation for the analysis of major organic components. *Am. J. Enol. and Vitic.* 35:28–29.

McCord, J.M. 1985. Oxygen-derived free radicals in postischemic tissue injury. *N. Engl. J. Med.* 312:159–163.

McCord, J.M. and Omar, B.A. 1993. Sources of free radicals. *Toxicol. Ind. Health* 9:23–37.

McDaniel, M.R., Henderson, L.A., Watson, B.T., and Heatherbell, D.A. 1988. Sensory panel training and descriptive analysis. Gewurtztraminer clonal wines. *Proceedings of the Second International Cool Climate Vitic. and Oenol. Symposium.* pp. 346–349.

McKinnon, A.J., Cattrall, R.W., and Scollary, G.R. 1993. Aluminum in wine—its measurement and identification of major sources. *Am. J. Enol. Vitic.* 43:166–170.

McKinnon, T. 1993. Some aspects of calcium tartrate precipitation. *Australian Grapegrower and Winemaker* 352:89–91.

McLaren, A.D., Peterson, G.H., and Barshad, I. 1958. The adsorption and reactions of enzymes and proteins on clay minerals. IV. Kaolinite and Monmorillonite. In: *Proceedings of Soil Science Society of America* 22:239–244.

McWilliams, D.J. and Ough, C.S. 1974. Measurement of ammonia in must and wine using a selective electrode. *Am. J. Enol. and Vitic.* 25(2):67–72.

Mehlhorn, R.J. and Cole, G. 1985. The free radical theory of aging: a critical review. *Free Rad. Biol. Med.* 1:165–223.

Mehlhorn, R.J. and Swanson, C.E. 1992. Nitroxide-stimulated H_2O_2 decomposition by peroxidases and pseudoperoxidases. *Free Rad. Res. Commun.* 17:157–175.

Meidell, J. 1987. Unsuitability of fluorescence microscopy for rapid detection of small numbers of yeast cells on a membrane filter. *Am. J. Enol. and Vitic.* 38(2): 159–160.

Meilgaard, M., Civille, G.V., and Carr, B.T. 1991. *Sensory Evaluation Techniques,* 2nd ed. Boston, MA: CRC Press, Inc.

Meister, A. 1973. On the enzymology of amino acid transport. *Science* 180:33–39.

Meister, A. 1992. On the antioxidant effects of ascorbic acid and glutathione. *Biochem. Pharmacol.* 44:1905–1915.

Melamed, N. 1962. Determination des sucres residuels des vins. Leur relation avec la fermentation malolactique. *Ann. Technol. Agric.* 11:5–31.

Mello-Filho, A.C., Hoffmann, M.E., and Meneghini, R. 1984. Cell killing and DNA damage by hydrogen peroxide mediated by intracellular iron. *Biochem. J.* 218:273–275.

Melnick, D. and Luckmann, F.H. 1954. Sorbic acid as a fungistatic agent in foods III. Spectrophotometric determination of sorbic acid in cheese and cheese wrappers. *Food Res.* 19:20–21.

Mercer, W.A. and Sommers, I.I. 1957. Chlorine in food plant sanitation. *Adv. Food Res.* 7:129–168.

MESROB, B., GORINOVA, N., and TZAKOV, D. 1983. Characterization of the electrical properties and of the molecular weights of the proteins of white wines. *Nahrung* 27:727–733.

MILLER, C.V. and HEILMANN, A.S. 1952. Ascorbic acid and physiological breakdown in the fruits of the pineapple. *Science* 116:505.

MILLER, J. 1966. Viewpoint: Quality carbonation of wine. *Wine and Vines,* June.

MILLER, N.J., RICE-EVANS, C., and DAVIES, M.J. 1993. A new method for measuring antioxidant activity. *Biochem. Soc. Trans.* 21(Suppl.):95S.

MILLIES, K. 1975. Protein stabilization of wines using silica sol/gelatin fining. Mitteilungsblat der GDCH-Fachgruppe. *Lebensm. Gerich. Chemie.* 29:50–53.

MILLS, J.A. 1991. Aspirin, the ageless remedy? *N. Engl. J. Med.* 325:1303–1304.

MOBAY. 1976. Baykisol 30. Wine and Fruit Juice Fining with Baykisol 30. AC10022E.

MODRA, E.J. and WILLIAMS, P.J. 1988. Are proteases active in wines and juices? *Australian Grapegrower and Winemaker* 292:42–46.

MONCADA, S., HIGGS, E.A., HODSON, H.F., KNOWLES, R.G., LOPEZ-JARAMILLO, P., MCCALL, T., PALMER, R.M.J., RADOMSKI, M.W., REES, D.D., and SCHULZ, R. 1991a. The L-Arginine:Nitric Oxide Pathway. *J. Cardiovasc. Pharmacol.* (1991a) 17(Suppl.3):S1–S9.

MONCADA, S., PALMER, R.M.J., and HIGGS, E.A. 1989. Biosynthesis of nitric oxide from L-arginine. A pathway for the regulation of cell function and communication. *Biochem. Pharmacol.* 38:1709–1715.

MONCADA, S., PALMER, R.M.J., and HIGGS, E.A. 1991b. Nitric oxide: physiology, pathophysiology, and pharmacology. *Pharmacol. Rev.* 43:109–142.

MONK, R. 1986. Rehydration and propagation of active dry wine yeast. *Australian Wine Ind J.* 1(1):3–5.

MONTEIRO, F.F., TROUSDALE, E.K., and BISSON, L.F. 1989. Ethyl carabamate formation in wine: Use of radioactively labeled precursors to demonstrate the involvement of urea. *Am. J. Enol. and Vitic.* 40:1–8.

MORETTI, R.H. and BERG, H.W. 1965. Variability among wines to protein clouding. *Am. J. Enol. and Vitic.* 16:69–78.

MOROZ, L.A. 1977. Increased blood fibrinolytic activity after aspirin ingestion. *N. Engl. J. Med.* 296:525–9.

MORRIS, J.R., SIMS, C.A., and CAWTHON, D.L. 1983. Effects of excessive potassium levels on pH acidity and color of fresh and stored grape juice. *Am. J. Enol. and Vitic.* 34(1):35–39.

MOUSCHMOUSCH, B. and ABI-MANSOUR, P. 1991. Alcohol and the heart. The long-term effects of alcohol on the cardiovascular system. *Arch. Int. Med.* 151:36–42.

MULLER, C.J. and FUGELSANG, K.C. 1993. Gentisic acid: an aspirin-like constituent of wine. *Pract. Winery and Vineyard* Sep/Oct:45–46.

MULLER, C.J. and FUGELSANG, K.C. 1994a. Post bottling hydrogen sulfide and 'corkiness'—any relationship? *Pract. Winery and Vineyard* March/April:35–36.

MULLER, C.J., and FUGELSANG, K.C. 1994b. Take two glasses of wine and see me in the morning. *The Lancet* 343:1428–1429.

MULLER, C.J. and FUGELSANG, K.C. 1994c. Antioxidants in foods—another look. *Am. J. Clin. Nutrition* 60:456–57.

MULLER, C.J., FUGELSANG, K.C., and WAHLSTROM, V.L. 1993. Capture and use of volatile flavor constituents emitted during wine fermentations. In: *Beer and Wine Production: Analysis, Characterization and Technological Advances.* B.H. Gump (ed.). Am. Chem. Society Series 536 pp 219–34. Washington, D.C.

MULLER, C.J., STRIEGLER, R.K., FUGELSANG, K.C., and WINEMAN, D.R. 1994. Salicylic acid: rootstock defense? *Pract. Winery and Vineyard* March/April:17–19.

MULLER, T., WURDIG, G., SHOLTEN, G., and FRIEDRICH, G. 1990. Determination of the calcium tartrate saturation temperature of wines by means of conductivity measurements. *Mitteilung Klosterneuburg* 40:158–168.

MÜLLER-SPÄTH, H. 1992. POM TEST, phenolics detectable in an oxidizing medium. *Deutsche Weinbau* 47:1099–1100.

MULLER-SPATH, H., MOSCHTERT, N., and SCHAFER, G. 1978. Observations in winemaking: the present state of the art. Seitz Technical Communication. Reprinted from *Die Weinwertschaft,* 36.

MULLINS, M.G. and MEREDITH, C.P. 1989. The nature of clonal variation in winegrapes: A review. *Proceedings of the Seventh Australian Wine Industry Technical Conference.* (Adelaide) P.J. Williams, D.M. Davidson, and T.H. Lee (eds.). Austr. Ind. Publ. pp. 79–82.

MUNOZ, A.M., CIVILLE, G.V., and CARR, B.T. 1992. *Sensory Evaluation in Quality Control.* New York: Van Nostrand Rheinhold Pub.

MUNOZ, E. and INGLEDEW, W.M. 1989. Effect of yeast hulls on stuck and sluggish wine fermentations: Importance of the lipid component. *J. Applied and Environ. Microbiology* 55:1560–1564.

MUNOZ, E. and INGLEDEW, W.M. 1990. Yeast hulls in wine fermentations—a review. *J. Wine Res.* 1:197–210.

MUNZ, T. 1960. Die bildung des Ca-Doppelsalzes der Wein-und Apfelsaure die moglichkeiten seiner fallung durch $CaCO_3$ im Most. *Weinberg Keller* 7:239–247.

MUNZ, T. 1961. Methoden zur praktische Fallung der Wein-und Apfelsaure als Ca-Doppelsalz. *Weinberg Keller* 8:155–158.

MURAOKA, H., WATABE, Y., OGASAWARA, N., and TAKAHASHI, H. 1983. Trigger damage by oxygen deficiency to the acid production system during submerged acetic acid fermentation with *Acetobacter aceti. J. Ferm. Tech.* 61:89–93.

MURPHEY, J.M., SPAYD, S.E., and POWERS, J.R. 1989. Effect of grape maturation on soluble protein characteristics of Gewurztraminer and white riesling juice and wine. *Am. J. Enol. and Vitic.,* 40:199–207.

MURRELL, W.G. and RANKINE, B.C. 1979. Isolation and identification of a sporing *Bacillus* from bottled brandy. *Am. J. Enol. and Vitic.* 30:247–249.

NAES, H., UTKILEN, H.C., and POST, A. 1988. Factors influencing geosmin production by Cyanobacterium *Oscillatoria brevis*. *Water Sci. and Tech.* 20:125–131.

NAGEL, C.W. and GRABER, W.R. 1988. Effect of must oxidation on quality of white wines. *Am. J. Enol. and Vitic.* 39(1):1–4.

NAIR, N.G. 1985. Fungi associated with bunch rot grapes in the Hunter Valley. *Australian J Agric Res.* 36:435–442.

NEL, L., WINGFIELD, B.D., VAN DER MEER, L.S., and VAN VUUREN, H.J. 1987. Isolation and characterization of *Leuconostoc oenos* bacteriophages from wine and sugar cane. *FEMS Microbiol. Lett.* 44:63–67.

NELSON, K.E. 1951. Effect of humidity on infection of table grapes by *Botrytis cinerea*. *Phytopathology* 41:859–864.

NERADT, F. 1977. A reliable new tartrate stabilization process. Presented at the Annual Meeting of the Am. Soc. for Enol. and Vitic. Cornado, CA.

NEWMARK, H.L. 1987. Plant phenolics as inhibitors of mutational and precarcinogenic events. *Can. J. Physiol. Pharmacol.* 65:461–466.

NISHIMURA, K., and MASUDA, M. 1983. Compounds characteristic of Botrytized wines. *Proc. of the Am. Chem. Soc.* Seattle, WA.

NISHINO, H., MIYAZAKI, S., and TOHJO, K. 1985. Effect of osmotic pressure on the growth rate and fermentation activity of wine yeasts. *Am. J. Enol. and Vitic.* 36: 170–174.

NOBEL, A.C. 1984. Precision and communication: descriptive analysis of wine. *Wine Ind. Tech. Symposium* 33–41. San Jose, CA.

NOBEL, A.C. and SHANNON, M. 1987. Profiling Zinfandel wines by sensory and chemical analysis. *Am. J. Enol. and Vitic.* 38(1):1–8.

NOBEL, A.C., ARNOLD, R.A., BUECHENSTEIN, J., LEACH, E.J., SCHMIDT, J.O., and STERN, P.M. 1987. Research note: modification of a standardized system of aroma terminology. *Am. J. Enol. and Vitic.* 38(2):143–146.

NOBEL, A.C., ARNOLD, R.A., MASUDA, B.M., PECORE, S.D., SCHMIDT, J.O., and STERN, P.M. 1984. Progress towards a standardized system of wine aroma terminology. *Am. J. Enol. and Vitic* 35:107–109.

NORDSTROM, K. 1961. Formation of ethyl acetate in fermentation with Brewer's yeast. *J. Inst. Brew.* 67:173–181.

NORDSTROM, K. 1963. Formation of esters from acids by brewers yeast. I: Kinetic theory and basic experiments. *J. Inst. Brew.* 69:310–322.

NORDSTROM, K. 1964. formation of esters from acids by Brewer's yeast. *J. Inst. Brew.* 70:38–40.

NORDSTROM, K. 1965a. Formation of esters from lower fatty acids by various yeasts species. *J. Inst. Brew.* 72:38–40.

NORDSTROM, K. 1965b. Formation of volatile esters by brewers yeast. *Brewers Dig.* 40(11):60–67.

NORDSTROM, K. 1966. Formation of esters from lower fatty acids by various yeast species. *J. Inst. Brew.* 72:38–40.

OBATA, T., HOSOKAWA, H., and YAMANAKA, Y. 1993. Effect of ferrous iron on the generation of hydroxyl free radicals by liver microdialysis perfusion of salicylate. *Comp. Biochem. Physiol.* 106C:629–34.

OFFICE INTERNATIONAL DE LA VIGNE ET DU VIN (OIV), 1990. Official Analytical Methodology (English Translation).

OFFICIAL JOURNAL OF THE EUROPEAN COMMUNITIES (ENGLISH EDITION). 1978. Volume 21:28.

OFFICIAL METHODS OF ANALYSIS OF THE ASSOCIATION OF OFFICIAL ANALYTICAL CHEMISTS. 14TH ED. 1984. W. Horwitz (ed.). Washington, D.C.

OHLIN, H., BRATTSTROM, L., ISRAELSSON, B., BERGQVIST, D., and JERNTORP, P. 1991. Atherosclerosis and acetaldehyde metabolism in blood. *Biochem. Med. Metab. Biol.* 46:317–328.

OLIJVE, W. and KOK, J.J. 1979. Analysis of growth of *Gluconobacter oxydans* in glucose containing media. *Arch. Microbiol.* 121:283–290.

OLPHEN, H. 1963. *An Introduction to Clay Colloid Chemistry.* New York: Interscience Publishers.

OLSEN, E. 1948. Studies of bacteria in Danish fruit wines. *Antonie van Leeuwenhoek J. Microbiol. Serol.* 14:1–28.

ONG, B.Y. and NAGEL, C.W. 1978. Hydroxycinnamic acid-tartaric acid ester content in mature grapes and during maturation of White Riesling grapes. *Am. J. Enol. and Vitic.* 29(4):277–281.

OOGHE, W. and KASTELYN, H. 1988. Amenozuurpatroon van druinenmost aangewend voor de bereiding van wijenen van gegarandeede herkomst. *Belg. J. Food Chem. Biotechnol.* 43:15–21.

O'REILLY, P. 1993. A review of fining materials and practice. *Vineyard and Winery Management.* May/June.

ORGANIZATION OF AMERICAN STATES. 1975. Paper and thin-layer chromatography (Monograph No. 16). Regional Program of Scientific and Technological Development. Department of Scientific Affairs, Washington, D.C.

OSBORN, M.M., FUGELSANG, K.C. and MULLER, C.J. 1991. Impact of native yeast flora on grape wine quality. Presented at Brewers and Vintners Annual Conference, Shell Beach, CA.

OSZMINASKI, J., RONMEYER, F.M., SAPIS, J.C., and MACHEIX, J.J. 1986. Grape seed phenolics: extraction as effected by some conditions occurring during wine processing. *Am. J. Enol. and Vitic.* 37(1):7–12.

OUGH, C.S. 1960. Gelatin and polyvinylpyrrolidone compared for fining red wines. *Am. J. Enol. and Vitic.* 11:170–173.

OUGH, C.S. 1968. Proline content of grapes and wine. *Vitis* 7:321–331.

OUGH, C.S. 1969. Substances extracted during skin contact with white must. I. General wine composition and quality changes with contact time. *Am. J. Enol. and Vitic.* 20:93–100.

OUGH, C.S. 1993a. Lead in wines—a review of recent reports. *Am. J. Enol. and Vitic.* 44(4):464–467.

OUGH, C.S. 1993b. Report on ethyl carbamate for the Wine Institute. Ethyl carbamate/urease enzyme preparation. A compendium from June, 1993 seminars.

OUGH, C.S. and AMERINE, M.A. 1962. Studies with controlled fermentation. VII. Effect of ante-fermentation blending of red must and white juice on color, tannins, and quality of Cabernet Sauvignon wine. *Am. J. Enol. and Vitic.* 13:181–188.

OUGH, C.S. and AMERINE, M.A. 1988. *Methods for Analysis of Musts and Wines, 2nd ed.* NY: Wiley-Interscience.

OUGH, C.S. and ANELI, G. 1979. Zinfandel grape juice and amino acid makeup as affected by crop level. *Vitis* 30:8–10.

OUGH, C.S. and BELL, A.A. 1980. Effects of nitrogen fertilization of grape vines on amino acid metabolism and higher alcohol formation during grape juice fermentation. *Am. J. Enol. and Vitic.* 31:122–123.

OUGH, C.S. and BERG, H.W. 1974. The effect of two commercial pectic enzymes on grape musts and wines. *Am. J. Enol. and Vitic.* 25:208–211.

OUGH, C.S. and INGRAHAM, J.L. 1960. Use of sorbic acid and sulfur dioxide in sweet table wines. *Am. J. Enol. and Vitic.* 11:117–122.

OUGH, C.S. and KRIEL, A. 1985. Ammonia concentrations of musts of different grape cultivars and vineyards in the Stellenbosh area. *S. African J. Enol. Vitic.* 6:7–11.

OUGH, C.S. and KUNKEE, R.E. 1974. The effect of fumaric acid on malolactic fermentation in wine from warm areas. *Am. J. Enol. Vitic.* 25:188–190.

OUGH, C.S. and STASHAK, R.M. 1974. Further studies on proline concentration in grapes and wines. *Am. J. Enol. Vitic.* 25:7–12.

OUGH, C.S., CAPUTI, A., JR., and GROAT, M. 1979. A rapid colorimetric calcium method. *Am. J. Enol. and Vitic.* 30:8–10.

OUGH, C.S., CROWELL, E.A., and GUTLOVE, B.R. 1988. Carbamyl compound reaction with ethanol. *Am. J. Enol. and Vitic.* 39:239–242.

OUGH, C.S., FONG, D., and AMERINE, M.A. 1972. Glycerol in wine: determination and some factors affecting formation. *Am. J. Enol. and Vitic.* 23:1–5.

OUGH, C.S., GUYMON, J.F., and CROWELL, E.A. 1966. Formation of higher alcohols during the fermentation of grape juice at different temperatures. *J. Food Sci.* 31:620–625.

OUGH, C.S., STEVENS, D., SEDOVSKI, T., HUANG, Z., and AN, D. 1990. Factors contributing to urea formation in commercially fermented wines. *Am. J. Enol. Vitic.* 41:68–73.

OURA, E. 1977. Reaction products of yeast fermentation. Proc. *Biochem.* 12(3):19–23.

PALLOTTA, U. and CANTARELLI, C. 1979. Le Catechine: loro importanza sulla qualita dei vini bianchi. *Estratto da Vignevini, Gruppo Giornalistico Edagricole* 4:19–46.

PALMER, R.M.J. and MONCADA, S. 1989. A novel citrulline-forming enzyme implicated in the formation of nitric oxide by vascular endothelial cells. *Biochem. Biophys. Res. Commun.* 158:348–352.

PALMER, R.M.J., FERRIGE, A.G., and MONCADA, S. 1987. Nitric oxide release accounts for the biological activity of endothelium-derived relaxing factor. *Nature* (Letter) 327:524–526.

PARDO, I., GARCIA, M.I., ZUNIGA, M., and URUBURU, F. 1989. Dynamics of microbial populations during fermentation of wines from the Utiel-Requena region of Spain. *Applied and Environ. Microbiology* 55:539–541.

PELLETIER, O. 1975. Vitamin C and cigarette smokers. *Ann. NY Acad. Sci.* 258:156–168.

PENA, A., CINCO, G., GOMEZ-PUYON, A., and TUENA, M. 1972. Effect of the pH of the incubation medium on glycolysis and respiration in *Saccharomyces cerevisiae*. *Arch. Biochem. Biophys.* 153:413–425.

PERIN, J. 1977. Compte rendu de quelques essais de refrigeration des vins. *Le Vigneron Chapenois* 98(3):97–101.

PETERSON, R.G. 1976. Formation of reduced pressure during wine aging. *Am. J. Enol. Vitic.* 27:24–36.

PETERSON, R.G., JOSLYN, M.A., and DURBIN, P.W. 1958. Mechanism of copper formation in white table wines. III. Source of the sulfur sediment. *Food Res.* 23:518–524.

PEYNAUD, E. 1937. Etudes sur les phenomenes d'esterfication. *Rev. Viticult.* 86:209–475.

PEYNAUD, E. 1939–1940. Sur la formation et la dimunition des acides volatile pendant la fermentation alcoolique en anaerobiose. *Ann. Ferment.* 5:321–327, 385–402.

PEYNAUD, E. 1947. Contribution a l'etude biochimique de la maturation du raisin et de la composition des vins. *Inds. Agr. et Aliment.* 64:87–414.

PEYNAUD, E. 1956. New information concerning biological degradation of acids. *Am. J. Enol.* 7:150–156.

PEYNAUD, E. 1984. *Knowing and Making Wine.* New York: John Wiley and Sons, 391 pp.

PEYNAUD, E. and DOMERCO, S. 1953. Etude des levures de la gironde. *Ann. Technol. Agr.* 2:265–300.

PEYNAUD, E. and GUIMBERTEAU, G., 1961. Recerches sur la constitution et l'efficacite anticristallisante de l'acide metatartarique. *Ind. Alimant. Agric. (Paris)* 78:131–35, 413–18.

PEYNAUD, E. and MAURIE, A. 1953. Sur l'evolution de l'azote dans les differentes parties du raisin au cours de la maturation. *Ann. Technol. Agr.* 2:15–25.

PEYNAUD, E. and SUBRAUD, P. 1964. Utilisation de l'effet desacidifiant des *Schizosaccharomyces* en vinification de raisins acides. *Ann. Technol. Agr.* 13:309–328.

PEYNAUD, E., GIUMBERTEAU, G., and BLOUIN, J. 1964. Die Loslichkeitsgleichgewichte von Kalzium and Kalium in Wein. *Mitt. Rebe u. Wein. Serie A (Klosterneuburg)* 14:176–186.

PFEFFER, T.E., CLARY, C.D., and PETRUCCI, V.E. 1985. Adaptation of HPLC to Wine Grape Inspection. Report to the Wine Grape Inspection Advisory Committee, CDFA, Viticulture Research Center, California State University, Fresno.

PHAFF, H.J., MILLER, M.W., and MRAK, E.M. 1978. *The Life of Yeasts,* 2nd ed. Cambridge, MA: Harvard University Press, 154 pp.

PILONE, G.J. 1967. Effect of lactic acid on volatile acid determination of wine. *Am. J. Enol. and Vitic.* 18:149–156.

PILONE, G.J. 1979. Technical Note: Preservation of Wine Yeast and Lactic Acid Bacteria. *Am. J. Enol. and Vitic* 30(4):326.

PILONE, G.J. and BERG, H.W. 1965. Some factors affecting tartrate stability in wine. *Am. J. Enol. and Vitic.* 16:195–211.

PILONE, G.J. and KUNKEE, R.E. 1972. Characterization and energetics of *Leuconcstoc oenos* ML-34. *Am. J. Enol. and Vitic.* 23(2):61–69.

PILONE, G.J., CLAYTON, M.G., VAN DUIVENBODEN, R.J. 1991. Characterization of wine lactic acid bacteria: single broth culture tests of heterofermentation, mannitol from fructose and ammonia from arginine. *Am. J. Enol. and Vitic.* 42(2): 153–157.

PILONE, G.J., KUNKEE, R.E., and WEBB, A.D. 1966. Chemical characterization of wines fermented with various malolactic bacteria. *Appl. Microbiol.* 14:608–615.

PILONE, G.J., RANKINE, B.C., and PILONE, A.D. 1974. Inhibiting malolactic fermentations in Australian dry red wines by adding fumaric acid. *Am. J. Enol. and Vitic.* 25(2):99–107.

PITTILO, R.M. 1990. Cigarette smoking and endothelial injury: a review. In: *Tobacco Smoking and Atherosclerosis.* J.N. Diana (ed.). New York: Plenum Press, pp. 61–78.

PLANE, R.A., MATTICK, L.R., and WEIRS, L.D. 1980. An acidity index for the taste of wines. *Am. J. Enol. and Vitic.* 31:265–268.

PLANK, P.F.H. and ZENT, J.B. 1993. Use of enzymes in winemaking and grape processing. In: *Beer and Wine Production Analysis. Characterization and Technological Advances.* B.H. Gump (ed.). ACS symposium series 536, Washington, DC. pp. 181–196.

POCOCK, K.F. 1983. Analytical survey of commercial bentonites. *Aust. Wine Res. Inst. Tech. Rev.* 27:19–24.

POCOCK, K.F. and RANKINE, B.C. 1973. Heat test for detecting protein instability in wine. *Australian Wine Brew. and Spirit Rev.* 91(5):42–43.

POINTING, J.D. and JOHNSON, G. 1945. Determination of sulfur dioxide in fruits. *Ind. Eng. Chem. (Anal. Ed.)* 17:682–686.

POIRIER, J. and THIFFAULT, C. 1993. Are free radicals involved in the pathogenesis of idiopathic Parkinson's disease? *Eur. Neurol.* 33(Suppl.1):38–43.

POLI, G., PAROLA, M., LEONARDUZZI, and PINZANI, M. 1993. Modulation of hepatic fibrogenesis by antioxidants. *Molec. Aspects Med.* 14:259–264.

PONTALLIER, P. 1987. Pratiques actulles de l élevage en barriques des gran vins rouges. *Connaissance Vigne Vin.* Special issue 143–149.

PORTER, L.J. and OUGH, C.S. 1982. The effects of ethanol, temperature, and dimethyl dicarbonate on viability of *Saccharomyces cerevisiae* Montrachet No. 522 in wine. *Am. J. Enol. and Vitic.* 33(4):222–225.

POSTEL, W. 1983. La solubilite et la cinetique de crystalisation du tartrate de calcium dans de vin. *Bull. OIV* 629–630.

POSTEL, W. and PRASCH, E. 1977. Untersuchungen zur Weinsteinstabilizierung von Wein durh Elektrodialyse. I. Mitteilung. Absenkung der Kalium un Weinsaurekonzentration. *Weinwirtschaft* 113(45):1277–1283.

POWERS, J.R., SHIVELY, A., and NEGEL, C.W. 1980. Effect of Ethephon on the colour of Pinot noir fruit and wine. *Am. J. Enol. and Vitic.* 31:203–205.

PRATT, D.E. 1993. Antioxidants indigenous to foods. *Toxicol. Ind. Health* 9:63–75.

PRYOR, W.A. 1982. Free radical biology: xenobiotics, cancer, and aging. *Ann. NY Acad. Sci.* 393:1–22.

PRYOR, W.A. 1991. The antioxidant nutrients and disease prevention—what do we know and what do we need to find out? *Am. J. Clin. Nutr.* 53(Suppl.):391S–393S.

PRYOR, W.A. and STONE, K. 1993. Oxidants and cigarette smoke. Radicals, hydrogen peroxide, peroxinitrate and peroxynitrite. *Ann. NY Acad. Sci.* 686:12–28.

PUPPO, A. and HALLIWELL, B. 1988. Formation of hydroxyl radicals from hydrogen peroxide in the presence of iron. *Biochem. J.* 249:185–190.

PURCHEU-PLANTÉ, B. and MERCIER, M. 1983. Etude ultrastructural de l'interellation hôte-parasite entre le raisin et le champaignon *Botrytis cinerea*: example de la pourriture noble en Sauternes. *Can J Botany* 61:1785–1797.

QUINN, K. and SINGLETON, V.L. 1985. Isolation and identification of ellagitannins from white oak wood and an estimation of their roles in wine. *Am. J. Enol. and Vitic.* 36:148–155.

QUINSLAND, D. 1978. Technical Note. Identification of common sediments in wine. *Am. J. Enol. and Vitic* 29(1):70–71.

RADLER, F. 1965. The main constituents of the surace waxes of varieties and species of the genus *Vitis. Am. J. Enol. and Vitic.* 16:159–167.

RADLER, F. 1968. Bakterieller Apfelsaureabbau in Deutschen spitzenweinen. *Z. Lebensm. Untersuch. Forsch.* 138:35–39.

RADLER, F. 1989. Vitamine und Wuchsstoffe. In: *Chemie des Weins.* G. Wurdig and R. Woller (eds.). Stuttgart: Ulmer, pp. 119–20.

RADLER, F. and SCHUTZ, H. 1982. Glycerol production from various strains of *Saccharomyces. Am. J. Enol. and Vitic.* 33(1):36–40.

RADLER, F., PFEIFFER, P., and DENNERT, M. 1985. Killer toxins in new isolates of the yeast *Hanseniaspora uvarum* and *Pichia kluyveri. FEMS Microbiology Letters* 29:269–272.

RAINIERI, R. and WEISBURGER, J.H. 1975. Reduction of gastric carcinogens with ascorbic acid. *Ann. NY Acad. Sci.* 258:181–189.

RAMEY, D., BERTRAND, A., OUGH, C.S., SINGLETON, V.L., and SANDERS, E. 1982. Effects of temperature variation during skin contact on Chardonnay must and wine. *Am. J. Enol. and Vitic.* 27(2):1986.

RANJI, R.G., RODRIQUEZ, S.G., and THORNTON, R.J. 1988. Glycerol production by four common grape molds. *Am. J. Enol. and Vitic.* 39:77–82.

RANKINE, B.C. 1955. Quantitative differences in products of fermentation by different strains of wine yeasts. *Am. J. Enol. and Vitic.* 6(1):1–10.

RANKINE, B.C. 1962. Aluminum haze in wine. *Aust. Wine Brew. Spirits Rev.* 80(9): 14–16.

RANKINE, B.C. 1963. Nature, origin, and prevention of H_2S aroma in wines. *J. Sci. Food Agr.* 14:75–91.

RANKINE, B.C. 1966a. Decomposition of L-malic acid by wine yeasts. *J. Sci. Food Agric.* 17:312–316.

RANKINE, B.C. 1966b. *Pichia membrafaciens,* a yeast causing film formation and off flavor in table wine. *Am. J. Enol. and Vitic.* 17:302–307.

RANKINE, B.C. 1967. Formation of higher alcohols by wine yeasts, and relationship to taste thresholds. *J. Sci. Food. Agric.* 18:583–589.

RANKINE, B.C. 1972. Influence of yeast strain and malolactic fermentation on the composition and quality of table wines. *Am. J. Enol. and Vitic.* 22:152–158.

RANKINE, B.C. 1977. Developments in malolactic fermentations in Australian table wines. *Am. J. Enol. and Vitic* 28:27–33.

RANKINE, B.C. 1984. Use of isinglass to fine wines. *Australian Grapegrower and Winemaker* 249:16.

RANKINE, B.C. and BRIDSON, D.A. 1971. Glycerol in Australian wines and factors influencing its formation. *Am. J. Enol. and Vitic.* 22:6–12.

RANKINE, B.C. and POCOCK, K.F. 1969a. B-Phenethanol and n-hexanol in wines: influence of yeast strain, grape variety and other factors; and taste thresholds. *Vitis* 8:23–37.

RANKINE, B.C. and POCOCK, K.F. 1969b. Influence of yeast strain on binding of sulfur dioxide in wines and on its formation during fermentation. *J. Sci. Food Agric.* 20:104–109.

RANKINE, B.C. and POCOCK, K.F. 1971. A new method for detecting protein instability in white wines. *Australian Wine. Brew. Spirits Rev.* 89:61.

RANKINE, B.C., FORNACHON, M.C.M., BOEHM, E.N., and CELLIER, K.M. 1971. The influence of grape variety, climate and soil on grape composition and quality of table wines. *Vitis* 10:33–50.

RAO, M.R.R. 1957. The acetic acid bacteria. *Ann. Rev. Microbiol.* 11:317–338.

RAPP, A. 1988. Wine aroma substances from gas chromatographic analysis. In: *Modern Methods of Plant Analysis: Wine Analysis.* H.F. Linskens and J.F. Jackson (eds.). Springer-Verlag, NY. pp 29–61.

RAPP, A. and VERSINI, G. 1991. Influence of nitrogen compounds in grapes on aroma compounds of wine. In: *Proc. of the International Symposium on Nitogen in Grapes and Wines* (Seattle, WA.) Am. Soc. for Enol. and Vitic. J. Rantz (ed.). pp 156–64.

RAPP, A., KNIPSER, W., HASTRICH, H., and ENGEL, L. 1982. Possibilities of characterizing wine quality and vine varieties by means of capillary chromatography. A.D. Webb, (ed.). *Grape and wine centennial symposium proceedings;* June 18–21, 1980; Davis, CA: University of California, pp. 304–316.

RASKIN, I. 1992a. Role of salicylic acid in plants. *Annu. Rev. Plant Physiol. Mol. Biol.* 43:439–463.

RASKIN, I. 1992b. Salicylate, a new plant hormone. *Plant Physiol.* 99:799–803.

RAUHUT, D. 1990. Trace analysis of sulfurous off-flavors in wine caused by extremely volatile S-containing metabolites of pesticides e.g. Orthene. In: *Actualities Oenologiques 89.* Symposium International d'Oenologie, Bordeaux, 15–17 June, 1989. P. Ribereau-Gayon and A. Lonvaud (eds.). pp 482–487.

RAUHUT, D., and SPONHOLZ, W.R. 1992. Sulfur off-odors in wine. Wine aroma defects. In: *Proceedings of the ASEV/ES Workshop.* T. Henick-Kling (ed.). *Am. Soc. Enol. Vitic.* pp 44–76.

READ, N., FRENCH, S., and CUNNINGHAM, K. 1994. The role of the gut in regulating food intake by man. *Nutr. Rev.* 52:1–10.

REAZIN, G., SCALES, H., and ANDREASE, A. 1970. Mechanism of major cogener formation in alcoholic grain fermentations. *J. Agric. Food* 18(4):585–588.

RECHT, J.A. 1992. Enological materials and ingredients Part I. *Wine East* Sept/Oct: 22–27.

RECHT, J.A. 1993. Wine stabilization: chemical and physical stability. *Wine East* Jan/Feb:16–24.

RECHT, J. 1993. The *visual aspect of wine. Wine East*:8–14.

REED, G. 1966. *Enzymes in Food Processing.* York: Academic Press.

REED, G. and CHEN, S.L. 1978. Evaluating commercial active dry wine yeasts by fermentation activity. *Am. J. Enol. and Vitic.* 29:165–168.

REED, G. and NAGODAWITHANA, T.W. 1991. *Yeast Technology,* 2nd ed. New York: Van Nostrand Reinhold, 454 pp.

REED, G. and PEPPLER, H.J. 1973. *Yeast Technology.* Westport, CT: AVI Publishing Co.

REHAN, A., JOHNSON, K.J., WIGGINS, R.C., KUNKEL, R.G., and WARD, P.A. 1984. Evidence for the role of oxygen radicals in acute nephrotoxic nephritis. *Lab. Invest.* 51:396–403.

RENAUD, S. and DE LORGERIL, M. 1992. Wine, alcohol, platelets, and the French paradox for coronary heart disease. *Lanet* 339:1523–1526.

RENAUD, S. and DE LORGERIL, M. 1993. The French paradox: dietary factors and cigarette smoking-related health risks. *Ann. NY Acad. Sci.* 686:299–309.

RENTSCHLER, H. and TANNER, H. 1951a. Das Bitterwerden der Rotweine. Beitsag zur kennfinis des vorkommens von Acroleins in Getranken und seine Bezihung zum Bitterwerden der Weine. *Mitt. Gebiete Lebenson Hyg.* 42:463–475.

RENTSCHLER, H. and TANNER, H. 1951b. Uber die Kupfersulfittrubung von Weiss Weinen, und Sussmosten. *Schweiz. Zeit Obst-und Weinbau* 60:298–301.

RHEIN, O. and NERADT, F. 1979. Tartrate stabilization by the contact process. *Am. J. Enol. and Vitic.* 30(4):265–271.

RIBEREAU-GAYON, J. 1933. Contribution a l'etude des oxidations et reductions dans les vin. Application a l'etude du vieillissement et des casses. 2nd ed. Bordeaux, France.

RIBEREAU-GAYON, J. 1961. La composition chimique des vins. In: *Traite d'Oenologie.* Librairie Polytechnique Ch. Beranger, Paris.

RIBEREAU-GAYON, J. 1963. Phenomenon of oxidation and reduction in wines and applications. *Am. J. Enol. and Vitic.* 14:139–143.

RIBEREAU-GAYON, J. 1965. Identification d'esters des acides cinnamiques et l'acide tartique dans les limbes et les baies de *V. vinifera. C.R. Acad. Sci. Paris* 260:341–343.

RIBEREAU-GAYON, J. 1972. Evolution des composes phenoliques au cours de la maturation du raisin. II. Discussion des resultats obtenus en 1969, 1970 et 1971. *Connaissance Vigne Vin* 2:161–175.

RIBEREAU-GAYON, J. 1974. The chemistry of red wine color. In: *The Chemistry of Winemaking.* A.D. Webb (ed.). Advances in Chemistry 137. Washington, DC: American Chemical Society.

RIBEREAU-GAYON, P. 1988. *Botrytis:* Advantages and disadvantages for producing quality wines. In: *Proceedings of the Second Int. Symposium for Cool Climate Viticulture.* R. Smart, R. Thornton, J. Rodriquez and J. Young (eds.). New Zealand Soc. Vitic. Oenol. Auckland, N.Z. pp 319–323.

RIBEREAU-GAYON, J. and PEYNAUD, E. 1958. *Analyses et Controle des Vins,* 2nd ed. Paris and Liege: Libraire Polytechnique.

RIBEREAU-GAYON, J. and PEYNAUD, E. 1961. *Traite d'Oenologie.* Paris: Libraire Polytechnique.

RIBEREAU-GAYON, J., PEYNAUD, E., and LAFON, M. 1956a. Investigations on the origin of secondary products of alcoholic fermentation, Part I. *Am. J. Enol. and Vitic.* 7:53–61.

RIBEREAU-GAYON, J., PEYNAUD, E., and LAFON, M. 1956b. Investigations on the origin of secondary products of alcoholic fermentation. Part II. *Am. J. Enol. and Vitic* 7:112–118.

RIBEREAU-GAYON, J., PEYNAUD, E., SUDRAUD, P., and RIBEREAU-GAYON, P. 1972. *Sciences et Techniques du Vin.* Vol. I. Paris: Dunod, pp. 471–514.

RIBEREAU-GAYON, J., PEYNAUD, P., RIBEREAU-GAYON, P., and SUDRAUD, P. 1976. *Sciences et Technique du Vin.* Tome 3. *Vinifications Transformations du Vin.* Paris: Dunod.

RIBEREAU-GAYON, P. 1982. Incidence oenologiques de la pourriture du raisin. *Bull. OEPP* 12:201–214.

RIBEREAU-GAYON, P. 1985. New developments in wine microbiology. *Am. J. Enol. Vitic.* 36:1–10.

RIBEREAU-GAYON, P. and SUDRAUD, P. 1981. *Actualites Oenologiques et Viticoles.* Paris: Dunod, 395 pp.

RIBEREAU-GAYON, P., LAFON-LAFOURCADE, S., DUBOURDIEU, D., LUCMARET, V., and LARU, F. 1979. Metabolisme de *Saccharomyces cerevisiae* dans le mout de raisins parasites par *Botrytis cinerea.* Inhibition de la fermentation, formation d'acide acetique et de glycerol. *Comptes Rendus Academie des Sciences Paris* 289D:441–44.

RIBEREAU-GAYON, P. and GLORIES Y. 1986. Phenolics in grapes and wines. In: *Proceedings of the Sixth Australian Wine Industry Technical Conference,* T. Lee (ed.). (Adelaide) South Australia, Austr. Ind. Publ. pp. 247–256.

RIBEREAU-GAYON, P., BOIDRON, J.N., and TERRIER, A. 1975. Aroma of muscat grape varieties. *J. Agric. food Chem.* 23:1042–1047.

RIBEREAU-GAYON, J., RIBEREAU-GAYON, P., and SEGUIN, G. 1980. *Botrytis cinerea* in enology. In: *The Biology of Botrytis.* J.R. Cooley-Smith, K. Verhoeff, and W.R. Jarvis (eds.). London: Academic Press, 262 pp.

RIBOULET, J-M. 1991. Cork tastes. *Pract. Winery Vineyard* July/Aug:16–19.

RICE, A.C. 1965. The malolactic fermentation in New York State wines. *Am. J. Enol. and Vitic.* 16:62–68.

RICE-EVANS, C., MCCARTHY, P., HALLINAN, T., GREEN, N.A., GOR, J., and DIPLOCK, A.T. 1989. Iron overload and the predisposition of cells to antioxidant consumption and peroxidative damage. *Free Rad. Res. Commun.* 7:307–313.

RICHARDSON, J.S. 1993. Free radicals in the genesis of Alzheimer disease. *Ann. NY Acad. Sci.* 695:73–76.

RIDDLE, C. and TUREK, A. 1977. An indirect method for the sequential determination of silicon and phosphorus in rock analysis by atomic absorption spectrometry. *Anal. Chim. Acta* 92, 49–53.

RIDKER, P.M., VAUGHAN, D.E., STAMPFER, M.J., GLYNN, R.J., and HENNEKENS, C.H. 1994. Association of Moderate Alcohol Consumption and Plasma Concentration of Endogenous Tissue-Type Plasminogen Activator. *J. Am. Med. Assoc.* 272:929–33.

RIESE, H. 1980. Crystal-flow, a new process for rapid and continuous tartrate stabilization of wine. Presented at the 31st Annual Meeting of the American Society for Enology and Viticulture. Los Angeles, CA.

RIGAUD, J., ISSANCHOU, S., SARRIS, J., and LANGLOIS, D. 1984. Incidence des composes volatiles issues du liege sur le gout de bouchon des vins. *Sci. Alli.* 4:81–93.

RIMM, E.B., GIOVANNUCCI, E.L., WILLETT, W.C., COLDITZ, G.A., ASCHERIO, A., ROSNER, B., and STAMPFER, M.J. 1991. Prospective study of alcohol consumption and risk of coronary disease in man. *Lancet* 338:464–468.

ROBERTS, S. 1988. Personal communication.

ROBICHAUD, J.L. and NOBLE, A.C. 1990. Astringency and bitterness of selected phenolics in wine. *J. Sci. Food Agric.* 53:343–353.

ROCHA, S., DELGADILLO, I., FERRER CORREIA, A.J., BASTOS, A.M., ROSEIRA, I., and GUIMARAES, A. 1993. Elimination of cork taint in wine in the course of the cork manufacturing process. *Australian and New Zealand Wine Ind. J.* 8(3):223–227.

ROCHE, E. and ROMERO-ALVIRA, D. Oxidative stress in some dementia types. *Med. Hypoth.* 40:342–350.

RODRIQUEZ, S.B. 1987. A system for identifying spoilage yeast in packaged wine. *Am. J. Enol. and Vitic.* 38(4):237–276.

RODRIQUEZ, S. and THORNTON, R. 1988. Rapid utilization of malic acid by a mutant *Schizosaccharomyces malvidorans.* In: *Proc. of the Second Cool Climate Viticulture and Oenology Symposium.* Auckland, NZ. pp 313–315. New Zealand Soc. for Vitic. and Enol.

RODRIQUEZ, S. and THORNTON, R. 1990. Factors influencing the utilization of L-malate by yeasts. *FEMS Microbiology Letters* 72:17–22.

ROKHLENKO, S.G., VANYUSHKINA, L.D., PANIKHINA, S.L., TOKHMAKHCHI, N.S. 1980. Effect of some enzyme preparations on wine stability. *Prikl. Biokhim. Mikrobiol.* 16:291–295.

ROSE, A.H. and HARRISON, J.S. (eds.). 1970 and 1971. *The Yeasts.* Vol. II and III. London and New York: Academic Press.

ROSE, R.C. and BODE, A.M. 1993. Biology of free radical scavengers: an evaluation of ascorbate. *FASEB J.* 7:1135–1142.

ROSELL, P.F., OFRIA, H.V., and PALLERONI, N.J. 1968. Production of acetic acid from ethanol by wine yeasts. *Am. J. Enol. and Vitic.* 19(1):13–16.

ROSENQUIST, J.K. and MORRISON, J.C. 1989. Some factors affecting cuticle and wax accumulation on grapes. *Am. J. Enol. and Vitic.* 40:241–244.

ROSSI, J. and CLEMENTI, F. 1984. L-malic acid catabolism by polyacrylamide gel entraped *Leuconostoc oenos. Am. J. Enol. and Vitic.* 35(2):100–102.

ROSSI, J.A., JR. and SINGLETON, V.L. 1966. Flavor effects and adsorptive properties of purified fractions of grape seed phenols. *Am. J. Enol. and Vitic.* 17:240–246.

ROTH, G.J. and MAJERUS, P.W. 1975. The mechanism of the effect of aspirin on human platelets I. Acetylation of a particulate fraction protein. *J. Clin. Invest.* 56:624–32.

ROTH, G.J., STANFORD, N., and MEJERUS, P.W. 1975. Acetylation of prostaglandin synthase by aspirin. *Proc. Natl. Acad. Sci. USA* 72:3073–6.

ROWLAND, M., RIEGELMAN, S., HARRIS, P.A., andSHOLKOFF, S.D. 1972. Absorption Kinetics of Aspirin in Man following Oral Administration of an Aqueous Solution. *J. Pharm. Sci.* 61:379–85.

RYAN, T.P. and AUST, S.D. 1992. The role of iron in oxygen-mediated toxicities. *Crit. Rev. Toxicol.* 22:119–141.

SAAVEDRA, I.J. and GARRIDO, J.M. 1963. La levadura de "flor" en la crianza del vino. *Rev. Cienc. Apl.* 17:497–501.

SAGONE, A.L. and HUSNEY, R.M. 1987. Oxidation of salicylates by stimulated granulocytes: evidence that these drugs act as free radical scavengers in biological systems. *J. Immunol.* 138:2177–2183.

SAITO, K. and KASAI, Z. 1969. Tartaric acid synthesis from L-ascorbic acid-1-^{14}C in grape berries. *Phytochemistry* 8:2177–2182.

SALAGOITY-AUGUSTE, M-H. and BERTRAND, A. 1984. Wine phenolics—analysis of low molecular weight components by high performance liquid chromatography. *J. Sci. Food Agr.* 35:1241–1247.

SALL, M.A.,TEVIOTDALE, B.L., and SAVAGE, S.D. 1982. Bunch Rot Grape Pest Management. Div. Agric. Sciences, Univ. of California. Pub. no. 4105.

SALLER, W. 1957. Die Spontane-Sprosspilzflora frische gepresster Traubensafe und die Reinhefegarung. *Mitt. Rebe u. Wein Serie A (Klosterneuburg)* 7:130–138.

SANFEY, H., SARR, M.G., BULKLEY, G.B., and CAMERON, J.L. 1986. Oxygen-derived free radicals in pancreatitis: a review. *Acta Physiol. Scand.* Suppl.548:109–118.

SAVOLAINEN, M.J., BARAONA, E., and LIEBER, C.S. 1987. Acetaldehyde binding increases the catabolism of rat serum low-density lipoproteins. *Life Sci.* 40:841–846.

SCHANDERL, H. 1955. Uber Storungsfaktoren bei umgarungen einschlie blesh schaumweingarungen. *Weinberg und Keller* 2:313–330.

SCHANDERL, H. 1959. *Die Mikrobiologie des Mostes und Weines.* 2nd ed. Stuttgart: Eugen Ullmer.

SCHANDERL, H. 1962. Der Einfluss von polyphenolen un Gerbstoffen auf die physiologie der weinhefe und der wert des pH-7-test für des Auswahl von Sektgrundweinen. *Mitt. Rebe u. Wein, Serie A (Klosterneuburg)* 12:265–274.

SCHANDERL, H. 1971. "Korkgeschmack" von Weinen. *Alg. Deut. Wein* 107:3333–3336.

SCHANDERL, H. and DRACZINSKY, M. 1952. *Brettanomyces,* eine lustige Hefegattung in flashenvergorenen Schaumwein. *Wein und Rebe* 20:462–464.

SCHANDERL, H. and STAUDENMEYER, T. 1964. Uber den einfluss den schwefligen Saure auf die Acetaldehydbildung verschiedener hefen bei Most- und Schaumweiningarungen. *Mitt. Rebe u. Wein. Serie A (Klosterneuburg)* 14:267–281.

SCHMIDT, T.R. 1987. Potassium sorbate or sodium benzoate. *Wine and Vines* (Nov.): 42–44.

SCHMITT, A. 1987. Untersuchung des chemischen Verhaltrens von Tetramethyl thiuramdisulfid in Lebensmittein. Univ. Karlsruhe, Thesis.

SCHMITTHENNER, J. 1950. die Werkung der Kohlensaure aus Heen und Bakererien. *Seitz-Werke, Bad Kreuznach.*

SCHNEYDER, J. and VLCEK, G. 1977. Mitt Hocheren Bundeslehr-Versuchsanst. *Wein-Obstbau (Klosterneuburg)* 27:87–88.

SCHOLTEN, G. and KACPROWSKI, M. 1993. Zur Analytik von Polyphenolen in Wein. *Weinwiss.* 48(1):33–38.

SCHRADER, V.E., LEMPERLE, N., BECKER, J., and K.G. BERGNER. 1976. Der Aminosaure-Zucker, Saure und Mineralgehalt von Weenbeenin Abhangizkeit vom Kleinklima des Standortes de Rebe Mitteilung. *Wein Wiss* 31:9–24.

SCHREIER, P., DRAWERT, F., and JUNKER, A. 1976. Gaschromatographicsshmassenspektrometrische differenzierung der traubenaromastoffe verschiedener rebsorten von *Vitis vinifera. Chem. Mikrobiol. Technol. Lebensm.* 4:154–157.

SCHROETER, L.C. 1966. *Sulfur Dioxide: Applications in Foods, Beverages, and Pharmaceuticals.* New York: Pergamon Press.

SCHUG, W. 1982. Vinification of Fine Wine in California. Institute of Masters of Wine's International Symposium on Viticulture, Vinification, and the Treatment and Handling of Wine. Oxford, England.

SCHUTZ, M. and KUNKEE, R.E. 1977. Formation of hydrogen sulfide during fermentation by wine yeasts. *Am. J. Enol. and Vitic.* 28:137–140.

SCHUTZ, H. and RADLER, F. 1984. Anaerobic reduction of glycerol to 1,3-propandiol by *Lactobacillus bunchneri. Systematic and Applied Microbiology* 203:1–10.

SCHWIMMER, S. 1981. *Source Book of Food Enzymes.* Westport, CT: Avi Publishing Co., pp. 296,511–551.

SCOTT, B.C., BUTLER, J., HALLIWELL, B., and ARUOMA, O.I. 1993. Evaluation of the antioxidant actions of ferulic acid and catechins. *Free Rad. Res. Commun.* 19:241–253.

SCOTT, B.C., BUTLER, J., HALLIWELL, B., and ARUOMA, O.I. 1993. Evaluation of the antioxidant actions of ferulic acid and catechins. *Free Rad. Res. Commun.* 19:241–253.

SCOTT, R.S. 1967. Clarification—the better half of filtration. *Wine and Vines* 48(10): 29–30.

SCOTT, R.S., ANDERS, T.G., and HUMS, N. 1981. Rapid cold stabilization of wine by filtration. *Am. J. Enol. and Vitic.* 32(2):138–143.

SCUDAMORE-SMITH, P.D., HOOPER, R.L., and MCLARN, E.D. 1990. Color and phenolic changes of Cabernet Sauvignon wine made by simultaneous yeast/ bacterial fermentation and extended pomace contact. *Am. J. Enol. and Vitic.* 41:57–67.

SEFTON, M.A., FRANCIS, L.L. and WILLIAMS, P.J. 1990. Volatile flavor compounds of oakwood. In: *Proc. Seventh Australian Wine Ind. Tech. Conf.* (Adelaide) P.J. Williams, D.M. Davidson and T.H. Lee (eds.). Austr. Ind. Publs. pp 107–112.

SEGUIN, G. 1986. Terroirs and pedology of wine growing. *Experientia* 42:861–873.

SEIGNEUR, M., BONNET, J., DORIAN, B., BENCHIMOL, D., DROUILLET, F., GOUVERNEUR, G., LARRUE, J., CROCKETT, R., BOISSEAU, M.-R., RIBEREAU-GAYON, P., and BICAUD, H. 1990. Effect of the consumption of alcohol, white wine, and red wine on platelet function and serum lipids. *J. Appl. Cardiol.* 5:215–222.

SEIMANN, E.H. and L.L. CREASY. 1992. concentration of the phytoalexin resneratrol in wine. *Am. J. Enol. and Vitic.* 43:49–52.

SEITZ-WERKE GMBH (SWK MACHINES, INC) TECHNICAL FILE. 1978. Methods to determine and evaluate KHT stability by means of the contact method on a laboratory scale. Bath, New York.

SEMPOS, C.T., LOOKER, A.C., GILLUM, R.F., and MAKUC, D.M. 1994. Body iron stores and the risk of coronary heart disease. *N. Engl. J. Med.* 330:1119–1124.

SHAHIDI, F., JANITHA, P.K., and WANASUNDARA, P.D. 1992. Phenolic antioxidants. *Crit. Rev. Food Sci. Nutr.* 32:67–103.

SHARF, R. and MARGALITH, P. 1983. The effect of temperature on spontaneous wine fermentation. *Eur. J. Appl. Micobiol. Biotech.* 17:311–313.

SHARP, D. 1993. Coronary disease—when wine is red. *Lancet* (Lett.) 341:27–28.

SHIMAZU, Y. and WATANABE, M. 1981. Determination of oganic acids in grape must and wine by HPLC. *Nippon Jozo Kyokai Zasshi* 76:418–423.

SHIMIZU, K., ADACHI, T., KITANO, K., SHIMAZAKI, T., TOTSUKA, A., HARA, S., and DITTRICH, H.H. 1985. Killer properties of wine yeasts and characterization of killer wine yeasts. *J. Ferment. Technol.* 63:421–429.

SHOSEYOV, O., BRAVDO, B.A., SIEGEL, D., GOLDMAN, A., COHEN, S., SHOSEYOV, L., and IKAN, R. 1990. Immobilized endo-*beta*-glucosidase enriches flavor of wine and passion fruit juice. *J. Agric. Food Chem.* 38:1387–1390.

SIEMAN, E.H. and CREASY, L.L. 1992. Concentration of the phytoalexin resveratrol in wine. *Am. J. Enol. Vitic.* 43:49–52.

SIES, H. and DE GROOT, H. 1992. Role of reactive oxygen species in cell toxicity. *Toxicol. Lett.* 64/65:547–551.

SILVER, J. and LEIGHTON, T. 1981. Control of malolactic fermentation in wine. II. Isolation and characterization of a new malolactic organism. *Am. J. Enol. and Vitic.* 32:64–72.

SIMPSON, E. and TRACEY, R.P. 1986. Microbiological quality shelf life and fermentation activity of active dried yeast. *S. African J. Enol. Vitic* 7(2):61–65.

SIMPSON, R.F. and LEE, T.H. 1990. The microbiology and taints of cork and oak. In: *Proceedings of the Ninth International Oenology Symposium,* E. Lemperle and E. Figlestahler (eds.). pp. 653–667.

SIMPSON, R.F., AMON, J.M., and DAW, A.J. 1986. Off-flavour in wine caused by guaiacol. *Food Tech. Austr.* 38:31–33.

SIMS, C.A., JOHNSON, R.P., and BATES, R.P. 1988. Response of a hard-to-press *Vitis rotundifolia* cultivar and a hard-to-clarify *Euvitis* hybrid to commercial enzyme preparations. *Am. J. Enol. and Vitic.* 39:341–343.

SINCLAIR, A.J., BARNETT, A.H., and LUNEC, J. 1990. Free radicals and antioxidant systems in health and disease. *Br. J. Hosp. Med.* 43:334–344.

SINGLETON, V.L. 1967. Adsorption of natural phenols from beer and wine. Technical quarterly of the Masters Brewers Assn. of America. 4(4):245–253.

SINGLETON, V.L. 1982a. Grape and wine phenolics: background and prospects. In: *Proceedings of the University of California, Davis, Grape and Wine Centennial Sympo-*

sium. A.D. Webb, (ed.). pp. 215–227. Dept. of Viticulture and Enology, University of California, Davis.

SINGLETON, V.L. 1982b. Oxidation of wine. In: *Proceedings of the International Symposium on Viticulture, Vinification, and Treatment and Handling of Wine.* Oxford, England.

SINGLETON, V.L. 1985. Recent conclusions on wine oxidation. In: *Proceedings of the Eleventh Wine Industry Technical Symposium.* pp. 17–24. Wine Ind. Symposium, San Francisco.

SINGLETON, V.L. 1987. Oxygen with phenols and related reactions in musts, wines, and model system: observations and practical implications. *Am. J. Enol. and Vitic.* 38(1):69–77.

SINGLETON, V.L. 1988. Wine phenols. In: *Modern methods of plant analysis. New series Vol. 6 (Wine Analysis).* H.F. Linskens and J.F. Jackson (eds.). pp. 173–218.

SINGLETON, V.L. and DRAPER, D.E. 1962. Adsorbents and wines. I. Selection of activated charcoals for treatment of wines. *Am. J. Enol. and Vitic.* 13:114–125.

SINGLETON, V.L. and ESAU, P. 1969. *Phenolic Substances in Grapes and Wines and Their Significance.* New York: Academic Press.

SINGLETON, V.L. and GUYMON, J.F. 1963. A test of fractional addition of wine spirits to red and white port wines. *Am. J. Enol. and Vitic.* 14:129–146.

SINGLETON, V.L. and KRAMLING, T.E. 1976. Browning in white wines and an accelerated test for browning capacity. *Am. J. Enol. and Vitic.* 27:157–160.

SINGLETON, V.L. and NOBLE, A.C. 1976. Wine flavor and phenolic substances. In: *Phenolic, Sulfur, and Nitrogen Compounds in Food Flavors.* G. Charalambous and I. Katz (Eds.). A.C.S. Symposium Series 26:47–70.

SINGLETON, V.L. and ROSSI, J.A., JR. 1965. Colorimetry of total phenolics with phosphomolybdic-phosphotungstic acid reagents. *Am. J. Enol and Vitic.* 16:144–158.

SINGLETON, V.L. and TROUSDALE, E.K. 1992. Anthocyanin-tannin interactions explaining differences in polymeric phenols between white and red wines. *Am. J. Enol. and Vitic.* 43:63–70.

SINGLETON, V.L., BERG, H.W., and GUYMON, J.F. 1964. Anthocyanin color level in port-type wines as affected by the use of wine spirits containing aldehydes. *Am. J. Enol. and Vitic.* 15:75–81.

SINGLETON, V.L., SALGUES, M., ZAYA, J., and TROUSDALE, E. 1985. Caftaric acid disappearance and conversion to products of enzymic oxidation in grape must and wine. *Am. J. Enol. and Vitic.* 15:75–81.

SINGLETON, V.L., SULLIVAN, A.R., and KRAMER, C. 1971. An analysis of wine to indicate aging in wood or treatment with wood chips or tannic acid. *Am. J. Enol. and Vitic.* 22(3):161–166.

SINGLETON, V.L., ZAYA, J., and TROUSDALE, E. 1980. White table wine quality and polyphenol composition as effected by must sulfur dioxide content and pomace contact time. *Am. J. Enol. and Vitic.* 31(1):14–20.

SKOOG, D.A. and WEST, D.M. 1971. *Principles of Instrumental Analysis.* New York: Holt, Reinhart and Winston.

SLINIGER, P.J., BOTHAST, R.J., and SMILEY, K.L. 1983. Production of 3-hydroxypropionaldehyde from glycerol. *Appl. Environ. Microbiol.* 46:62–67.

SMART, R.E. 1976. Implication of radiation microclimate for productivity of vineyards. Ph.D. thesis. Cornell Univ.

SMART, R.E. 1985. Principles of grapevine canopy microclimate manipulation with implications of yield and quality. A review. *Am. J. Enol. Vitic.* 36:230–239.

SMART, R. and ROBINSON, M. 1991. *Sunlight into Wine.* A Handbook for Winegrape Canopy Management. Adelaide, South Australia: Winetitles, p. 88.

SMART, R.E. and YOUNG, J. (EDS.). 1987. *Proceedings of Vintage '87 Seminar.* Te Kauwhata, NZ: Te Kawhata Research Station, pp. 115–133.

SMART, R.E., DRY, P.R., and BRIEN, D.R.G. 1977. Field temperature of grape berries and implications for fruit composition. *Int. Symp. on Quality of the Vintage,* pp. 227–231, Capetown, South Africa.

SMITH, C. 1982. Review of basics on sulfur dioxide—part II. *Enology Briefs* 1(2):1–3. Cooperative Extension, University of California.

SMITH, C. 1992. Finding the stopper for cork aroma taint. *Pract. Winery and Vineyard* Nov/Dec:49–50.

SMITH, C. 1993. Studies on Sulfur Dioxide Toxicity for Two Wine Yeasts: M.S. Thesis, University of California, Davis.

SMITH, C., HALLIWELL, B., and ARUOMA, O.I. 1992. Protection by albumin against the pro-oxidant actions of phenolic dietary components. *Food Chem. Toxicol.* 30:483–489.

SMITH, E.B. and THOMPSON, W.D. 1994. Fibrin as a factor in athergenesis. *Thromb. Res.* 73:1–19.

SNOW, R. 1979. Toward genetic improvement of wine yeasts. *Am. J. Enol. and Vitic.* 30(1):33–36.

SNOW, P.G. and GALLANDER, J.F. 1979. Deacidification of white table wines through partial fermentation with *Schizosaccharomyces pombe. Am. J. Enol. and Vitic.* 30(1): 45–48.

SNYDER, S.H. and BREDT, D.S. 1991. Nitric oxide as a neuronal messenger. *Trends Pharmacol. Sci.* 12:125–128.

SNYDER, S.H. and BREDT, D.S. 1992. Biological roles of nitric oxide. *Sci. Amer.* May:68–77.

SOBOLESKY, W.J. 1968. The Rodac plate—a useful tool for sanitarians. *J. Environ. Health.* 30(5):525–527.

SOHAL, R.S. and ORR, W.C. 1992. Relationship between antioxidants, prooxidants, and the aging process. *Ann. NY Acad. Sci.* 663:74–84.

SOLES, R.M., OUGH, C.S., KUNKEE, R.E. 1982. Ester concentration differences in wines fermented by various species and strains of yeast. *Am. J. Enol. Vitic.* 33:94–98.

SOLS, A., GANCEDO, C., and DELAFUENTE, G. 1971. Energy yielding metabolism in yeasts. In: *The Yeasts II.* A.H. Rose and J.S. Harrison (eds.). New York: Academic Press, pp. 271–303.

SOMATTMADJA, D., POWERS, J.J., and HANDY, M.K. 1964. Anthocyanins VI. Chelation studies on anthocyanins and other related compounds. *J. Food Sci.* 29:655–660.

SOMERS, T.C. 1978. Interpretation of color composition in young red wines. *Vitis* 17:161–167.

SOMERS, T.C. 1984. *Botrytes cinerea*—consequences for red wines. *Australian Grapegrower Winemaker* 244:80, 83, 85.

SOMERS, T.C. 1987. Assessment of phenolic components in viticulture and oenology. In: *Proceedings of the Sixth Australian Wine Industry Technical Conference.* T. Lee (ed.). Adelaide, Australia: Australian Industrial Publishers, pp. 257–261.

SOMERS, T.C. and EVANS, M.E. 1977. Spectral evaluation of young red wines: anthocyanin equilibria, total phenolics, free and molecular sulfur dioxide. *J. Sci. Food Agr.* 28:279–287.

SOMERS, T.C. and EVANS, M.E. 1979. Grape pigment phenomena: interpretation of major colour losses during vinification. *J. Sci. Food Agric.* 30:623–633.

SOMERS, T.C., EVANS, M.E., and CELLIER, K.M. 1983. Red wine quality and style: diversities of composition and adverse influences from free sulfur dioxide. *Vitis* 22:384–53.

SOMERS, T.C. and VERETTE, E. 1988. Phenolic composition of natural wine types. *In: Wine Analyses. Modern Methods of Plant Analysis.* Vol. 8. H.F. Linskens and J.F. Jackson (eds.). pp. 219–257.

SOMERS, T.C. and ZIEMELIS, G. 1972. Interpretations of ultraviolet absorption in white wines. *J. Sci. Food Agr.* 23:441–453.

SOMERS, T.C. and ZIEMELIS, G. 1973a. The use of gel column analysis in evaluation of bentonite fining procedures. *Am. J. Enol. and Vitic.* 24(2):51–54.

SOMERS, T.C. and ZIEMELIS, G. 1973b. Direct determination of wine proteins. *Am. J. Enol. and Vitic.* 24(2):47–50.

SOMERS, T.C. and ZIEMELIS, G. 1985. Spectral evaluation of total phenolic components in *Vitis vinifera* grapes and wine. *J. Sci. Food Agric.* 36:1275–1284.

SPACEK, P. and JELINKOVA M. 1991. Sorption of flavonoids from wine and its investigation by HPLC in the reverse phase. *J. Liq. Chrom.* 14:237–251.

SPECTOR, A., WANG, G.-M., WANG, R.-R., GARNER, W.H., and MOLL, H. 1993. The prevention of cataract caused by oxidative stress in cultured rat lenses. I. H_2O_2 and photochemically induced cataract. *Curr. Eye Res.* 12:163–179.

SPEDDING, D.J. and RAUT, P. 1982. The influence of dimethyl sulfide and carbon disulfide on bouquet of wines. *Vitis* 21:240–246.

SPENCER, J.F.T. and SALLANS, H.R. 1956. Production of polyhydric alcohols by osmophilic yeasts. *Canadian J. Microbiol.* 2:72–79.

SPENCER, J.F.T. and SPENCER, D.M. 1980. Production of polyhydric alcohols by osmotolerant yeasts. In: *The Biochemistry of Plants, Vol 3.* J. Preiss (ed.). pp 393–425. New York: Academic Press.

SPETTOLI, P., BOTTACIN, A., NUTI, M.P., and ZAMORANI, A. 1982. Immobilization of *Leuconostoc oenos* ML-34 in calcium alginate gels and its application to wine technology. *Am. J. Enol. and Vitic.* 33(1):1–5.

SPLITTSTOESSER, D.F. 1981. Preservation of fresh grape juice. *Proceedings of the Ohio Grape and Wine Short Course.*

SPLITTSTOESSER, D.F. and WILKINSON, M. 1973. Some factors affecting the activity of DEPC as a sterilant. *Appl. Microbiol.* 25:853–857.

SPONHOLZ, W.R. 1988. Der Wein 4.3. Fehlerhafte und unerwunschte Erscheinungen im Wein. In: *Chemie des Weines.* G. Wurdig and R. Woller (eds.). Stuttgart: Ulmer.

SPONHOLZ, W.R. 1991. Nitrogen compounds in grapes must and wine. In: *Proceedings of the International Symposium on Nitrogen in Grapes and Wine.* Seattle, WA. Am. J. Enol. Vitic. J. Rantz (ed.) pp. 67–77.

SPONHOLZ, W.R. and DITTRICH, H.H. 1974. die Bildung von SO_2-bindenden Garungs-Nebenprodukten, hoheren Alkoholen und Estern bein einigen Reinzuchthefestammen und einigen fur die Weinbereitung wichtigen "wilden" Heffen. *Wein Wiss* 29:301–314.

SPONHOLZ, W.R. and DITTRICH, H.H. 1985. Uber die herkunft von Gluconsaure, 2- und 5-oxo Gluconsaure sowie Glucuron- und Galacturonsaure in Mosten un Weinen. *Vitis* 24:51–58.

SPONHOLZ, W.R., DITTRICH, H.H., and HAN, K. 1990. Die Beeinflussung der Garung und der Essigsaureethylester-bildung durch *Hanseniaspora uvarum. Vitic. and Enol. Sci.* 45:65–72.

SPONHOLZ, W.R., LACHER, M., and DITTRICH, H.H. 1986. Die Bildung von Alditolen durch die Hefen des Weines. *Chemie Mikrobiologie Technologie Lebensmittel* 9:19–24.

STAFFORD, P.A. and OUGH, C.S. 1976. Formation of methanol and ethyl methyl carbonate by dimethyldicarbonate in wine and model solutions. *Am. J. Enol. Vitic.* 27:7–11.

STAHL, E. 1967. *Thin-Layer Chromatography. A Lab Handbook.* Singapore: Toppar Printing Co.

STAMLER, J.S., SINGEL, D.J., and LOSCALZO, J. 1992a. Biochemistry of nitric oxide and its redox-activated forms. *Science* 258:1898–1902.

STAMLER, J.S., JARAKI, O., OSBORNE, J., SIMON, D.I., KEANEY, J., VITA, J., SINGEL, D., VALERI, C.R., and LOSCALZO, J. 1992b. Nitric oxide circulates in mammalian plasma primarily as an S-nitroso adduct of serum albumin. *Proc. Natl. Acad. USA* 89:7674–7677.

STATE OF CALIFORNIA. 1973. California Administrative Code, Title 17: "Public Health." Bureau of Printing (Documents Division) pp. 431–443.

STEELE, J.T. and KUNKEE, R.E. 1978. Deacidification of musts from western United States by calcium double salt precipitation process. *Am. J. Enol. and Vitic.* 29(3): 153–160.

STEELE, J.T. and KUNKEE, R.E. 1979. Deacidification of high acid California wines by calcium double salt precipitation. *Am. J. Enol. and Vitic.* 30(3):227–231.

STEFFAN, H., ZIEGLER, A., and RAPP, A. 1988. N-Salicyloyl-Asparaginsaure: Eine neue phenolische Verbindung aus Reben. *Vitis* 27:79–86.

STEINSCHREIBER, P. 1984. Diacetyl in wines. A varietal, pH and inhibition study. Masters Thesis, C.S.U. Fresno, Fresno, CA.

STERN, P. 1983. Technical Projects Workshop. American Society of Enologists Annual Meeting, Reno, NV.

STERNS, G.F. 1987. Extraction of colour during fermentation of Pinot Noir wines and its stability on aging. Dissertation, Dipl. Hort. Sci., Lincoln College, University of Canterbury, New Zealand.

STEVENS, D.F. and OUGH, C.S. 1993. Ethyl carbamate formation: reaction of urea and citrulline with ethanol in wine under low to normal temperature conditions. *Am. J. Enol. and Vitic.,* 44:309–312.

STICH, H.F. 1991. The beneficial and hazardous effects of simple phenolic compounds. *Mutation Res.* 259:307–324.

STICH, H.F. and ROSIN, M.P. 1984. Naturally occurring phenolics as antimutagenic an anticarcinogenic agents. *Adv. Exp. Biol. Med.* 177:1–29.

STOCKER, R., MCDONAGH, A.F., GLAZER, A.N., and AMES, B.N. 1990. Antioxidant activities of bile pigments: billiverdin and bilirubin. *Meth. Enzymol.* 186:301–309.

STONE, H., and SIDEL, J.L. 1985. *Sensory Evaluation Practices.* New York: Academic Press.

STUCKEY, W., ILAND, P., HENSCHKE, P.A., and GAWEL, R. 1991. The effect of lees contact time on chardonnay wine composition. *Proceedings of the International Symposium on Nitrogen in Grapes in Wines.* (Seattle, WA) J.M. Rantz (ed.). The American Society of Enology and Viticultures, pp. 325–319.

STUTZ, C. 1993. The use of enzymes in ultrafiltration. International Fruit Juice Week, Wiesbaden, Germany.

SU, C.T. and SINGLETON, V.L. 1969. Identification of three flavan-3-ols from grapes. *Phytochem* 8:1553–1558.

SUDRAUD, P. 1958. Interpretation des courbes de'absorption des vin rouges. *Ann. Technol. Agric.* 7:203–208.

SUOMALAINEN, H. and OURA, E. 1971. Yeast nutrition and solute uptake. In: *The Yeasts,* Vol. II. A.H. Rose and J.S. Harrison, (eds.). London: Academic Press.

SURDIN-KERJAN, Y., CHEREST, H., and DE ROBICHON-SZULMAJSTER, H. 1976. Regulation of methionine synthesis in *Saccharomyces cerevisiae* operates through independent signals: methionyl-tRNA^{met} and S-adenosylmethionine. *Acta Microbiol. Acad. Scintarum Hungaricae* 23:252–258.

SUZUKAWA, Y., ISHIKAWA, T., YOSHIDA, H., HOSOAI, K., NISHIO, E., YAMASHITA, T., NAKAMURA, H., HASHIZUME, N., and SUZUKI, K. 1994. Effects of alcohol consumption on antioxidant content and susceptibility of low-density lipoprotein to oxidative modification. *J. Am. Coll. Nutr.* 13:237–242.

SUZZI, G., ROMANO, P., and ZAMBONELLI, C. *Saccharomyces* strain selection in minimizing sulfur dioxide requirements during vinification. *Am. J. Enol. and Vitic.* 36(3):199–202.

TANNER, H. 1969. Der Weinbockser, Entstehung und Beseitigung. *Zeitschrit fur Obst- und Weinbau.* 105:252–258.

TANNER, H. and VETSCH, U. 1956. How to characterize cloudiness in beverages. *Am. J. Enol.* 7:142–149.

TANNER, H. and ZANIER, C. 1981. Zur analytischen Differenzierung von Muffton und Korkgeschmack in Weinen. *Schweiz. Zietschrift Obst-und Weinbau* 117:752–757.

TAYEH, M.A. and MARLETTA, M.A. 1989. Macrophage oxidation of L-arginine to nitric oxide, nitrite, and nitrate. Tetrahydrobiopterin is required as a cofactor. *J. Biol. Chem.* 264:19654–19658.

TAYLOR, A., JACQUES, P.F., and DOREY, C.K. 1993. Oxidation and aging: Impact on vision. *Toxicol. Ind. Health* 9:349–371.

TEEL, R.W. and CASTONGUAY, A. 1992. Antimutagenic effects of polyphenolic compounds. *Cancer Lett.* 66:107–113.

TEMPERLI, A., KUNSCH, U., MAYER, K., BUSCH, I. 1965. Reinigung und Eigenschaften der Malatdehydrogenase aus Hefe. *Biochemica et Biophysica Acta* 110: 630–632.

TERRELL, F.R., MORRIS, J.R., JOHNSON, M.G., GBUR, E.E., and MAKUS, D.J. 1993. Yeast inhibition in grape juice containing sulfur dioxide, sorbic acid, and dimethyldicarbonate. *J. Food Science* 58(5):1132–34.

TERRIER, A., BOIDRON, J.N., RIBÉREAU-GAYON, P. 1972. Teneurs' en composés terpéniques des raisins de *V. vinifera. C.R. Hebd. Seances Acad. Sci. Ser. D* 275: 941–944.

THOMAS, S. and DAVENPORT, R.R. 1985. *Zygosaccharomyces bailii*—a profile of characteristics and spoilage activities. *Food Microbiol.* 2:157–169.

THORNGATE, J.H. 1993. Flavan-3-ol and their polymers: analytical techniques and sensory considerations. *In: Beer and Wine Production: Characterization, Analysis and Technical Advances.* B.H. Gump (ed.). Am. Chem. Soc. Symp. Series 536. ACS, Washington D.C. pp. 57–63.

THOUKIS, G. 1958. The mechanism of isoamyl alcohol formation using tracer techniques. *Am. J. Enol. Vitic.* 9161–9166.

THOUKIS, G. and AMERINE, M.A. 1956. Fate of copper and iron during fermentation of grape musts. *Am. J. Enol.* 7:45–52.

THOUKIS, G. and STERN, L.A. 1962. A review of some studies of the effect of sulfur of formation of off odors in wine. *Am. J. Enol. and Vitic.* 13:133–140.

THOUKIS, G., UEDA, M., and WRIGHT, D. 1965. Formation of succinic acid during alcoholic fermentation. *Am. J. Enol. and Vitic.* 16:1–8.

TIMBERLAKE, C.F. and BRIDLE, P. 1967. Flavylium salts, anthocyanidins and anthocyanins. II. Reactions with sulfur dioxide. *J. Sci. Food. Agr.* 18:479–485.

TIMBERLAKE, C.F. and BRIDLE, P. 1976. Interactions between anthocyanins, phenolic compounds, and acetaldehyde and their significance in red wines. *Am. J. Enol. and Vitic.* 27:97–105.

TINSDALE, R.C. 1987. Shipping container floors: a potential source of chloroanisole contamination in dried fruit. *Chem. Ind. (London)* July 6:458–459.

TOMADA, Y. 1927. On the production of glycerine by fermentation. IV: dissociation of acetaldehyde-bisulfite complex in alkaline solution. *J. Soc. Chem. Ind. (Japan)* 30:747–759.

TORLE, J., CILLARD, J., and CILLARD, P. 1986. Antioxidant activity of flavonoids and reactivity with peroxy radical. *Phytochemistry* 25:383–385.

TRAPPEL, A.L. 1968. Will antioxidant nutrients slow the aging process? *Geriatrics* 23:97–105.

TRIOLI, G. and OUGH, C.S. 1989. Causes for inhibition of an acid urease from *Lactobacillus fermentum. Am. J. Enol. and Vitic.* 40:245–252.

TROMP, A. and AGENBACH, W.A. 1981. Sorbic acid as a preservative. Its efficacy and organoleptic thresholds. *S. Afr. J. Enol. and Vitic.* 2:1–5.

TROOST, G. 1972. *Die Technologie des Weines.* 4th ed. Ullmer: Stuttgart.

TROOST, G. and FETTER, K. 1960. Zur praxis de eiweisstabilisiesung der weine durch bentonite. *Weinberg Keller* 7:444–459.

TROUP, G.J., HUTTON, D.R., HEWITT, D.G., and HUNTER, C.R. 1994. How red wine is good for you. *Wine Industry J. (Austr.)* 9:145–147.

TSUKAMOTO, H. 1993. Oxidative stress, antioxidants, and alcoholic liver fibrogenesis. *Alcohol* 10:465–467.

TYSON, P.J., LUIS, E.S., DAY, W.R., and LEE, T.H. 1981. Estimation of soluble proteins in must and wine by high performance liquid chromatography. *Am. J. Enol. and Vitic.* 32(3):241–243.

UNITED STATES BUREAU OF ALCOHOL, TOBACCO and FIREARMS. Wine: Part 240 of Title 27, Code of Federal Regulations.

URLAUB, R. 1985. Benefits of combined use of a protease and a pectic enzyme in white wine processing. In: *Proceedings of the Australian Society of Viticulture and Oenology.* Glen Osmond, Australia. Australian Ind. Publishers.

URLAUB, R. 1989. *Enzymes for Winemaking—A Technical Bulletin.* Darmstadt, Germany: Rohm GmbH.

U.S. BUREAU OF STANDARDS. 1924. Standard Density and Volumetric Tables (Circ. 19). U.S. Govt. Printing Office, Washington, DC.

U.S. INTERNAL REVENUE SERVICE. 1945. Regulations No. 7, Wine. U.S. Govt. Printing office, Washington, DC.

VAHL, J.M. 1979. Relative density-extract-ethanol nomograph for table wines. *Am. J. Enol. and Vitic.* 30(3):262–263.

VAN BUREN, J.P., HRAZDINA, G., and ROBINSON, W.B. 1974. Color of anthocyanin solutions expressed in lightness and chromaticity terms. Effect of pH and the type of anthocyanin. *J. Food Sci.* 39:325–328.

VANELLA, A., DI GIACOMO, C., SORRENTI, V., RUSSO, A., CASTORINA, C., CAMPISI, A., RENIS, M., and PEREZ-POLO, J.R. 1993. Free radical scavenger depletion in post-ischemic reperfusion brain damage. *Neurochem. Res.* 18:1337–1340.

VAN ESCH, F. 1992. Yeast in soft drinks and concentrated fruit juices. *Brygmesteren—NR* 4:9–20.

VAN ZYL, J.A. 1962. Turbidity in South African dry wines caused by the development of *Brettanomyces* yeasts. *Bull. Vitic. and Enol. Res. Institute, Stellenbosch* 381:11–41.

VARMA, S.D. 1991. Scientific basis for medical therapy of cataracts by antioxidants. *Am. J. Clin. Nutr.* 53(Suppl.):335S–345S.

VAS, K. and INGRAHAM, M. 1949. Preservation of fruit juices with less sulfur dioxide. *Food Manuf.* 24:414–416.

VAUGHN, R.H. 1938. Some effects of association and competition on *Acetobacter*. *J. Bact.* 36:357–367.

VAUGHN, R.H. 1955. Bacterial spoilage of wine with special reference to California conditions. In: *Adv. Food Res.* 6:67–108.

VAUGHN, R.H. and DOUGLAS, H.C. 1938. Some Lactobacilli encountered in abnormal musts. *J. Bact.* 36:318–319.

VAUGHN, R.H. and TCHELISTCHEFF, A. 1957. Studies on the malic acid fermentation of California table wines. I. An introduction to the problem. *Am. J. Enol. and Vitic.* 8:74–79.

VEENSTRA, J. 1991. Moderate alcohol use and coronary heart disease: a U-shaped curve? *World Rev. Nutr. Diet.* 65:38–71.

VERDUYN, C., POSTMA, E., SCHEFFERS, W.A., and VAN DIJKEN, J.P. 1990. Physiology of *Saccharomyces cerevisiae* in anaerobic glucose-limited chemostat cultures. *J. Gen. Microbiology.* 136:359–403.

VERSINI. 1985. Sull'aroma del vino Traminer aromatico o Gewurztraminer. *Vigne vini N.* 1–2:57–65.

VETCH, V. and LUTHI, H. 1964. Farbstoffverlustewahrend des Biologischen Saureabbaues Schweiz. *Z. Obst-Weinbau* 73:124–126.

VILLETTAZ, J.-C. 1986. A new method for production of low alcohol wines and better balanced wines. In: *Proceedings of Sixth Australian Wine Industry Technical Conference.* (Adelaide) T.H. Lee (ed.) Australian Wine Research Institute, pp. 125–128.

VILLETTAZ, J.C., AMADO, R., NEUKOM, H., HORISBERGER, M., and HORMAN, T. 1980. Comparative structural studies of the D-mannans from a rosé wine and *Saccharomyces uvarum*. *Carbohydrate Res.* 81:341–344.

VILLETTAZ, J.-C., STEINER, D., TROGUS, H. 1984. The use of *beta*-glucanase as an enzyme in wine clarification and filtration. *Am. J. Enol. Vitic.* 34:253–256.

VOS, P.J.A. 1981. Assimilable nitrogen—A factor influencing the quality of wines. In: *International Assoc. for Modern Winery Technology and Management, Sixth International Oenological Symposium.* Mainz, Germany. pp 163–90. April 28–30, 1981.

VOS, P.J.A. and GRAY, R.S. 1979. The origin and control of H_2S during fermentation of grape must. *Am. J. Enol. and Vitic.* 30(3):187–196.

VOS, P.J.A., CROUS, E., and SWART, L. 1980. Fermentation and the optimal nitrogen balance of musts. *Wynboer.* 582:58–62.

VOS, P.J.A., ZEEMAN, W., HEYMANN, H. 1979. The effect on wine quality of diammonium phosphate additions to musts. *Proc. South African Enol. Soc.* 87–109.

WAHAB, A., WITZKE, W., and CRUESS, W.V. 1949. Experiments with ester-forming yeast. *Fruit Prod. J.* 28:198–219.

WAHLBERG, I. and ENZELL, C.R. 1987. Tobacco isoprenoids. *Nat. Prod. Rep.* 4(3): 237–276.

WAHLSTROM, V.L. and FUGELSANG, K.C. 1988. Utilization of yeast hulls in winemaking. California Agricultural Tech. Institute Bull. 880103.

WAHLSTROM, V.E., BURR, S., and FUGELSANG, K.C. 1992. Sensitivity of wine yeast to several levels of fluoride present during fermentation. Presented at the Annual Meeting of the American Society for Enology and Viticulture. Seattle, WA.

WALLACE, J.R. 1980. The uses of gas in winemaking. *East. Grapegrower-Winemaker* Oct.:47–49.

WANG, L.-F. 1985. Off-flavor development in white by *Brettanomyces* and *Dekkera.* Masters Thesis, California State University Fresno, Fresno, CA.

WARD, P.A., JOHNSON, K.J., and TILL, G.O. 1986. Oxygen radicals and microvascular injury of lungs and kidney. *Acta Physiol. Scand.* Suppl.548:79–85.

WARKENTINE, H. and NURY, M.S. 1963. Alcohol losses during fermentation of grape juice in closed containers. *Am. J. Enol. and Vitic.* 14:68–74.

WARNER, C.R., BRUMLEY, W.C., DANIELS, D.H., JOE, JR., F.L., and FAZIO, T. 1986. Reactions of antioxidants in foods. *Food Chem. Toxicol.* 24:1015–1019.

WATERHOUSE, A.L. and FRANKEL, E.N. 1993. *Wine Antioxidants May Reduce Heart Disease and Cancer.* Proc. O.I.V. 73rd Gen. Assem., San Francisco, California, Aug. 29–Sept. 3.

WATERS, E.J., WALLACE, W., and WILLIAMS, P.J. 1990. Peptidases in winemaking. In: *Proc. of the Seventh Australian Wine Industry Technical Conference* (Adelaide) P.J. Williams, P.M. Davidson and T.H. Lee, (eds.) Australian Industrial Publishers. pp. 186–191.

WEEKS, C. 1969. Production of sulfur dioxide-binding compounds and sulfur dioxide by two *Saccharomyces* yeasts. *Am. J. Enol. and Vitic.* 20:32–39.

WEENK, G., OLIJVE, W., and HARDER, W. 1984. Ketogluconate formation by *Gluconobacter* species. *Appl. Microbiol. Biotech.* 20:400–405.

WEETHALL, H.W., ZELKO, J.T., and BAILEY, L.F. 1984. A new method for stabilization of white wine. *Am. J. of Enol. and Vitic.* 35(4):212–215.

WEILLER, H.G. and RADLER, F. 1972. Vitamin-und Aminosaure-Bedarf von Milchsaurenbakterien aus Wein und Rebenblatterm. Mitt. Hoecheren Bundeslehr Versuchsanst. *Wein Obst. Kloesterneuburg* 22:4–18.

WENZEL, K., DITTRICH, H.H., SEYFFARDT, H.P., and BOHNERT, J. 1980. Schwefelruckstande auf Trauben und im Most und ihr Einfluss auf die H_2S-Bildung. *Wein-Wissenschaft* 35:414–420.

WESTRICH, M. 1993. Managing oxygen pickup during bottling of white wines. *Vineyard and Winery Mgmt.* 20–21.

WHITE, B.B. and OUGH, C.S. 1973. Oxygen uptake studies in grape juice. *Am. J. Enol. and Vitic.* 24(4):148–152.

WHITE, C.R., BROCK, T.A., CHANG, L-Y., CRAPO, J., BRISCOE, P., KU, D., BRADLEY, W.A., GIANTURCO, S.H., GORE, J., FREEMAN, B.A., and TARPEY, M.M. 1994. Superoxide and peroxynitrite in atherosclerosis. *Proc. Natl. Acad. Sci USA* 91:1044–1048.

WHITING, G.C. 1976. Organic acid metabolism of yeasts during fermentation of alcoholic beverages: a review. *J. of the Institute of Brewing* 82:84–93.

WHITING, G.C. and COGGINS, R.A. 1971. The role of quinate and skimate in metabolism of lactobacilli. *Antonie van Leeuwenhoek J. Microbiol. Serol.* 37:33–49.

WHITNEY, P.A., and COOPER, T.G. 1972. Urea carboxylase and allophanate hydrolase: Two components of adenosine triphosphate:urea amidolyase in *Saccharomyces cerevisae. J. Biol. Chem.* 247:1349–1353.

WIBOWO, D., ESCHENBRUCH, R., DAVIS, C.R., FLEET, G.H., and LEE, T.H. 1985. Occurrence and growth of lactic acid bacteria in wine. A review. *Am. J. Enol. and Vitic* 36:302–313.

WILDENRADT, H.L. and SINGLETON, V.L. 1974. The production of aldehydes as a result of oxidation of polyphenolic compounds and its relation to wine aging. *Am. J. Enol. and Vitic.* 25(2):119–126.

WILLIAMS, J.T., OUGH, C.S., and BERG, H.W. 1978. White wine composition and quality as influenced by methods of must clarification. *Am. J. Enol. and Vitic.* 29(1):92–96.

WILLIAMS, P.J. 1990. Hydrolytic flavor release from non-volatile precursors. ACS Short Course, 200th National Meeting of the American Chemical Society. Washington, DC.

WILLIAMS, P.J., STRAUSS, C.R., and WILSON, B. 1980. New linalool derivatives in Muscat of Alexandria grapes and wines. *Phytochemistry* 19:1137–1139.

WILLIAMS, P.J., STRAUSS, C.R., and WILSON, B. 1981. Classification of the monoterpenoid composition of muscat grapes. *Am. J. Enol. and Vitic.* 32:230–235.

WILLIAMS, P.J., STRAUSS, C.R., and WILSON, B. 1988. Developments in flavor research on premium varieties. In: *Proceedings of the Second International Cool Climate Viticulture and Oenology Symposium.* R.E. Smart, R.J. Thornton, S.B. Rodriquez, and J.E. Young (eds.). pp. 331–332. New Zealand. Society of Viticulture and Oenology, Auchland.

WILLIAMS, P.J., STRAUSS, C.R., WILSON, B., and DIMITRIADIS, E. 1985. Recent studies into grape terpene glycosides 1984. In: *Progress in Flavor Research.* pp. 349–357.

WILLIAMS, P.J., STRUASS, C.R., WILSON, B., and MASSY-WESTROPP, R.A. 1982. Studies on the hydrolysis of *Vitis vinifera* monoterpene precursor compounds and model monoterpenes *beta*-D-glycosides rationalizing the monoterpene composition of grapes. *J. Agric. Food Chem.* 30:1219–1223.

WILSON, B., STRAUSS, C.R., and WILLIAMS, P.J. 1984. Changes in free and glycosidically bound monoterpenes in developing Muscat grapes. *J. Agric. Food Chem.* 32:919–924.

WILSON, K.S., WALKER, W.O., MARS, C.V., and RINELLI, W.R. 1943. Liquid sulfur dioxide in the fruit industry. *Fruit Prods. J.* 23:72–82.

WINKLER, A.J., COOK, J.A., KLIEWER, W.M., and LIDER, A.L. 1974. *General Viticulture.* Berkeley, CA: University of California Press.

WOIDICH, H. and PFANNHAUSER, W. 1974. Zur Gaschromatograpahischen Analyse von Branntweinen: Quatitative bestimming von Acetaldehyd, Essigsauremethylester, Essigsaureanthylester, Methanol, Butanol-(1), Butanol-(2), Propanol-(1), 2-Methylpropanol-(1), "Amylalkoholen" und Hexanol-(1). Mitt. Hoehesen Bundeslehr-Versuchsanst. *Wein Obstbau. Klosterneuburg* 24:155–156.

WOLLER, R., DODIE, I., and THUL, B. 1981. Histamin im Wein. Bestimmungsmethoden und Gehalte. *Dipl. Arbeit Thul.*

WOLPERT, J.A. and VILAS, E.P. 1991. Estimating vineyard yield: Introduction to a simple two-step method. *Am. J. Enol. Vitic.* 43:384–388.

WONG, G. and CAPUTI, A., JR. 1966. A new indicator for total acid determination in wines. *Am. J. Enol and Vitic.* 17(3):174–177.

WRIGHT, J.M. and PARLE, J.N. 1974. *Brettanomyces* in The New Zealand Wine Industry *New Zealand J. Agric. Res.* 17:273–278.

WUCHERPFENNIG, K. 1978. Possible applications of procedures using membranes for stabilization of wines. *Ann. Technol. Agric.* 27(1):319–331.

WUCHERPFENNING, K. and DITTRICH, H.H. 1984. Composition, behaviour and enzymatic degration of special colloids of wine. Proceedings of the Symposium for Australian Wine and Citrus Ind. Adelaide, South Australia.

WUCHERPFENNING, K. and RATZKA, P. 1967. Veber die verzogerung der weinsteinausscheidung durch polymere substanzen des weines. *Weinberg Keller* 14:499–509.

WUCHERPFENNING, K. and SEMMLER, G. 1973. Uber den SO_2-bedarf der Wein aus verschiedenen Weinbau gebieten der Welt und dessen Abhangigkeit von der bildung von Acetaldehyd im verlauf der garung. *Deut. Weinbau* 28:851–855.

WURDIG, G. 1976. Mucic acid: a constituent of wines made from *Botrytis* infected grapes. *Weinwissenschaft* 112:16–17.

WURDIG, G., SCHLOTTER, H.A., and KLEIN, E. 1974. Uber die Ursachen des sogenannten Geranientones. *Allg. Deut. Weinfachztg.* 110:578–583.

YALPANI, N. and RASKIN, I. 1993. Salicylic acid: a systemic signal in induced plant disease resistance. *Trends Microbiol.* 1:88–92.

YAMADA, H. and DESOTO, R.T. 1963. Relationship of solubility products to long-range tartrate stability. *Am. J. Enol. and Vitic.* 14:43–51.

YAMADA, S., NABE, K., IZUO, M., and CHIBATA, I. 1979. Enzymic production of dihydroxyacetone by *Acetobacter suboxydans* ATCC 621. *J. Ferment. Tech.* 57:321–326.

YANG, G., CANDY, T.E.G., BOARO, M., WILKIN, H.E., JONES, P., NAZHAT, N.B., SAADALLA, R.A., and BLAKE, D.R. 1992. Free radical yields from the homolysis of peroxinitrous acid. *Free Rad. Biol. Med.* 12:327–330.

YANTIS, J.E. 1992. The role of Sensory Analysis in Quality Control. ASTM MNL 14. Philadelphia, PA: The American Society for Testing and Materials.

YOSHIKAWA, T. 1993. Free radicals and their scavengers in Parkinson's disease. *Eur. J. Neurol.* 33(Suppl.1):60–68.

YOSHIOKA, K., HASHIMOTO, N. 1981. Ester formation by alcohol acetyltransferase from Brewer's yeast. *Agric. Biol. Chem.* 45:2183–2190.

YOSHIZAWA, K., and TAKAHASHI, K. 1988. Utilization of urease for decomposition of urea in sake. *J. Brew. Soc. Japan.* 83:142–144.

YU, B.P. 1994. Cellular defenses against damage from reactive oxygen species. *Physiol. Rev.* 74:139–162.

YORK, G. 1986. More on chlorine sanitizers—Review of basics VI enology briefs 5(1). *Cooperative Extension Bull.* University of California, Davis.

ZEEMAN, W., SNYMAN, J.P., and VAN WYK, C.J. 1982. The influence of yeast strain and malolactic fermentation on some volatile bouquet substances and on quality of table wines. In: *Proceedings of the Univ. of California, Davis Grape and Wine Centennial Symp.* A.D. Webb (ed.). Dept. of Vitic. and Enology. U.C. Davis pp. 79–90.

ZENT, J.B. and INAMA, S. 1992. Influence of macerating enzymes on the quality and composition of wines obtained from red Valpolicella wine grapes. *Am. J. Enol. and Vitic.* 43:311.

ZHU, L., GUNN, C., and BECKMAN, J.S. 1992. Bactericidal activity of peroxynitrite. *Arch. Biochem. Biophys.* 298:452–457.

ZIEGLER, B. 1990. Untersuchung von Trubruckstanden der Weinbereitung auf Nahrstoff- und Schwermetallgehalte. *Wein-Wiss.* 45(1):24–26.

ZIEMELIS, T. and SOMERS, T.C. 1978. Rapid determination of sorbic acid in wine. *Am. J. Enol. and Vitic.* 29:217–219.

ZIMMERMAN, H.W. 1963. Studies on the dichromate method of alcohol determination. *Am. J. Enol. and Vitic.* 14:205–213.

ZOECKLEIN, B.W. 1984. Bentonite fining. *The Practical Winery* 3(5) May/June:84–91.

ZOECKLEIN, B.W. 1986. Vineyard sampling. Virginia Polytechnic Inst. Cooperative Extension Bulletin.

ZOECKLEIN, B.W. 1987. pH imbalance in Cabernet sauvignon. In: *Proc. of the Am. Soc. Enol. Vitic. Eastern Sec. Regional Meeting.* pp. 26–37.

ZOECKLEIN, B.W. 1988. A review of potassium bitartrate stabilization of wines. Virginia Cooperative Extension. VPI-SU No. 463–013.

ZOECKLEIN, B.W. 1994. Red wine quality components. Vineyard and Winery Information Series. Virginia Cooperative Extension 9(3).

ZOECKLEIN, B.W., FUGELSANG, K.C., GUMP, B.H., and NURY, F.S. 1990. *Production Wine Analysis.* Van Nostrand Reinhold Publs. New York.

ZOECKLEIN, B.W., WOLF, T.K., DUNCAN, N.W., JUDGE, J.M. and COOK, M.K. 1992. Effects of fruit zone leaf removal on yield, fruit composition and fruit rot incidence of Chardonnay and White Riesling grapes. *Am. J. Enol. Vitic.* 43:139–148.

INDEX

CPSIA information can be obtained at www.ICGtesting.com
Printed in the USA
LVOW04s1422051014

407338LV00005B/16/P

9 781475 769807